Lecture Notes in Mathematics

Volume 2304

This series reports on new developments in all areas of mathematics and their applications - quickly, informally and at a high level. Mathematical texts analysing new developments in modelling and numerical simulation are welcome. The type of material considered for publication includes:

1. Research monographs
2. Lectures on a new field or presentations of a new angle in a classical field
3. Summer schools and intensive courses on topics of current research.

Texts which are out of print but still in demand may also be considered if they fall within these categories. The timeliness of a manuscript is sometimes more important than its form, which may be preliminary or tentative.

Titles from this series are indexed by Scopus, Web of Science, Mathematical Reviews, and zbMATH.

Cédric Arhancet • Christoph Kriegler

Riesz Transforms, Hodge-Dirac Operators and Functional Calculus for Multipliers

Springer

Cédric Arhancet
Albi, France

Christoph Kriegler
Laboratoire de Mathematiques Blaise
Pascal (LMBP), CNRS
Université Clermont Auvergne
Clermont-Ferrand, France

ISSN 0075-8434   ISSN 1617-9692   (electronic)
Lecture Notes in Mathematics
ISBN 978-3-030-99010-7   ISBN 978-3-030-99011-4   (eBook)
https://doi.org/10.1007/978-3-030-99011-4

Funding Information: Agence Nationale de la Recherche anr-18-ce40-0021 anr-17-ce40-0021
(http://dx.doi.org/10.13039/501100001665)

Mathematics Subject Classification: 46L51, 46L07, 47D03, 58B34

This Springer imprint is published by the registered company Springer Nature Switzerland AG
The registered company address is: Gewerbestrasse 11, 6330 Cham, Switzerland

We dedicate this book to our families: Clément, Lise, Nina, Christine, Raphael and Clara.

Preface

The starting point of the writing of this book is an open problem of functional calculus of operators explicitly stated in a remarkable recent paper of Junge, Mei and Parcet. This problem is connected to Riesz transforms defined by semigroups of operators acting on spaces associated to discrete groups. In this book, we give a solution for a large class of groups relying in particular on some approximation properties of groups.

Along the way, we also present a significant number of topics which are new and of independent interest: a transference principle between Fourier multipliers and Fourier multipliers on crossed products, Khintchine type equivalences for deformed Gaussians in spaces associated to crossed products, and Hodge decompositions. We think that these tools will be used in the next developments of this theory. Indeed, Riesz transforms and functional calculus are still evolving mathematical fields. It is transparent for us that a lot of things remain to be done in this area. Consequently, we describe an incomplete story.

In this book, we also want to popularize the fact that noncommutative spaces appear naturally in the classical and old topics of Riesz transforms and functional calculus. This is an astonishing and really striking fact linked to cohomology. Furthermore, we will show in the last chapter that the language of noncommutative geometry is natural and allows us to illuminate these topics with a new geometric point of view. In some sense, there is some "hidden noncommutative geometry" in some semigroups of operators acting on classical spaces. Moreover, we also describe and define a Banach space generalization of the notion of spectral triple of noncommutative geometry. We believe that this generalization will be used in a large number of settings in the future.

We also prove similar results suitable for semigroups of Schur multipliers, and we wanted to show the analogies and the differences between those two settings. It is known that Fourier multipliers and Schur multipliers are connected, and we continue in this book to display interesting features and links.

We hope we made the topic accessible. Our guiding principle has been to give all details of computations and approximation arguments. We think that it is easier for a reader to have complete proofs than proofs left to the reader. We equally wanted to close some gaps in the literature. We have tried to reduce the required prerequisites to a minimum. We also hope that our effort will consolidate the beautiful setting of modern noncommutative harmonic analysis which is a mixture of probabilities, analysis, algebra and geometry.

Albi, France   Cédric Arhancet
Clermont-Ferrand, France   Christoph Kriegler
February 2022

Acknowledgements

The authors are supported by the grant of the French National Research Agency ANR-18-CE40-0021 (project HASCON). The first author would like to thank Françoise Lust-Piquard and Cyril Lévy for some discussions. We are grateful to Marek Bożejko for the reference [41], Quanhua Xu for the reference [100] and to Frédéric Latrémolière for his confirmation of Proposition 5.1. Finally, we are indebted to Markus Haase for his help to devise rigorously the end of the proof of Lemma 5.6. The second author acknowledges support by the grant ANR-17-CE40-0021 (project Front).

Contents

Chapter 1
Introduction

Abstract In this chapter, we give an overview of the book and its contents. We start by giving a short introduction to Riesz transforms and their L^p-boundedness in various settings in that they have been studied in the literature. We also explain the emergence of noncommutative L^p-spaces and noncommutative geometry in this context. Furthermore, we describe our main results concerning Riesz transforms, functional calculus of Hodge-Dirac operators and spectral triples. We equally present examples that can used with our results and end with an overview of the contents of the other chapters.

The continuity of the Hilbert transform H on $L^p(\mathbb{R})$ by Riesz [199] is known as one of the greatest discoveries in analysis of the twentieth century. This transformation is defined by the principal value

$$(Hf)(x) \stackrel{\text{def}}{=} \frac{1}{\pi} \text{p.v.} \int_{\mathbb{R}} \frac{f(y)}{x-y} \, dy, \quad f \in \mathscr{S}(\mathbb{R}), \text{ a.e. } x \in \mathbb{R}. \tag{1.1}$$

It is at the heart of many areas: complex analysis, harmonic analysis, Banach space geometry, martingale theory and signal processing. We refer to the thick books [109, 110, 137, 138] and references therein for more information. Directional Riesz transforms R_j are higher-dimensional generalizations of the Hilbert transform defined by the formula

$$R_j \stackrel{\text{def}}{=} \partial_j \circ (-\Delta)^{-\frac{1}{2}}, \quad j = 1, \ldots, n \tag{1.2}$$

© The Author(s), under exclusive license to Springer Nature Switzerland AG 2022
C. Arhancet, C. Kriegler, *Riesz Transforms, Hodge-Dirac Operators and Functional Calculus for Multipliers*, Lecture Notes in Mathematics 2304,
https://doi.org/10.1007/978-3-030-99011-4_1

where Δ is the Laplacian on $\mathbb{R}^n$. Generalizing Riesz's result, Calderón and Zygmund proved in [47] that these operators are bounded on $L^p(\mathbb{R}^n)$ if $1 < p < \infty$ by developing a theory of singular integrals. Indeed, each Riesz transform R_j is defined by the following integral

$$(R_j f)(x) = \frac{\Gamma(\frac{n+1}{2})}{\pi^{\frac{n+1}{2}}} \, \mathrm{p.\,v.} \int_{\mathbb{R}^n} f(y) \frac{x_j - y_j}{|x - y|^{n+1}} \, dy, \quad f \in \mathscr{S}(\mathbb{R}), \text{ a.e. } x \in \mathbb{R}^n.$$

It is known that the L^p-norms do not depend on the dimension n by Banuelos and Wang [30] and Iwaniec and Martin [112]. More precisely, we have

$$\left\| R_j \right\|_{L^p(\mathbb{R}^n) \to L^p(\mathbb{R}^n)} = \begin{cases} \tan \frac{\pi}{2p} & \text{if } 1 < p \leqslant 2 \\ \cot \frac{\pi}{2p} & \text{if } 2 \leqslant p < \infty \end{cases}. \tag{1.3}$$

The case $n = 1$ is a classical result of Pichorides [186], see [91, 92] for a short proof. In [211] (see also [30, 112]), Stein showed that the vectorial Riesz transform $\nabla(-\Delta)^{-\frac{1}{2}}$ satisfies

$$\left\| \nabla(-\Delta)^{-\frac{1}{2}} f \right\|_{L^p(\mathbb{R}^n, \ell^2)} \lesssim_p \|f\|_{L^p(\mathbb{R}^n)}, \text{ i.e. } \left\| \left(\sum_{j=1}^n |R_j f|^2 \right)^{\frac{1}{2}} \right\|_{L^p(\mathbb{R}^n)} \lesssim_p \|f\|_{L^p(\mathbb{R}^n)} \tag{1.4}$$

with free dimensional bound where $\nabla f \overset{\text{def}}{=} (\partial_1 f, \ldots, \partial_n f)$ is the gradient of a function f belonging to some suitable subspace of $L^p(\mathbb{R}^n)$. It is an open problem to find the best constants in (1.4), explicitly stated in [28, Problem 6]. Furthermore, by duality we have by e.g. [67, Proposition 2.1], [204, p. 5] an equivalence of the form

$$\left\| (-\Delta)^{\frac{1}{2}} f \right\|_{L^p(\mathbb{R}^n)} \approx_p \|\nabla f\|_{L^p(\mathbb{R}^n, \ell_n^2)}, \text{ i.e. } \left\| (-\Delta)^{\frac{1}{2}} f \right\|_{L^p} \approx_p \left\| \left(\sum_{j=1}^n |\partial_j f|^2 \right)^{\frac{1}{2}} \right\|_{L^p(\mathbb{R}^n)}. \tag{1.5}$$

Note that the case $p = 2$ is easy since an integration by parts gives

$$\left\| (-\Delta)^{\frac{1}{2}} f \right\|_{L^2(\mathbb{R}^n)} = \langle -\Delta f, f \rangle_{L^2(\mathbb{R}^n)}^{\frac{1}{2}} = \left(\sum_{j=1}^n \|\partial_j f\|_{L^2(\mathbb{R}^n)}^2 \right)^{\frac{1}{2}} = \|\nabla f\|_{L^2(\mathbb{R}^n, \ell_n^2)}.$$

Moreover, we can show that the domain of the operator $(-\Delta)^{\frac{1}{2}}$ is the Sobolev space

$$\mathrm{W}^{1,2}(\mathbb{R}^n) \overset{\text{def}}{=} \left\{ f \in L^2(\mathbb{R}^n) : \partial_i f \in L^2(\mathbb{R}^n) \text{ for any } 1 \leqslant i \leqslant n \right\}.$$

Note that the case $p = 2$ can be seen as a very particular case of the famous Kato square root problem solved in [21, 23], see also [106, 217]. In this context, the Laplacian $-\Delta$ is replaced by an operator $L \overset{\text{def}}{=} -\text{div}(A\nabla)$ where $A \colon \mathbb{R}^n \to M_n(\mathbb{C})$ is a measurable bounded function satisfying the ellipticity condition

$$\lambda|\xi|^2 \leqslant \text{Re}\langle A\xi, \xi\rangle_{\ell_n^2} \quad \text{and} \quad |\langle A\xi, \zeta\rangle_{\ell_n^2}| \leqslant \Lambda|\xi||\zeta|, \quad \xi, \zeta \in \mathbb{C}^n$$

for some constants $0 < \lambda \leqslant \Lambda$. If $A(x) = I$ for any $x \in \mathbb{R}^n$ we recover the Laplacian $-\Delta$. Note that the precise definition of L relies on the theory of sesquilinear forms on Hilbert spaces, see [136, Chapter 6]. The square root problem essentially introduced by Kato and refined by some other authors consisted to establish the estimate

$$\left\|L^{\frac{1}{2}}f\right\|_{L^2(\mathbb{R}^n)} \approx \|\nabla f\|_{L^2(\mathbb{R}^n, \ell_n^2)}$$

with constants depending only on n, λ and Λ, and the equality $\text{dom}\, L^{\frac{1}{2}} = W^{1,2}(\mathbb{R}^n)$. Some authors investigated the L^p-version of this problem. The corresponding L^p-equivalence is not necessarily true. We refer to the memoir [20] and references therein.

The study of Riesz transforms and variants in many contexts has a long history and was the source of many fundamental developments such as the Calderón-Zygmund theory [47] or the classical work of Stein [210] on Littlewood-Paley Theory. Nowadays, Riesz transforms associated to various geometric structures is a recurrent theme in analysis and geometry. For example, the singular integral

$$(Bf)(z) = -\frac{1}{\pi}\, \text{p.v.} \int_{\mathbb{C}} \frac{f(w)}{(z-w)^2}\, dw, \quad f \in C_c^\infty, \text{ a.e. } z \in \mathbb{C}$$

where the integration is with respect to the Lebesgue measure on the plane, is known as the complex Hilbert transform or the Beurling-Ahlfors transform. This operator is fundamental in complex analysis, since it intertwines the two partial differential operators

$$\partial_z \overset{\text{def}}{=} \frac{1}{2}(\partial_x - i\partial_y), \quad \partial_{\bar{z}} \overset{\text{def}}{=} \frac{1}{2}(\partial_x + i\partial_y)$$

in the sense that

$$B \circ \partial_{\bar{z}} = \partial_z.$$

For any $1 < p < \infty$, this integral induces a bounded operator $B \colon L^p(\mathbb{C}) \to L^p(\mathbb{C})$. In the remarkable works of Donaldson and Sullivan [77] and Iwaniec and Martin [111], it has been observed that the knowledge of the exact value or a good estimate of the L^p-norm of this operator will lead to important applications in the study of quasi-conformal mappings and related nonlinear geometric partial differential equations as well as in the L^p-Hodge decomposition theory. The computation of the exact value is an important open problem known as Iwaniec's conjecture.

In the field of harmonic analysis in the discrete setting, a discrete Hilbert transform was introduced by Hilbert at the beginning of twentieth century. This operator $\mathcal{H}$ maps a finite support sequence $a = (a_n)_{n \in \mathbb{Z}}$ of complex numbers to the sequence $\mathcal{H}a$ defined by

$$(\mathcal{H}a)_n = \frac{1}{\pi} \sum_{m \in \mathbb{Z} \setminus \{0\}} \frac{a_{n-m}}{m}, \quad n \in \mathbb{Z}.$$

Riesz and Titchmarsh proved that for any $1 < p < \infty$, $\mathcal{H}$ induces a bounded operator $\mathcal{H} \colon \ell^p_{\mathbb{Z}} \to \ell^p_{\mathbb{Z}}$ where $\|a\|_{\ell^p_{\mathbb{Z}}} \overset{\text{def}}{=} \left(\sum_{n \in \mathbb{Z}} |a_n|^p \right)^{\frac{1}{p}}$. Recently, the exact computation of the norm of this operator was given in [29], a century after its introduction. The result is identical with the norm (1.3) of the continuous Hilbert transform H defined in (1.1).

Riesz transforms associated with Ornstein-Uhlenbeck operators were studied and play a fundamental role in the Malliavin calculus on the Wiener space [161, 178]. Let n be a fixed positive integer and suppose that γ_n is the standard Gaussian measure on $\mathbb{R}^n$, i.e.

$$d\gamma_n(x) = \frac{1}{(2\pi)^{\frac{n}{2}}} e^{-\frac{|x|}{2}} \, dx, \quad x \in \mathbb{R}^n.$$

Recall that the Ornstein-Uhlenbeck operator L on $L^2(\mathbb{R}^n, \gamma_n)$ is a negative operator defined by

$$L \overset{\text{def}}{=} \Delta - x \cdot \nabla = \sum_{j=1}^{n} \left(\frac{\partial^2}{\partial x_j^2} - x_j \frac{\partial}{\partial x_j} \right).$$

This operator generates the Ornstein-Uhlenbeck semigroup in n dimensions. In [171], Meyer introduced Riesz transforms associated with the Ornstein-Uhlenbeck operator L by

$$R_L \overset{\text{def}}{=} \nabla \circ (-L)^{-\frac{1}{2}}.$$

He proved that for any $1 < p < \infty$, the Riesz transform R_L induces a bounded operator on the space $L^p(\mathbb{R}^n, \gamma_n)$. This result remains true on the infinite-dimensional Wiener space, see e.g. [178, Proposition 1.5.2]. In [187], Pisier gave a different proof of Meyer's result by using the L^p-boundedness of the Hilbert transform H of (1.1).

Arcozzi considered in [6] the Riesz transform

$$R_{S^n} \overset{\text{def}}{=} \nabla_{S^n} \circ (-\Delta_{S_n})^{-\frac{1}{2}}$$

on the n-dimensional unit sphere $S^n \overset{\text{def}}{=} \{x \in \mathbb{R}^{n+1} : \|x\| = 1\}$. Here, Δ_{S_n} denotes the Laplace-Beltrami operator on S^n and ∇_{S^n} is the gradient assocated to S^n. For any $1 < p < \infty$, he showed that

$$\|R_{S^n}\|_{L^p(S^n) \to L^p(S^n, TS^n)} \leqslant 2 \begin{cases} \frac{1}{p-1} & \text{if } 1 < p \leqslant 2 \\ p-1 & \text{if } 2 \leqslant p < \infty \end{cases}. \tag{1.6}$$

We refer to [7] for other Riesz transforms on spheres.

A well-known fact is that many geometric objects on the $(n-1)$-dimensional sphere of radius $\sqrt{n}$ pass in the limit to the corresponding objects on the infinite-dimensional Wiener space, see e.g. [166]. This is often referred to as Poincaré's limit or Poincaré's observation, though the argument can be traced back to the work of Mehler [168]. Combining this observation and the estimates (1.6), Arcozzi obtained for any $1 < p < \infty$ that

$$\|R_L\|_{L^p(\mathbb{R}^n, \gamma_n) \to L^p(\mathbb{R}^n, \gamma_n, \ell_n^2)} \leqslant 2 \begin{cases} \frac{1}{p-1} & \text{if } 1 < p \leqslant 2 \\ p-1 & \text{if } 2 \leqslant p < \infty \end{cases}$$

and the same estimate also for the Riesz transforms acting on the infinite-dimensional Wiener space. The exact value of the norm is an open question. See [142] for a study of the growth of the constants when $p \to 1$ and when $p \to \infty$.

Generalizing the cases of $\mathbb{R}^n$ and the spheres S^n, Strichartz raised in [213] the question concerning the structure of a complete Riemannian manifold M which guarantees that the Riesz transform

$$R_M \overset{\text{def}}{=} \nabla_M \circ (-\Delta_M)^{-\frac{1}{2}}$$

induces a bounded operator on the space $L^p(M)$ for $1 < p < \infty$ where Δ_M is the Laplace-Beltrami operator of M and ∇_M the gradient. A classical result of Bakry [25] is that it is true for any complete Riemannian manifold with nonnegative Ricci curvature and any $1 < p < \infty$. Note that there exist some manifolds for which R_M does not induce a bounded operator on $L^p(M)$ for some (or all) $p > 2$, see e.g. [4] or [50].

Is it conjectured in [67, Conjecture 1.1] that for any $1 < p < 2$ there exists a constant $C_p > 0$ such that

$$\|R_M\|_{L^p(M) \to L^p(M, TM)} \leqslant C_p$$

for all complete Riemannian manifolds M. By Coulhon and Duong [66, Theorem 1.1], we know that the Riesz transform R_M is bounded on $L^p(M)$ for $1 < p \leqslant 2$ as soon as the manifold M satisfies the doubling property

$$V(x, 2r) \lesssim V(x, r), \quad x \in M, r > 0$$

and a diagonal estimate

$$p_t(x, x) \lesssim \frac{1}{V(x, \sqrt{t})}, \quad x \in M, t > 0$$

on the kernel $p_t(x, y)$ of the heat semigroup $(e^{t\Delta})_{t \geqslant 0}$, where $V(x, r) \overset{\text{def}}{=} \mu(B(x, r))$ is the Riemannian volume of the geodesic ball $B(x, r)$ of center x and radius $r > 0$. See the survey [65] and the recent paper [115] and references therein for more information. We equally refer to the surveys [63, 64] for a deeper discussion of heat kernels.

Some authors investigated some generalizations of (1.2) to Lie groups. Let G be a unimodular connected Lie group equipped with a Haar measure μ_G and $\mathfrak{g}$ its Lie algebra which can be identified with the Lie algebra of left invariant vector fields on G where the product is the Lie bracket $(X, Y) \mapsto [X, Y]$. Consider a family $X = (X_1, \ldots, X_m)$ of left-invariant vector fields on G satisfying the Hörmander condition, which means that the Lie subalgebra generated by the X_k's is $\mathfrak{g}$, and such that the vectors $X_1(e), \ldots, X_m(e)$ are linearly independent. We can consider the sublaplacian Δ_G defined by[1]

$$\Delta_G \overset{\text{def}}{=} -\sum_{j=1}^{m} X_j^2.$$

Suppose that the Lie group G has polynomial volume growth, i.e. that there exists an integer $D \geqslant 0$ such that

$$\mu_G(B(x, r)) \approx r^D, \quad r \geqslant 1$$

where $B(x, r)$ is the open ball centred at $x \in G$ and of radius r with respect to the Carnot-Carathéodory metric. Alexopoulos proved in [1, Theorem 2] that in this case the Riesz transforms

$$R_{G,j} \overset{\text{def}}{=} X_j \Delta_G^{-\frac{1}{2}}, \quad j = 1, \ldots, n$$

induce bounded operators on $L^p(G)$ for any $1 < p < \infty$. His approach relies on some ideas inspired by homogenization theory. Recall that a connected nilpotent

---

[1] From now on, the Laplacian type operators incorporate a minus sign and thus are positive definite.

Lie group or a connected compact Lie group have polynomial volume growth (with $D = 0$ in the compact case).

We continue with the case of the hypercube $\{-1, 1\}^n$ where $n \in \mathbb{N}$. For any function $f: \{-1, 1\}^n \to \mathbb{R}$ and any index $j \in \{1, \ldots, n\}$ we can consider the hypercube partial derivative $\partial_j f: \{-1, 1\}^n \to \mathbb{R}$ defined by

$$(\partial_j f)(\varepsilon) \overset{\text{def}}{=} f(\varepsilon) - f(\varepsilon_1, \ldots, \varepsilon_{j-1}, -\varepsilon_j, \varepsilon_{j+1}, \ldots, \varepsilon_n), \quad \varepsilon \in \{-1, 1\}^n. \quad (1.7)$$

We recall the usual Fourier–Walsh expansion of a function $f: \{-1, 1\}^n \to \mathbb{R}$. For any subset A of $\{-1, 1\}^n$ consider the corresponding Walsh function $w_A: \{-1, 1\}^n \to \mathbb{R}$ given by

$$w_A(\varepsilon) \overset{\text{def}}{=} \prod_{j \in A} \varepsilon_j, \quad \varepsilon \in \{-1, 1\}^n \quad \text{and denote} \quad \hat{f}(A) \overset{\text{def}}{=} \frac{1}{2^n} \sum_{\varepsilon \in \{-1,1\}^n} f(\varepsilon) w_A(\varepsilon).$$

$$(1.8)$$

Then we have

$$f(\varepsilon) = \frac{1}{2^n} \sum_{A \subset \{-1,1\}^n} \hat{f}(A) w_A(\varepsilon), \quad \varepsilon \in \{-1, 1\}^n.$$

For any function $f: \{-1, 1\}^n \to \mathbb{R}$ we equally define the discrete Laplacian $\Delta f: \{-1, 1\}^n \to \mathbb{R}$ by

$$(\Delta f)(\varepsilon) \overset{\text{def}}{=} \sum_{A \subset \{1,\ldots,n\}} |A| \, \hat{f}(A) w_A(\varepsilon), \quad \varepsilon \in \{-1, 1\}^n \quad (1.9)$$

where $|A|$ is the cardinal of A. This operator is sometimes called the number operator since $\Delta(w_A) = |A| w_A$. Then we can introduce the hypercube Riesz transforms $\mathsf{R}_1, \ldots, \mathsf{R}_n$ by

$$\mathsf{R}_j \overset{\text{def}}{=} \frac{1}{2} \partial_j \circ \Delta^{-\frac{1}{2}}, \quad j \in \{-1, 1\}^n.$$

More concretely, we can show that

$$(\mathsf{R}_j f)(\varepsilon) = \sum_{\substack{A \subset \{1,\ldots,n\} \\ j \in A}} \frac{\hat{f}(A)}{\sqrt{|A|}} w_A(\varepsilon), \quad \varepsilon \in \{-1, 1\}^n. \quad (1.10)$$

Similarly to (1.4), Lust-Piquard proved[2] in [154, Theorem 0.1 (a)] that for any $2 \leqslant p < \infty$ and any function $f \in \{-1, 1\}^n \to \mathbb{R}$ of mean 0,[3] we have

$$\frac{1}{p^{3/2}} \|f\|_{L^p(\{-1,1\}^n)} \lesssim \left\| \left(\sum_{j=1}^n (\mathsf{R}_j f)^2 \right)^{\frac{1}{2}} \right\|_{L^p(\{-1,1\}^n)} \lesssim p \|f\|_{L^p(\{-1,1\}^n)} .$$

(1.11)

where the L^p norm is taken with respect to the normalized counting measure on the discrete hypercube $\{-1, 1\}^n$, i.e.

$$\|f\|_{L^p(\{-1,1\}^n)} \overset{\text{def}}{=} \left(\frac{1}{2^n} \sum_{\varepsilon \in \{-1,1\}^n} |f(\varepsilon)|^p \right)^{\frac{1}{p}} .$$

We can rewrite these inequalities as

$$\frac{1}{p^{3/2}} \|\Delta^{\frac{1}{2}} f\|_{L^p(\{-1,1\}^n)} \lesssim \left\| \left(\sum_{j=1}^n (\partial_j f)^2 \right)^{\frac{1}{2}} \right\|_{L^p(\{-1,1\}^n)} \lesssim p \|\Delta^{\frac{1}{2}} f\|_{L^p(\{-1,1\}^n)}$$

(1.12)

which are analogues of (1.5). Note that the obtained constants are dimension independent. In sharp contrast with the case of $\mathbb{R}^n$ or the Wiener space, these inequalities are false for $1 < p < 2$ as showed in [154], see also [82, Section 5.5]. Indeed, Lust-Piquard shows that we can use a transference trick to the fermion algebra in order to reduce the problem for $2 \leqslant p < \infty$ to some noncommutative estimates. This striking dichotomy between the cases $1 < p < 2$ and $2 \leqslant p < \infty$ is very natural in noncommutative analysis due to noncommutative Khintchine inequalities [153, 158] (see Theorem 1.5 for a generalization to q-gaussians and crossed products) which imply in general two formulations of the results in noncommutative analysis according to the value of p. Actually, Lust-Piquard showed a more complicated substitute for (1.11) in the case $1 < p < 2$:

$$\|f\|_p \lesssim_p \inf_{\mathsf{R}_j(f) = g_j + h_j} \left\{ \left\| \left(\sum_{j=1}^n |g_j|^2 \right)^{\frac{1}{2}} \right\|_p + \left\| \left(\sum_{j=1}^n |\mathcal{T}_{e_j}(h_j)|^2 \right)^{\frac{1}{2}} \right\|_p \right\} \lesssim_p \|f\|_p$$

(1.13)

---

[2] The constants proved by Lust-Piquard were worse than the ones of (1.11).
[3] That means that

$$\frac{1}{2^n} \sum_{\varepsilon \in \{-1,1\}^n} f(\varepsilon) = 0.$$

where the infimum is taken over all decompositions with $g_j, h_j \colon \{-1, 1\}^n \to \mathbb{R}$ and where $\mathcal{T}_{e_j}$ is the translation by $e_j = (1, \ldots, -1, 1, \ldots, 1)$ where -1 occurs at coordinate j.

For the study of Hilbert transforms and Riesz transforms in other contexts, the interested reader can consult the papers [79, 152, 209, 210, 218] (Lie groups), [5] (symmetric spaces), [154] (fermion algebras), [155] (deformed Gaussian algebras), [76, 156] (abelian groups), [157] (generalized Heisenberg groups), [24, 55, 203] (graphs), [19, 73, 103] (Schrödinger operators), [101, 102] (fields of p-adic numbers), [114, 163, 164, 173, 220] (fractal sets and measures), [169] (von Neumann algebras of free groups), [51] (free Araki-Woods factors), [119, 184, 192] (von Neumann algebras), [52] (quantum groups) and [224] (free probability). Indeed, Riesz transforms have become a cornerstone of analysis and the literature is quite huge and it would be impossible to give complete references here. We refer to [28, 204] for nice surveys.

An important generalization of (1.5) was given by Meyer [171]. It consists in replacing the Laplacian Δ by the L^p-realization A_p of the negative infinitesimal generator A of a Markov semigroup $(T_t)_{t \geqslant 0}$ of operators acting on the L^p-spaces of a measure space Ω and to replace the gradient ∇ by the "carré du champ" Γ introduced by Roth [202] (see also [105]) defined[4] by

$$\Gamma(f, g) \overset{\text{def}}{=} \frac{1}{2}\big[A(\overline{f})g + \overline{f}A(g) - A(\overline{f}g)\big]. \tag{1.14}$$

In the case of the Heat semigroup $(e^{-t\Delta})_{t \geqslant 0}$ with generator Δ, we recover the gradient form $\langle \nabla f, \nabla g \rangle_{\ell_n^2}$. Meyer was interested in the equivalence

$$\big\| A_p^{\frac{1}{2}}(f) \big\|_{L^p(\Omega)} \approx_p \big\| \Gamma(f, f)^{\frac{1}{2}} \big\|_{L^p(\Omega)} \tag{1.15}$$

on some suitable subspace (ideally dom $A_p^{\frac{1}{2}}$) of $L^p(\Omega)$. Meyer proved such equivalence for the Ornstein-Uhlenbeck semigroup. Nevertheless, with sharp contrast, if $1 < p < 2$ these estimates are surprisingly false for the Poisson semigroup on $L^p(\mathbb{R}^n)$ which is a Markov semigroup of Fourier multipliers, see [131, Appendix D]. Actually, as we said, other examples of semigroups illustrating this phenomenon are already present in the papers of Lust-Piquard [156, Proposition 2.9] and [154, p. 283] relying on an observation of Lamberton.

Of course, when something goes wrong with a mathematical problem it is rather natural to change slightly the formulation of the problem in order to obtain a natural positive statement. By introducing some gradients with values in a *noncommutative* space, Junge, Mei and Parcet obtained in [131] dimension free estimates for Riesz transforms associated with arbitrary Markov semigroups $(T_t)_{t \geqslant 0}$ of Fourier multipliers acting on classical L^p-spaces $L^p(\hat{G})$ where G is for example

---

[4] Here, the domain of A must contain a suitable involutive algebra.

an abelian discrete group with (compact) dual group $\hat{G}$ (and more generally on the noncommutative L^p-spaces $L^p(\text{VN}(G))$ associated with a nonabelian group G). We denote by $\psi \colon G \to \mathbb{C}$ the symbol of the (negative) infinitesimal generator A of the semigroup. In the spirit of (1.5), the previous authors proved estimates of the form

$$\left\| A_p^{\frac{1}{2}}(f) \right\|_{L^p(\hat{G})} \approx_p \left\| \partial_{\psi,1,p}(f) \right\|_{L^p(L^\infty(\Omega) \rtimes_\alpha G)} \tag{1.16}$$

where $\partial_{\psi,1,p}$ is some kind of gradient defined on a dense subspace of the classical L^p-space $L^p(\hat{G})$. It takes values in a closed subpace $\Omega_{\psi,1,p}$ of a *noncommutative* L^p-space $L^p(L^\infty(\Omega) \rtimes_\alpha G)$ associated with some crossed product $L^\infty(\Omega) \rtimes_\alpha G$ where Ω is a probability space and where $\alpha \colon G \to \text{Aut}(L^\infty(\Omega))$ is an action of G on $L^\infty(\Omega)$ determined by the semigroup. Let us explain the simplest case, i.e. the case where α is trivial. In this *non-crossed and very particular situation*, we have an identification of $L^p(L^\infty(\Omega) \rtimes_\alpha G)$ with the classical L^p-space $L^p(\Omega \otimes \hat{G})$ and the map $\partial_{\psi,1,p}$ is defined on the span of characters $\langle s, \cdot \rangle_{G,\hat{G}}$ in $L^p(\hat{G})$ with values in $L^p(\Omega \otimes \hat{G})$. It is defined by

$$\partial_{\psi,1,p}\big(\langle s, \cdot \rangle_{G,\hat{G}} \big) \stackrel{\text{def}}{=} \text{W}(b_\psi(s)) \otimes \langle s, \cdot \rangle_{G,\hat{G}}. \tag{1.17}$$

where $\text{W} \colon H \to L^0(\Omega)$ is an H-isonormal *Gaussian process*[5] for some real Hilbert space H and where $b_\psi \colon G \to H$ is a specific function satisfying

$$\psi(s) = \left\| b_\psi(s) \right\|_H^2, \quad s \in G. \tag{1.18}$$

We refer to Sect. 2.6 for the (crossed) general situation where the action α is obtained by second quantization from an orthogonal representation $\pi \colon G \to \text{B}(H)$ associated to the semigroup, see (2.83).

The approach by Junge, Mei and Parcet highlights an intrinsic noncommutativity since $\Omega_{\psi,1,p}$ is in general a highly noncommutative object *although* the group G may be abelian. It is fair to say that this need of noncommutativity was first noticed and explicitly written by Lust-Piquard in [154, 156] in some particular cases under a somewhat different but essentially equivalent form of (1.16), see (1.13). Moreover, it is remarkable that the estimates of [156] were exploited in a decisive way by Naor [172] to understand subtle geometric phenomena. Finally, note that the existence of gradients suitable for arbitrary Markov semigroups of linear operators appears already in the work of Sauvageot and Cipriani, see [57, 205] and the survey [56]. Finally, we refer to [17, 128, 139, 191] and references therein for more information on noncommutative L^p-spaces.

In the context of Riesz transforms, the authors of the classical and remarkable paper [23] were the first to introduce suitable Hodge-Dirac operators. The L^p-

---

[5] In particular, for any $h \in H$ the random variable $\text{W}(h)$ is a centred real Gaussian.

boundedness of the H^∞ calculus of this unbounded operator allows everyone to obtain immediately the L^p-boundedness of Riesz transforms. The authors of [131] introduced a similar operator in the context of Markov semigroups $(T_t)_{t \geqslant 0}$ of Fourier multipliers acting on classical L^p-spaces and more generally on noncommutative L^p-spaces $L^p(\mathrm{VN}(G))$ associated with group von Neumann algebras $\mathrm{VN}(G)$ where $1 < p < \infty$ and where G is a discrete group. We refer to the papers [56, Definition 10.4], [58, 107, 108, 160, 165, 176] for Hodge-Dirac operators in related contexts.

Recall that if G is a discrete group then the von Neumann algebra $\mathrm{VN}(G)$, whose elements are bounded operators acting on the Hilbert space ℓ_G^2, is generated by the left translation unitaries $\lambda_s \colon \ell_G^2 \to \ell_G^2$, $\delta_r \mapsto \delta_{sr}$ where $r, s \in G$. If G is abelian, then $\mathrm{VN}(G)$ is $*$-isomorphic to the algebra $L^\infty(\hat{G})$ of essentially bounded functions on the dual group $\hat{G}$ of G. In this case, we can see the functions of $L^\infty(\hat{G})$ as multiplication operators on $L^2(\hat{G})$. As basic models of quantum groups, these von Neumann algebras play a fundamental role in operator algebras. Moreover, we can equip $\mathrm{VN}(G)$ with a normalized trace (=noncommutative integral) and if $1 \leqslant p \leqslant \infty$ we have a canonical identification

$$L^p(\mathrm{VN}(G)) = L^p(\hat{G}). \tag{1.19}$$

A Markov semigroup $(T_t)_{t \geqslant 0}$ of Fourier multipliers on $\mathrm{VN}(G)$ is characterized by a conditionally negative length $\psi \colon G \to \mathbb{C}$ such that the symbol of each operator T_t of the semigroup is $e^{-t\psi}$. Moreover, the symbol of the (negative) infinitesimal generator A_p on the noncommutative L^p-space $L^p(\mathrm{VN}(G))$ of the semigroup is ψ.

Introducing the Banach space $L^p(\mathrm{VN}(G)) \oplus_p \Omega_{\psi,1,p}$, the authors of [131] define the Hodge-Dirac operator

$$D_{\psi,1,p} \overset{\text{def}}{=} \begin{bmatrix} 0 & (\partial_{\psi,1,p})^* \\ \partial_{\psi,1,p} & 0 \end{bmatrix} \tag{1.20}$$

which is an unbounded operator defined on a dense subspace. In [131, Problem C.5], the authors ask for dimension free estimates for the operator $\operatorname{sgn} D_{\psi,1,p} \overset{\text{def}}{=} D_{\psi,1,p}|D_{\psi,1,p}|^{-1}$. We affirmatively answer this question for a large class of groups including all amenable discrete groups and free groups by showing the following result in the spirit of [23] (see also [22]). Here we use the bisector $\Sigma_\omega^\pm \overset{\text{def}}{=} \Sigma_\omega \cup (-\Sigma_\omega)$ where $\Sigma_\omega \overset{\text{def}}{=} \{z \in \mathbb{C} \backslash \{0\} : |\arg z| < \omega\}$ for any angle $\omega \in (0, \frac{\pi}{2})$.

Theorem 1.1 (see Theorems 4.3, 4.4 and 4.6 and Remark 4.3) *Suppose* $1 < p < \infty$. *Let* G *be a weakly amenable discrete group such that the crossed product* $L^\infty(\Omega) \rtimes_\alpha G$ *has QWEP. The Hodge-Dirac operator* $D_{\psi,1,p}$ *is bisectorial on* $L^p(\mathrm{VN}(G)) \oplus_p \Omega_{\psi,1,p}$ *and admits a bounded* $H^\infty(\Sigma_\omega^\pm)$ *functional calculus on a* *bisector* $\Sigma_\omega^\pm$. *Moreover, the norm of the functional calculus is bounded by a constant* $K_{\omega,p}$ *which depends neither on G nor on the semigroup.*[6]

We refer to [83, 110] for more information on bisectorial operators. Roughly speaking, our result says that

$$\left\| f(D_{\psi,1,p}) \right\|_{L^p(\mathrm{VN}(G))\oplus_p \Omega_{\psi,1,p} \to L^p(\mathrm{VN}(G))\oplus_p \Omega_{\psi,1,p}} \leqslant K_{\omega,p} \, \|f\|_{H^\infty(\Sigma_\omega^\pm)} \qquad (1.21)$$

for any suitable function f of $H^\infty(\Sigma_\omega^\pm)$. Using the function sgn defined by $\mathrm{sgn}(z) \stackrel{\mathrm{def}}{=} 1_{\Sigma_\omega}(z) - 1_{-\Sigma_\omega}(z)$, we obtain the estimate for the operator sgn $D_{\psi,1,p}$. Our result can be seen as a strengthening of the dimension free estimates (1.16) of Riesz transforms of the previous authors since it is almost immediate that the boundedness of the H^∞ functional calculus implies the equivalence (1.16), see Remark 4.1. Note that the H^∞ functional calculus of bisectorial Hodge-Dirac operators plays an important role in the geometry of Riemannian manifolds, notably for the regularity properties of geometric flows, the Riesz continuity of the Atiyah-Singer Dirac operator or even for boundary value problems of elliptic operators. We refer to the recent survey [27] for more information. We expect that our H^∞ functional calculus result from Theorem 4.4 and its dimension free estimate from Remark 4.3 will have similar geometric consequences for the noncommutative manifolds (i.e. spectral triples) investigated in Chap. 5.

Our argument relies in part on a new transference argument between Fourier multipliers on crossed products and classical Fourier multipliers (see Proposition 2.8) which is of independent interest and which needs that a crossed product has QWEP, which is an approximation property, see [181]. So we need a QWEP assumption on the von Neumann algebra $L^\infty(\Omega) \rtimes_\alpha G$ of Theorem 1.1. Note that this assumption is satisfied for amenable groups by Ozawa [181, Proposition 4.1] and for free groups $\mathbb{F}_n$ by the same reasoning as used in the proof of [10, Proposition 4.8].

Moreover, we assume that the discrete group G is weakly amenable. Indeed, we need in the proof some form of L^p-summability of noncommutative Fourier series in order to work with elements whose Fourier series have finite support. The weak amenability assumption and our transference result allow us to have uniformly bounded approximations of arbitrary elements of noncommutative L^p-spaces associated to crossed products, see the discussion at the end of Sect. 2.4.

We also show a q-gaussian version of Theorem 1.1, see Theorem 4.3. This kind of generalization will be useful in Chap. 5. In this introduction, we will explain its interest in the discussion following Theorem 1.4.

With the help of an extension of this result (Theorem 4.4), we obtain a Hodge decomposition (see Theorem 4.5). In particular, we are able to deduce functional calculus for an extension of $D_{\psi,1,p}$ on the whole space $L^p(\mathrm{VN}(G)) \oplus_p L^p(L^\infty(\Omega) \rtimes_\alpha G)$.

---

[6] In particular, it is independent of the dimension of the Hilbert space H associated with the 1-cocycle by Proposition 2.3.

Below, we describe concrete semigroups in which Theorem 1.1 applies.

Semigroups on Abelian Groups Recall that a particular case of [33, Corollary 18.20] says that a function $\psi \colon G \to \mathbb{R}$ on a discrete abelian group G is a conditionally negative length if and only if there exists a quadratic form[7] $q \colon G \to \mathbb{R}^+$ and a symmetric positive measure μ on $\hat{G} - \{0\}$ such that

$$\int_{\hat{G}-\{0\}} \left(1 - \operatorname{Re}\chi(s)\right) d\mu(\chi) < \infty$$

for any $s \in G$ satisfying

$$\psi(s) = q(s) + \int_{\hat{G}-\{0\}} \left(1 - \operatorname{Re}\chi(s)\right) d\mu(\chi), \quad s \in G. \tag{1.22}$$

In this case, μ is the so called Lévy measure of ψ and q is determined by the formula $q(s) = \lim_{n\to+\infty} \frac{\psi(ns)}{n^2}$. This is the Lévy-Khintchine representation of ψ as a continuous sum of elementary conditionally negative lengths.[8]

(a) If $G = \mathbb{Z}^n$, we recover the semigroups on the L^p-spaces $L^p(\mathrm{VN}(\mathbb{Z}^n)) \overset{(1.19)}{=} L^p(\mathbb{T}^n)$ of the torus $\mathbb{T}^n$. For example, taking $\mu = 0$ and $\psi(k_1, \dots, k_n) = q(k_1, \dots, k_n) = k_1^2 + \cdots + k_n^2$, we obtain the function defining the heat semigroup $(e^{-t\Delta})_{t \geqslant 0}$. By choosing $q = 0$ and the right measure μ, we can obtain the Poisson semigroup $(e^{-t\Delta^{\frac{1}{2}}})_{t \geqslant 0}$ or more generally the semigroups associated with the fractional Laplacians Δ^α with $0 \leqslant \alpha < 2$ [200, 201]. Note that in the particular case of the Poisson semigroup on the L^p-space $L^p(\mathbb{T})$ associated with the torus $\mathbb{T}$, the Lévy measure is given by

$$d\mu(e^{ix}) = \operatorname{Re}\left[\frac{-2e^{ix}}{(1 - e^{ix})^2}\right] d\mu_{\mathbb{T}}(e^{ix}),$$

where $\mu_{\mathbb{T}}$ denotes the normalized Haar measure on the torus.

(b) Fix some integer $n \geqslant 1$. We consider the group G of Walsh functions w_A defined in (1.8) where $\varepsilon = (\varepsilon_1, \dots, \varepsilon_n)$ belongs to the discrete abelian group $\hat{G} = \{-1, 1\}^n$. For any $1 \leqslant i \leqslant n$, we let $e_i \overset{\text{def}}{=} (1, \dots, 1, -1, 1, \dots, 1)$. If we consider the atomic measure $\mu = \frac{1}{2}\sum_{i=1}^n \delta_{e_i}$ on $\hat{G} - \{(1, \dots, 1)\}$ and

---

[7] That means that $2q(s) + 2q(t) = q(s+t) + q(s-t)$ for any $s, t \in G$.
[8] Recall that a quadratic form $q \colon G \to \mathbb{R}^+$ and a function $G \to \mathbb{R}$, $s \mapsto 1 - \operatorname{Re}\chi(s)$ are conditionally negative lengths by Berg and Forst [33, Proposition 7.19, Proposition 7.4 (ii), and Corollary 7.7].

$q = 0$, we obtain[9] $\psi(w_A) = |A|$. So we recover the discrete Heat semigroup[10] of [110, p. 19] whose generator is the discrete Laplacian (1.9). It is also related to [82, 154, 156, 172].

Semigroups on Finitely Generated Groups Let G be a finitely generated group and S be a generating set for G such that $S^{-1} = S$ and $e \notin S$. Any element s admits a decomposition

$$s = s_1 s_2 \cdots s_n \tag{1.23}$$

where $s_1, \ldots, s_n$ are elements of S. The word length $|s|$ of s with respect to the generating set S is defined to be the minimal integer n of such a decomposition and is a basic notion in geometric group theory. As a special case, the neutral element e has length zero.

(a) **Coxeter groups** Here, we refer to [45] and references therein for more information. Recall that a group $G = W$ is called a Coxeter group if W admits the following presentation:

$$W = \left\langle S \mid (s_1 s_2)^{m(s_1, s_2)} = e : s_1, s_2 \in S, m(s_1, s_2) \neq \infty \right\rangle$$

where $m: S \times S \to \{1, 2, 3, \ldots, \infty\}$ is a function such that $m(s_1, s_2) = m(s_2, s_1)$ for any $s_1, s_2 \in S$ and $m(s_1, s_2) = 1$ if and only if $s_1 = s_2$. The pair (W, S) is called a Coxeter system. In particular, every generator $s \in S$ has order two. By Bożejko [41, Theorem 7.3.3], the word length $|s|$ is a conditionally negative length and our results can be used with the semigroup generated by this function. Recall that dihedral groups

$$D_n = \left\langle s_1, s_2 \mid s_1^2 = s_2^2 = (s_1 s_2)^2 = e \right\rangle,$$

product groups $\mathbb{Z}_2 \times \cdots \times \mathbb{Z}_2$, symmetric groups S_n with $S = \{(n, n+1) : n \in \mathbb{N}\}$ and the infinite symmetric group S_∞ of all finite permutations of the set $\mathbb{N}$ with $S = \{(n, n+1) : n \in \mathbb{N}\}$ are examples of Coxeter groups. In the case of symmetric groups, the length $|\sigma|$ of σ is the number of crossings in the diagram which represents the permutation σ.

---

[9] For any $1 \leqslant i \leqslant n$, note that $w_A(e_i) \overset{(1.8)}{=} \prod_{j \in A}(e_i)_j$ is equal to 1 if $i \notin A$ and to -1 if $i \in A$. So we have

$$\psi(w_A) \overset{(1.22)}{=} \frac{1}{2} \sum_{i=1}^n (1 - w_A(e_i)) = \frac{1}{2}\left(n - \sum_{i=1}^n w_A(e_i)\right) = |A|.$$

[10] By analogy with the case of $\mathbb{R}^n$, the name "discrete Poisson semigroup" seems more appropriate.

Consider a Coxeter group W. If $s \in W$, note that the sequence $s_1, \ldots, s_n$ in (1.23) chosen in such a way that n is minimal is not unique in general. However, the set of involved generators is unique, i.e. if $s = s_1 s_2 \cdots s_n = s_1' s_2' \cdots s_n'$ are minimal words of $s \in W$ then $\{s_1, s_2, \ldots, s_n\} = \{s_1', s_2', \ldots, s_n'\}$. This subset $\{s_1, s_2, \ldots, s_n\}$ of S is denoted S_s and is called the colour of s, following [45, p. 585]. We define the colour-length of s putting $\|s\| \stackrel{\text{def}}{=} \text{card } S_s$. We always have $\|w\| \leqslant |w|$. By Bożejko et al. [45, Theorem 4.3 and Corollary 5.4], if $0 \leqslant \alpha \leqslant 1$ the functions $|\cdot|^{\alpha}$ and $\|\cdot\|$ are conditionally negative lengths on S_∞. Finally, see also [129, p. 1971] for other examples for S_n for $n < \infty$.

(b) **Free groups** Our results can be used with the noncommutative Poisson semigroup [93], [128, Definition 10.1] on free groups $\mathbb{F}_n$ ($1 \leqslant n \leqslant \infty$) whose negative generator is the length $|\cdot|$. Moreover, we can use the characterization [95, Theorem 1.2] of radial functions $\psi \colon \mathbb{F}_n \to \mathbb{C}$ with $\psi(e) = 0$ which are conditionally negative definite. If $\varphi_z \colon \mathbb{F}_n \to \mathbb{C}$ denotes the spherical function[11] of parameter $z \in \mathbb{C}$, these functions can be written

$$\psi(s) = \int_{-1}^{1} \psi_z(s) \, d\nu(z), \quad s \in \mathbb{F}_n$$

for some finite positive Borel measure ν on $[-1, 1]$ where $\psi_z(s) \stackrel{\text{def}}{=} \frac{1 - \varphi_z(s)}{1 - z}$, $z \in \mathbb{C} \setminus \{1\}$ and $\psi_1(s) \stackrel{\text{def}}{=} \lim_{z \to 1} \frac{1 - \varphi_z(s)}{1 - z}$. Finally, [130] contains (but without proof) examples of weighted forms of the word length which are conditionally negative.

We can write $\mathbb{F}_n = *_{i=1}^{n} G_i$ with $G_i = \mathbb{Z}$. Every element s of $\mathbb{F}_n - \{e\}$ has a unique representation $s = s_{i_1} s_{i_2} \cdots s_{i_m}$, where $s_{i_k} \in G_{i_k}$ are distinct from the corresponding neutral elements and $i_1 \neq i_2 \neq \cdots \neq i_k$. The number m is called the block length of s and denoted $\|s\|$. By Junge and Zeng [126, Example 6.14], this function is a conditionally negative length.

(c) **Cyclic groups** The word length on $\mathbb{Z}_n$ is given by $|k| = \min\{n, n - k\}$. It is known that this length is negative definite, see for example [131, p. 553], [130, Appendix B], [126, Section 5.3] and [127, Example 5.9] for more information. By Junge et al. [132, p. 925], the function ψ_n defined on $\mathbb{Z}_n$ by $\psi_n(k) = \frac{n^2}{2\pi^2} \left(1 - \cos\left(\frac{2\pi k}{n}\right)\right)$ is another example of conditionally negative length on $\mathbb{Z}_n$.

Semigroups on the Discrete Heisenberg Group Let $\mathbb{H} = \mathbb{Z}^{2n+1}$ be the discrete Heisenberg group with group operations

$$(a, b, t) \cdot (a', b', t') = (a + a', b + b', t + t' + ab') \quad \text{and}$$

$$(a, b, t)^{-1} = (-a, -b, -t + ab) \tag{1.24}$$

---

[11] In the case $n = \infty$, we have $\varphi_z(s) = z^{|s|}$ for any $s \in \mathbb{F}_\infty$ and any $z \in \mathbb{C}$.

where $a, b, a', b' \in \mathbb{Z}^n$ and $t, t' \in \mathbb{Z}$. By Junge and Zeng [126, Proposition 5.13] and [127, p. 261], the map $\psi \colon \mathbb{H} \to \mathbb{R}$, $(a, b, t) \mapsto |a| + |b|$ is a conditionally negative length.

We also prove in this paper an analogue of the equivalences (1.16) for markovian semigroups $(T_t)_{t \geqslant 0}$ of Schur multipliers acting on Schatten spaces $S_I^p \overset{\text{def}}{=} S^p(\ell_I^2)$ for $1 < p < \infty$ where I is an index set. In this case, by Arhancet [10, Proposition 5.4], the Schur multiplier symbol $[a_{ij}]$ of the negative generator A of $(T_t)_{t \geqslant 0}$ is given by $a_{ij} = \|\alpha_i - \alpha_j\|_H^2$ for some family $\alpha = (\alpha_i)_{i \in I}$ of vectors of a real Hilbert space H. We define a gradient operator $\partial_{\alpha,1,p}$ as the closure of the unbounded linear operator $M_{I,\text{fin}} \to L^p(\Omega, S_I^p)$, $e_{ij} \mapsto W(\alpha_i - \alpha_j) \otimes e_{ij}$ where $W \colon H \to L^0(\Omega)$ is an H-isonormal Gaussian process, Ω is the associated probability space and where $M_{I,\text{fin}}$ is the subspace of S_I^p of matrices with a finite number of non null entries. Then the result reads as follows.

Theorem 1.2 (see Theorem 3.3 and (3.88)) *Let I be an index set and A be the negative generator of a markovian semigroup $(T_t)_{t \geqslant 0}$ of Schur multipliers on $B(\ell_I^2)$. Suppose $1 < p < \infty$. For any $x \in M_{I,\text{fin}}$, we have*

$$\left\| A_p^{\frac{1}{2}}(x) \right\|_{S_I^p} \approx_p \left\| \partial_{\alpha,1,p}(x) \right\|_{L^p(\Omega, S_I^p)}. \tag{1.25}$$

We also obtain an analogue of Theorem 1.1. With this result, we are equally able to obtain a Hodge decomposition, see Theorem 4.10.

Theorem 1.3 (see Theorems 4.9 and 4.11) *Suppose $1 < p < \infty$. The unbounded operator* $\mathcal{D}_{\alpha,1,p} \overset{\text{def}}{=} \begin{bmatrix} 0 & (\partial_{\alpha,1,p^*})^* \\ \partial_{\alpha,1,p} & 0 \end{bmatrix}$ *on the Banach space $S_I^p \oplus L^p(\Omega, S_I^p)$ is bisectorial and admits a bounded $H^\infty(\Sigma_\omega^\pm)$ functional calculus on a bisector $\Sigma_\omega^\pm$. Moreover, the norm of the functional calculus is bounded by a constant $K_{\omega,p}$ which depends neither on I nor on the semigroup. In particular it is independent of the dimension of H.*

Moreover, we also relate the equivalences (1.16) and (1.25) with the ones of Meyer's formulation (1.15). To achieve this, we define and study in the spirit of (1.14) a carré du champ Γ (see (2.78) and (2.91)) and its closed extension in the sense of Definition 2.7 and we connect this notion to some approximation properties of groups. It leads us to obtain alternative formulations of (1.16) and (1.25). Note that some carrés du champ were studied in the papers [56], [57, Section 9], [119, 126, 205] mainly in the σ-finite case and for L^2-spaces (see [70] for related things) but unfortunately their approach does not suffice for our work on L^p-spaces. By the way, it is rather surprising that even in the *commutative* setting, no one has examined the carré du champ on L^p-spaces with $p \neq 2$. The following is an example of result that we have achieved, and which can be compared with (1.15).

Theorem 1.4 (see Theorem 3.5) *Suppose* $2 \leqslant p < \infty$. *Let* A *be the negative generator of a markovian semigroup of Schur multipliers on* $\mathrm{B}(\ell_I^2)$. *For any* $x \in$ dom $A_p^{\frac{1}{2}}$, *we have*

$$\left\| A_p^{\frac{1}{2}}(x) \right\|_{S_I^p} \approx_p \max \left\{ \left\| \Gamma(x,x)^{\frac{1}{2}} \right\|_{S_I^p}, \left\| \Gamma(x^*,x^*)^{\frac{1}{2}} \right\|_{S_I^p} \right\}. \tag{1.26}$$

The maximum is natural in noncommutative analysis due to the use of noncommutative Khintchine inequalities.

It is remarkable that the point of view of Hodge-Dirac operators fits perfectly into the setting of noncommutative geometry if $p = 2$. If G is a discrete group, the Hilbert space $H \overset{\text{def}}{=} \mathrm{L}^2(\mathrm{VN}(G)) \oplus_2 \Omega_{\psi,1,2}$, the $*$-algebra $A \overset{\text{def}}{=} \mathrm{span}\{\lambda_s : s \in G\}$ of trigonometric polynomials and the Hodge-Dirac operator $D_{\psi,1,2}$ on $\mathrm{L}^2(\mathrm{VN}(G)) \oplus_2 \Omega_{\psi,1,2}$ define a triple $(A, H, D_{\psi,1,2})$ in the spirit of noncommutative geometry [60, 90, 222]. Recall that the notion of spectral triple (A, H, D) (= noncommutative manifold) à la Connes covers a huge variety of different geometries such as Riemmannian manifolds, fractals, quantum groups or even non-Hausdorff spaces. We refer to [62] for an extensive list of examples and to [53, 61, 141, 208, 222] for some surveys. From here, it is apparent that we can see Markov semigroups of Fourier multipliers as geometric objects. The same observation is true for Markov semigroups of Schur multipliers. Nevertheless, the Hilbert space setting of the noncommutative geometry is too narrow to encompass our setting on L^p-spaces. So, we develop in Sect. 5.7 a natural Banach space variant (A, X, D) of a spectral triple where the selfadjoint operator D acting on the Hilbert space H is replaced by a bisectorial operator D acting on a (reflexive) Banach space X, allowing us to use (noncommutative) L^p-spaces $(1 < p < \infty)$.

It is well-known that Gaussian variables are not bounded, i.e. do not belong to $\mathrm{L}^\infty(\Omega)$. From the perspective of noncommutative geometry, this is problematic under technical aspects as the boundedness of the commutators $[D, \pi(a)]$ of a spectral triple (A, H, D). Indeed, the noncommutative gradients (1.17) appear naturally in the commutators of our spectral triples and these gradients are defined with Gaussian variables. Fortunately, the noncommutative setting is very flexible and allows us to introduce a continuum of gradients $\partial_{\psi,q,p}$ and $\partial_{\alpha,q,p}$ indexed by a new parameter $-1 \leqslant q \leqslant 1$ replacing Gaussian variables $(q = 1)$ by *bounded* noncommutative q-deformed Gaussian variables $(q < 1)$ and $\mathrm{L}^\infty(\Omega)$ by the von Neumann algebra $\Gamma_q(H)$ of [43, 44]. Note that $\Gamma_{-1}(H)$ is the fermion algebra and that $\mathrm{L}^\infty(\Omega)$ can be identified with the boson algebra $\Gamma_1(H)$. Our main theorems on Hodge-Dirac operators admit extensions in these cases, see Theorems 4.3 and 4.6. We expect some differences of behaviour when q varies. We also prove that

Theorem 1.2 and its counterpart for markovian semigroups of Fourier multipliers
are valid under the forms

$$\left\| A_p^{\frac{1}{2}}(x) \right\|_{S_I^p} \approx_p \left\| \partial_{\alpha,q,p}(x) \right\|_{L^p(\Gamma_q(H) \overline{\otimes} B(\ell_I^2))} \tag{1.27}$$

$$\left\| A_p^{\frac{1}{2}}(x) \right\|_{L^p(\mathrm{VN}(G))} \approx_p \left\| \partial_{\psi,q,p}(x) \right\|_{L^p(\Gamma_q(H) \rtimes_\alpha G)} \tag{1.28}$$

with gradients $\partial_{\alpha,q,p}$ and $\partial_{\psi,q,p}$ taking values in L^p spaces over amplifications of the
q-deformed Gaussian von Neumann algebra $\Gamma_q(H)$. A major ingredient for (1.27)
and (1.28) are Khintchine inequalities for q-Gaussians for that we give full proofs.
They read in the following form.

Theorem 1.5 (Khintchine Inequalities, see Theorem 3.1 and Lemma 3.19)
Consider $-1 \leqslant q \leqslant 1$ *and* $1 < p < \infty$.

1. Let G be a discrete group. Denote the conditional expectation $\mathbb{E}: \Gamma_q(H) \rtimes_\alpha G \to$
*$\mathrm{VN}(G)$, $x \rtimes \lambda_s \mapsto \tau_{\Gamma_q(H)}(x)\lambda_s$. Then for any finite sum $f = \sum_{s,h} f_{s,h} s_q(h) \rtimes \lambda_s$,
we have in case* $1 < p < 2$

$$\|f\|_{L^p(\Gamma_q(H) \rtimes_\alpha G)} \approx_p \inf_{f=g+h} \left\{ \left\| \left(\mathbb{E}(g^*g)\right)^{\frac{1}{2}} \right\|_{L^p(\mathrm{VN}(G))}, \left\| \left(\mathbb{E}(hh^*)\right)^{\frac{1}{2}} \right\|_{L^p(\mathrm{VN}(G))} \right\}$$

and in case $2 < p < \infty$

$$\|f\|_{L^p(\Gamma_q(H) \rtimes_\alpha G)} \approx_p \max \left\{ \left\| \left(\mathbb{E}(f^*f)\right)^{\frac{1}{2}} \right\|_{L^p(\mathrm{VN}(G))}, \left\| \left(\mathbb{E}(ff^*)\right)^{\frac{1}{2}} \right\|_{L^p(\mathrm{VN}(G))} \right\}.$$

2. Let I be an index set. Denote the conditional expectation $\mathbb{E}: \Gamma_q(H) \overline{\otimes} B(\ell_I^2) \to$
*$B(\ell_I^2)$, $x \otimes y \mapsto \tau_{\Gamma_q(H)}(x)y$. Then for any finite sum $f = \sum_{i,j,h} f_{i,j,h} s_q(h) \otimes e_{ij}$,
we have in case* $1 < p < 2$

$$\|f\|_{L^p(\Gamma_q(H) \overline{\otimes} B(\ell_I^2))} \approx_p \inf_{f=g+h} \left\{ \left\| \left(\mathbb{E}(g^*g)\right)^{\frac{1}{2}} \right\|_{S_I^p}, \left\| \left(\mathbb{E}(hh^*)\right)^{\frac{1}{2}} \right\|_{S_I^p} \right\}$$

and in case $2 < p < \infty$

$$\|f\|_{L^p(\Gamma_q(H) \overline{\otimes} B(\ell_I^2))} \approx_p \max \left\{ \left\| \left(\mathbb{E}(f^*f)\right)^{\frac{1}{2}} \right\|_{S_I^p}, \left\| \left(\mathbb{E}(ff^*)\right)^{\frac{1}{2}} \right\|_{S_I^p} \right\}.$$

In Chap. 5, we give a first elementary study of the triples associated to Markov
semigroups of multipliers. In particular, we give sufficient conditions for the
verification of axioms of noncommutative geometry and more generally the axioms
of our Banach spectral triples. One of our main results in this part reads as follows
and can be used with arbitrary amenable groups or free groups. Here $C_r^*(G)$ denotes
the reduced C*-algebra of the discrete group G.

Theorem 1.6 (see Theorem 5.4) *Suppose* $1 < p < \infty$ *and* $-1 \leqslant q < 1$. *Let* G *be a weakly amenable discrete group. Let* $(T_t)_{t \geqslant 0}$ *be a markovian semigroup of Fourier multipliers on the group von Neumann algebra* $\mathrm{VN}(G)$. *Consider the associated function* $b_\psi \colon G \to H$ *from* (1.18). *Assume that the Hilbert space* H *is finite-dimensional, that* b_ψ *is injective and that*

$$\mathrm{Gap}_\psi \overset{\text{def}}{=} \inf_{b_\psi(s) \neq b_\psi(t)} \left\| b_\psi(s) - b_\psi(t) \right\|_H^2 > 0.$$

Finally assume that the crossed product von Neumann algebra $\Gamma_q(H) \rtimes_\alpha G$ *has* QWEP. *Let*

$$\pi \colon \mathrm{C}_r^*(G) \to \mathrm{B}(\mathrm{L}^p(\mathrm{VN}(G)) \oplus_p \mathrm{L}^p(\Gamma_q(H) \rtimes_\alpha G))$$

be the Banach algebra homomorphism such that $\pi(a)(x, y) \overset{\text{def}}{=} (ax, (1 \rtimes a)y)$ *where* $x \in \mathrm{L}^p(\mathrm{VN}(G))$ *and* $y \in \mathrm{L}^p(\Gamma_q(H) \rtimes_\alpha G)$. *Let* $\mathcal{D}_{\psi,q,p}$ *be the Hodge-Dirac operator on the Banach space* $\mathrm{L}^p(\mathrm{VN}(G)) \oplus_p \mathrm{L}^p(\Gamma_q(H) \rtimes_\alpha G)$ *defined in* (5.45). *Then* $(\mathrm{C}_r^*(G), \mathrm{L}^p(\mathrm{VN}(G)) \oplus_p \mathrm{L}^p(\Gamma_q(H) \rtimes_\alpha G), \mathcal{D}_{\psi,q,p})$ *is a Banach spectral triple in the sense of Definition 5.10.*

We are equally interested in the metric aspect [147] of noncommutative geometry, see Sect. 5.1 for background on quantum (locally) compact metric spaces.

We introduce new quantum (locally) compact metric spaces in the sense of [143, 147, 197] associated with these spectral triples. It relies on L^p-variants of the seminorms of [119, Section 1.2]. Here, we check carefully the axioms taking into account all problems of domains required by this theory. Note that it is not clear how to do the same analysis at the level $p = \infty$ considered in [119, Section 1.2] since we cannot hope for the boundedness of the Riesz transform on L^∞ which is an important tool in this part. So, on this point, L^p-seminorms seem more natural.

We observe significant differences between the case of Fourier multipliers (Theorem 5.1) and the one of Schur multipliers (Theorem 5.3). For example, we need to use C*-*algebras* for semigroups of Fourier multipliers and which produces quantum *compact* metric spaces contrarily to the case of semigroups of Schur multipliers which requires *order-unit spaces* and that leads to quantum *locally* compact metric spaces if I is infinite. Furthermore, our analysis with quantum compact metric spaces relying on noncommutative L^p-spaces makes appear a new phenomenon when the value of the parameter p changes, see Remark 5.3.

Note that the combination of our spectral triples and our quantum (locally) compact metric spaces is in the spirit of the papers [32, 150] (see also [59]) but it is more subtle here since the link between the norms of the commutators and the seminorms of our quantum metric spaces is not as direct as the ones of [32, 150].

Finally, we are interested in other Dirac operators related to Riesz transforms. We study the associated noncommutative geometries in Sects. 5.9 and 5.11 and their properties of functional calculus in Sect. 5.12.

The paper is organized as follows.

Chapter 2 gives background and preliminary results. In Sect. 2.1, we collect some elementary information on operator theory, functional calculus and semigroup theory. In Sect. 2.2, we recall some important information on isonormal Gaussian processes and more generally q-Gaussian functors. These notions are fundamental for the construction of the noncommutative gradients. In Sect. 2.3, we develop a short theory of vector-valued unbounded bilinear forms on Banach spaces that we need for our carré du champ with values in a noncommutative L^p-space. In Sect. 2.4, we introduce a new transference result (Proposition 2.8) between Fourier multipliers on crossed products and classical Fourier multipliers. In Sect. 2.5, we discuss Hilbertian valued noncommutative L^p-spaces which are fundamental for us. We complete and clarify some technical points of the literature. In Sect. 2.6, we introduce the carré du champ Γ and the gradients with values in a noncommutative L^p-space for semigroups of Fourier multipliers. In Sect. 2.7, we discuss the carré du champ Γ and the gradients associated to Markov semigroups of Schur multipliers.

In Chap. 3, we investigate dimension free Riesz estimates for semigroups of Schur multipliers and we complement the results of [131] for Riesz transforms associated to semigroups of Markov Fourier multipliers. In Sect. 3.1, we obtain Khintchine inequalities for q-Gaussians in crossed products generalizing some result of [131]. In Sect. 3.2, we extend the equivalences (1.16) of [131] to q-deformed Gaussians variables on a larger (maximal) domain, see (3.31). We also examine the constants. In Sect. 3.3, we are interested in Meyer's equivalences (1.15) in the context of semigroups of Fourier multipliers in the case $p \geqslant 2$, see Theorem 3.2. In Sect. 3.4, we show dimension free estimates for Riesz transforms associated to semigroups of Markov Schur multipliers, see Theorem 3.3 and (3.88). Moreover, we examine carefully the obtained constants in this equivalence. In Sect. 3.5, we turn again to the formulation of these equivalences in the spirit of Meyer's equivalence (1.15) in connection with the carré du champ. We also obtain concrete equivalences similar to the ones of Lust-Piquard [154, 156] and related to (1.4). Note that for the case $1 < p < 2$, the statement becomes more involved, see Corollary 3.1.

In the next Chap. 4, we show our main results on functional calculus of Hodge-Dirac operators. In Sect. 4.1, we start with the case of the Hodge-Dirac operator for semigroups of Fourier multipliers. We show the boundedness of the functional calculus on $L^p(\mathrm{VN}(G)) \oplus_p \overline{\mathrm{Ran}\, \partial_{\psi,q,p}}$, see Theorems 4.2 and 4.3. In Sect. 4.2, we extend this boundedness to the larger space $L^p(\mathrm{VN}(G)) \oplus L^p(\Gamma_q(H) \rtimes_\alpha G)$. In Sect. 4.3, we obtain the boundedness on $L^p(\mathrm{VN}(G)) \oplus \Omega_{\psi,q,p}$. In Sect. 4.4, we examine the bimodule properties of $\Omega_{\psi,q,p}$. In Sect. 4.5, we change the setting and we consider the Hodge-Dirac operator associated to semigroups of Schur multipliers and we show the boundedness of the functional calculus on $S_I^p \oplus_p \overline{\mathrm{Ran}\, \partial_{\alpha,q,p}}$, see Theorem 4.8. In a second step, we extend in Sect. 4.6 this boundedness to $S_I^p \oplus L^p(\Gamma_q(H) \overline{\otimes} \mathrm{B}(\ell_I^2))$ in Corollary 4.2 and Theorem 4.9. In Sect. 4.7, we show that the constants are independent from H and α.

In Chap. 5, we examine the noncommutative geometries induced by the Hodge-Dirac operators in both settings and another two Hodge-Dirac operators. In Sect. 5.1, we give background on quantum (locally) compact metric spaces. In Sect. 5.2, we introduce L^p-versions of the quantum compact metric spaces of [119]. In Sect. 5.3, we establish fundamental estimates for the sequel and we introduce the constant Gap_α which plays an important role. In Sect. 5.4, we introduce the seminorm which allows us to define the new quantum (locally) compact metric spaces in the next Sect. 5.5. In Sect. 5.6, we compute the gaps of some explicit semigroups and we compare the notions in both of our settings (Fourier multipliers and Schur multipliers), see Proposition 5.8. In Sect. 5.7, we define a Banach space generalization of the notion of spectral triple. In Sect. 5.8, we give sufficient conditions to make sure that the triple induced by the Hodge-Dirac operator for a semigroup of Fourier multipliers gives birth to a Banach spectral triple satisfying axioms of noncommutative geometry and more generally the axioms of Sect. 5.7. In Sect. 5.9, we introduce a second Hodge-Dirac operator for Fourier multipliers and we study the noncommutative geometry induced by this operator. In Sect. 5.10, we give sufficient conditions to ensure that the triple induced by the Hodge-Dirac operator for a semigroup of Schur multipliers gives rise to a spectral triple. In Sect. 5.11, we introduce a second related Hodge-Dirac operator and we study the noncommutative geometry induced by this operator. In Sect. 5.12, in the bosonic case $q = 1$ we establish the bisectoriality and the boundedness of the functional calculus of Dirac operators introduced in Sects. 5.9 and 5.11.

Finally, in the short Appendix A, we discuss how the Lévy measure of a continuous conditionally of negative type function ψ on a locally compact abelian group G induces a 1-cocycle.

Chapter 2
Preliminaries

Abstract In this chapter, we start by recalling the used properties of the notions used in this book: operators, semigroups, H$^\infty$ functional calculus, noncommutative L^p-spaces and probabilities. In particular, the construction of our markovian semigroups of Fourier and Schur multipliers is a standing assumption in the rest of the book. We equally investigate vector-valued unbounded bilinear forms on Banach spaces which will be used as a framework for (the domain of) the carré du champ. We also show a transference result between Fourier multipliers on group von Neumann algebras and Fourier multipliers on crossed product von Neumann algebras. Then we give useful results on Hilbertian valued noncommutative L^p-spaces for the sequel of the book. Finally, we examine in detail the carré du champ and the first order differential calculus for semigroups of Fourier multipliers and semigroups of Schur multipliers.

2.1 Operators, Functional Calculus and Semigroups

In this short section, we describe various notions that play a role in this book at different places.

Closed and Weak* Closed Operators An operator $T: \operatorname{dom} T \subset X \to Y$ is closed if

$$\text{for any sequence } (x_n) \text{ of } \operatorname{dom} T \text{ with } x_n \to x \text{ and } T(x_n) \to y \text{ with } x \in X \tag{2.1}$$

and $y \in Y$ we have $x \in \operatorname{dom} T$ and $T(x) = y$.

By Kato [136, p. 165], an operator $T: \operatorname{dom} T \subset X \to Y$ is closed if and only if its domain $\operatorname{dom} T$ is a complete space with respect to the graph norm

$$\|x\|_{\operatorname{dom} T} \overset{\text{def}}{=} \|x\|_X + \|T(x)\|_Y. \tag{2.2}$$

© The Author(s), under exclusive license to Springer Nature Switzerland AG 2022

C. Arhancet, C. Kriegler, *Riesz Transforms, Hodge-Dirac Operators and Functional Calculus for Multipliers*, Lecture Notes in Mathematics 2304, https://doi.org/10.1007/978-3-030-99011-4_2

A linear subspace C of dom T is a core of T if C is dense in dom T for the graph norm, that is

for any $x \in$ dom T there is (x_n) of C s. th. $x_n \to x$ in X and $T(x_n) \to T(x)$ in Y.
(2.3)

If T is closed, a subspace C of dom T is a core for T if and only if T is the closure of its restriction $T|C$. Recall that if $T:$ dom $T \subset X \to Y$ is a densely defined unbounded operator then dom T^* is equal to

$$\{y^* \in Y^* : \text{there exists } x^* \in X^* \text{ s. th. } \langle T(x), y^*\rangle_{Y,Y^*} = \langle x, x^*\rangle_{X,X^*} \text{ for all } x \in \text{dom } T\}.$$
(2.4)

If $y^* \in$ dom T^*, the previous $x^* \in X^*$ is determined uniquely by y^* and we let $T^*(y^*) = x^*$. For any $x \in$ dom T and any $y^* \in$ dom T^*, we have

$$\langle T(x), y^*\rangle_{Y,Y^*} = \langle x, T^*(y^*)\rangle_{X,X^*}.$$
(2.5)

Recall that the unbounded operator $T:$ dom $T \subset X \to Y$ is closable [136, p. 165] if and only if

$$x_n \in \text{dom } T, \ x_n \to 0 \text{ and } T(x_n) \to y \text{ imply } y = 0.$$
(2.6)

If the unbounded operator $T:$ dom $T \subset X \to Y$ is closable then

$x \in$ dom $\overline{T}$ iff there exists $(x_n) \subset$ dom T s. th. $x_n \to x$ and $T(x_n) \to y$ for some y.
(2.7)

In this case $\overline{T}(x) = y$. Finally, if T is a densely defined closable operator then $T^* = \overline{T}^*$ by Kato [136, Problem 5.24]. If T is densely defined, by Kato [136, Problem 5.27 p. 168], we have

$$\text{Ker } T^* = (\text{Ran } T)^{\perp}.$$
(2.8)

If T is a densely defined unbounded operator and if $T \subset R$, by Kato [136, Problem 5.25 p. 168] we have

$$R^* \subset T^*.$$
(2.9)

If TR and T are densely defined, by Kato [136, Problem 5.26 p. 168] we have

$$R^*T^* \subset (TR)^*.$$
(2.10)

In Sect. 5, we need the following lemma. To this end, an operator $T:$ dom $T \subset X \to Y$ between dual spaces X and Y is called weak* closed if (2.1) holds, when the convergences are in the weak* sense.

Lemma 2.1 *Let X, Y be Banach spaces and $T:$ dom $T \subset X^* \to Y^*$ be a weak* closed operator with a weak*-core[1] $D \subset$ dom T. Then any $a \in$ dom T admits a bounded net (a_j) of elements of D such that the net $(T(a_j))$ is also bounded, $a_j \to a$ and $T(a_j) \to T(a)$ both for the weak* topology.*

Proof We consider the Banach space $X^* \oplus Y^*$ which is canonically the dual space of $X \oplus Y$ and thus carries a weak* topology. Let $N = \{(a, T(a)) : a \in D\} \subset X^* \oplus Y^*$ and $N_1 = N \cap B_1$ where B_1 is the closed unit ball of $X^* \oplus Y^*$.

We claim that $\overline{\mathbb{C} N_1}^{w^*} = \overline{N}^{w^*}$. The inclusion "$\subset$" is clear. For the inclusion "$\supset$" it suffices to show that $\overline{\mathbb{C} N_1}^{w^*}$ contains N and is weak* closed. That it contains N is easy to see.

For the weak*-closedness, we will employ the Krein-Smulian Theorem [167, Theorem 2.7.11]. So it suffices to show that $\overline{\mathbb{C} N_1}^{w^*} \cap B_r$ is weak* closed for any $r > 0$. Let (x_j) be a net in $\overline{\mathbb{C} N_1}^{w^*} \cap B_r$ converging to some $x \in X^* \oplus Y^*$. Since B_r is weak* closed (by Alaoglu Theorem), we have $x \in B_r$. Moreover, for any j we can write $x_j = \lambda_j y_j$ with $y_j \in \overline{N_1}^{w^*}$ and $\lambda_j \in \mathbb{C}$. We have a freedom in choice of both factors and can take $\|y_j\|_{X^* \oplus Y^*} = 1$. Thus, $|\lambda_j| \leqslant r$. By Alaoglu Theorem, (y_j) admits a weak* convergent subnet (y_{j_k}) with a limit thus in $\overline{N_1}^{w^*}$. Moreover, we can suppose that the net (λ_{j_k}) is convergent. Thus, $x_{j_k} = \lambda_{j_k} y_{j_k}$ converges for the weak* topology with limit a fortiori equal to x. Thus $x \in \overline{\mathbb{C} N_1}^{w^*}$. We infer that $x \in \overline{\mathbb{C} N_1}^{w^*} \cap B_r$. Consequently $\overline{\mathbb{C} N_1}^{w^*} \cap B_r$ is weak*-closed and hence $\overline{\mathbb{C} N_1}^{w^*}$ is weak*-closed. So we have shown $\overline{\mathbb{C} N_1}^{w^*} = \overline{N}^{w^*}$.

Since D is a weak*-core of T, the closure $\overline{N}^{w^*}$ equals the graph of T. Now if $a \in$ dom T, the elements $(a, T(a))$ belongs to $\overline{N}^{w^*} = \overline{\mathbb{C} N_1}^{w^*}$, so that we can choose a net (a_j) of graph norm less than 1 and $\lambda \in \mathbb{C}$ such that $a_j \in D$, $\lambda a_j \to a$ and $T(\lambda a_j) \to T(a)$ both for the weak* topology. $\square$

The following is [219, Corollary 5.6 p. 144].

Theorem 2.1 *Let T be a densely defined closed operator on a Hilbert space H. Then the operator T^*T on $(\text{Ker } T)^\perp$ is unitarily equivalent to the operator TT^* on $(\text{Ker } T^*)^\perp$.*

Semigroups and Sectorial Operators If $(T_t)_{t \geqslant 0}$ is a strongly continuous semigroup on a Banach space X with (negative) infinitesimal generator A, we have

$$\frac{1}{t}(T_t(x) - x) \xrightarrow[t \to 0^+]{} -A(x), \qquad x \in \text{dom } A \tag{2.11}$$

by Engel and Nagel [85, Definition 1.2 p. 49] and $\text{Ker } A = \{x \in X : T_t(x) = x \text{ for all } t \geqslant 0\}$ by Engel and Nagel [85, p. 337]. Moreover, by Engel and Nagel

---

[1] For any $a \in$ dom T there exists a net (a_j) in D such that $a_j \to a$ and $T(a_j) \to T(a)$ both for the weak* topology.

[85, Corollary 5.5 p. 223], for any $t \geqslant 0$, we have

$$\left(\text{Id} + \frac{t}{n}A\right)^{-n} x \xrightarrow[n \to +\infty]{} T_t(x), \quad x \in X. \tag{2.12}$$

Furthermore, by Engel and Nagel [85, (1.5) p. 50], if $x \in \text{dom } A$ and $t \geqslant 0$, then $T_t(x)$ belongs to dom A and

$$T_t A(x) = A T_t(x). \tag{2.13}$$

For any angle $\sigma \in (0, \pi)$, we will use the open sector symmetric around the positive real half-axis with opening angle 2σ

$$\Sigma_\sigma^+ = \Sigma_\sigma \overset{\text{def}}{=} \left\{ z \in \mathbb{C} \backslash \{0\} : |\arg z| < \sigma \right\}.$$

It will be convenient to set $\Sigma_0^+ \overset{\text{def}}{=} (0, \infty)$. We refer to [83, 98, 110, 128] for background on sectorial and bisectorial operators and their H^∞ functional calculus, as well as R-boundedness. A sectorial operator $A \colon \text{dom } A \subset X \to X$ of type σ on a Banach space X is a closed densely defined operator such that $\sigma(A) \subset \overline{\Sigma_\sigma}$ for some $\sigma \in [0, \pi)$ and such that $\sup\{\|zR(z, A)\| : z \notin \overline{\Sigma_{\sigma'}}\} < \infty$ for any $\sigma' \in (\sigma, \pi)$. We introduce the angle of sectoriality

$$\omega(A) = \inf \sigma. \tag{2.14}$$

An operator A is called R-sectorial of type σ if it is sectorial of type σ and if for any $\theta \in (\sigma, \pi)$ the set

$$\left\{ zR(z, A) : z \notin \overline{\Sigma_\theta} \right\} \tag{2.15}$$

is R-bounded. If A is the negative generator of a bounded strongly continuous semigroup $(T_t)_{t \geqslant 0}$ on a Banach space X then A is sectorial of type $\frac{\pi}{2}$ by Hytönen [110, Example 10.1.2].

If A is sectorial operator on a *reflexive* Banach space X, we have by Haase [98, Proposition 2.1.1 (h)] or [110, Proposition 10.1.9] a decomposition

$$X = \text{Ker } A \oplus \overline{\text{Ran } A}. \tag{2.16}$$

If in addition A is the negative generator of a bounded strongly continuous semigroup $(T_t)_{t \geqslant 0}$, this means that $(T_t)_{t \geqslant 0}$ is mean ergodic [8, Definition 4.3.3], [85, p. 338]. In particular, the map $P \colon X \mapsto X, x \mapsto \lim_{t \to \infty} \frac{1}{t} \int_0^t T_s(x) \, ds$ is the bounded projection on Ker A associated to (2.16). By Arendt et al. [8, Corollary 4.3.2] (see also [110, (10.5) and p. 367]), it is also given by

$$P(x) = \lim_{\lambda \to 0} \lambda(\lambda + A)^{-1} x, \quad x \in X. \tag{2.17}$$

Let A be the negative generator of a bounded strongly continuous semigroup $(T_t)_{t \geqslant 0}$ on X. For any $x \in X$ and any $z \notin \overline{\Sigma_{\frac{\pi}{2}}}$, we have by e.g. [85, p. 55] the following expression of the resolvent as a Laplace transform

$$(z - A)^{-1}x = -\int_0^\infty e^{zt} T_t(x)\, dt. \tag{2.18}$$

The following is [110, Proposition G.2.4 p. 526].

Lemma 2.2 *Let $(T_t)_{t \geqslant 0}$ be a strongly continuous semigroup of bounded operators on a Banach space X with (negative) generator A. If Y is a subspace of $\mathrm{dom}\, A$ which is dense in X and invariant under each operator T_t, then Y is a core of A.*

Bisectorial Operators In a similar manner, for $\sigma \in [0, \frac{\pi}{2})$, we consider the open bisector $\Sigma_\sigma^\pm \overset{\mathrm{def}}{=} \Sigma_\sigma \cup (-\Sigma_\sigma)$ and we say that a closed densely defined operator A is $(R$-)bisectorial of type σ if $\sigma(A) \subset \overline{\Sigma_\sigma^\pm}$ for some $\sigma \in [0, \frac{\pi}{2})$ and if $\{zR(z, A) : z \notin \overline{\Sigma_{\sigma'}^\pm}\}$ is $(R$-)bounded for any $\sigma' \in (\sigma, \frac{\pi}{2})$. By Hytönen [110, p. 447] (see also [110, Proposition 10.3.3]), a linear operator is $(R$-)bisectorial if and only if

$$i\mathbb{R} - \{0\} \subset \rho(A) \quad \text{and} \quad \sup_{t \in \mathbb{R}^+ - \{0\}} \|tR(it, A)\| < \infty \tag{2.19}$$

(resp. $\{tR(it, A) : t \in \mathbb{R}^+ - \{0\}\}$ is R-bounded). Self-adjoint operators are bisectorial of type 0. If A is bisectorial of type σ then by Hytönen [110, Proposition 10.6.2 (2)] the operator A^2 is sectorial of type 2σ and we have

$$\overline{\mathrm{Ran}\, A^2} = \overline{\mathrm{Ran}\, A} \quad \text{and} \quad \mathrm{Ker}\, A^2 = \mathrm{Ker}\, A. \tag{2.20}$$

Functional Calculus of Sectorial and Bisectorial Operators For any $0 < \theta < \pi$, let $\mathrm{H}^\infty(\Sigma_\theta)$ be the algebra of all bounded analytic functions $f : \Sigma_\theta \to \mathbb{C}$, equipped with the supremum norm $\|f\|_{\mathrm{H}^\infty(\Sigma_\theta)} \overset{\mathrm{def}}{=} \sup\{|f(z)| : z \in \Sigma_\theta\}$. Let $\mathrm{H}_0^\infty(\Sigma_\theta) \subset \mathrm{H}^\infty(\Sigma_\theta)$ be the subalgebra of bounded analytic functions $f : \Sigma_\theta \to \mathbb{C}$ for which there exist $s, c > 0$ such that $|f(z)| \leqslant \frac{c|z|^s}{(1+|z|)^{2s}}$ for any $z \in \Sigma_\theta$.

Given a sectorial operator A of type $0 < \omega < \pi$, a bigger angle $\theta \in (\omega, \pi)$, and a function $f \in \mathrm{H}_0^\infty(\Sigma_\theta)$, one may define a bounded operator $f(A)$ by means of a Cauchy integral on Σ_θ (see e.g. [94, Section 2.3] or [140, Section 9]). The resulting mapping $\mathrm{H}_0^\infty(\Sigma_\theta) \to \mathrm{B}(X)$, $f \mapsto f(A)$ is an algebra homomorphism. By definition, A has a bounded $\mathrm{H}^\infty(\Sigma_\theta)$ functional calculus provided that this homomorphism is bounded, that is if there exists a positive constant C such that $\|f(A)\|_{X \to X} \leqslant C\|f\|_{\mathrm{H}^\infty(\Sigma_\theta)}$ for any $f \in \mathrm{H}_0^\infty(\Sigma_\theta)$. In the case where A has a dense range, the latter boundedness condition allows a natural extension of $f \mapsto f(A)$ to the full algebra $\mathrm{H}^\infty(\Sigma_\theta)$.

For a bisectorial operator A, we can make a similar definition by integrating on the boundary of a bisector. In particular, A is said to have a bounded $\mathrm{H}^\infty(\Sigma_\theta^\pm)$ calcu-

lus, if there is some $C > 0$ such that $\left\| f(A) \right\|_{X \to X} \leqslant C \left\| f \right\|_{H^\infty(\Sigma_\theta^\pm)}$ for any f of the
space $H_0^\infty(\Sigma_\theta^\pm) \overset{\text{def}}{=} \left\{ g \in H^\infty(\Sigma_\theta^\pm) : \exists c, s > 0 \, \forall z \in \Sigma_\theta^\pm : |g(z)| \leqslant c|z|^s/(1 + |z|)^{2s} \right\}$.
The following is a particular case of [176, Proposition 2.3], see also [110, Theorem
10.6.7].

Proposition 2.1 *Suppose that A is an R-bisectorial operator on a Banach space
X of finite cotype. Then A^2 is R-sectorial and for each $\omega \in (0, \frac{\pi}{2})$ the following
assertions are equivalent.*

1. *The operator A admits a bounded $H^\infty(\Sigma_\omega^\pm)$ functional calculus.*
2. *The operator A^2 admits a bounded $H^\infty(\Sigma_{2\omega})$ functional calculus.*

Fractional Powers We refer to [8, 98, 99, 140, 162] for more information on
fractional powers. Let A be a sectorial operator of type σ on a Banach space X.
If $\alpha \in (0, \frac{\pi}{\sigma})$, then by Haase [98, Proposition 3.1.2] the operator A^α is sectorial of
angle $\alpha\sigma$. For all α, β with $\operatorname{Re}\alpha, \operatorname{Re}\beta > 0$ we have $A^\alpha A^\beta = A^{\alpha+\beta}$. By Haase [98,
page 62], [98, Corollary 3.1.11] and [162, p. 142], for any $\alpha \in \mathbb{C}$ with $\operatorname{Re}\alpha > 0$ we
have

$$\overline{\operatorname{Ran} A^\alpha} = \overline{\operatorname{Ran} A} \quad \text{and} \quad \operatorname{Ker} A^\alpha = \operatorname{Ker} A. \tag{2.21}$$

If A is densely defined and $0 < \operatorname{Re}\alpha < 1$, then the space $\operatorname{dom} A$ is a core of A^α by
Haase [98, page 62].

For $t, s > 0$, we let $f_t(s) \overset{\text{def}}{=} \frac{t}{\sqrt{4\pi s^3}} e^{-\frac{t^2}{4s}}$. It is well-known [8, Lemma 1.6.7] that

$$\int_0^{+\infty} e^{-s\lambda} f_t(s) \, \mathrm{d}s = e^{-t\sqrt{\lambda}}. \tag{2.22}$$

Let A be the negative generator of a bounded semigroup $(e^{-tA})_{t \geqslant 0}$ acting on a
Banach space X. By Haase [98, Example 3.4.6], for any $x \in X$ we have

$$e^{-tA^{\frac{1}{2}}} x = \int_0^{+\infty} f_t(s) e^{-sA}(x) \, \mathrm{d}s. \tag{2.23}$$

Compact Operators and Complex Interpolation Recall the following result [69,
Theorem 9] (see also [135, Theorem 5.5]) which allows to obtain compactness via
complex interpolation.

Theorem 2.2 *Suppose that (X_0, X_1) and (Y_0, Y_1) are Banach couples and that X_0
is a UMD-space. Let $T: X_0 + X_1 \to Y_0 + Y_1$ such that its restrictions $T_0: X_0 \to Y_0$
resp. $T_1: X_1 \to Y_1$ are compact resp. bounded. Then for any $0 < \theta < 1$ the map
$T: (X_0, X_1)_\theta \to (Y_0, Y_1)_\theta$ is compact.*

Noncommutative L^p-Spaces We give a description of noncommutative L^p-spaces
associated with a semifinite von Neumann algebra. We refer the reader to [191] and
the references therein for further information on these spaces.

Let M be a semifinite von Neumann algebra equipped with a normal semifinite faithful trace τ. We let M_+ denote the positive part of M. Let $\mathcal{S}_+$ be the set of all $x \in M_+$ whose support projection has a finite trace. Then any $x \in \mathcal{S}_+$ has a finite trace. Let $\mathcal{S} \subset M$ be the linear span of $\mathcal{S}_+$, then $\mathcal{S}$ is a weak* dense $*$-subalgebra of M.

Let $0 < p < \infty$. For any $x \in \mathcal{S}$, the operator $|x|^p$ belongs to $\mathcal{S}_+$ and we set

$$\|x\|_{\mathrm{L}^p(M)} \overset{\text{def}}{=} \left(\tau(|x|^p)\right)^{\frac{1}{p}}, \qquad x \in \mathcal{S}$$

where $|x| \overset{\text{def}}{=} (x^*x)^{\frac{1}{2}}$ denotes the modulus of x. It turns out that $\|\cdot\|_{\mathrm{L}^p(M)}$ is a norm on $\mathcal{S}$ if $p \geqslant 1$, and a p-norm if $p < 1$. By definition, the noncommutative L^p-space associated with (M, τ) is the completion $\mathrm{L}^p(M)$ of $(\mathcal{S}, \|\ \|_p)$. For convenience, we also set $\mathrm{L}^\infty(M) \overset{\text{def}}{=} M$ equipped with its operator norm. Note that by definition, $\mathrm{L}^p(M) \cap M$ is dense in $\mathrm{L}^p(M)$ for any $1 \leqslant p < \infty$.

Assume that M acts on some Hilbert space H. We denote by $\mathrm{L}^0(M)$ the set of all measurable operators which is a $*$-algebra under suitable operations. For any x in $\mathrm{L}^0(M)$ and any $0 < p < \infty$, the operator $|x|^p = (x^*x)^{\frac{p}{2}}$ belongs to $\mathrm{L}^0(M)$. Let $\mathrm{L}^0(M)_+$ be the positive part of $\mathrm{L}^0(M)$, that is, the set of all selfadjoint positive operators in $\mathrm{L}^0(M)$. Then the trace τ extends to a positive tracial functional on $\mathrm{L}^0(M)_+$, still denoted by τ, in such a way that for any $0 < p < \infty$, we have an isometric identification

$$\mathrm{L}^p(M) = \left\{x \in \mathrm{L}^0(M) : \tau(|x|^p) < \infty\right\},$$

where the latter space is equipped with $\|x\|_p = \left(\tau(|x|^p)\right)^{\frac{1}{p}}$. Furthermore, τ uniquely extends to a bounded linear functional on $\mathrm{L}^1(M)$, still denoted by τ. Indeed we have $|\tau(x)| \leqslant \tau(|x|) = \|x\|_1$ for any $x \in \mathrm{L}^1(M)$.

For any $0 < p \leqslant \infty$ and any $x \in \mathrm{L}^p(M)$, the adjoint operator x^* belongs to $\mathrm{L}^p(M)$ as well, with $\|x^*\|_p = \|x\|_p$. Clearly, we also have that $x^*x \in \mathrm{L}^{\frac{p}{2}}(M)$ and $|x| \in \mathrm{L}^p(M)$, with $\|\,|x|\,\|_p = \|x\|_p$. We let $\mathrm{L}^p(M)_+ = \mathrm{L}^0(M)_+ \cap \mathrm{L}^p(M)$ denote the positive part of $\mathrm{L}^p(M)$. The space $\mathrm{L}^p(M)$ is spanned by $\mathrm{L}^p(M)_+$.

We recall the noncommutative Hölder inequality. If $0 < p, q, r \leqslant \infty$ satisfy $\frac{1}{p} + \frac{1}{q} = \frac{1}{r}$, then

$$\|xy\|_r \leqslant \|x\|_p \|y\|_q, \qquad x \in \mathrm{L}^p(M), \ y \in \mathrm{L}^q(M). \tag{2.24}$$

Conversely for any $z \in \mathrm{L}^r(M)$, there exist $x \in \mathrm{L}^p(M)$ and $y \in \mathrm{L}^q(M)$ such that $z = xy$ with $\|z\|_r = \|x\|_p \|y\|_q$. For any $x \in \mathrm{L}^{2p}(M)$ and any $y \in \mathrm{L}^p(M)_+$, we have if $\alpha > 0$

$$\|x^*x\|_p = \|x\|_{2p}^2 \quad \text{and} \quad \|y^\alpha\|_{\frac{p}{\alpha}} = \|y\|_p^\alpha. \tag{2.25}$$

For any $1 \leqslant p < \infty$, let $p^* \overset{\text{def}}{=} \frac{p}{p-1}$ be the conjugate number of p. Applying (2.24) with $q = p^*$ and $r = 1$, we may define a duality pairing between $L^p(M)$ and $L^{p^*}(M)$ by

$$\langle x, y \rangle = \tau(xy), \qquad x \in L^p(M), \ y \in L^{p^*}(M). \tag{2.26}$$

This induces an isometric isomorphism

$$L^{p^*}(M) = (L^p(M))^*, \qquad 1 \leqslant p < \infty, \quad \frac{1}{p} + \frac{1}{p^*} = 1. \tag{2.27}$$

In particular, we may identify $L^1(M)$ with the unique predual M_* of the von Neumann algebra M.

By means of the natural embeddings of $L^\infty(M) = M$ and $L^1(M) = M_*$ into $L^0(M)$, one may regard $(L^\infty(M), L^1(M))$ as a compatible couple of Banach spaces. Then we have

$$(L^\infty(M), L^1(M))_{\frac{1}{p}} = L^p(M), \qquad 1 \leqslant p \leqslant \infty, \tag{2.28}$$

where $(\ ,\ )_\theta$ stands for the interpolation space obtained by the complex interpolation method, see [35].

Finally, for any $1 < p < \infty$ note that the Banach space $L^p(M)$ is a UMD Banach space and has finite cotype, see [191, Section 7].

The space $L^2(M)$ is a Hilbert space, with inner product given by $(x, y) \mapsto \langle x, y^* \rangle = \tau(xy^*)$. Note that the identity (2.27) provided by (2.26) for $p = 2$ differs from the canonical (antilinear) identification of a Hilbert space with its dual space.

Let $T \colon L^2(M) \to L^2(M)$ be any bounded operator. We will denote by T^* the adjoint of T provided by (2.27) and (2.26) defined by

$$\tau\big(T(x)y\big) = \tau\big(x T^*(y)\big), \qquad x, y \in L^2(M).$$

For any $1 \leqslant p \leqslant \infty$ and any $T \colon L^p(M) \to L^p(M)$, we can consider the map $T^\circ \colon L^p(M) \to L^p(M)$ defined by

$$T^\circ(x) \overset{\text{def}}{=} T(x^*)^*, \qquad x \in L^p(M). \tag{2.29}$$

If $p = 2$ and if we denote by $T^\dagger$ the adjoint of T in the usual sense of Hilbertian operator theory, that is

$$\tau\big(T(x)y^*\big) = \tau\big(x(T^\dagger(y))^*\big), \qquad x, y \in L^2(M),$$

we see that

$$T^{\dagger} = T^{*\circ}. \tag{2.30}$$

In particular, the selfadjointness of $T : L^2(M) \to L^2(M)$ means that $T^* = T^{\circ}$.

Selfadjoint Maps Let $T : M \to M$ be a contraction. We say that T is selfadjoint if

$$\tau\big(T(x)y^*\big) = \tau\big(xT(y)^*\big), \qquad x, y \in M \cap L^1(M). \tag{2.31}$$

In this case, for any x, y in $M \cap L^1(M)$, we have $\big|\tau(T(x)y)\big| = \big|\tau(xT(y^*)^*)\big| \leqslant \|x\|_1 \|T(y^*)^*\|_\infty \leqslant \|x\|_1 \|y\|_\infty$. Hence the restriction of T to $M \cap L^1(M)$ uniquely extends to a contraction $T_1 : L^1(M) \to L^1(M)$. Then by (2.28), it also extends to a contraction $T_p : L^p(M) \to L^p(M)$ for any $1 \leqslant p < \infty$. We write $T_\infty = T$ by convention. If T is in addition weak* continuous, we obtain from (2.31) that

$$(T_p)^* = (T_{p^*})^{\circ}, \qquad 1 \leqslant p < \infty, \qquad \frac{1}{p} + \frac{1}{p^*} = 1.$$

By (2.30), this implies that the operator $T_2 : L^2(M) \to L^2(M)$ is selfadjoint.

If $T : M \to M$ is in addition positive, then each operator T_p is positive. In particular $(T_p)^{\circ} = T_p$. Thus in this case, we have $T_p^* = T_{p^*}$ for any $1 \leqslant p < \infty$.

Markovian Semigroups of Fourier Multipliers In this paragraph, we recall the basic theory of markovian semigroups of Fourier multipliers. The following definition and properties of a markovian semigroup are fundamental for us. Thus the assumptions and notations which follow these lines are standing for all the book. A companion definition of equal importance is Definition 2.2 together with Proposition 2.3 which follow.

Definition 2.1 Let M be a von Neumann algebra equipped with a normal semifinite faithful trace. We say that a weak* continuous semigroup $(T_t)_{t \geqslant 0}$ of operators on M is a markovian semigroup if each T_t is a weak* continuous selfadjoint completely positive unital contraction.

For any $1 \leqslant p < \infty$, such a semigroup induces a strongly continuous semigroup of operators on each $L^p(M)$ satisfying

1. each T_t is a contraction on $L^p(M)$,
2. each T_t is selfadjoint on $L^2(M)$,
3. each T_t is completely positive on $L^p(M)$,
4. $T_t(1) = 1$.

By Hellmich [104, Corollary 1.27], if M is finite, then there exists a normal conditional expectation $\mathbb{E}$ on the fixed point subalgebra $\{x \in M : T_t(x) = x$ for all $t \geqslant 0\}$. A careful reading of the proof shows that $\mathbb{E}$ is trace preserving since it is easy to check that each T_t is[2] trace preserving.

If $\{x \in M : T_t(x) = x$ for all $t \geqslant 0\} = \mathbb{C}1$, the conditional expectation is defined by $\mathbb{E}(x) = \tau(x)1$. We use the notation $L_0^p(M)$ for the subspace $\mathrm{Ker}\,\mathbb{E}_p$ of $L^p(M)$. We have $L_0^p(M) = \overline{\mathrm{Ran}\,A_p}$ if A_p is the L^p-realization of the generator of the semigroup. The following is [119, Theorem 1.1.7]. Note that the proof of this result uses [119, Lemma 1.1.6] which seems false in the light of [135, Problem 5.4]. However, [119, Lemma 1.1.6] can be replaced by Theorem 2.2.

Proposition 2.2 *Let M be a finite von Neumann algebra equipped with a normal finite faithful trace. Let $(T_t)_{t \geqslant 0}$ be a weak* continuous markovian semigroup on $L^\infty(M)$ with $\{x \in M : T_t(x) = x$ for all $t \geqslant 0\} = \mathbb{C}1$ satisfying $\left\| T_t : L_0^1(M) \to L^\infty(M) \right\| \lesssim \frac{1}{t^{\frac{1}{2}}}$ and such that A^{-1} is compact on $L_0^2(M)$. Then for all $1 \leqslant p < q \leqslant \infty$ such that $\frac{2\,\mathrm{Re}\,z}{n} > \frac{1}{p} - \frac{1}{q}$ we have a well-defined compact operator $A^{-z} : L_0^p(M) \to L_0^q(M)$.*

Let G be a discrete group and consider its left regular representation $\lambda : G \to B(\ell_G^2)$ over the Hilbert space ℓ_G^2 given by left translations $\lambda_s : \delta_r \mapsto \delta_{sr}$ where $r, s \in G$. Then the group von Neumann algebra $\mathrm{VN}(G)$ is the von Neumann subalgebra of $B(\ell_G^2)$ generated by these left translations. We also recall that $C_r^*(G)$ stands for the reduced group C*-algebra sitting inside $\mathrm{VN}(G)$ and generated by these λ_s. Let us write

$$\mathcal{P}_G \overset{\text{def}}{=} \mathrm{span}\,\{\lambda_s : s \in G\} \tag{2.32}$$

for the space of "trigonometric polynomials". The von Neumann algebra $\mathrm{VN}(G)$ is equipped with the tracial faithful normal state $\tau(\lambda_s) \overset{\text{def}}{=} \delta_{s=e} = \langle \delta_e, \lambda_s(\delta_e) \rangle_{\ell_G^2}$. Now, we introduce the main class of multipliers in which we are interested.

Definition 2.2 Let G be a discrete group. A Fourier multiplier on $\mathrm{VN}(G)$ is a weak* continuous linear map $T : \mathrm{VN}(G) \to \mathrm{VN}(G)$ such that there exists a (unique) complex function $\phi : G \to \mathbb{C}$ such that for any $s \in G$ we have $T(\lambda_s) = \phi_s \lambda_s$. In this case, we let $M_\phi = T$ and we say that ϕ is the symbol of T.

We have the following folklore characterization of markovian semigroups of Fourier multipliers. Proposition 2.3 is central for the remainder of the book and the mappings π and $b_\psi = b$ will be used throughout tacitly.

Recall that a function $\psi : G \to \mathbb{C}$ is conditionally negative definite if $\psi(s) = \psi(s^{-1})$ for any $s \in G$, $\psi(e) \geqslant 0$ and if the condition $\sum_{i=1}^n c_i = 0$ implies $\sum_{i,j=1}^n \overline{c_i} c_j \psi(s_j^{-1} s_i) \leqslant 0$. If H is a real Hilbert space, $O(H)$ stands for the orthogonal group.

---

[2] If $x \in M$, we have $\tau(T_t(x)) = \tau(T_t(x)1) = \tau(x T_t(1)) = \tau(x)$.

Proposition 2.3 *Let G be a discrete group and $(T_t)_{t \geqslant 0}$ be a family of weak* continuous operators on* $\mathrm{VN}(G)$. *Then the following are equivalent.*

1. $(T_t)_{t \geqslant 0}$ *is a markovian semigroup of Fourier multipliers.*
2. *There exists a (unique) real-valued conditionally negative definite function* $\psi : G \to \mathbb{R}$ *satisfying* $\psi(e) = 0$ *such that* $T_t(\lambda_s) = \exp(-t\psi(s))\lambda_s$ *for any* $t \geqslant 0$ *and any* $s \in G$.
3. *There exists a (unique) function* $\psi : G \to \mathbb{R}$ *such that* $T_t(\lambda_s) = \exp(-t\psi(s))\lambda_s$ *for any* $t \geqslant 0$ *and any* $s \in G$ *with the following property: there exists a real Hilbert space* H *together with a mapping* $b_\psi : G \to \mathbb{C}$ *and a homomorphism* $\pi : G \to \mathrm{O}(H)$ *such that the 1-cocycle law holds*

$$\pi_s(b_\psi(t)) = b_\psi(st) - b_\psi(s), \quad i.e. \quad b_\psi(st) = b_\psi(s) + \pi_s(b_\psi(t)) \tag{2.33}$$

for any $s, t \in G$ *and such that*

$$\psi(s) = \|b_\psi(s)\|_H^2, \quad s \in G. \tag{2.34}$$

Under these conditions, we say that ψ is a conditionally negative length. Note also the equality

$$b_\psi(s^{-1}) = -\pi_{s^{-1}} b_\psi(s) \tag{2.35}$$

which is immediate from (2.33). We refer to Appendix A for a link between Lévy measures and 1-cohomology strongly related to this result.

Truncations of Matrices See [9, p. 488] for the following definition.

Definition 2.3 We denote by $\mathrm{M}_{I,\mathrm{fin}}$ the space of matrices indexed by $I \times I$ with a finite number of non null entries. Given a set I, the set $\mathcal{P}_f(I)$ of all finite subsets of I is directed with respect to set inclusion. For $J \in \mathcal{P}_f(I)$ and $A \in \mathrm{M}_I$ a matrix indexed by $I \times I$, we write $\mathcal{T}_J(A)$ for the matrix obtained from A by setting each entry to zero if its row and column index are not both in J. We call $\big(\mathcal{T}_J(A)\big)_{J \in \mathcal{P}_f(I)}$ the net of finite submatrices of A.

If J is a finite subset of I, recall that the truncation $\mathcal{T}_J : S_I^p \to S_I^p$ is a complete contraction. It is well-known that $\mathcal{T}_J(x) \to x$ in S_I^p as $J \to I$ where $1 \leqslant p \leqslant \infty$.

Markovian Semigroups of Schur Multipliers Let I be any non-empty index set. Let $A = [a_{ij}]_{i,j \in I}$ be a matrix of M_I. By definition, the Schur multiplier on $\mathrm{B}(\ell_I^2)$ associated with this matrix is the unbounded linear operator M_A whose domain $\mathrm{dom}\, M_A$ is the space of all $B = [b_{ij}]_{i,j \in I}$ of $\mathrm{B}(\ell_I^2)$ such that $[a_{ij}b_{ij}]_{i,j \in I}$ belongs to $\mathrm{B}(\ell_I^2)$, and whose action on $B = [b_{ij}]_{i,j \in I}$ is given by $M_A(B) \overset{\mathrm{def}}{=} [a_{ij}b_{ij}]_{i,j \in I}$. In the literature, it is often written $A * B$ instead of $M_A(B)$ for the Schur product $[a_{ij}b_{ij}]_{i,j \in I}$.

The following description [10, Proposition 5.4], [15] of a markovian semigroup consisting of Schur multipliers is central for our paper. See also [11] for a generalization to semigroups of non positive Schur multipliers.

Proposition 2.4 *Let I be some non-empty index set. Then the following are equivalent.*

1. *If $(T_t)_{t \geqslant 0}$ is a weak* continuous semigroup of selfadjoint unital completely positive Schur multipliers on $\mathrm{B}(\ell_I^2)$ then there exists a real Hilbert space H and a family*

$$\alpha = (\alpha_i)_{i \in I} \tag{2.36}$$

of elements of H such that for any $t \geqslant 0$, the Schur multiplier $T_t : \mathrm{B}(\ell_I^2) \to \mathrm{B}(\ell_I^2)$ is associated to the matrix

$$\left[\mathrm{e}^{-t\|\alpha_i - \alpha_j\|_H^2} \right]_{i,j \in I}. \tag{2.37}$$

2. *If there exists a real Hilbert space H and a family $\alpha = (\alpha_i)_{i \in I}$ of elements of H then (2.37) defines a weak* continuous semigroup $(T_t)_{t \geqslant 0}$ of selfadjoint unital completely positive Schur multipliers on $\mathrm{B}(\ell_I^2)$*

In this case, the weak* (negative) infinitesimal generator A of this markovian semigroup of Schur multipliers acts by $A(e_{ij}) = \|\alpha_i - \alpha_j\|_H^2 e_{ij}$.

2.2 q-Gaussian Functors, Isonormal Processes and Probability

The definitions (2.84) and (2.95) of the noncommutative gradients associated with our markovian semigroups of Schur and Fourier multipliers need q-deformed Gaussian variables from [44]. We recall here several facts about the associated von Neumann algebras. We denote by S_n the symmetric group. If σ is a permutation of S_n we denote by $|\sigma|$ the number card $\{(i, j) \mid 1 \leqslant i, j \leqslant n, \sigma(i) > \sigma(j)\}$ of inversions of σ. Let H be a real Hilbert space with complexification $H_{\mathbb{C}}$. If $-1 \leqslant q < 1$ the q-Fock space over H is

$$\mathcal{F}_q(H) \stackrel{\text{def}}{=} \mathbb{C}\Omega \oplus \bigoplus_{n \geqslant 1} H_{\mathbb{C}}^{\otimes n}$$

where Ω is a unit vector, called the vacuum and where the scalar product on $H_{\mathbb{C}}^{\otimes n}$ is given by

$$\langle h_1 \otimes \cdots \otimes h_n, k_1 \otimes \cdots \otimes k_n \rangle_q = \sum_{\sigma \in S_n} q^{|\sigma|} \langle h_1, k_{\sigma(1)} \rangle_{H_{\mathbb{C}}} \cdots \langle h_n, k_{\sigma(n)} \rangle_{H_{\mathbb{C}}}.$$

If $q = -1$, we must first divide out by the null space, and we obtain the usual antisymmetric Fock space. The creation operator $\ell(e)$ for $e \in H$ is given by

$$\ell(e): \quad \mathcal{F}_q(H) \quad \longrightarrow \quad \mathcal{F}_q(H)$$
$$h_1 \otimes \cdots \otimes h_n \longmapsto e \otimes h_1 \otimes \cdots \otimes h_n.$$

They satisfy the q-relation

$$\ell(f)^*\ell(e) - q\ell(e)\ell(f)^* = \langle f, e \rangle_H \mathrm{Id}_{\mathcal{F}_q(H)}.$$

We denote by $s_q(e): \mathcal{F}_q(H) \to \mathcal{F}_q(H)$ the selfadjoint operator $\ell(e) + \ell(e)^*$. The q-von Neumann algebra $\Gamma_q(H)$ is the von Neumann algebra over $\mathcal{F}_q(H)$ generated by the operators $s_q(e)$ where $e \in H$. It is a finite von Neumann algebra with the trace τ defined by $\tau(x) = \langle \Omega, x(\Omega) \rangle_{\mathcal{F}_q(H)}$ where $x \in \Gamma_q(H)$.

Let H and K be real Hilbert spaces and $T: H \to K$ be a contraction with complexification $T_\mathbb{C}: H_\mathbb{C} \to K_\mathbb{C}$. We define the following linear map

$$\mathcal{F}_q(T): \quad \mathcal{F}_q(H) \quad \longrightarrow \quad \mathcal{F}_q(K)$$
$$h_1 \otimes \cdots \otimes h_n \longmapsto T_\mathbb{C}(h_1) \otimes \cdots \otimes T_\mathbb{C}(h_n).$$

Then there exists a unique map $\Gamma_q(T): \Gamma_q(H) \to \Gamma_q(K)$ such that for every $x \in \Gamma_q(H)$ we have

$$\big(\Gamma_q(T)(x)\big)\Omega = \mathcal{F}_q(T)(x\Omega). \tag{2.38}$$

This map is weak* continuous, unital, completely positive and trace preserving. If $T: H \to K$ is an isometry, $\Gamma_q(T)$ is an injective *-homomorphism. If $1 \leqslant p < \infty$, it extends to a contraction $\Gamma_q^p(T): L^p(\Gamma_q(H)) \to L^p(\Gamma_q(K))$.

Moreover, we need the following Wick formula, (see [42, p. 2] and [80, Corollary 2.1]). In order to state this, we denote, if $k \geqslant 1$ is an integer, by $\mathcal{P}_2(2k)$ the set of 2-partitions of the set $\{1, 2, \ldots, 2k\}$. If $\mathcal{V} \in \mathcal{P}_2(2k)$ we let $c(\mathcal{V})$ the number of crossings of $\mathcal{V}$, which is given by the number of pairs of blocks of $\mathcal{V}$ which cross (see [80, p. 8630] for a precise definition). Then, if $f_1, \ldots, f_{2k} \in H$ we have

$$\tau\big(s_q(f_1)s_q(f_2)\cdots s_q(f_{2k})\big) = \sum_{\mathcal{V} \in \mathcal{P}_2(2k)} q^{c(\mathcal{V})} \prod_{(i,j) \in \mathcal{V}} \langle f_i, f_j \rangle_H \tag{2.39}$$

and for an odd number of factors of q-Gaussians, the trace vanishes,

$$\tau\big(s_q(f_1)s_q(f_2)\cdots s_q(f_{2k-1})\big) = 0. \tag{2.40}$$

In particular, for any $e, f \in H$, we have

$$\tau\big(s_q(e)s_q(f)\big) = \langle e, f \rangle_H. \tag{2.41}$$

Recall that if $e \in H$ has norm 1, then the operator $s_{-1}(e)$ satisfies

$$s_{-1}(e)^2 = \mathrm{Id}_{\mathcal{F}_{-1}(H)}. \tag{2.42}$$

The q-Gaussian functor for $q = 1$ identifies to an H-isonormal process on a probability space (Ω, μ) [175, Definition 6.5], [178, Definition 1.1.1], that is a linear mapping $\mathrm{W} \colon H \to \mathrm{L}^0(\Omega)$ with the following properties:

for any $h \in H$ the random variable $\mathrm{W}(h)$ is a centered real Gaussian, $\tag{2.43}$

for any $h_1, h_2 \in H$ we have $\mathbb{E}\big(\mathrm{W}(h_1)\mathrm{W}(h_2)\big) = \langle h_1, h_2 \rangle_H$. $\tag{2.44}$

The linear span of the products $\mathrm{W}(h_1)\mathrm{W}(h_2) \cdots \mathrm{W}(h_m)$, with $m \geqslant 0$ $\tag{2.45}$

and $h_1, \ldots, h_m$ in H, is dense in the real Hilbert space $\mathrm{L}^2_{\mathbb{R}}(\Omega)$.

Here $\mathrm{L}^0(\Omega)$ denotes the space of measurable functions on Ω and we make the convention that the empty product, corresponding to $m = 0$ in (2.45), is the constant function 1. Recall that the span of elements $\mathrm{e}^{\mathrm{i}\mathrm{W}(h)}$ is dense in $\mathrm{L}^p(\Omega)$ by Janson [113, Theorem 2.12] if $1 \leqslant p < \infty$ and weak* dense if $p = \infty$.

If $1 \leqslant p \leqslant \infty$ and if $T \colon H \to H$ is a contraction, we denote by $\Gamma_1^p(T) \colon \mathrm{L}^p(\Omega) \to \mathrm{L}^p(\Omega)$ the (symmetric) second quantization of T acting on the *complex* Banach space $\mathrm{L}^p(\Omega)$. Recall that the map $\Gamma_1^\infty(T) \colon \mathrm{L}^\infty(\Omega) \to \mathrm{L}^\infty(\Omega)$ preserves the integral.[3] If $T \colon H_1 \to H_2$ is an isometry between Hilbert spaces then $\Gamma_1^\infty(T) \colon \mathrm{L}^\infty(\Omega_{H_1}) \to \mathrm{L}^\infty(\Omega_{H_2})$ is a trace preserving injective weak* continuous unital *-homomorphism which is surjective if T is surjective. Moreover, we have if $1 \leqslant p < \infty$

$$\Gamma_1^p(T)\mathrm{W}(h) = \mathrm{W}(T(h)), \quad h \in H, \tag{2.46}$$

and

$$\Gamma_1^\infty(T)\mathrm{e}^{\mathrm{i}\mathrm{W}(h)} = \mathrm{e}^{\mathrm{i}\mathrm{W}(T(h))}, \quad h \in H. \tag{2.47}$$

Finally, note that if $T \colon H \to H$ is a surjective isometry then $\Gamma_1^\infty(T) \colon \mathrm{L}^\infty(\Omega) \to \mathrm{L}^\infty(\Omega)$ is a *-automorphism of the von Neumann algebra $\mathrm{L}^\infty(\Omega)$.

Rosenthal's Inequalities Recall the following notion of conditional independence, which was introduced in [75, Definition 6.1] and which is similar to the one of [124, (I) p. 233]. Let M be a semifinite von Neumann algebra equipped with a normal semifinite faithful trace τ. Suppose that (N_k) is a family of von Neumann subalgebras of M and that N is a common von Neumann subalgebra of the N_k such that $\tau|N$ is semifinite. Let $\mathbb{E}_N \colon M \to N$ be the canonical trace preserving faithful

---

[3] That means that for any $f \in \mathrm{L}^\infty(\Omega)$ we have $\int_\Omega \Gamma_1^\infty(T)f \, \mathrm{d}\mu = \int_\Omega f \, \mathrm{d}\mu$.

normal conditional expectation with respect to N. We say that the family (N_k) is independent[4] with respect to N if for every k we have

$$\mathbb{E}_N(xy) = \mathbb{E}_N(x)\mathbb{E}_N(y), \quad x \in N_k, y \in W^*((N_j)_{j \neq k})$$

where $W^*((N_j)_{j \neq k})$ denotes the von Neumann subalgebra generated by the N_j with $j \neq k$. A sequence (x_k) of elements of $L^p(M)$ is said to be faithfully independent with respect to a von Neumann subalgebra N if there exists a family (N_k) with $x_k \in L^p(N_k)$ for any k such that (N_k) is independent with respect to N.

Let $(N_k)_{k \geqslant 1}$ be an independent sequence of von Neumann subalgebras of M with respect to N. In [125, Theorem 0.4 and (3.1)], it is shown in the case where M is finite that if $2 \leqslant p < \infty$ and if (x_k) is a sequence such that $x_k \in L^p(N_k)$ and $\mathbb{E}_N(x_k) = 0$ for all k, then

$$\left\| \sum_{k=1}^n x_k \right\|_{L^p(M)} \lesssim \max \left\{ \sqrt{p} \left\| \left(\sum_{k=1}^n \mathbb{E}_N(|x_k|^2) \right)^{\frac{1}{2}} \right\|_{L^p(M)}, \right. \tag{2.48}$$

$$\left. \sqrt{p} \left\| \left(\sum_{k=1}^n \mathbb{E}_N(|x_k^*|^2) \right)^{\frac{1}{2}} \right\|_{L^p(M)}, p \left(\sum_{k=1}^n \|x_k\|_{L^p(M)}^p \right)^{\frac{1}{p}} \right\}.$$

We refer to [75, Theorem 6.3] for a variant of this result. Using reduction [97], it is easy to extend this inequality to the non-finite case.

2.3 Vector-Valued Unbounded Bilinear Forms on Banach Spaces

We need some notions on vector-valued unbounded bilinear forms on Banach spaces. The reason is that these notions give an appropriate framework for the carré du champ operators defined in (2.78) and (2.91) and their domains associated with the markovian semigroups of Fourier multipliers or Schur multipliers. In the case of scalar-valued unbounded forms on Hilbert spaces, we refer to [136, 159, 180].

Definition 2.4 Let X be a Banach space and Z be an ordered-$*$-quasi-Banach space so carrying a positive cone and an involution. Assume that $D \subset X$ is a subspace and that $a : D \times D \to Z$ is an $\mathbb{R}$-bilinear map. We say that

1. a is densely defined if D is dense in X.
2. a is symmetric if for any $x, y \in D$ we have $a(x, y)^* = a(y, x)$.
3. a is positive if for any $x \in D$ we have $a(x, x) \geqslant 0$.

---

[4] The authors of [125] say fully independent.

4. a satisfies the Cauchy-Schwarz inequality if for any $x, y \in D$

$$\|a(x, y)\|_Z \leqslant \|a(x, x)\|_Z^{\frac{1}{2}} \|a(y, y)\|_Z^{\frac{1}{2}}. \tag{2.49}$$

5. A net (x_i) of elements of D is called a-convergent to $x \in X$ if $x_i \to x$ in X and if $a(x_i - x_j, x_i - x_j) \to 0$ for $i, j \to \infty$. We denote this by $x_i \overset{a}{\to} x$.

Remark 2.1 If a satisfies the Cauchy-Schwarz inequality and if Z is p-normed,[5] then for any $x, y \in D$ we have

$$\|a(x + y, x + y)\|_Z^{\frac{p}{2}} = \left(\|a(x + y, x + y)\|_Z^p \right)^{\frac{1}{2}}$$

$$= \left(\|a(x, x) + a(x, y) + a(y, x) + a(y, y)\|_Z^p \right)^{\frac{1}{2}} \tag{2.50}$$

$$\leqslant \left(\|a(x, x)\|_Z^p + \|a(x, y)\|_Z^p + \|a(y, x)\|_Z^p + \|a(y, y)\|_Z^p \right)^{\frac{1}{2}}$$

$$\overset{(2.49)}{\leqslant} \left(\|a(x, x)\|_Z^p + 2 \|a(y, y)\|_Z^{\frac{p}{2}} \|a(y, y)\|_Z^{\frac{p}{2}} + \|a(y, y)\|_Z^p \right)^{\frac{1}{2}}$$

$$= \|a(x, x)\|_Z^{\frac{p}{2}} + \|a(y, y)\|_Z^{\frac{p}{2}}.$$

Hence, if $x_i \overset{a}{\to} x$ and $y_j \overset{a}{\to} y$ and if $\alpha, \beta \in \mathbb{R}$ then $\alpha x_i + \beta y_j \overset{a}{\to} \alpha x + \beta y$.

In the sequel, we suppose that Z is an ordered $*$-Banach space.

Definition 2.5 Let $a: D \times D \to Z$ be a densely defined $\mathbb{R}$-bilinear map satisfying the Cauchy-Schwarz inequality. We say that a is closed if D is complete for the norm

$$\|x\|_a \overset{\text{def}}{=} \sqrt{\|a(x, x)\|_Z + \|x\|_X^2}. \tag{2.51}$$

Proposition 2.5 *Let $a: D \times D \to Z$ be a densely defined $\mathbb{R}$-bilinear map satisfying the Cauchy-Schwarz inequality. The following are equivalent.*

1. *a is closed*
2. *For any net (x_i), $x_i \overset{a}{\to} x$ implies $x \in D$ and $a(x_i - x, x_i - x) \to 0$*
3. *For any sequence (x_n), $x_n \overset{a}{\to} x$ implies $x \in D$ and $a(x_n - x, x_n - x) \to 0$.*

Proof $1 \Rightarrow 2$: Suppose that $x_i \overset{a}{\to} x$. This implies that $a(x_i - x_j, x_i - x_j) \to 0$ and $\|x_i - x_j\|_X \to 0$ for $i, j \to \infty$, so by (2.51) that $\|x_i - x_j\|_a \to 0$ as $i, j \to \infty$. By the completeness of D there is a $x_0 \in D$ such that $\|x_i - x\|_a \to 0$. Hence by (2.51), we have $a(x_i - x_0, x_j - x_0) \to 0$ and $\|x_i - x_0\|_X \to 0$. Hence we must have $x = x_0 \in D$ and $a(x_i - x, x_i - x) \to 0$.

$\quad 2 \Rightarrow 3$: This is obvious.

---

[5] That means that $\|y + z\|_Z^p \leqslant \|y\|_Z^p + \|z\|_Z^p$ for any $y, z \in Z$.

$3 \Rightarrow 1$: Let (x_n) be a Cauchy sequence in D, that is $\|x_n - x_m\|_a \to 0$ for $n, m \to \infty$. By (2.51), the sequence (x_n) is also Cauchy in X, so that there is an $x \in X$ such that $x_n \to x$. Since $a(x_n - x_m, x_n - x_m) \to 0$ by (2.51), the sequence (x_n) is a-convergent and the assumption implies that $x \in D$ and $a(x_n - x, x_n - x) \to 0$. By (2.51), we have $\|x_n - x\|_a \to 0$, which shows that D is complete. $\qquad\square$

Proposition 2.6 *Let $a: D \times D \to Z$ be a densely defined $\mathbb{R}$-bilinear form satisfying the Cauchy-Schwarz inequality.*

1. *Let (x_i) and (y_i) be nets in D. If $x_i \xrightarrow{a} x$ and $y_i \xrightarrow{a} y$ such that the nets $(a(x_i, x_i))$ and $(a(y_i, y_i))$ are bounded, then $\lim_i a(x_i, y_i)$ exists. If a is closed, then $\lim_i a(x_i, y_i) = a(x, y)$.*
2. *Let (x_n) be a sequence in D. If $x_n \xrightarrow{a} x$ then the sequence $(a(x_n, x_n))$ is bounded.*

Proof

1. In order to see that $\lim_i a(x_i, y_i)$ exists, we write:

$$\big\|a(x_i, y_i) - a(x_j, y_j)\big\|_Z = \big\|a(x_i - x_j, y_i) + a(x_j, y_i - y_j)\big\|_Z$$

$$\leqslant \big\|a(x_i - x_j, y_i)\big\|_Z + \big\|a(x_j, y_i - y_j)\big\|_Z$$

$$\overset{(2.49)}{\leqslant} \big\|a(x_i - x_j, x_i - x_j)\big\|_Z^{\frac{1}{2}} \big\|a(y_i, y_i)\big\|_Z^{\frac{1}{2}}$$

$$+ \big\|a(x_j, x_j)\big\|_Z^{\frac{1}{2}} \big\|a(y_i - y_j, y_i - y_j)\big\|_Z^{\frac{1}{2}}.$$

It follows from the previous inequality that $(a(x_i, y_i))$ is a Cauchy net, hence a convergent net.

If a is in addition closed, note by Proposition 2.5 that $x, y \in D$, $a(x_i - x, x_i - x) \to 0$ and $a(y_i - y, y_i - y) \to 0$. We conclude with

$$\|a(x_i, y_i) - a(x, y)\|_Z = \|a(x_i - x, y_i) + a(x, y_i - y)\|_Z$$

$$\leqslant \|a(x_i - x, y_i)\|_Z + \|a(x, y_i - y)\|_Z$$

$$\overset{(2.49)}{\leqslant} \|a(x_i - x, x_i - x)\|_Z^{\frac{1}{2}} \|a(y_i, y_i)\|_Z^{\frac{1}{2}}$$

$$+ \|a(x, x)\|_Z^{\frac{1}{2}} \|a(y_i - y, y_i - y)\|_Z^{\frac{1}{2}}.$$

2. We have $a(x_n - x_m, x_n - x_m) \to 0$ and $\|x_n - x_m\|_X \to 0$ for $n, m \to \infty$ so by (2.51) that $\|x_n - x_m\|_a \to 0$ as $n, m \to \infty$. So (x_n) is a Cauchy sequence for the norm (2.51), hence a bounded sequence. Hence the sequence $(a(x_n, x_n))$ is bounded by (2.51). $\qquad\square$

Definition 2.6 A densely defined $\mathbb{R}$-bilinear map a satisfying the Cauchy-Schwarz inequality is called closable if there exists a closed map $b: D' \times D' \to Z$ such that $D \subset D' \subset X$ with $a(x, y) = b(x, y)$ for any $x, y \in D$.

If a is closable, we let

$$\operatorname{dom}\overline{a} \stackrel{\text{def}}{=} \left\{ x \in X \text{ such that there exists a sequence } (x_n) \text{ of } D \text{ satisfying } x_n \stackrel{a}{\to} x \right\}. \tag{2.52}$$

Proposition 2.7 *Let $a: D \times D \to Z$ be a densely defined (symmetric positive) $\mathbb{R}$-bilinear form satisfying the Cauchy-Schwarz inequality. Then a is closable if and only if $x_n \stackrel{a}{\to} 0$ implies $a(x_n, x_n) \to 0$. In this case, if $x_n \stackrel{a}{\to} x$ and if $y_n \stackrel{a}{\to} y$ where $x, y \in \operatorname{dom}\overline{a}$ then the limit*

$$\overline{a}(x, y) \stackrel{\text{def}}{=} \lim_{n \to \infty} a(x_n, y_n) \tag{2.53}$$

exists and $\overline{a}$ is a well-defined, densely defined (symmetric positive) closed form and satisfies the Cauchy-Schwarz inequality. In addition, every closed extension of a is also an extension of $\overline{a}$.

Proof $\Rightarrow$: Assume that a is closable and let $b: D' \times D' \to Z$ be a closed extension. Suppose $x_n \stackrel{a}{\to} 0$. Then $x_n \stackrel{b}{\to} 0$. Hence $a(x_n, x_n) = b(x_n, x_n) = b(x_n - 0, x_n - 0) \to 0$ since b is closed.

$\Leftarrow$: By Proposition 2.6, the limit (2.53) exists. We will prove that $\lim_{n \to \infty} a(x_n, y_n)$ is independent of the chosen sequences (x_n) and (y_n). Indeed, if in addition $x'_n \stackrel{a}{\to} x$ and $y'_n \stackrel{a}{\to} y$, then

$$\left\| a(x_n, y_n) - a(x'_n, y'_n) \right\|_Z = \left\| a(x_n - x'_n, y_n) + a(x'_n, y_n - y'_n) \right\|_Z$$

$$\leqslant \left\| a(x_n - x'_n, y_n) \right\|_Z + \left\| a(x'_n, y_n - y'_n) \right\|_Z$$

$$\stackrel{(2.49)}{\leqslant} \left\| a(x_n - x'_n, x_n - x'_n) \right\|_Z^{\frac{1}{2}} \left\| a(y_n, y_n) \right\|_Z^{\frac{1}{2}}$$

$$+ \left\| a(x'_n, x'_n) \right\|_Z^{\frac{1}{2}} \left\| a(y_n - y'_n, y_n - y'_n) \right\|_Z^{\frac{1}{2}}.$$

Note that $x_n - x'_n \stackrel{a}{\to} 0$ and $y_n - y'_n \stackrel{a}{\to} 0$. So by the assumption $a(x_n - x'_n, x_n - x'_n) \to 0$ and $a(y_n - y'_n, y_n - y'_n) \to 0$. With the point 2 of Proposition 2.6, we obtain $\left\| a(x_n, y_n) - a(x'_n, y'_n) \right\|_Z \to 0$ as $n \to \infty$.

For the closedeness of $\overline{a}$, consider a Cauchy sequence (x_n) of elements of D for the norm $\|x\|_{\overline{a}}$. It converges to some $x \in X$. From the definition (2.52) of $\operatorname{dom}\overline{a}$, we infer that x belongs to $\operatorname{dom}\overline{a}$. We infer that

$$\overline{a}(x_n - x, x_n - x) \stackrel{(2.53)}{=} \lim_{n \to \infty} \lim_{m \to \infty} a(x_n - x_m, x_n - x_m) = 0$$

which means that the sequence (x_n) is convergent for the norm $\|x\|_{\overline{a}}$. Finally, note that D is dense in $\operatorname{dom}\overline{a}$. So by a classical argument $\operatorname{dom}\overline{a}$ is also complete.

Now we will show that $\overline{a}$ is the smallest closed extension of a. Let b be a closed extension of a. Let $x \in \operatorname{dom}\overline{a}$. By (2.52), there exists a sequence (x_n) of D such that $x_n \overset{a}{\to} x$. Then $x_n \overset{b}{\to} x$. So x is an element of $\operatorname{dom}b$. We conclude that $\operatorname{dom}\overline{a} \subset \operatorname{dom}b$ and $a(x, y) = b(x, y)$ if $x, y \in D$ by (2.53) and Proposition 2.6.   □

Definition 2.7 If the form a is closable, then $\overline{a}$ defined by (2.53) with domain $\operatorname{dom}(\overline{a})$ is called the closure of the bilinear map a.

2.4   Transference of Fourier Multipliers on Crossed Product Von Neumann Algebras

In the statement of our results on Riesz transforms associated with markovian semigroups of Fourier multipliers of Chap. 3 and later, we will need to use crossed products which is a basic method of construction of a new von Neumann algebra $M \rtimes_\alpha G$ from a von Neumann algebra M on which acts a discrete group G. In our work, M will be a q-Gaussian von Neumann algebra $\Gamma_q(H)$ of Sect. 2.2 and the action will be constructed in (2.83) by means of the second quantization from the orthogonal representation of Proposition 2.3. We refer to [94, 98, 212, 215, 216] for more information on crossed products.

Let M be a von Neumann algebra acting on a Hilbert space H. Let G be a locally compact group equipped with some left Haar measure μ_G. Let $\alpha \colon G \to \operatorname{Aut}(M)$ be a representation of G on M which is weak* continuous, i.e. for any $x \in M$ and any $y \in M_*$, the map $G \to M, s \mapsto \langle \alpha_s(x), y \rangle_{M, M_*}$ is continuous. For any $x \in M$, we define the operators $\pi(x) \colon \mathrm{L}^2(G, H) \to \mathrm{L}^2(G, H)$ [212, (2) p. 263] by

$$\big(\pi(x)\xi\big)(s) \overset{\text{def}}{=} \alpha_s^{-1}(x)\xi(s), \qquad \xi \in \mathrm{L}^2(G, H), s \in G. \tag{2.54}$$

These operators satisfy the following commutation relation [212, (2) p. 292]:

$$(\lambda_s \otimes \operatorname{Id}_H)\pi(x)(\lambda_s \otimes \operatorname{Id}_H)^* = \pi(\alpha_s(x)), \quad x \in M, s \in G. \tag{2.55}$$

Recall that the crossed product of M and G with respect to α is the von Neumann algebra $M \rtimes_\alpha G = (\pi(M) \cup \{\lambda_s \otimes \operatorname{Id}_H : s \in G\})''$ on the Hilbert space $\mathrm{L}^2(G, H)$ generated by the operators $\pi(x)$ and $\lambda_s \otimes \operatorname{Id}_H$ where $x \in M$ and $s \in G$. By Stratila [212, p. 263] or [221, Proposition 2.5], π is a normal injective $*$-homomorphism from M into $M \rtimes_\alpha G$ (hence σ-strong* continuous).

In the sequel, we suppose that G is discrete. For any $s \in G$ and any $x \in M$, we let $x \rtimes \lambda_s \overset{\text{def}}{=} \pi(x)(\lambda_s \otimes \operatorname{Id}_H)$. We recall the rules of product and adjoint:

$$(x \rtimes \lambda_s)(y \rtimes \lambda_t) = x\alpha_s(y) \rtimes \lambda_{st}, \quad (s, t \in G, x, y \in M) \tag{2.56}$$

and

$$(x \rtimes \lambda_s)^* = \alpha_{s^{-1}}(x^*) \rtimes \lambda_{s^{-1}} \quad (s \in G, x \in M). \tag{2.57}$$

These relations will be used frequently, as well as the following definition. Namely, for $1 \leqslant p \leqslant \infty$ and M a semifinite von Neumann algebra, we let

$$\mathcal{P}_{p,\rtimes,G} = \operatorname{span}\{x \rtimes \lambda_s : x \in \mathrm{L}^p(M), s \in G\} \subset \mathrm{L}^p(M \rtimes_\alpha G), \tag{2.58}$$

and write in short $\mathcal{P}_{\rtimes,G} = \mathcal{P}_{\infty,\rtimes,G}$, which will be the most used of these spaces. Finally, if M is equipped with a normal finite faithful trace τ_M, then $M \rtimes_\alpha G$ is equipped with the normal finite faithful trace defined by $\tau_{M \rtimes_\alpha G}(x \rtimes \lambda_s) = \tau_M(x)\delta_{s=e}$ if τ_M is G-invariant.

Lemma 2.3 *Assume M to be a finite von Neumann algebra. Let $1 < p < \infty$. Then $\mathcal{P}_{\rtimes,G}$ and $\mathcal{P}_{p,\rtimes,G}$ are dense subspaces of $\mathrm{L}^p(M \rtimes_\alpha G)$. In particular, $\mathcal{P}_G$ is dense in $\mathrm{L}^p(\mathrm{VN}(G))$.*

Proof Since M is finite, we have $M \subset \mathrm{L}^p(M)$, and thus, $\mathcal{P}_{\rtimes,G} \subset \mathcal{P}_{p,\rtimes,G}$. It thus suffices to prove that $\mathcal{P}_{\rtimes,G}$ is dense. Let $y \in \mathrm{L}^{p^*}(M \rtimes_\alpha G)$ such that $\langle x, y \rangle = 0$ for all $x \in \mathcal{P}_{\rtimes,G}$. By the Hahn-Banach theorem, it suffices to show that $y = 0$. But $y \in \mathrm{L}^1(M \rtimes_\alpha G) = (M \rtimes_\alpha G)_*$. By the rule (2.56), $\mathcal{P}_{\rtimes,G}$ contains span $(\pi(M) \cup \{\lambda_s \otimes \mathrm{Id}_H : s \in G\})$, and therefore by construction, $\mathcal{P}_{\rtimes,G}$ is weak* dense in $M \rtimes_\alpha G$. So $\langle x, y \rangle = 0$ for all $x \in M \rtimes_\alpha G$. Thus, $y = 0$. $\qquad\qquad\square$

Given a discrete group G, we can consider the "fundamental unitary" $W_H : \ell_G^2 \otimes_2 H \otimes_2 \ell_G^2 \to \ell_G^2 \otimes_2 H \otimes_2 \ell_G^2$ defined by

$$W_H(\varepsilon_t \otimes \xi \otimes \varepsilon_r) = \varepsilon_t \otimes \xi \otimes \varepsilon_{tr}. \tag{2.59}$$

Lemma 2.4 *For any $s \in G$ and any $x \in M$, we have*

$$W_H\big((x \rtimes \lambda_s) \otimes \mathrm{Id}_{\ell_G^2}\big)W_H^* = (x \rtimes \lambda_s) \otimes \lambda_s.$$

Proof On the one hand, for any $s, t, r \in G$, we have

$$W_H\big((x \rtimes \lambda_s) \otimes \mathrm{Id}_{\ell_G^2}\big)(\varepsilon_t \otimes \xi \otimes \varepsilon_r)$$

$$= W_H\big(x(\lambda_s \otimes \mathrm{Id}_H) \otimes \mathrm{Id}_{\ell_G^2}\big)(\varepsilon_t \otimes \xi \otimes \varepsilon_r)$$

$$= W_H\big(x \otimes \mathrm{Id}_{\ell_G^2}\big)(\varepsilon_{st} \otimes \xi \otimes \varepsilon_r) = W_H(\varepsilon_{st} \otimes \alpha_{st}^{-1}(x)\xi \otimes \varepsilon_r)$$

$$= \varepsilon_{st} \otimes \alpha_{st}^{-1}(x)\xi \otimes \varepsilon_{str}.$$

On the other hand, we have

$$((x \rtimes \lambda_s) \otimes \lambda_s) W_H(\varepsilon_t \otimes \xi \otimes \varepsilon_r) = ((x(\lambda_s \otimes \mathrm{Id}_H)) \otimes \lambda_s)(\varepsilon_t \otimes \xi \otimes \varepsilon_{tr})$$

$$= (x \otimes \mathrm{Id}_{\ell_G^2})(\varepsilon_{st} \otimes \xi \otimes \varepsilon_{str})$$

$$= \varepsilon_{st} \otimes \alpha_{st}^{-1}(x)\xi \otimes \varepsilon_{str}.$$

We conclude by linearity and density.   □

As a composition of an amplification and a spatial isomorphism, the map $M \rtimes_\alpha G \to \mathrm{VN}(G)\overline{\otimes}(M \rtimes_\alpha G)$, $z \mapsto W_H(z \otimes \mathrm{Id}_{\ell_G^2})W_H^*$ is a unital normal injective *-homomorphism. This map is a well-defined kind of "crossed coproduct" defined by

$$\Delta: M \rtimes_\alpha G \longrightarrow \mathrm{VN}(G)\overline{\otimes}(M \rtimes_\alpha G) \tag{2.60}$$
$$x \rtimes \lambda_s \longmapsto \lambda_s \otimes (x \rtimes \lambda_s)$$

This homormorphism is trace preserving since for any $s \in G$ and any $x \in M$

$$\tau_{\mathrm{VN}(G)\overline{\otimes}(M \rtimes_\alpha G)}\big(\Delta(x \rtimes \lambda_s)\big) \overset{(2.60)}{=} \tau_{\mathrm{VN}(G)\overline{\otimes}(M \rtimes_\alpha G)}(\lambda_s \otimes (x \rtimes \lambda_s))$$

$$= \tau_M(x)\delta_{s=e}\tau_{\mathrm{VN}(G)}(\lambda_s)$$

$$= \tau_M(x)\delta_{s=e} = \tau_{M \rtimes_\alpha G}(x \rtimes \lambda_s).$$

By Arhancet [9, Lemma 4.1], it admits a completely isometric L^p-extension $\Delta_p: L^p(M \rtimes_\alpha G) \to L^p(\mathrm{VN}(G)\overline{\otimes}(M \rtimes_\alpha G))$ for any $1 \leqslant p < \infty$.

If $x \in L^p(M)$ and $s \in G$ and if $\phi: G \to \mathbb{C}$, we let

$$\big(\mathrm{Id}_{L^p(M)} \rtimes M_\phi\big)(x \rtimes \lambda_s) \overset{\mathrm{def}}{=} \phi(s)x \rtimes \lambda_s. \tag{2.61}$$

The main result of this section is the following transference result. See [16] for a larger perspective of this result. The assumptions are satisfied in the case where M has QWEP and where the action $\alpha: G \to \mathrm{Aut}(M)$ is amenable, see [181, Proposition 4.1 (vi)]. See also [10, Proposition 4.8].

Proposition 2.8 *Suppose $1 \leqslant p \leqslant \infty$. Let $\phi: G \to \mathbb{C}$ a function which induces a completely bounded Fourier multiplier $M_\phi: L^p(\mathrm{VN}(G)) \to L^p(\mathrm{VN}(G))$. If $1 \leqslant p < \infty$, assume in addition that $M \rtimes_\alpha G$ has QWEP. Then (2.61) induces a completely bounded map $\mathrm{Id}_{L^p(M)} \rtimes M_\phi: L^p(M \rtimes_\alpha G) \to L^p(M \rtimes_\alpha G)$ and*

$$\big\|\mathrm{Id}_{L^p(M)} \rtimes M_\phi\big\|_{\mathrm{cb}, L^p(M \rtimes_\alpha G) \to L^p(M \rtimes_\alpha G)} \leqslant \big\|M_\phi\big\|_{\mathrm{cb}, L^p(\mathrm{VN}(G)) \to L^p(\mathrm{VN}(G))}. \tag{2.62}$$

Moreover, we have

$$\Delta_p\big(\mathrm{Id}_{\mathrm{L}^p(M)} \rtimes M_\phi\big) = \big(M_\phi \otimes \mathrm{Id}_{\mathrm{L}^p(M \rtimes_\alpha G)}\big)\Delta_p. \qquad (2.63)$$

Proof On the one hand, for any $x \in \mathrm{L}^p(M)$ and any $s \in G$, we have

$$\Delta_p\big(\mathrm{Id}_{\mathrm{L}^p(M)} \rtimes M_\phi\big)(x \rtimes \lambda_s) \overset{(2.61)}{=} \phi(s)\Delta_p(x \rtimes \lambda_s) \overset{(2.60)}{=} \phi(s)\big(\lambda_s \otimes (x \rtimes \lambda_s)\big).$$

On the other hand, we have

$$\big(M_\phi \otimes \mathrm{Id}_{\mathrm{L}^p(M \rtimes_\alpha G)}\big)\Delta_p(x \rtimes \lambda_s) \overset{(2.60)}{=} \big(M_\phi \otimes \mathrm{Id}_{\mathrm{L}^p(M \rtimes_\alpha G)}\big)\big(\lambda_s \otimes (x \rtimes \lambda_s)\big)$$

$$= \phi(s)\big(\lambda_s \otimes (x \rtimes \lambda_s)\big).$$

We conclude that (2.63) is true on $\mathcal{P}_{p,\rtimes,G}$. Since $M \rtimes_\alpha G$ has QWEP, the von Neumann algebra $\mathrm{VN}(G)$ is also QWEP. So, by Junge [117, (iii) p. 984], the map $M_\phi \otimes \mathrm{Id}_{\mathrm{L}^p(M \rtimes_\alpha G)}\colon \mathrm{L}^p(\mathrm{VN}(G)\overline{\otimes}(M \rtimes_\alpha G)) \to \mathrm{L}^p(\mathrm{VN}(G)\overline{\otimes}(M \rtimes_\alpha G))$ is completely bounded (in the case $p = \infty$, note that this assumption is useless) with

$$\big\| M_\phi \otimes \mathrm{Id}_{\mathrm{L}^p(M \rtimes_\alpha G)} \big\|_{\mathrm{cb},\mathrm{L}^p(\mathrm{VN}(G)\overline{\otimes}(M \rtimes_\alpha G)) \to \mathrm{L}^p(\mathrm{VN}(G)\overline{\otimes}(M \rtimes_\alpha G))} \qquad (2.64)$$

$$\leqslant \big\| M_\phi \big\|_{\mathrm{cb},\mathrm{L}^p(\mathrm{VN}(G)) \to \mathrm{L}^p(\mathrm{VN}(G))}.$$

Note that $\mathrm{Id}_{S^p} \otimes \Delta_p\colon \mathrm{L}^p(\mathrm{B}(\ell^2)\overline{\otimes}(M \rtimes_\alpha G)) \to \mathrm{L}^p(\mathrm{B}(\ell^2)\overline{\otimes}\mathrm{VN}(G)\overline{\otimes}(M \rtimes_\alpha G))$ is a complete isometry. If $z \in S^p \otimes \mathcal{P}_{p,\rtimes,G}$, we have

$$\big\| \big(\mathrm{Id}_{S^p} \otimes (\mathrm{Id}_{\mathrm{L}^p(M)} \rtimes M_\phi)\big)(z) \big\| = \big\| \big(\mathrm{Id}_{S^p} \otimes \big(\Delta_p(\mathrm{Id}_{\mathrm{L}^p(M)} \rtimes M_\phi)\big)\big)(z) \big\|$$

$$\overset{(2.63)}{=} \big\| \big(\mathrm{Id}_{S^p} \otimes \big(M_\phi \otimes \mathrm{Id}_{\mathrm{L}^p(M \rtimes_\alpha G)}\big)\Delta_p\big)(z) \big\|$$

$$\leqslant \big\| M_\phi \otimes \mathrm{Id}_{\mathrm{L}^p(M \rtimes_\alpha G)} \big\|_{\mathrm{cb}} \big\| \Delta_p \big\|_{\mathrm{cb}} \|z\|$$

$$\overset{(2.64)}{\leqslant} \big\| M_\phi \big\|_{\mathrm{cb},\mathrm{L}^p(\mathrm{VN}(G)) \to \mathrm{L}^p(\mathrm{VN}(G))} \|z\|.$$

So (2.63) is proved.   □

Recall that a discrete group has AP if there exists a net (φ_j) of finitely supported functions on G such that $M_{\varphi_j} \otimes \mathrm{Id}_{\mathrm{B}(H)} \to \mathrm{Id}_{\mathrm{C}^*_r(G)\otimes\mathrm{B}(H)}$ in the point-norm topology for any Hilbert space H by Haagerup and Kraus [96, Theorem 1.9]. A discrete group G is weakly amenable if there exists a net (φ_j) of finitely supported functions on G such that $\varphi_j \to 1$ pointwise and $\sup_j \big\| M_{\varphi_j} \big\|_{\mathrm{cb},\mathrm{VN}(G)\to\mathrm{VN}(G)} < \infty$. Recall that a weakly amenable discrete group has AP by Haagerup and Kraus [96, p. 677]. Finally note that amenable groups, free groups, $\mathrm{SL}(2, \mathbb{Z})$ and hyperbolic groups are weakly amenable. The first three parts of the following result are [48, Lemma 2.1].

The part 4 is a straightforward consequence of Proposition 2.8 and the proof of [96, Theorem 1.12].

Proposition 2.9 *Let G be a discrete group with AP and $\alpha \colon G \to \mathrm{Aut}(M)$ be a weak* continuous representation of G on a von Neumann algebra M. Then there exist a net (φ_j) of finitely supported functions on G with the following properties:*

1. *For each $s \in G$, we have $\lim_j \varphi_j(s) = 1$.*
2. *Each M_{φ_j} induces a normal completely bounded map $\mathrm{Id} \rtimes M_{\varphi_j} \colon M \rtimes_\alpha G \to M \rtimes_\alpha G$.*
3. *The net $(\mathrm{Id} \rtimes M_{\varphi_j})$ converges to $\mathrm{Id}_{M \rtimes_\alpha G}$ for the point weak* topology.*
4. *If G is in addition weakly amenable then $\sup_j \left\| \mathrm{Id} \rtimes M_{\varphi_j} \right\|_{\mathrm{cb}, M \rtimes_\alpha G \to M \rtimes_\alpha G} < \infty$.*

An operator space E has CBAP [81, p. 205] (see also [46, p. 365] for the particular case of C*-algebras) when there exists a net (T_j) of finite-rank linear maps $T_j \colon E \to E$ satisfying the properties:

1. for any $x \in E$, we have $\lim_j \|T_j(x) - x\|_E = 0$,
2. $\sup_j \|T_j\|_{\mathrm{cb}, E \to E} < \infty$.

Then E has in addition CCAP if the supremum in the second point can be chosen to be 1. By Junge and Ruan [121, Proposition 4.3], if M is a finite von Neumann algebra then $L^p(M)$ has CCAP if and only if $L^p(M)$ has CBAP. In the case where E is a noncommutative L^p-space associated to a group von Neumann algebra of a discrete group G, we can use an average construction of Haagerup [94, proof of Lemma 2.5] to replace the net (T_j) by a net of Fourier multipliers. Indeed, we have the following observation which is contained in the proof of [121, Theorem 4.4].

Lemma 2.5 *Let G be a discrete group such that $\mathrm{VN}(G)$ has QWEP. Suppose $1 \leqslant p < \infty$. Then the operator space $L^p(\mathrm{VN}(G))$ has CCAP if and only if there exists a net (φ_j) of functions $\varphi_j \colon G \to \mathbb{C}$ with finite support which converge pointwise to 1 such that the net (M_{φ_j}) converges to $\mathrm{Id}_{L^p(\mathrm{VN}(G))}$ in the point-norm topology with*

$$\sup_j \left\| M_{\varphi_j} \right\|_{\mathrm{cb}, L^p(\mathrm{VN}(G)) \to L^p(\mathrm{VN}(G))} \leqslant 1.$$

Recall that by Junge and Ruan [121, Proposition 3.5] if G is a weakly amenable discrete group then $L^p(\mathrm{VN}(G))$ has CBAP for any $1 < p < \infty$. A careful reading of the proofs of [121, Proposition 3.2] and [121, Proposition 3.5] shows the existence of an approximating net (M_{φ_j}) as in Lemma 2.5 which converges to $\mathrm{Id}_{L^p(\mathrm{VN}(G))}$ for *all* $1 \leqslant p < \infty$ *and* in $\mathrm{C}^*_r(G)$ in the point-norm topologies with in addition $\sup_j \left\| M_{\varphi_j} \right\|_{\mathrm{cb}, \mathrm{C}^*_r(G) \to \mathrm{C}^*_r(G)} < \infty$ (and $\sup_j \left\| M_{\varphi_j} \right\|_{\mathrm{cb}, \mathrm{C}^*_r(G) \to \mathrm{C}^*_r(G)} = 1$ if its Cowling-Haagerup constant is equal to one). If $M \rtimes_\alpha G$ has QWEP, then using [84, Proposition C.18], Proposition 2.8 and the density of $\mathcal{P}_{\rtimes, G}$ in $L^p(M \rtimes_\alpha G)$, it is easy to check that the net $(\mathrm{Id} \rtimes M_{\varphi_j})$ of completely bounded maps converges to $\mathrm{Id}_{L^p(M \rtimes_\alpha G)}$ for the point norm topology of $L^p(M \rtimes_\alpha G)$.

Consider finally a discrete group G with AP such that $VN(G)$ has QWEP. It is stated in [121, Theorem 1.2] that if $1 < p < \infty$ then $L^p(VN(G))$ has CCAP. But in this case, it is unclear if the previous paragraph is true for this class of groups and it is an interesting open question.

We will use the following result which is stated in [221, Proposition 3.13 and its proof], see also [97, Theorem 4.1] for complements. Then the rather easy proof of Lemma 2.6 follows the same lines.

Lemma 2.6 *Let* $T: M \to M$ *be a completely bounded normal map. Let* G *be a discrete group and* $\alpha: G \to \mathrm{Aut}(M)$ *be a weak* continuous action. If* $T\alpha(s) = \alpha(s)T$ *for any* $s \in G$, *then there is a linear normal mapping* $T \rtimes \mathrm{Id}_{VN(G)}: M \rtimes_\alpha G \to M \rtimes_\alpha G$ *of completely bounded norm that of* T *such that*

$$\left(T \rtimes \mathrm{Id}_{VN(G)}\right)(x \rtimes \lambda_s) = T(x) \rtimes \lambda_s.$$

If T *is in addition a (faithful) normal conditional expectation then* $T \rtimes \mathrm{Id}_{VN(G)}$ *is a (faithful) normal conditional expectation.*

2.5   Hilbertian Valued Noncommutative L^p-Spaces

We shall use various Hilbert-valued noncommutative L^p-spaces. These are some kind of noncommutative Bochner spaces which will be invaluable when we will use noncommutative Khintchine inequalities and some variants. We refer to [128, Chapter 2] for more information on these spaces. However, note that our notations are slightly different since we want that our notations remain compatible with vector-valued noncommutative L^p-spaces in the hyperfinite case in the spirit of the notations of [132].

Let M be a semifinite von Neumann algebra. Let X be a complex vector space X which is a right M-module. If X is equipped with an $L^{\frac{p}{2}}(M)$-valued inner product $\langle \cdot, \cdot \rangle_X$[6] where $0 < p \leqslant \infty$, recall that for any $x, y \in X$, we have by Junge and Sherman [123, (2.2)]

$$\|\langle x, y \rangle_X\|_{L^{\frac{p}{2}}(M)} \leqslant \|x\|_X \|y\|_X \tag{2.65}$$

---

[6] By an $L^{\frac{p}{2}}$-valued inner product on X we mean a sesquilinear mapping $\langle \cdot, \cdot \rangle : X \times X \to L^{\frac{p}{2}}(M)$, conjugate linear in the first variable, satisfying for any $x, y \in X$ and any $z \in M$

1. $\langle x, yz \rangle = \langle x, y \rangle z$
2. $\langle x, y \rangle = \langle y, x \rangle^*$
3. $\langle x, x \rangle \geqslant 0$
4. $\langle x, x \rangle = 0 \iff x = 0$.

where

$$\|x\|_X \overset{\text{def}}{=} \left\| \langle x, x \rangle_X \right\|_{L^{\frac{p}{2}}(M)}^{\frac{1}{2}}. \tag{2.66}$$

Note that the inner product is antilinear in the first and linear in the second variable. The following is [123, Proposition 2.2]. We include a short proof for the sake of completeness.

Lemma 2.7 *Suppose* $0 < p \leqslant \infty$. *Let* X *be a complex vector space which is a right M-module equipped with an* $L^{\frac{p}{2}}(M)$-*valued inner product* $\langle \cdot, \cdot \rangle_X$ *on* X. *Then (2.66) defines a norm on* X *if* $2 \leqslant p \leqslant \infty$ *and a* $\frac{p}{2}$-*norm if* $0 < p \leqslant 2$.

Proof For any $x, y \in X$, we have

$$\left\| \langle x, y \rangle_X \right\|_{L^{\frac{p}{2}}(M)} \overset{(2.65)}{\leqslant} \|x\|_X \|y\|_X \overset{(2.66)}{=} \left\| \langle x, x \rangle_X \right\|_{L^{\frac{p}{2}}(M)}^{\frac{1}{2}} \left\| \langle y, y \rangle_X \right\|_{L^{\frac{p}{2}}(M)}^{\frac{1}{2}}.$$

Hence the inequality (2.49) is valid with $Z = L^{\frac{p}{2}}(M)$ and with $a(x, y) = \langle x, y \rangle_X$. By Remark 2.1, the proof is finished. $\qquad\square$

Suppose $1 \leqslant p < \infty$. Let H be a Hilbert space. For any elements $\sum_{k=1}^{n} x_k \otimes a_k, \sum_{j=1}^{m} y_j \otimes b_j$ of $L^p(M) \otimes H$, we define the $L^{\frac{p}{2}}(M)$-valued inner product

$$\left\langle \sum_{k=1}^{n} x_k \otimes a_k, \sum_{j=1}^{m} y_j \otimes b_j \right\rangle_{L^p(M,H_{c,p})} \overset{\text{def}}{=} \sum_{k,j=1}^{n,m} \langle a_k, b_j \rangle_H x_k^* y_j. \tag{2.67}$$

For any element $\sum_{k=1}^{n} x_k \otimes a_k$ of $L^p(M) \otimes H$, we set [128, (2.9)]

$$\left\| \sum_{k=1}^{n} x_k \otimes a_k \right\|_{L^p(M,H_{c,p})} \overset{\text{def}}{=} \left\| \left(\left\langle \sum_{k=1}^{n} x_k \otimes a_k, \sum_{j=1}^{n} x_j \otimes a_j \right\rangle_{L^p(M,H_{c,p})} \right)^{\frac{1}{2}} \right\|_{L^p(M)} \tag{2.68}$$

$$= \left\| \left(\sum_{k,j=1}^{n} \langle a_k, a_j \rangle_H x_k^* x_j \right)^{\frac{1}{2}} \right\|_{L^p(M)}.$$

The space $L^p(M, H_{c,p})$ is the completion of $L^p(M) \otimes H$ for this norm [128, p. 10]. It is folklore that $L^p(M, H_{c,p})$ equipped with (2.67) is a right L^p-M-module [123, Definition 3.3], see also [120, Section 1.3].

If $(e_1, \ldots, e_n)$ is an orthonormal family of H and if $x_1, \ldots, x_n$ belong to $L^p(M)$, it follows from (2.68) (see [128, (2.10)]) that

$$\left\| \sum_{k=1}^n x_k \otimes e_k \right\|_{L^p(M, H_{c,p})} = \left\| \left(\sum_{k=1}^n |x_k|^2 \right)^{\frac{1}{2}} \right\|_{L^p(M)} = \left\| \sum_{k=1}^n e_{k1} \otimes x_k \right\|_{S^p(L^p(M))}.$$

We define similarly $L^p(M, H_{r,p})$ and we let $L^p(M, H_{\mathrm{rad},p}) \overset{\mathrm{def}}{=} L^p(M, H_{r,p}) \cap L^p(M, H_{c,p})$ if $p \geqslant 2$ and $L^p(M, H_{\mathrm{rad},p}) \overset{\mathrm{def}}{=} L^p(M, H_{r,p}) + L^p(M, H_{c,p})$ if $1 \leqslant p \leqslant 2$. Recall that [128, (2.25)]

$$L^p(M, H_{\mathrm{rad},p})^* = L^{p^*}(M, \overline{H}_{\mathrm{rad},p^*}). \tag{2.69}$$

If $T \colon L^p(M) \to L^p(M)$ is completely bounded then by Junge et al. [128, p. 22] the map $T \otimes \mathrm{Id}_H$ induces a bounded operator on $L^p(M, H_{c,p})$, on $L^p(M, H_{r,p})$ and on $L^p(M, H_{\mathrm{rad},p})$ and we have

$$\left\| T \otimes \mathrm{Id}_H \right\|_{L^p(M, H_{c,p}) \to L^p(M, H_{c,p})} \leqslant \| T \|_{\mathrm{cb}, L^p(M) \to L^p(M)}. \tag{2.70}$$

(and similarly for the others, and also similarly for $T \colon L^p(M) \to L^p(N)$ completely bounded between distinct spaces). If $T \colon H \to K$ between Hilbert spaces is bounded then the map $\mathrm{Id}_{L^p(M)} \otimes T$ induces a bounded operator from $L^p(M, H_{\mathrm{rad},p})$ into $L^p(M, K_{\mathrm{rad},p})$ (and on $L^p(M, H_{c,p})$ and $L^p(M, H_{r,p})$) [128, p. 14] and we have

$$\left\| \mathrm{Id}_{L^p(M)} \otimes T \right\|_{L^p(M, H_{\mathrm{rad},p}) \to L^p(M, K_{\mathrm{rad},p})} = \| T \|_{H \to K} \tag{2.71}$$

(and similarly for the others).

We need some semifinite variant of spaces introduced in [116] in the σ-finite case. Let $\mathbb{E} \colon \mathcal{N} \to M$ be a trace preserving normal faithful conditional expectation between semifinite von Neumann algebras equipped with normal semifinite faithful traces. If $2 \leqslant p \leqslant \infty$, for any $f, g \in L^p(\mathcal{N})$, using the boundedness[7] of the conditional expectation $\mathbb{E} \colon L^{\frac{p}{2}}(\mathcal{N}) \to L^{\frac{p}{2}}(M)$ we let $\langle f, g \rangle_{L^p_c(\mathbb{E})} \overset{\mathrm{def}}{=} \mathbb{E}(f^* g)$. It is clear that this defines an $L^{\frac{p}{2}}(M)$-valued inner product and the associated right L^p-M-module[8] is denoted by $L^p_c(\mathbb{E})$. This means that $L^p_c(\mathbb{E})$ is the completion of $L^p(\mathcal{N})$ with respect to the norm

$$\| f \|_{L^p_c(\mathbb{E})} \overset{\mathrm{def}}{=} \left\| \mathbb{E}(f^* f) \right\|_{L^{\frac{p}{2}}(M)}^{\frac{1}{2}}. \tag{2.72}$$

---

[7] In the case $0 < p < 2$, note that the conditional expectation $\mathbb{E} \colon L^{\frac{p}{2}}(\mathcal{N}) \to L^{\frac{p}{2}}(M)$ is not bounded in general.

[8] In the case $p = \infty$, we obtain a right W*-module.

We still denote by $\langle f, g \rangle_{L_c^p(\mathbb{E})}$ the extension of the bracket on this space. Similarly, we define $\| f \|_{L_r^p(\mathbb{E})} \overset{\text{def}}{=} \left\| \mathbb{E}(ff^*)^{\frac{1}{2}} \right\|_{L^{\frac{p}{2}}(M)}$. The proof of [116, Remark 2.7], gives the following result. For the sake of completeness, we give the short computation.

Lemma 2.8 *Suppose* $2 \leqslant p \leqslant \infty$. *We have a contractive injective inclusion of* $L^p(\mathcal{N})$ *into* $L_c^p(\mathbb{E})$ *and into* $L_r^p(\mathbb{E})$.

Proof For any $f \in L^p(\mathcal{N})$, we have

$$
\| f \|_{L_c^p(\mathbb{E})} \overset{(2.72)}{=} \left\| \mathbb{E}(f^* f)^{\frac{1}{2}} \right\|_{L^{\frac{p}{2}}(M)} \leqslant \left\| f^* f \right\|_{L^{\frac{p}{2}}(\mathcal{N})}^{\frac{1}{2}} = \| f \|_{L^p(\mathcal{N})} .
$$

We conclude by density. The row result is similar. □

In the case $0 < p < 2$, the definition of these spaces is unclear in the *semifinite* case.[9] However, if $N = M \overline{\otimes} N$ is a product with N finite, we can use the conditional expectation $\mathbb{E} = \mathrm{Id}_M \otimes \tau \colon M \overline{\otimes} N \to M$ at the level $p = \infty$. Indeed, for any $f, g \in (M \cap L^p(M)) \otimes N$, we can consider the element $\langle f, g \rangle_{L_c^p(\mathbb{E})} \overset{\text{def}}{=} \mathbb{E}(f^* g)$ of $M \cap L^{\frac{p}{2}}(M)$. It is clear that this defines an $L^{\frac{p}{2}}(M)$-valued inner product and we can consider the associated right L^p-M-module $L_c^p(\mathbb{E})$.

Lemma 2.9 *Suppose* $0 < p < \infty$. *Assume that the von Neumann algebra N is finite and that M is semifinite. Consider the canonical trace preserving normal faithful conditional expectation* $\mathbb{E} \colon M \overline{\otimes} N \to M$. *If f and g are elements of* $(M \cap L^p(M)) \otimes N$ *then we have*

$$
\mathbb{E}(f^* g) = \langle f, g \rangle_{L^p(M, L^2(N)_{c,p})}. \tag{2.73}
$$

and

$$
\| f \|_{L_c^p(\mathbb{E})} = \| f \|_{L^p(M, L^2(N)_{c,p})} . \tag{2.74}
$$

So, we have a canonical isometric isomorphism $L_c^p(\mathbb{E}) = L^p(M, L^2(N)_{c,p})$.

Proof We can suppose that $f = \sum_{k=1}^n x_k \otimes a_k$ and $g = \sum_{j=1}^m y_j \otimes b_j$. Note that $N \subset L^p(N)$ for any $0 < p \leqslant \infty$. If τ is the normalized trace of N, we have

$$
\mathbb{E}(f^* g) = (\mathrm{Id}_M \otimes \tau) \left(\sum_{k,j=1}^{n,m} (x_k^* \otimes a_k^*)(y_j \otimes b_j) \right) = \sum_{k,j=1}^{n,m} \tau(a_k^* b_j) x_k^* y_j
$$

$$
= \sum_{k,j=1}^{n,m} \langle a_k, b_j \rangle_{L^2(N)} x_k^* y_j \overset{(2.67)}{=} \langle f, g \rangle_{L^p(M, L^2(N)_{c,p})}.
$$

---

[9] In the finite case, we can use the conditional expectation at the level $p = \infty$ and define $\|\cdot\|_{L_c^p(\mathbb{E})}$ on the vector space $\mathcal{N}$. See below.

Moreover, we have

$$\|f\|_{L_c^p(\mathbb{E})} \overset{(2.72)}{=} \left\|\mathbb{E}(f^*f)\right\|_{L^{\frac{p}{2}}(M)}^{\frac{1}{2}} \overset{(2.73)}{=} \left\|\langle f, f\rangle_{L^p(M,L^2(N)_{c,p})}\right\|_{L^{\frac{p}{2}}(M)}^{\frac{1}{2}}$$

$$\overset{(2.68)}{=} \|f\|_{L^p(M,L^2(N)_{c,p})} .$$

Finally, if $0 < p < 2$ note that $(M \cap L^p(M)) \otimes N$ is dense in $L_c^p(\mathbb{E})$ by definition. If $2 \leqslant p \leqslant \infty$, note that $(M \cap L^p(M)) \otimes N$ is dense in $L^p(M \overline{\otimes} N)$, hence in $L_c^p(\mathbb{E})$ by Lemma 2.8. Moreover, $(M \cap L^p(M)) \otimes N$ is dense in $L^p(M, L^2(N)_{c,p})$ since $\|\cdot\|_{L^p(M,L^2(N)_{c,p})}$ is a cross norm by (2.68). $\qquad\square$

Let $\mathbb{E}: N \to M$ be a trace preserving normal faithful conditional expectation between finite von Neumann algebras equipped with normal finite faithful traces. For any $f, g \in N$, we can consider the element $\langle f, g\rangle_{L_c^p(\mathbb{E})} \overset{\text{def}}{=} \mathbb{E}(f^*g)$ of $M \subset L^{\frac{p}{2}}(M)$. It is clear that this defines an $L^{\frac{p}{2}}(M)$-valued inner product and we can consider the associated right L^p-M-module $L_c^p(\mathbb{E})$. If N is a crossed product, we have the following result. See Sect. 2.4 for background on crossed products. We denote here in accordance with (2.58) below $\mathcal{P}_{\rtimes,G} = \mathrm{span}\{x \rtimes \lambda_s : x \in N, s \in G\}$. Note that in (2.76), we interpret f to be equal to $\sum_{k=1}^n x_k \rtimes \lambda_{s_k}$ on the left hand side, and reinterpret this as $f = \sum_{k=1}^n \lambda_{s_k} \otimes x_k$ on the right hand side.

Lemma 2.10 *Suppose* $0 < p < \infty$. *Let* G *be a discrete group. Assume that the von Neumann algebra* N *is finite. Let* $\alpha: G \to N$ *be an action of* G *on* N *by trace preserving $*$-automorphisms. Consider the canonical trace preserving normal faithful conditional expectation* $\mathbb{E}: N \rtimes_\alpha G \to \mathrm{VN}(G)$. *If* $f = \sum_{k=1}^n x_k \rtimes \lambda_{s_k}$ *and* $g = \sum_{j=1}^m y_j \rtimes \lambda_{s_j}$ *are elements of* $\mathcal{P}_{\rtimes,G}$, *then we have*

$$\mathbb{E}(f^*g) = \left\langle \sum_{k=1}^n \lambda_{s_k} \otimes x_k, \sum_{j=1}^m \lambda_{s_j} \otimes y_j \right\rangle_{L^p(\mathrm{VN}(G),L^2(N)_{c,p})}, \tag{2.75}$$

and

$$\|f\|_{L_c^p(\mathbb{E})} = \|f\|_{L^p(\mathrm{VN}(G),L^2(N)_{c,p})} . \tag{2.76}$$

So, we have a canonical isometric isomorphism $L_c^p(\mathbb{E}) = L^p(\mathrm{VN}(G), L^2(N)_{c,p})$.

Proof Note that $N \subset L^p(N)$ for any $0 < p \leqslant \infty$. If τ is the normalized trace of N, we have

$$\mathbb{E}(f^*g) = \sum_{k,j=1}^{n,m} \mathbb{E}\big((x_k \rtimes \lambda_{s_k})^*(y_j \rtimes \lambda_{s_j})\big) \overset{(2.57)}{=} \sum_{k,j=1}^{n,m} \mathbb{E}\big((\alpha_{s_k^{-1}}(x_k^*) \rtimes \lambda_{s_k}^{-1})(y_j \rtimes \lambda_{s_j})\big)$$

$$\overset{(2.56)}{=} \sum_{k,j=1}^{n,m} \mathbb{E}\big((\alpha_{s_k^{-1}}(x_k^*)\alpha_{s_k^{-1}}(y_j) \rtimes \lambda_{s_k}^{-1}\lambda_{s_j})\big) = \sum_{k,j=1}^{n,m} \tau\big(\alpha_{s_k^{-1}}(x_k^* y_j)\big) \rtimes \lambda_{s_k}^{-1}\lambda_{s_j}$$

$$= \sum_{k,j=1}^{n,m} \tau(x_k^* y_j)\lambda_{s_k}^* \lambda_{s_j} \overset{(2.67)}{=} \Big\langle \sum_{k=1}^{n} \lambda_{s_k} \otimes x_k, \sum_{j=1}^{m} \lambda_{s_j} \otimes y_j \Big\rangle_{L^p(\mathrm{VN}(G),L^2(N)_{c,p})}.$$

Moreover, we have

$$\|f\|_{L_c^p(\mathbb{E})} \overset{(2.72)}{=} \big\|\mathbb{E}(f^*f)\big\|_{L^{\frac{p}{2}}(\mathrm{VN}(G))}^{\frac{1}{2}}$$

$$\overset{(2.75)}{=} \big\|\langle f, f\rangle_{L^p(\mathrm{VN}(G),L^2(N)_{c,p})}\big\|_{L^{\frac{p}{2}}(\mathrm{VN}(G))}^{\frac{1}{2}} \overset{(2.68)}{=} \|f\|_{L^p(\mathrm{VN}(G),L^2(N)_{c,p})}.$$

Finally, if $0 < p < 2$ note that $\mathcal{P}_{\rtimes,G}$ is dense in $L_c^p(\mathbb{E})$ by definition. If $2 \leqslant p \leqslant \infty$, note that $\mathcal{P}_{\rtimes,G}$ is dense in $L^p(N \rtimes_\alpha G)$, hence in $L_c^p(\mathbb{E})$. Moreover, $\mathrm{VN}(G) \otimes N$ is dense in $L^p(\mathrm{VN}(G), L^2(N)_{c,p})$ since $\|\cdot\|_{L^p(\mathrm{VN}(G),L^2(N)_{c,p})}$ is a cross norm by (2.68). $\qquad\square$

Remark 2.2 It is easy to see that (2.74) holds in the row case $\|f\|_{L_r^p(\mathbb{E})} = \|f\|_{L^p(M,L^2(N)_{r,p})}$, too. Note however that it in general the row analogue of (2.76) does not hold true. Indeed, an elementary calculation reveals that

$$\Big\|\sum_k x_k \rtimes \lambda_{s_k}\Big\|_{L_r^p(\mathbb{E})} = \Big\|\Big(\sum_{k,l} \tau(x_k \alpha_{s_k s_l^{-1}}(x_l^*))\lambda_{s_k s_l^{-1}}\Big)^{\frac{1}{2}}\Big\|_{L_r^p(\mathbb{E})},$$

whereas according to the row version of (2.68)

$$\Big\|\sum_k \lambda_{s_k} \otimes x_k\Big\|_{L^p(\mathrm{VN}(G),L^2(N)_{r,p})} = \Big\|\Big(\sum_{k,l} \tau(x_k x_l^*)\lambda_{s_k s_l^{-1}}\Big)^{\frac{1}{2}}\Big\|_{L^p(\mathrm{VN}(G))}.$$

For the rest of the section, we shall show that in the situations of Lemmas 2.9 and 2.10, the Banach spaces $L_c^p(\mathbb{E})$ and $L_r^p(\mathbb{E})$ define an interpolation couple and that $L_c^p(\mathbb{E})$ and $L_r^p(\mathbb{E})$ admit $L_c^{p^*}(\mathbb{E})$ and $L_r^{p^*}(\mathbb{E})$ as dual spaces.

First we consider the easy situation of Lemma 2.9. According to Lemma 2.9 (resp. Remark 2.2), the spaces $L_c^p(\mathbb{E})$ and $L^p(M, L^2(N)_{c,p})$ (resp. $L_r^p(\mathbb{E})$ and $L^p(M, L^2(N)_{r,p})$) are isometric. According to [128, p. 13], the spaces $L_c^p(\mathbb{E})$ and $L_r^p(\mathbb{E})$ embed into a common Banach space Z (one can take the injective tensor product of $L^p(M)$ and $L^2(N)$). In the sequel, we will always consider these embeddings.

Lemma 2.11 *Suppose* $1 < p < \infty$. *Let N be a finite von Neumann algebra and M be a semifinite von Neumann algebra. Consider the conditional expectation* $\mathbb{E}: M \overline{\otimes} N \to M$ *from Lemma 2.9.*

1. *The embeddings respect* $(M \cap L^p(M)) \otimes L^2(N)$, *which is dense in* $L_c^p(\mathbb{E})$, *in* $L_r^p(\mathbb{E})$ *and in* $L_c^p(\mathbb{E}) \cap L_r^p(\mathbb{E})$.
2. *The dual space of* $L_c^p(\mathbb{E})$ *(resp.* $L_r^p(\mathbb{E})$*) is* $L_c^{p^*}(\mathbb{E})$ *(resp.* $L_r^{p^*}(\mathbb{E})$*) with duality bracket* $\langle f, g \rangle = \tau_{M \overline{\otimes} N}(f^*g)$ *for* $f \in (M \cap L^p(M)) \otimes L^2(N)$ *and* $g \in (M \cap L^{p^*}(M)) \otimes L^2(N)$.

Proof

1. The previous mentioned embedding into the injective tensor product respects the subspace $(M \cap L^p(M)) \otimes L^2(N)$ which is dense in both $L_c^p(\mathbb{E})$ and $L_r^p(\mathbb{E})$. We prove the third density. Let (p_i) be an increasing net of orthogonal projections over $L^2(N)$ with finite-dimensional range, converging strongly to the identity. Then by (2.71), (2.74) and Remark 2.2, the net $(\text{Id}_{L^p(M)} \otimes p_i)$ is uniformly bounded as operators from $L_c^p(\mathbb{E})$ into $L_c^p(\mathbb{E})$ and from $L_r^p(\mathbb{E})$ into $L_r^p(\mathbb{E})$. Moreover, by density of $(M \cap L^p(M)) \otimes L^2(N)$ separately in both $L_c^p(\mathbb{E})$ and $L_r^p(\mathbb{E})$, the net $(\text{Id}_M \otimes p_i)$ converges strongly to the identity of both $L_c^p(\mathbb{E})$ and $L_r^p(\mathbb{E})$. Moreover, both mappings are compatible as restrictions of a common mapping $\text{Id}_{L^p(M)} \otimes p_i : Z \to Z$. But if $f \in L_c^p(\mathbb{E}) \cap L_r^p(\mathbb{E})$, then $(\text{Id}_M \otimes p_i)(f) = \sum_k f_k \otimes e_k$, where the sum is finite, $f_k \in L^p(M)$, and $e_k \in L^2(N)$ span the range of p_i. Taking $g_k \in M \cap L^p(M)$ with $\|g_k - f_k\|_{L^p(M)} < \varepsilon$, we see that $(M \cap L^p(M)) \otimes L^2(N)$ is dense in the intersection $L_c^p(\mathbb{E}) \cap L_r^p(\mathbb{E})$.
2. For the duality result, see [128, (2.14)] (note that the duality bracket is bilinear there, so we have to reinterpret it).

$\square$

Now we consider the situation of Lemma 2.10. Concerning the compatibility, we have the following result.

Proposition 2.10 *Suppose* $1 < p < \infty$. *Let N be a finite von Neumann algebra. Consider the conditional expectation* $\mathbb{E}: N \rtimes_\alpha G \to \text{VN}(G)$ *from Lemma 2.10. There exist continuous injective embeddings* $w_c: L_c^p(\mathbb{E}) \hookrightarrow \ell_G^\infty(L^2(N))$ *and* $w_r: L_r^p(\mathbb{E}) \hookrightarrow \ell_G^\infty(L^2(N))$ *such that* $w_c(x) = w_r(x)$ *for all* $x \in \mathcal{P}_{\rtimes, G}$.

Proof For $x = \sum_s x_s \rtimes \lambda_s$ belonging to $\mathcal{P}_{\rtimes, G}$, we consider the map $w(x) \overset{\text{def}}{=} (x_s)_{s \in G}$. Let us show that w extends to continuous mappings $w_c: L_c^p(\mathbb{E}) \to \ell_G^\infty(L^2(N))$ and $w_r: L_r^p(\mathbb{E}) \to \ell_G^\infty(L^2(N))$.

Fix some $t \in G$. Since the trace τ is unital, we have a completely[10] contractive mapping $\phi_t \colon L^p(\mathrm{VN}(G)) \to \mathbb{C}$, $x \mapsto x_t \overset{\text{def}}{=} \tau(x\lambda_{t^{-1}})$. Thus according to (2.70), $\phi_t \otimes \mathrm{Id}_{L^2(N)}$ extends to a contraction $L^p(\mathrm{VN}(G), L^2(N)_{c,p}) \to L^p(\mathbb{C}, L^2(N)_{c,p}) = L^2(N)$ where the last equality is [128, Remark 2.13]. We deduce the inequality

$$\|x_t\|_{L^2(N)} \leqslant \|x\|_{L^p(\mathrm{VN}(G), L^2(N)_{c,p})} \overset{(2.76)}{=} \|x\|_{L_c^p(\mathbb{E})}.$$ Taking the supremum over all $t \in G$ on the left hand side, since $\mathcal{P}_{\rtimes, G}$ is dense in $L_c^p(\mathbb{E})$, this shows that w extends to a contraction $w_c \colon L_c^p(\mathbb{E}) \to \ell_G^\infty(L^2(N))$.

Next we show that w extends to a continuous mapping $w_r \colon L_r^p(\mathbb{E}) \to \ell_G^\infty(L^2(N))$. According to Remark 2.2, we have $\|x\|_{L_r^p(\mathbb{E})} = \|y\|_{L^p(\mathrm{VN}(G), L^2(N)_{r,p})}$ with $y = \sum_s \lambda_s \otimes \alpha_{s^{-1}}(x_s)$. For any $t \in G$, note that since $\phi_t \otimes \mathrm{Id}_{L^2(N)}$ equally extends to a contraction $L^p(\mathrm{VN}(G), L^2(N)_{r,p}) \to L^p(\mathbb{C}, L^2(N)_{r,p}) = L^2(N)$ and $\alpha_{t^{-1}} \colon L^2(N) \to L^2(N)$ is an isometry, we also have $\|x_t\|_{L^2(N)} = \|\alpha_{t^{-1}}(x_t)\|_{L^2(N)} \leqslant \|y\|_{L^p(\mathrm{VN}(G), L^2(N)_{r,p})} = \|x\|_{L_r^p(\mathbb{E})}.$ Taking the supremum over all $t \in G$ on the left hand side, since $\mathcal{P}_{\rtimes, G}$ is dense in $L_r^p(\mathbb{E})$, this shows that w extends to a contraction $w_r \colon L_r^p(\mathbb{E}) \to \ell_G^\infty(L^2(N))$.

Next we show that w_c is injective. Assume that $w_c(x) = 0$ for some $x \in L_c^p(\mathbb{E})$. If $\Psi \colon L_c^p(\mathbb{E}) \to L^p(\mathrm{VN}(G), L^2(N)_{c,p})$ is the isometry from (2.76), the previous proof of continuity shows that $w_c(x) = \big((\phi_s \otimes \mathrm{Id}_{L^2(N)}) \circ \Psi(x)\big)_{s \in G}$. We infer that $(\phi_s \otimes \mathrm{Id}_{L^2(N)}) \circ \Psi(x) = 0$ for all $s \in G$. Thus,[11] $\Psi(x) = 0$, hence $x = 0$.

Finally we show that w_r is injective. Assume that $w_r(x) = 0$ for some $x \in L_r^p(\mathbb{E})$. If $\Phi \colon L_r^p(\mathbb{E}) \to L^p(\mathrm{VN}(G), L^2(N)_{r,p})$ is the isometry from Remark 2.2, we have $w_r(x) = \big(\alpha_s \circ (\phi_s \otimes \mathrm{Id}_{L^2(N)}) \circ \Phi(x)\big)_{s \in G}$. We infer that $\alpha_s \circ (\phi_s \otimes \mathrm{Id}_{L^2(N)}) \circ \Phi(x) = 0$ for all $s \in G$. Consequently, we have $(\phi_s \otimes \mathrm{Id}_{L^2(N)}) \circ \Phi(x) = 0$ for all $s \in G$. Hence $\Phi(x) = 0$, so $x = 0$.   □

We keep the notations from Proposition 2.10. As we have observed in the proof of Lemma 2.10, $\mathcal{P}_{\rtimes, G}$ is dense in $L_c^p(\mathbb{E})$, and thus also in $L_r^p(\mathbb{E})$, since $\mathcal{P}_{\rtimes, G}$ is stable by taking adjoints. We shall also need that $\mathcal{P}_{\rtimes, G}$ is dense in the intersection $L_c^p(\mathbb{E}) \cap L_r^p(\mathbb{E})$. This is implicitly used in a similar situation in [131, p. 541][12] and [131, p. 543][13] in the case of a crossed product $\mathcal{M} \overline{\otimes} L^\infty(\Omega, \Sigma, \mu) \rtimes G$ where $\mathcal{M}$ is a finite von Neumann algebra. We provide a proof with some assumptions in Lemma 2.13 which is needed in the proof of Theorem 3.1 below. Note that the approach to define

---

[10] Recall that a contractive linear form is completely contractive by Effros and Ruan [81, Corollary 2.2.3] and that $L^p(M) \subset L^1(M)$.

[11] If $(\phi_s \otimes \mathrm{Id}_{L^2(N)})(y) = 0$ for some $y \in L^p(\mathrm{VN}(G), L^2(N)_{c,p})$ and all $s \in G$, then to conclude that $y = 0$, by duality, it suffices to check that $\langle y, y^* \rangle = 0$ for y^* belonging to some total subset of $L^{p^*}(\mathrm{VN}(G), L^2(N)_{c,p})$. We can choose the total subset $\mathcal{P}_G \otimes L^2(N)$ and conclude by $\langle y, \lambda_s \otimes h \rangle = \langle (\phi_s \otimes \mathrm{Id}_{L^2(N)})(y), h \rangle = 0$ (the first equality is true on a dense subset, hence correct).

[12] In line -3 of [131, p. 541], it is stated that f is a finite sum $\sum_{g,h}(B(h) \otimes f_{g,h}) \rtimes \lambda(g)$.

[13] In the calculations line 3 and 4 of [131, p. 543], it is used that the duality brackets between $L_{rc}^p(E_{\mathcal{M} \rtimes G})$ and $L_{rc}^q(E_{\mathcal{M} \rtimes G})$, and between $L^p(\mathcal{A})$ and $L^q(\mathcal{A})$ are compatible, which a priori only makes sense on common subspaces. It is also used that the Gaussian projection is bounded $\widehat{Q} \colon L_{rc}^q(E_{\mathcal{M} \rtimes G}) \to G_q(\mathcal{M}) \rtimes G$ which is a priori only clear on the subspace of f as in footnote 12.

compatibility via *non-constructive* Hilbert module theory seems *ineffective* to obtain this kind of results. We first record, also for later use, the following lemma.

Lemma 2.12 *Let* $1 < p < \infty$ *and* G *be a discrete group acting on* N *by trace preserving* $*$-*automorphisms, where* N *is some finite von Neumann algebra. Let* M_φ *be a completely bounded Fourier multiplier on* $\mathrm{L}^p(\mathrm{VN}(G))$. *Then*

$$\left\| (\mathrm{Id}_N \rtimes M_\varphi)(f) \right\|_{\mathrm{L}_c^p(\mathbb{E})} \lesssim_p \|f\|_{\mathrm{L}_c^p(\mathbb{E})} \ \text{and} \ \left\| (\mathrm{Id}_N \rtimes M_\varphi)(f) \right\|_{\mathrm{L}_r^p(\mathbb{E})} \lesssim_p \|f\|_{\mathrm{L}_r^p(\mathbb{E})}.$$
(2.77)

Proof By the isometry in Lemma 2.10 and the fact that M_φ is completely bounded, the column part of (2.77) follows immediately from (2.70). For the row part of (2.77), we need a little bit of work, due to Remark 2.2. Let $f = \sum_k x_k \rtimes \lambda_{s_k}$ be an element of $\mathcal{P}_{\rtimes,G}$ which is a dense subspace of $\mathrm{L}_r^p(\mathbb{E})$. Using the notation $\varphi^*(s) \overset{\mathrm{def}}{=} \overline{\varphi(s^{-1})}$ and [17, Lemma 6.4] in the last equality, we have

$$\left\| (\mathrm{Id}_N \rtimes M_\varphi)(f) \right\|_{\mathrm{L}_r^p(\mathbb{E})} = \left\| \left((\mathrm{Id}_N \rtimes M_\varphi)(f) \right)^* \right\|_{\mathrm{L}_c^p(\mathbb{E})}$$

$$= \left\| \left(\sum_k \varphi(s_k) x_k \rtimes \lambda_{s_k} \right)^* \right\|_{\mathrm{L}_c^p(\mathbb{E})} \overset{(2.57)}{=} \left\| \sum_k \overline{\varphi(s_k)} \alpha_{s_k^{-1}}(x_k^*) \rtimes \lambda_{s_k^{-1}} \right\|_{\mathrm{L}_c^p(\mathbb{E})}$$

$$= \left\| (\mathrm{Id}_N \rtimes M_{\varphi^*}) \left(\sum_k \alpha_{s_k^{-1}}(x_k^*) \rtimes \lambda_{s_k^{-1}} \right) \right\|_{\mathrm{L}_c^p(\mathbb{E})}$$

$$= \left\| (\mathrm{Id}_N \rtimes M_{\check{\varphi}})(f^*) \right\|_{\mathrm{L}_c^p(\mathbb{E})} \overset{(2.77)}{\leq} \left\| M_{\varphi^*} \right\|_{\mathrm{cb}, \mathrm{L}^p(\mathrm{VN}(G)) \to \mathrm{L}^p(\mathrm{VN}(G))} \left\| f^* \right\|_{\mathrm{L}_c^p(\mathbb{E})}$$

$$= \left\| M_{\varphi^*} \right\|_{\mathrm{cb}, \mathrm{L}^p(\mathrm{VN}(G)) \to \mathrm{L}^p(\mathrm{VN}(G))} \|f\|_{\mathrm{L}_r^p(\mathbb{E})}$$

$$= \left\| M_\varphi \right\|_{\mathrm{cb}, \mathrm{L}^p(\mathrm{VN}(G)) \to \mathrm{L}^p(\mathrm{VN}(G))} \|f\|_{\mathrm{L}_r^p(\mathbb{E})}.$$

$\square$

With this lemma at hand, we get the following result.

Lemma 2.13 *Let* $1 < p < \infty$ *and keep the assumptions of Proposition 2.10. Assume in addition that* $\mathrm{L}^p(\mathrm{VN}(G))$ *has* CCAP *and that* $\mathrm{VN}(G)$ *has* QWEP. *Consider* $\mathrm{L}_c^p(\mathbb{E})$ *and* $\mathrm{L}_r^p(\mathbb{E})$ *injected into* $\ell_G^\infty(\mathrm{L}^2(N))$ *as in Proposition 2.10. Then* $\mathcal{P}_{\rtimes,G}$ *is dense in* $\mathrm{L}_c^p(\mathbb{E}) \cap \mathrm{L}_r^p(\mathbb{E})$.

Proof Consider the approximating net (M_{φ_j}) of completely bounded Fourier multipliers provided by the CCAP assumption by Lemma 2.5 with

$$\left\| M_{\varphi_j} \right\|_{\mathrm{cb}, \mathrm{L}^p(\mathrm{VN}(G)) \to \mathrm{L}^p(\mathrm{VN}(G))} \leq 1.$$

By Arhancet and Kriegler [17, Lemma 6.5, Lemma 6.6], note that (φ_j) is a bounded net of ℓ_G^∞.

A direct estimate shows that the mappings $M_{\varphi_j} \otimes \mathrm{Id}_{\mathrm{L}^2(N)} \colon \ell_G^\infty(\mathrm{L}^2(N)) \to \ell_G^\infty(\mathrm{L}^2(N))$, $(x_s)_{s \in G} \mapsto (\varphi_j(s)x_s)_{s \in G}$ define a bounded net of operators. Note that according to Lemma 2.12, the extensions $\mathrm{Id}_N \rtimes M_{\varphi_j}$ are uniformly bounded as operators $\mathrm{L}_c^p(\mathbb{E}) \to \mathrm{L}_c^p(\mathbb{E})$ and $\mathrm{L}_r^p(\mathbb{E}) \to \mathrm{L}_r^p(\mathbb{E})$.

Thus, by density of $\mathcal{P}_{\rtimes,G}$ in $\mathrm{L}_c^p(\mathbb{E})$ and in $\mathrm{L}_r^p(\mathbb{E})$, the net $(\mathrm{Id}_N \rtimes M_{\varphi_j})$ converges strongly to $\mathrm{Id}_{\mathrm{L}_c^p(\mathbb{E})}$ resp. $\mathrm{Id}_{\mathrm{L}_r^p(\mathbb{E})}$. By compatibility of this mapping on both spaces as a restriction of $M_{\varphi_j} \otimes \mathrm{Id}_{\mathrm{L}^2(N)} \colon \ell_G^\infty(\mathrm{L}^2(N)) \to \ell_G^\infty(\mathrm{L}^2(N))$, the net $(\mathrm{Id}_N \rtimes M_{\varphi_j})$ converges also to the identity on $\mathrm{L}_c^p(\mathbb{E}) \cap \mathrm{L}_r^p(\mathbb{E})$. It thus suffices to show density of $\mathcal{P}_{\rtimes,G}$ in $\mathrm{Ran}(\mathrm{Id}_N \rtimes M_{\varphi_j} \colon \mathrm{L}_c^p(\mathbb{E}) \cap \mathrm{L}_r^p(\mathbb{E}) \to \mathrm{L}_c^p(\mathbb{E}) \cap \mathrm{L}_r^p(\mathbb{E}))$ for some fixed j.

For $f \in \mathrm{L}_c^p(\mathbb{E}) \cap \mathrm{L}_r^p(\mathbb{E})$, we have $(\mathrm{Id}_N \rtimes M_{\varphi_j})(f) = \sum_{s \in \mathrm{supp}\,\varphi_j} \varphi_j(s)f_s \rtimes \lambda_s$ for some $f_s \in \mathrm{L}^2(N)$. Since the previous sum is finite, we are left to show that $f \rtimes \lambda_s$ can be approximated by elements of $\mathcal{P}_{\rtimes,G}$ in $\mathrm{L}_c^p(\mathbb{E}) \cap \mathrm{L}_r^p(\mathbb{E})$. But this is immediate since both column and row norms are cross norms and N is dense in $\mathrm{L}^2(N)$. □

By Lemma 2.11 (resp. Proposition 2.10), $(\mathrm{L}_c^p(\mathbb{E}), \mathrm{L}_r^p(\mathbb{E}))$ is an interpolation couple. Therefore, it is meaningful to let $\mathrm{L}_{cr}^p(\mathbb{E}) \stackrel{\text{def}}{=} \mathrm{L}_c^p(\mathbb{E}) + \mathrm{L}_r^p(\mathbb{E})$ if $1 < p < 2$ and $\mathrm{L}_{cr}^p(\mathbb{E}) \stackrel{\text{def}}{=} \mathrm{L}_c^p(\mathbb{E}) \cap \mathrm{L}_r^p(\mathbb{E})$ if $2 \leqslant p < \infty$. In the situation of Lemma 2.11, we get immediately with [35, 2.7.1 Theorem] that for $1 < p < \infty$, $\mathrm{L}_{cr}^p(\mathbb{E})$ and $\mathrm{L}_{cr}^{p^*}(\mathbb{E})$ are dual spaces with duality bracket $\langle f, g \rangle = \tau_{M \overline{\otimes} N}(f^* g)$.

In the situation of Proposition 2.10, we have the following duality result, see [35, 2.7.1 Theorem] and [116, Corollary 2.12]. Note that there is a separability assumption in [116, Corollary 2.12]. So our discrete groups considered in this paper are implicitly countable when we use this result. Nevertheless, this assumption seems removable with the transparent argument of [126, Remark 6.11] (see also the discussion before [122, Lemma 3.4]).

Lemma 2.14 *For $1 < p < \infty$, $\mathrm{L}_c^p(\mathbb{E})$ and $\mathrm{L}_c^{p^*}(\mathbb{E})$ (resp. $\mathrm{L}_r^p(\mathbb{E})$ and $\mathrm{L}_r^{p^*}(\mathbb{E})$, resp. $\mathrm{L}_{cr}^p(\mathbb{E})$ and $\mathrm{L}_{cr}^{p^*}(\mathbb{E})$) are dual spaces with respect to each other. Here, the duality pairing is $\langle f, g \rangle = \tau_\rtimes(f^* g)$ for all $f, g \in N \rtimes_\alpha G$.*

2.6 Carré Du Champ Γ and First Order Differential Calculus for Fourier Multipliers

Let G be a discrete group. Given a markovian semigroup $(T_t)_{t \geqslant 0}$ of Fourier multipliers on the von Neumann algebra $\mathrm{VN}(G)$ as given in Proposition 2.3, we shall introduce the noncommutative gradient $\partial_{\psi,q}$ and the carré du champ Γ, and interrelate them together with the square root $A^{\frac{1}{2}}$ of the negative generator A. For any $1 \leqslant p \leqslant \infty$, note that the space $\mathcal{P}_G$ from (2.32) is a subspace of $\mathrm{L}^p(\mathrm{VN}(G))$.

For $x, y \in \mathcal{P}_G$, we define the element

$$\Gamma(x, y) \stackrel{\text{def}}{=} \frac{1}{2}\left[A(x^*)y + x^*A(y) - A(x^*y)\right] \tag{2.78}$$

of span $\{\lambda_s : s \in G\}$. For any $s, t \in G$, note that

$$\Gamma(\lambda_s, \lambda_t) = \frac{1}{2}\left[\psi(s^{-1}) + \psi(t) - \psi(s^{-1}t)\right]\lambda_{s^{-1}t}. \tag{2.79}$$

The form Γ is usually called carré du champ, associated with the markovian semigroup $(T_t)_{t \geqslant 0}$. The first part of the following result shows how to construct the carré du champ directly from the semigroup $(T_t)_{t \geqslant 0}$.

We refer to the books [26, 40] for the general case of semigroups acting on classical L^p-spaces and to [170]. For the semigroups acting on noncommutative L^p-spaces, the carré du champ and its applications were studied in the papers [56], [57, Section 9], [119, 126, 205] mainly in the σ-finite case and for L^2-spaces (see [70] for related things). Unfortunately, these papers do not suffice for our setting. So, in the following, we describe an elementary, independent and more concrete approach. However, some results of this section could be known, at least in the case $p = 2$.

Lemma 2.15 *Let G be a discrete group.*

1. *Suppose $1 \leqslant p \leqslant \infty$. For any $x, y \in \mathcal{P}_G$, we have*

$$\Gamma(x, y) = \lim_{t \to 0^+} \frac{1}{2t}\left(T_t(x^*y) - (T_t(x))^*T_t(y)\right) \tag{2.80}$$

   in $L^p(\mathrm{VN}(G))$.
2. *For any $x \in \mathcal{P}_G$, we have $\Gamma(x, x) \geqslant 0$.*
3. *For any $x, y \in \mathcal{P}_G$, the matrix $\begin{bmatrix} \Gamma(x, x) & \Gamma(x, y) \\ \Gamma(y, x) & \Gamma(y, y) \end{bmatrix}$ is positive.*
4. *For any $x, y \in \mathcal{P}_G$, we have*

$$\|\Gamma(x, y)\|_{L^p(\mathrm{VN}(G))} \leqslant \|\Gamma(x, x)\|_{L^p(\mathrm{VN}(G))}^{\frac{1}{2}} \|\Gamma(y, y)\|_{L^p(\mathrm{VN}(G))}^{\frac{1}{2}}. \tag{2.81}$$

   This inequality even holds for $0 < p \leqslant \infty$.

Proof

1. For $x, y \in \mathrm{span}\{\lambda_s : s \in G\}$, using the joint continuity of multiplication on bounded sets for the strong operator topology on $L^p(\mathrm{VN}(G))$ [84, Proposition C.19] (if $p = \infty$ we use that $x, y \in \mathrm{span}\{\lambda_s : s \in G\}$, so all operators below belong to some finite dimensional space $\subset L^p(\mathrm{VN}(G)) \cap \mathrm{VN}(G)$ on which

convergence in VN(G) norm follows from convergence in $L^p(\text{VN}(G))$ norm), we have

$$\frac{1}{2t}\big(T_t(x^*y) - (T_t(x))^*T_t(y)\big)$$

$$= \frac{1}{2t}\big(T_t(x^*y) - x^*y + x^*y - (T_t(x))^*T_t(y)\big)$$

$$= \frac{1}{2t}\big(T_t(x^*y) - x^*y + x^*y - (T_t(x))^*y + (T_t(x))^*y - (T_t(x))^*T_t(y)\big)$$

$$= \frac{1}{2t}\big(T_t(x^*y) - x^*y + (x^* - T_t(x^*))y + (T_t(x))^*(y - T_t(y))\big)$$

$$\xrightarrow[t \to 0^+]{(2.11)} \frac{1}{2}\big(-A(x^*y) + A(x^*)y + x^*A(y)\big) \overset{(2.78)}{=} \Gamma(x, y).$$

2. Each operator $T_t : \text{VN}(G) \to \text{VN}(G)$ is completely positive and unital. Hence by the Schwarz inequality [185, Proposition 3.3] (or [183, Proposition 9.9.4]), we have $T_t(x)^*T_t(x) \leqslant T_t(x^*x)$. Recall that the positive cone of VN(G) is weak* closed (see e.g. [17, (2.3)]). Using the point 1, we conclude that $\Gamma(x, x) \geqslant 0$.

3. For $x, y \in \text{span}\{\lambda_s : s \in G\}$, using the point 1 in the first equality and the point 2 in the last inequality, we obtain

$$\begin{bmatrix} \Gamma(x, x) & \Gamma(x, y) \\ \Gamma(y, x) & \Gamma(y, y) \end{bmatrix}$$

$$\overset{(2.80)}{=} \lim_{t \to 0^+} \frac{1}{2t} \begin{bmatrix} T_t(x^*x) - (T_t(x))^*T_t(x) & T_t(x^*y) - (T_t(x))^*T_t(y) \\ T_t(y^*x) - (T_t(y))^*T_t(x) & T_t(y^*y) - (T_t(y))^*T_t(y) \end{bmatrix}$$

$$= \lim_{t \to 0^+} \frac{1}{2t} \left(\begin{bmatrix} T_t(x^*x) & T_t(x^*y) \\ T_t(y^*x) & T_t(y^*y) \end{bmatrix} - \begin{bmatrix} (T_t(x))^*T_t(x) & (T_t(x))^*T_t(y) \\ (T_t(y))^*T_t(x) & (T_t(y))^*T_t(y) \end{bmatrix} \right)$$

$$= \lim_{t \to 0^+} \frac{1}{2t} \left((\text{Id}_{M_2} \otimes T_t) \left(\begin{bmatrix} x^*x & x^*y \\ y^*x & y^*y \end{bmatrix} \right) - \begin{bmatrix} (T_t(x))^* & 0 \\ (T_t(y))^* & 0 \end{bmatrix} \begin{bmatrix} T_t(x) & T_t(y) \\ 0 & 0 \end{bmatrix} \right)$$

$$= \lim_{t \to 0^+} \frac{1}{2t} \left((\text{Id}_{M_2} \otimes T_t) \left(\begin{bmatrix} x & y \\ 0 & 0 \end{bmatrix}^* \begin{bmatrix} x & y \\ 0 & 0 \end{bmatrix} \right) \right.$$

$$\left. - \left((\text{Id}_{M_2} \otimes T_t) \left(\begin{bmatrix} x & y \\ 0 & 0 \end{bmatrix} \right) \right)^* \left((\text{Id}_{M_2} \otimes T_t) \begin{bmatrix} x & y \\ 0 & 0 \end{bmatrix} \right) \right)$$

$$\geqslant 0.$$

4. By Arhancet and Kriegler [17, Lemma 2.11],

$$\|\Gamma(x, y)\|_{L^p(\text{VN}(G))} \leqslant \frac{1}{2^{\frac{1}{p}}} \left(\|\Gamma(x, x)\|_{L^p(\text{VN}(G))}^p + \|\Gamma(y, y)\|_{L^p(\text{VN}(G))}^p \right)^{\frac{1}{p}}.$$

Since we have $\Gamma(\lambda x, \frac{1}{\lambda}y) = \Gamma(x, y)$ for $\lambda > 0$, we deduce

$$\|\Gamma(x, y)\|_{L^p(\mathrm{VN}(G))} \leqslant \frac{1}{2^{\frac{1}{p}}} \inf_{\lambda > 0} \left(\lambda \|\Gamma(x, x)\|_{L^p(\mathrm{VN}(G))}^p + \frac{1}{\lambda} \|\Gamma(y, y)\|_{L^p(\mathrm{VN}(G))}^p \right)^{\frac{1}{p}}.$$

Ruling out beforehand the easy cases $\Gamma(x, x) = 0$ or $\Gamma(y, y) = 0$, we choose

$$\lambda = \left(\|\Gamma(y, y)\|_{L^p(\mathrm{VN}(G))}^p / \|\Gamma(x, x)\|_{L^p(\mathrm{VN}(G))}^p \right)^{\frac{1}{2}}$$

and obtain

$$\|\Gamma(x, y)\|_{L^p(\mathrm{VN}(G))} \leqslant \frac{1}{2^{\frac{1}{p}}} \left(2 \|\Gamma(x, x)\|_{L^p(\mathrm{VN}(G))}^{\frac{p}{2}} \|\Gamma(y, y)\|_{L^p(\mathrm{VN}(G))}^{\frac{p}{2}} \right)^{\frac{1}{p}}$$

(2.82)

$$= \|\Gamma(x, x)\|_{L^p(\mathrm{VN}(G))}^{\frac{1}{2}} \|\Gamma(y, y)\|_{L^p(\mathrm{VN}(G))}^{\frac{1}{2}}.$$

□

Suppose that $(T_t)_{t \geqslant 0}$ is associated to the length $\psi: G \to \mathbb{R}_+$ with associated cocycle $b = b_\psi: G \to H$ and the orthogonal representation $\pi: G \to B(H)$, $s \mapsto \pi_s$ of G on H from Proposition 2.3. Suppose $-1 \leqslant q \leqslant 1$. For any $s \in G$, we will use the second quantization from (2.38) by letting

$$\alpha_s \stackrel{\text{def}}{=} \Gamma_q^\infty(\pi_s): \Gamma_q(H) \to \Gamma_q(H) \tag{2.83}$$

which is trace preserving. We obtain an action $\alpha: G \to \mathrm{Aut}(\Gamma_q(H))$. So we can consider the crossed product $\Gamma_q(H) \rtimes_\alpha G$ as studied in Sect. 2.4, which comes equipped with its canonical normal finite faithful trace $\tau_\rtimes$.

Suppose $-1 \leqslant q \leqslant 1$ and $1 \leqslant p < \infty$. We introduce the map $\partial_{\psi,q}: \mathcal{P}_G \to L^p(\Gamma_q(H) \rtimes_\alpha G)$ defined by

$$\partial_{\psi,q}(\lambda_s) = s_q(b_\psi(s)) \rtimes \lambda_s. \tag{2.84}$$

which is a slight generalization of the map of [131, p. 535].

Note that $L^p(\Gamma_q(H) \rtimes_\alpha G)$ is a VN(G)-bimodule with left and right actions induced by

$$\lambda_s(z \rtimes \lambda_t) \stackrel{\text{def}}{=} \alpha_s(z) \rtimes \lambda_{st} \quad \text{and} \quad (z \rtimes \lambda_t)\lambda_s \stackrel{\text{def}}{=} z \rtimes \lambda_{ts}, \quad z \in \Gamma_q(H), s, t \in G. \tag{2.85}$$

The following is stated in the particular case $q = 1$ without proof in [131, p. 544]. For the sake of completness, we give a short proof.

Lemma 2.16 *Suppose* $-1 \leqslant q \leqslant 1$. *Let* G *be a discrete group. For any* $x, y \in \mathcal{P}_G$, *we have*

$$\partial_{\psi,q}(xy) = x\partial_{\psi,q}(y) + \partial_{\psi,q}(x)y. \tag{2.86}$$

Proof For any $s, t \in G$, we have

$$
\lambda_s \partial_{\psi,q}(\lambda_t) + \partial_{\psi,q}(\lambda_s)\lambda_t \overset{(2.84)}{=} \lambda_s\big(s_q(b_\psi(t)) \rtimes \lambda_t\big) + \big(s_q(b_\psi(s)) \rtimes \lambda_s\big)\lambda_t
$$

$$
\overset{(2.85)}{=} \alpha_s\big(s_q(b_\psi(t))\big) \rtimes \lambda_{st} + s_q(b_\psi(s)) \rtimes \lambda_{st}
$$

$$
= s_q\big(\pi_s(b_\psi(t))\big) \rtimes \lambda_{st} + s_q(b_\psi(s)) \rtimes \lambda_{st}
$$

$$
\overset{(2.84)}{=} \big(s_q(\pi_s(b_\psi(t)) + s_q(b_\psi(s)))\big) \rtimes \lambda_{st}
$$

$$
\overset{(2.33)}{=} \big(s_q(b_\psi(st))\big) \rtimes \lambda_{st} = \partial_{\psi,q}(\lambda_{st}) = \partial_{\psi,q}(\lambda_s\lambda_t).
$$

□

The following is a slight generalization of [131, Remark 1.3]. The proof is not difficult.[14] Here $\mathbb{E}\colon L^p(\Gamma_q(H) \rtimes_\alpha G) \to L^p(\mathrm{VN}(G))$ denotes the canonical conditional expectation.

---

[14] On the one hand, for any $s, t \in G$, we have

$$
\Gamma(\lambda_s, \lambda_t) \overset{(2.79)}{=} \frac{1}{2}\big[\psi(s^{-1}) + \psi(t) - \psi(s^{-1}t)\big]\lambda_{s^{-1}t}
$$

$$
\overset{(2.34)}{=} \frac{1}{2}\Big[\big\|b_\psi(s^{-1})\big\|^2 + \big\|b_\psi(t)\big\|^2 - \big\|b_\psi(s^{-1}t)\big\|^2\Big]\lambda_{s^{-1}t}
$$

$$
\overset{(2.33)}{=} \frac{1}{2}\Big[\big\|b_\psi(s^{-1})\big\|^2 + \big\|b_\psi(t)\big\|^2
$$

$$
- \big\|b_\psi(s^{-1})\big\|^2 - 2\langle b_\psi(s^{-1}), \pi_{s^{-1}}(b_\psi(t))\rangle_H - \big\|b_\psi(t)\big\|^2\Big]\lambda_{s^{-1}t}
$$

$$
= -\langle b_\psi(s^{-1}), \pi_{s^{-1}}(b_\psi(t))\rangle_H\lambda_{s^{-1}t}.
$$

On the other hand, we have

$$
\mathbb{E}\big[(\partial_{\psi,q}(\lambda_s))^*\partial_{\psi,q}(\lambda_t)\big] \overset{(2.84)}{=} \mathbb{E}\big[\big(s_q(b_\psi(s)) \rtimes \lambda_s\big)^*\big(s_q(b_\psi(t)) \rtimes \lambda_t\big)\big]
$$

$$
\overset{(2.57)}{=} \mathbb{E}\big[\big(\alpha_{s^{-1}}\big(s_q(b_\psi(s))\big) \rtimes \lambda_{s^{-1}}\big)\big(s_q(b_\psi(t)) \rtimes \lambda_t\big)\big]
$$

$$
= \mathbb{E}\big[\big(s_q(\pi_{s^{-1}}(b_\psi(s))) \rtimes \lambda_{s^{-1}}\big)\big(s_q(b_\psi(t)) \rtimes \lambda_t\big)\big]
$$

$$
\overset{(2.33)}{=} \mathbb{E}\big[\big(s_q(b_\psi(e) - b_\psi(s^{-1})) \rtimes \lambda_{s^{-1}}\big)\big(s_q(b_\psi(t)) \rtimes \lambda_t\big)\big]
$$

$$
= -\mathbb{E}\big[\big(s_q(b_\psi(s^{-1})) \rtimes \lambda_{s^{-1}}\big)\big(s_q(b_\psi(t)) \rtimes \lambda_t\big)\big]
$$

Proposition 2.11 *Suppose* $-1 \leqslant q \leqslant 1$. *For any* $x, y \in \mathcal{P}_G$, *we have*

$$\Gamma(x, y) = \mathbb{E}\big[\big(\partial_{\psi,q}(x)\big)^* \partial_{\psi,q}(y)\big]. \tag{2.87}$$

Suppose $1 \leqslant p < \infty$. The following equalities are in [132, pp. 930-931] for $q = 1$. For any $x, y \in \mathcal{P}_G$, we have

$$\Gamma(x, y) = \big\langle \partial_{\psi,q}(x), \partial_{\psi,q}(y)\big\rangle_{\mathrm{L}^p(\mathrm{VN}(G), \mathrm{L}^2(\Gamma_q(H))_{c,p})}, \tag{2.88}$$

and

$$\big\|\Gamma(x, x)^{\frac{1}{2}}\big\|_{\mathrm{L}^p(\mathrm{VN}(G))} = \big\|\partial_{\psi,q}(x)\big\|_{\mathrm{L}^p(\mathrm{VN}(G), \mathrm{L}^2(\Gamma_q(H))_{c,p})}. \tag{2.89}$$

We shall also need the following Riesz transform norm equivalence for markovian semigroup of Fourier multipliers from [131, Theorem A2, Remark 1.3].

Proposition 2.12 *Let G be a discrete group and $(T_t)_{t \geqslant 0}$ a markovian semigroup of Fourier multipliers with symbol ψ of the negative generator A. Then for $2 \leqslant p < \infty$, we have the norm equivalence*

$$\big\|A^{\frac{1}{2}}(x)\big\|_{\mathrm{L}^p(\mathrm{VN}(G))} \approx_p \max\Big\{\big\|\Gamma(x, x)^{\frac{1}{2}}\big\|_{\mathrm{L}^p(\mathrm{VN}(G))}, \big\|\Gamma(x^*, x^*)^{\frac{1}{2}}\big\|_{\mathrm{L}^p(\mathrm{VN}(G))}\Big\} \tag{2.90}$$

for any $x \in \mathcal{P}_G$.

2.7 Carré Du Champ Γ and First Order Differential Calculus for Schur Multipliers

This section on markovian semigroups of Schur multipliers is the analogue of Sect. 2.6 on Fourier multipliers. We suppose here that we are given a markovian semigroup of Schur multipliers as in Proposition 2.4 and we fix the associated family $(\alpha_i)_{i \in I}$ from (2.36).

$$\overset{(2.56)}{=} -\mathbb{E}\Big[\Big(s_q(b_\psi(s^{-1}))\alpha_{s-1}(s_q(b_\psi(t)))\Big) \rtimes \lambda_{s^{-1}t}\Big]$$

$$\overset{(2.33)}{=} -\mathbb{E}\Big[\Big(s_q(b_\psi(s^{-1}))s_q\big(\pi_{s-1}(b_\psi(t))\big)\Big) \rtimes \lambda_{s^{-1}t}\Big]$$

$$\overset{(2.41)}{=} -\big\langle b_\psi(s^{-1}), \pi_{s-1}(b_\psi(t))\big\rangle_H \lambda_{s^{-1}t}.$$

For any $x, y \in M_{I,\text{fin}}$, we define the element

$$\Gamma(x, y) \overset{\text{def}}{=} \frac{1}{2}\left[A(x^*)y + x^*A(y) - A(x^*y)\right].\tag{2.91}$$

of $M_{I,\text{fin}}$. As in (2.78) in the case of a markovian semigroup of Fourier multipliers, Γ is called the carré du champ, associated with the semigroup $(T_t)_{t \geqslant 0}$. For any $i, j, k, l \in I$, note that

$$\Gamma(e_{ij}, e_{kl}) = \frac{1}{2}\left[A(e_{ji})e_{kl} + e_{ji}A(e_{kl}) - A(e_{ji}e_{kl})\right] = \frac{1}{2}\delta_{i=k}\left[a_{ji} + a_{kl} - a_{jl}\right]e_{jl}.\tag{2.92}$$

The first part of the following result shows how to construct the carré du champ directly from the semigroup $(T_t)_{t \geqslant 0}$. The proof is similar to the one of Lemma 2.15 to which we refer.

Lemma 2.17

1. Suppose $1 \leqslant p \leqslant \infty$. For any $x, y \in M_{I,\text{fin}}$, we have

$$\Gamma(x, y) = \lim_{t \to 0^+} \frac{1}{2t}\left(T_t(x^*y) - (T_t(x))^*T_t(y)\right)\tag{2.93}$$

in S_I^p.

2. For any $x \in M_{I,\text{fin}}$, we have $\Gamma(x, x) \geqslant 0$.

3. For any $x, y \in M_{I,\text{fin}}$, the matrix $\begin{bmatrix} \Gamma(x, x) & \Gamma(x, y) \\ \Gamma(y, x) & \Gamma(y, y) \end{bmatrix}$ is positive.

4. For any $x, y \in M_{I,\text{fin}}$, we have

$$\|\Gamma(x, y)\|_{S_I^p} \leqslant \|\Gamma(x, x)\|_{S_I^p}^{\frac{1}{2}} \|\Gamma(y, y)\|_{S_I^p}^{\frac{1}{2}}.\tag{2.94}$$

This inequality even holds for $0 < p \leqslant \infty$.

We shall link, more profoundly in Sect. 3, the (square root of the) negative generator A_p, the carré du champ Γ and some noncommutative gradient from (2.95). Suppose $1 \leqslant p < \infty$ and $-1 \leqslant q \leqslant 1$. Note that the Bochner space $L^p(\Gamma_q(H)\overline{\otimes}B(\ell_I^2))$ is equipped with a canonical structure of S_I^∞-bimodule whose operations are defined by $(f \otimes x)y \overset{\text{def}}{=} f \otimes xy$ and $y(f \otimes x) \overset{\text{def}}{=} f \otimes yx$ where $x \in S_I^p$, $y \in S_I^\infty$ and $f \in L^p(\Gamma_q(H))$. If Γ_q is the q-Gaussian functor of Sect. 2.2 associated with the real Hilbert space H stemming from Proposition 2.4, we can consider the linear map $\partial_{\alpha,q} \colon M_{I,\text{fin}} \to \Gamma_q(H) \otimes M_{I,\text{fin}}$ (resp. $\partial_{\alpha,1} \colon M_{I,\text{fin}} \to L^0(\Omega) \otimes M_{I,\text{fin}}$ if $q = 1$) defined by

$$\partial_{\alpha,q}(e_{ij}) \overset{\text{def}}{=} s_q(\alpha_i - \alpha_j) \otimes e_{ij}, \qquad i, j \in I. \tag{2.95}$$

We have the following Leibniz rule.

Lemma 2.18 *Suppose* $-1 \leqslant q \leqslant 1$. *For any* $x, y \in M_{I,\text{fin}}$, *we have*

$$\partial_{\alpha,q}(xy) = x\partial_{\alpha,q}(y) + \partial_{\alpha,q}(x)y. \tag{2.96}$$

Proof On the one hand, for any $i, j, k, l \in I$, we have

$$\partial_{\alpha,q}(e_{ij}e_{kl}) = \delta_{j=k}\partial_{\alpha,q}(e_{il}) \overset{(2.95)}{=} \delta_{j=k}s_q(\alpha_i - \alpha_l) \otimes e_{il}.$$

On the other hand, for any $i, j, k, l \in I$, we have

$$
\begin{aligned}
e_{ij}&\partial_{\alpha,q}(e_{kl}) + \partial_{\alpha,q}(e_{ij})e_{kl} \\
&= e_{ij}\big(s_q(\alpha_k - \alpha_l) \otimes e_{kl}\big) + \big(s_q(\alpha_i - \alpha_j) \otimes e_{ij}\big)e_{kl} \\
&= s_q(\alpha_k - \alpha_l) \otimes e_{ij}e_{kl} + s_q(\alpha_i - \alpha_j) \otimes e_{ij}e_{kl} \\
&= s_q(\alpha_k - \alpha_l) \otimes \delta_{j=k}e_{il} + s_q(\alpha_i - \alpha_j) \otimes \delta_{j=k}e_{il} \\
&= \delta_{j=k}\big(s_q(\alpha_k - \alpha_l) + s_q(\alpha_i - \alpha_j)\big) \otimes e_{il} = \delta_{j=k}s_q(\alpha_i - \alpha_l) \otimes e_{il}.
\end{aligned}
$$

The result follows by linearity. $\square$

Now, we describe a connection between the carré du champ and the map $\partial_{\alpha,q}$ which is analogous to the equality of [205, Section 1.4] (see also [206]). For that, we introduce the canonical trace preserving normal faithful conditional expectation $\mathbb{E} \colon \Gamma_q(H)\overline{\otimes}B(\ell_I^2) \to B(\ell_I^2)$.

Proposition 2.13 *Suppose* $-1 \leqslant q \leqslant 1$ *and* $1 \leqslant p < \infty$.

1. For any $x, y \in M_{I,\text{fin}}$, *we have*

$$\Gamma(x, y) = \mathbb{E}\big((\partial_{\alpha,q}(x))^*\partial_{\alpha,q}(y)\big) = \big\langle \partial_{\alpha,q}(x), \partial_{\alpha,q}(y)\big\rangle_{S_I^p(L^2(\Gamma_q(H))_{c,p})}. \tag{2.97}$$

2. For any $x \in M_{I,\text{fin}}$, *we have*

$$\big\|\Gamma(x, x)^{\frac{1}{2}}\big\|_{S_I^p} = \big\|\partial_{\alpha,q}(x)\big\|_{S_I^p(L^2(\Gamma_q(H))_{c,p})}, \tag{2.98}$$

Proof

1. On the one hand, for any $i, j, k, l \in I$, we have

$$\Gamma(e_{ij}, e_{kl}) \overset{(2.92)}{=} \frac{1}{2}\delta_{i=k}\big[a_{ji} + a_{kl} - a_{jl}\big]e_{jl}$$

$$= \frac{1}{2}\delta_{i=k}\left(\|\alpha_j - \alpha_i\|_H^2 + \|\alpha_k - \alpha_l\|_H^2 - \|\alpha_j - \alpha_l\|_H^2\right)e_{jl}$$

$$= \frac{1}{2}\delta_{i=k}\left(2\|\alpha_k\|^2 - 2\langle\alpha_i, \alpha_k\rangle - 2\langle\alpha_j, \alpha_k\rangle + 2\langle\alpha_i, \alpha_j\rangle\right)e_{jl}$$

$$= \langle\alpha_k - \alpha_i, \alpha_k - \alpha_j\rangle_H e_{jl}.$$

On the other hand, we have

$$\mathbb{E}\big[(\partial_{\alpha,q}(e_{ij}))^*\partial_{\alpha,q}(e_{kl})\big] \overset{(2.95)}{=} \mathbb{E}\big[(s_q(\alpha_i - \alpha_j)\otimes e_{ij})^*(s_q(\alpha_k - \alpha_l)\otimes e_{kl})\big]$$

$$= \mathbb{E}\big[s_q(\alpha_i - \alpha_j)s_q(\alpha_k - \alpha_l)\otimes e_{ji}e_{kl}\big]$$

$$= \delta_{i=k}\tau(s_q(\alpha_i - \alpha_j)s_q(\alpha_k - \alpha_l))e_{jl}$$

$$= \delta_{i=k}\langle\alpha_i - \alpha_j, \alpha_k - \alpha_l\rangle_H e_{jl}$$

The second equality is a consequence of (2.73).

2. If $x \in M_{I,\mathrm{fin}}$, we have

$$\|\partial_{\alpha,q}(x)\|_{S_I^p(\mathrm{L}^2(\Gamma_q(H))_{c,p})} \overset{(2.68)}{=} \left\|\langle\partial_{\alpha,q}(x), \partial_{\alpha,q}(x)\rangle_{S_I^p(\mathrm{L}^2(\Gamma_q(H))_{c,p})}^{\frac{1}{2}}\right\|_{S_I^p}$$

$$\overset{(2.97)}{=} \left\|\Gamma(x, x)^{\frac{1}{2}}\right\|_{S_I^p}.$$

$\square$

Chapter 3
Riesz Transforms Associated to Semigroups of Markov Multipliers

Abstract In this chapter, we start by proving Khintchine inequalities for q-Gaussians in crossed products. As a consequence, we obtain boundedness of L^p-Riesz transforms associated with markovian semigroups of Fourier multipliers and defined over these crossed products with q-Gaussians. We also give dependence in p and independence of the group G and the markovian semigroup, of these L^p-inequalities. Then we examine in detail the domains of the operators related to Kato's square root problem for markovian semigroups of Fourier multipliers. We also show how to extend the carré du champ Γ associated with such a markovian semigroup to a closed form. Moreover, we solve the Kato square root problem for markovian semigroups of Schur multipliers. In particular, in its course, we again prove Khintchine inequalities for q-Gaussians. We also obtain the constants of the Kato square root problem independently of the markovian semigroup and discuss dependence in p. Finally, we also investigate Meyer's problem for semigroups of Schur multipliers and study L^p-boundedness of directional Riesz transforms.

3.1 Khintchine Inequalities for q-Gaussians in Crossed Products

In this section, we consider a markovian semigroup of Fourier multipliers on $\mathrm{VN}(G)$, where G is a discrete group satisfying Proposition 2.3 with cocycle (b_ψ, π, H). Moreover, we have seen in (2.83) that by second quantization we have an action $\alpha \colon G \to \mathrm{Aut}(\Gamma_q(H))$. The aim of the section is to prove Theorem 3.1 below which generalizes [131, Theorem 1.1]. In the following, the conditional expectation is $\mathbb{E} \colon \Gamma_q(H) \rtimes_\alpha G \to \Gamma_q(H) \rtimes_\alpha G$, $x \rtimes \lambda_s \mapsto \tau_{\Gamma_q(H)}(x)\lambda_s$. We let

$$\mathrm{Gauss}_{q,p,\rtimes}(L^p(\mathrm{VN}(G))) \overset{\text{def}}{=} \overline{\mathrm{span}\left\{s_q(h) \rtimes x : h \in H, x \in L^p(\mathrm{VN}(G))\right\}} \quad (3.1)$$

where the closure is taken in $L^p(\Gamma_q(H) \rtimes_\alpha G)$ (for the weak* topology if $p = \infty$ and $-1 \leqslant q < 1$).

Lemma 3.1 *Let $1 < p < \infty$. Let G be a discrete group. If $(e_k)_{k \in K}$ is an orthonormal basis of the Hilbert space H, then $\mathrm{Gauss}_{q,p,\rtimes}(\mathrm{L}^p(\mathrm{VN}(G)))$ equals the closure of the span of the $s_q(e_k) \rtimes \lambda_s$'s with $k \in K$ and $s \in G$.*

Proof We write temporarily G' for the closure of the span of the $s_q(e_k) \rtimes \lambda_s$. First note that it is trivial that $G' \subset \mathrm{Gauss}_{q,p}(\mathrm{L}^p(\mathrm{VN}(G)))$. For the reverse inclusion, by linearity and density and (3.1), it suffices to show that $s_q(h) \rtimes \lambda_s$ belongs to G' for any $h \in H$ and any $s \in G$. Let $\varepsilon > 0$ be fixed. Then there exists some finite subset F of K and $\alpha_k \in \mathbb{C}$ for $k \in F$ such that $\left\| h - \sum_{k \in F} \alpha_k e_k \right\|_H < \varepsilon$. We have $\sum_{k \in F} \alpha_k s_q(e_k) \rtimes \lambda_s \in G'$ and[1]

$$
\left\| s_q(h) \rtimes \lambda_s - \sum_{k \in F} \alpha_k s_q(e_k) \rtimes \lambda_s \right\|_{\mathrm{L}^p(\Gamma_q(H) \rtimes_\alpha G)} = \left\| s_q\left(h - \sum_{k \in F} \alpha_k e_k \right) \rtimes \lambda_s \right\|_{\mathrm{L}^p}
$$

$$
= \left\| s_q\left(h - \sum_{k \in F} \alpha_k e_k \right) \right\|_{\mathrm{L}^p(\Gamma_q(H))} \cong \left\| s_q\left(h - \sum_{k \in F} \alpha_k e_k \right) \right\|_{\mathrm{L}^2(\Gamma_q(H))} < \varepsilon.
$$

We conclude $\mathrm{Gauss}_{q,p,\rtimes}(\mathrm{L}^p(\mathrm{VN}(G))) \subset G'$ by closedness of G'. □

If $2 \leqslant p < \infty$, recall that we have a contractive injective inclusion

$$
\mathrm{Gauss}_{q,p,\rtimes}(\mathrm{L}^p(\mathrm{VN}(G))) \subset \mathrm{L}^p(\Gamma_q(H) \rtimes_\alpha G) \overset{\text{Lemma 2.8}}{\subset} \mathrm{L}^p_{cr}(\mathbb{E}) \tag{3.2}
$$

and that if $1 < p < 2$ we have an inclusion

$$
\mathrm{span}\left\{ s_q(h) \rtimes x : h \in H, \ x \in \mathrm{L}^p \right\} \subset \mathrm{L}^p(\mathrm{VN}(G), \mathrm{L}^2(\Gamma_q(H)_{c,p})) \tag{3.3}
$$

$$
\overset{(2.74)}{=} \mathrm{L}^p_c(\mathbb{E}) \subset \mathrm{L}^p_{cr}(\mathbb{E}).
$$

In Theorem 3.1 below, we shall need several operations on $\mathrm{L}^p_c(\mathbb{E})$, $\mathrm{L}^p_r(\mathbb{E})$ and $\mathrm{L}^p(\Gamma_q(H) \rtimes_\alpha G)$. To this end, we have the following result.

---

[1] Recall that α preserves the trace. If $s \in G$ and $k \in H$, we have

$$
\left\| s_q(h) \rtimes \lambda_s \right\|^p_{\mathrm{L}^p(\Gamma_q(H) \rtimes_\alpha G)} = \tau_\rtimes \left((s_q(h) \rtimes \lambda_s)^* (s_q(h) \rtimes \lambda_s) \right)^{\frac{p}{2}}
$$

$$
\overset{(2.57)}{=} \tau_\rtimes \left((\alpha_{s^{-1}}(s_q(h))) \rtimes \lambda_s ((s_q(h) \rtimes \lambda_s))^{\frac{p}{2}} \right)
$$

$$
\overset{(2.56)}{=} \tau_\rtimes \left((\alpha_{s^{-1}}(s_q(h)^2) \rtimes \lambda_e)^{\frac{p}{2}} \right) = \tau_\rtimes \left(\pi \left(\alpha_{s^{-1}}(s_q(h)^2) \right)^{\frac{p}{2}} \right) = \tau \left(\alpha_{s^{-1}}((s_q(h)^p)) \right)
$$

$$
= \tau \left(s_q(h)^p \right) = \left\| s_q(h) \right\|^p_{\mathrm{L}^p(\Gamma_q(H))}.
$$

Lemma 3.2 *Let* $-1 \leqslant q \leqslant 1$ *and* $1 < p < \infty$. *Consider* $P \colon L^2(\Gamma_q(H)) \to L^2(\Gamma_q(H))$ *the orthogonal projection onto* $\mathrm{Gauss}_{q,2}(\mathbb{C}) = \mathrm{span}\{s_q(e) : e \in H\}$. *Let* G *be a discrete group. Consider* $Q_p = P \rtimes \mathrm{Id}_{L^p(\mathrm{VN}(G))}$ *initially defined on* $\mathcal{P}_{\rtimes,G}$. *Then* Q_p *induces well-defined contractions*

$$Q_p \colon L_c^p(\mathbb{E}) \to L_c^p(\mathbb{E}) \quad and \quad Q_p \colon L_r^p(\mathbb{E}) \to L_r^p(\mathbb{E}). \tag{3.4}$$

Consequently, according to Proposition 2.10, Q_p equally extends to a contraction on $L_{cr}^p(\mathbb{E})$.

Proof By (2.71) for the column case and (2.76), the column part of (3.4) follows immediately. We turn to row part of (3.4). Note that the immediate analogue of (2.71) to the row case holds true, but not for (2.76), according to Remark 2.2. So we need to argue differently.

We start by showing that for any $h \in L^2(\Gamma_q(H))$ and any $s \in G$ we have

$$\alpha_s\big((P(h))^*\big) = P\big(\alpha_s(h^*)\big). \tag{3.5}$$

Indeed, by anti-linearity and $L^2(\Gamma_q(H))$ continuity of both sides, it suffices to show (3.5) for $h = w(e_{n_1} \otimes \cdots \otimes e_{n_N})$ the Wick word of $e_{n_1} \otimes \cdots \otimes e_{n_N}$, where (e_k) is an orthonormal basis of H. But then we have[2]

$$\alpha_s(P(h)^*) = \alpha_s(\delta_{N=1} s_q(e_{n_1})^*) = \delta_{N=1} \alpha_s(s_q(e_{n_1}))$$
$$\overset{(2.83)}{=} \delta_{N=1} s_q(\pi_s(e_{n_1})) = P(\alpha_s(h^*)).$$

We deduce that

$$\left\| (P \rtimes \mathrm{Id}_{L^p(\mathrm{VN}(G))}) \left(\sum_k h_k \rtimes \lambda_{s_k} \right) \right\|_{L_r^p(\mathbb{E})} \tag{3.6}$$

$$= \left\| \left((P \rtimes \mathrm{Id}_{L^p(\mathrm{VN}(G))}) \left(\sum_k h_k \rtimes \lambda_{s_k} \right) \right)^* \right\|_{L_c^p(\mathbb{E})}$$

$$= \left\| \left(\sum_k P(h_k) \rtimes \lambda_{s_k} \right)^* \right\|_{L_c^p(\mathbb{E})} \overset{(2.57)}{=} \left\| \sum_k \alpha_{s_k^{-1}}(P(h_k)^*) \rtimes \lambda_{s_k^{-1}} \right\|_{L_c^p(\mathbb{E})}$$

$$\overset{(3.5)}{=} \left\| \sum_k P(\alpha_{s_k^{-1}}(h_k^*)) \rtimes \lambda_{s_k^{-1}} \right\|_{L_c^p(\mathbb{E})}$$

---

[2] In the last equality, we use that $w(e_{n_1} \otimes \cdots \otimes e_{n_N}) = w(e_{n_N} \otimes \cdots \otimes e_{n_1})^*$ [177, p. 21].

$$\begin{aligned}
&= \left\| \left(P \rtimes \mathrm{Id}_{L^p(\mathrm{VN}(G))} \right) \left(\sum_k \alpha_{s_k^{-1}}(h_k^*) \rtimes \lambda_{s_k^{-1}} \right) \right\|_{L_c^p(\mathbb{E})} \\
&\overset{(2.57)}{=} \left\| \left(P \rtimes \mathrm{Id}_{L^p(\mathrm{VN}(G))} \right) \left(\left(\sum_k h_k \rtimes \lambda_{s_k} \right)^* \right) \right\|_{L_c^p(\mathbb{E})} \\
&\overset{(3.4)}{\leqslant} \left\| \left(\sum_k h_k \rtimes \lambda_{s_k} \right)^* \right\|_{L_c^p(\mathbb{E})} = \left\| \sum_k h_k \rtimes \lambda_{s_k} \right\|_{L_r^p(\mathbb{E})}.
\end{aligned}$$

We have proved the row part of (3.4). □

Now, the noncommutative Khintchine inequalities can be rewritten under the following form.

Theorem 3.1 *Consider $-1 \leqslant q \leqslant 1$ and $1 < p < \infty$. Let G be a discrete group.*

1. *Suppose $1 < p < 2$. For any element $f = \sum_{s,h} f_{s,h} s_q(h) \rtimes \lambda_s$ of span $\{ s_q(h) : h \in H \} \rtimes \mathcal{P}_G$, we have*

$$\| f \|_{\mathrm{Gauss}_{q,p,\rtimes}(L^p(\mathrm{VN}(G)))} \approx_p \inf_{f=g+h} \left\{ \left\| \left(\mathbb{E}(g^*g) \right)^{\frac{1}{2}} \right\|_{L^p(\mathrm{VN}(G))}, \left\| \left(\mathbb{E}(hh^*) \right)^{\frac{1}{2}} \right\|_{L^p} \right\}. \tag{3.7}$$

Here the infimum runs over all $g \in L_c^p(\mathbb{E})$ and $h \in L_r^p(\mathbb{E})$ such that $f = g + h$. Further one can restrict the infimum to all $g \in \mathrm{Ran}\, Q_p|_{L_c^p(\mathbb{E})}$, $h \in \mathrm{Ran}\, Q_p|_{L_r^p(\mathbb{E})}$, where Q_p is the mapping from Lemma 3.2.

   Finally, in case that $L^p(\mathrm{VN}(G))$ has CCAP and $\mathrm{VN}(G)$ has QWEP, the infimum can be taken over all $g, h \in$ span $\{ s_q(e) : e \in H \} \rtimes \mathcal{P}_G$.

2. *Suppose $2 \leqslant p < \infty$. For any element $f = \sum_{s,h} f_{s,h} s_q(h) \rtimes \lambda_s$ of the space $\mathrm{Gauss}_{q,p,\rtimes}(L^p(\mathrm{VN}(G)))$ with $f_{s,h} \in \mathbb{C}$, we have*

$$\max \left\{ \left\| \left(\mathbb{E}(f^*f) \right)^{\frac{1}{2}} \right\|_{L^p}, \left\| \left(\mathbb{E}(ff^*) \right)^{\frac{1}{2}} \right\|_{L^p} \right\} \leqslant \| f \|_{\mathrm{Gauss}_{q,p,\rtimes}(L^p(\mathrm{VN}(G)))} \tag{3.8}$$

$$\lesssim \sqrt{p} \underbrace{\max \left\{ \left\| \left(\mathbb{E}(f^*f) \right)^{\frac{1}{2}} \right\|_{L^p(\mathrm{VN}(G))}, \left\| \left(\mathbb{E}(ff^*) \right)^{\frac{1}{2}} \right\|_{L^p(\mathrm{VN}(G))} \right\}}_{\| f \|_{L_{cr}^p(\mathbb{E})}}.$$

Proof 2. Suppose $2 < p < \infty$. We begin by proving the upper estimate of (3.8). Fix $m \geqslant 1$. We have[3] an isometric embedding $J_m \colon H \to \ell_m^2(H)$ defined by

$$J_m(h) \overset{\text{def}}{=} \frac{1}{\sqrt{m}} \sum_{l=1}^m e_l \otimes h. \tag{3.9}$$

Hence we can consider the operator $\Gamma_q^p(J_m) \colon L^p(\Gamma_q(H)) \to L^p(\Gamma_q(\ell_m^2(H)))$ of second quantization (2.38) which is an isometric completely positive map. We define the maps $\pi_s^m \colon G \to \mathrm{B}(\ell_m^2(H))$, $s \mapsto (e_l \otimes h \mapsto e_l \otimes \pi_s(h))$ and $\alpha_s^m \overset{\text{def}}{=} \Gamma_q(\pi_s^m) \colon \Gamma_q(\ell_m^2(H)) \to \Gamma_q(\ell_m^2(H))$. For any $h \in H$, any $s \in G$, note that

$$\pi_s^m \circ J_m(h) \overset{(3.9)}{=} \pi_s^m \left(\frac{1}{\sqrt{m}} \sum_{l=1}^m e_l \otimes h \right) = \frac{1}{\sqrt{m}} \sum_{l=1}^m \pi_s^m(e_l \otimes h)$$

$$= \frac{1}{\sqrt{m}} \sum_{l=1}^m e_l \otimes \pi_s(h) \overset{(3.9)}{=} J_m \circ \pi_s(h). \tag{3.10}$$

We deduce that

$$\alpha_s^m \circ \Gamma_q(J_m) = \Gamma_q(\pi_s^m) \circ \Gamma_q(J_m) = \Gamma_q(\pi_s^m \circ J_m) \overset{(3.10)}{=} \Gamma_q(J_m \circ \pi_s) = \Gamma_q(J_m) \circ \alpha_s.$$

By Lemma 2.6, we obtain a trace preserving unital normal injective $*$-homomorphism $\Gamma_q(J_m) \rtimes \mathrm{Id}_{\mathrm{VN}(G)} \colon \Gamma_q(H) \rtimes_\alpha G \to \Gamma_q(\ell_m^2(H)) \rtimes_{\alpha^m} G$. This map induces an isometric map $\rho \overset{\text{def}}{=} \Gamma_q^p(J_m) \rtimes_\alpha \mathrm{Id}_{L^p(\mathrm{VN}(G))} \colon L^p(\Gamma_q(H) \rtimes_\alpha G) \to L^p(\Gamma_q(\ell_m^2(H)) \rtimes_{\alpha^m} G)$. For any finite sum $f = \sum_{s,h} f_{s,h} s_q(h) \rtimes \lambda_s$ of $\mathcal{P}_{\rtimes,G}$ with $f_{s,h} \in \mathbb{C}$, we obtain

$$\|f\|_{L^p(\Gamma_q(H) \rtimes_\alpha G)} = \left\| \sum_{s,h} f_{s,h} s_q(h) \rtimes \lambda_s \right\|_{L^p(\Gamma_q(H) \rtimes_\alpha G)} \tag{3.11}$$

$$= \left\| \rho \left(\sum_{s,h} f_{s,h} s_q(h) \rtimes \lambda_s \right) \right\|_{L^p(\Gamma_q(\ell_m^2(H)) \rtimes_{\alpha^m} G)}$$

---

[3] If $h \in H$, we have

$$\left\| \frac{1}{\sqrt{m}} \sum_{l=1}^m e_l \otimes h \right\|_{\ell_m^2(H)} = \frac{1}{\sqrt{m}} \left\| \sum_{l=1}^m e_l \right\|_{\ell_m^2} \|h\|_H = \|h\|_H.$$

$$= \left\| \sum_{s,h} f_{s,h} \Gamma_q(J_m)(s_q(h)) \rtimes_{\alpha^m} \lambda_s \right\|_{L^p(\Gamma_q(\ell_m^2(H)) \rtimes_{\alpha^m} G)}$$

$$\overset{(2.38)}{=} \left\| \sum_{s,h} f_{s,h} s_{q,m}(J_m(h)) \rtimes \lambda_s \right\|_{L^p(\Gamma_q(\ell_m^2(H)) \rtimes_{\alpha^m} G)}$$

$$\overset{(3.9)}{=} \left\| \sum_{s,h} f_{s,h} s_{q,m} \left(\frac{1}{\sqrt{m}} \sum_{l=1}^m e_l \otimes h \right) \rtimes \lambda_s \right\|_{L^p(\Gamma_q(\ell_m^2(H)) \rtimes_{\alpha^m} G)}$$

$$= \frac{1}{\sqrt{m}} \left\| \sum_{s,h,l} f_{s,h} s_{q,m}(e_l \otimes h) \rtimes \lambda_s \right\|_{L^p(\Gamma_q(\ell_m^2(H)) \rtimes_{\alpha^m} G)}. \tag{3.12}$$

For any $1 \leqslant l \leqslant m$, consider the element $f_l \overset{\text{def}}{=} \sum_{s,h} f_{s,h} s_{q,m}(e_l \otimes h) \rtimes \lambda_s$ and the conditional expectation $\mathbb{E} \colon L^p(\Gamma_q(\ell_m^2(H)) \rtimes_{\alpha^m} G) \to L^p(\mathrm{VN}(G))$. For any $1 \leqslant l \leqslant m$, note that

$$\mathbb{E}(f_l) = \mathbb{E}\left(\sum_{s,h} f_{s,h} s_{q,m}(e_l \otimes h) \rtimes \lambda_s \right) = \sum_{s,h} f_{s,h} \mathbb{E}\big(s_{q,m}(e_l \otimes h) \rtimes \lambda_s\big)$$

$$= \sum_{s,h} f_{s,h} \tau(s_{q,m}(e_l \otimes h)) \lambda_s \overset{(2.40)}{=} 0.$$

We deduce that the random variables f_l are mean-zero. Now, we prove in addition that the f_l are independent over $\mathrm{VN}(G)$, so that we will be able to apply the noncommutative Rosenthal inequality (2.48) to them. The rest of the proof follows page 72.   □

Lemma 3.3 *The f_l defined previously are independent over* $\mathrm{VN}(G)$. *More precisely, the von Neumann algebras generated by f_l are independent with respect to the von Neumann algebra* $\mathrm{VN}(G)$.

Proof Due to normality of $\mathbb{E}$, it suffices to prove

$$\mathbb{E}(xy) = \mathbb{E}(x)\mathbb{E}(y) \tag{3.13}$$

for x (resp. y) belonging to a weak* dense subset of $W^*(f_l)$ (resp. of $W^*((f_k)_{k \neq l})$). Moreover, by bilinearity in x, y of both sides of (3.13) and selfadjointness of $s_{q,m}(e_l \otimes h)$, it suffices to take $x = s_{q,m}(e_l \otimes h)^n \rtimes \lambda_s$ for some $n \in \mathbb{N}$ and $y = \prod_{t=1}^T s_{q,m}(e_{k_t} \otimes h)^{n_t} \rtimes \lambda_u$ for some $T \in \mathbb{N}$, $k_t \neq l$ and $n_t \in \mathbb{N}$. Since

$\alpha_s^m = \Gamma_q(\pi_s^m)$ is trace preserving, we have

$$\mathbb{E}(x)\mathbb{E}(y) = \tau(s_{q,m}(e_l \otimes h)^n)\lambda_s \, \tau\left(\prod_{t=1}^{T} s_{q,m}(e_{k_t} \otimes h)^{n_t}\right)\lambda_u$$

$$= \tau(s_{q,m}(e_l \otimes h)^n)\tau\left(\prod_{t=1}^{T} s_{q,m}(e_{k_t} \otimes h)^{n_t}\right)\lambda_{su}$$

$$= \tau(s_{q,m}(e_l \otimes h)^n)\tau\left(\alpha_s^m\left(\prod_{t=1}^{T} s_{q,m}(e_{k_t} \otimes h)^{n_t}\right)\right)\lambda_{su}$$

$$= \tau(s_{q,m}(e_l \otimes h)^n)\tau\left(\prod_{t=1}^{T} s_{q,m}(e_{k_t} \otimes \pi_s(h))^{n_t}\right)\lambda_{su}$$

and

$$\mathbb{E}(xy) = \mathbb{E}\left(\left(s_{q,m}(e_l \otimes h)^n \rtimes \lambda_s\right)\left(\prod_{t=1}^{T} s_{q,m}(e_{k_t} \otimes h)^{n_t} \rtimes \lambda_u\right)\right)$$

$$\overset{(2.56)}{=} \mathbb{E}\left(\left(s_{q,m}(e_l \otimes h)^n \alpha_s^m\left(\prod_{t=1}^{T} s_{q,m}(e_{k_t} \otimes h)^{n_t}\right) \rtimes \lambda_{su}\right)\right)$$

$$= \tau\left(s_{q,m}(e_l \otimes h)^n \alpha_s^m\left(\prod_{t=1}^{T} s_{q,m}(e_{k_t} \otimes h)^{n_t}\right)\right)\lambda_{su}$$

$$= \tau\left(s_{q,m}(e_l \otimes h)^n \prod_{t=1}^{T} s_{q,m}(e_{k_t} \otimes \pi_s(h))^{n_t}\right)\lambda_{su}.$$

We shall now apply the Wick formulae (2.39) and (2.40) to the pevious trace term.

Note that if $n + \sum_{t=1}^{T} n_t$ is odd, then according to the Wick formula (2.40) we have $\mathbb{E}(xy) = 0$. On the other hand, then either n or $\sum_{t=1}^{T} n_t$ is odd, so according to the Wick formula (2.40), either $\mathbb{E}(x) = 0$ or $\mathbb{E}(y) = 0$. Thus, (3.13) follows in this case.

Now suppose that $n + \sum_{t=1}^{T} n_t \overset{\text{def}}{=} 2k$ is even. Consider a 2-partition $\mathcal{V} \in \mathcal{P}_2(2k)$. If both n and $\sum_{t=1}^{T} n_t$ are odd, then we must have some $(i, j) \in \mathcal{V}$ such that one term $\langle f_i, f_j \rangle_{\ell_m^2(H)}$ in the Wick formula (2.39) equals $\langle e_l \otimes h, e_{k_t} \otimes \pi_s(h)\rangle = 0$, since $k_t \neq l$. Thus, $\mathbb{E}(xy) = 0$, and since n is odd, also $\mathbb{E}(x) = 0$, and therefore, (3.13) follows. If both n and $\sum_{t=1}^{T} n_t$ are even, then in the Wick formula (2.39), we only need to consider those 2-partitions $\mathcal{V}$ without a mixed term, since otherwise we have as previously $\langle f_i, f_j \rangle = 0$. Such a $\mathcal{V}$ is clearly the disjoint union $\mathcal{V} = \mathcal{V}_1 \cup \mathcal{V}_2$ of 2-partitions corresponding to n and to $\sum_{t=1}^{T} n_t$. Moreover, we have for the number of crossings, $c(\mathcal{V}) = c(\mathcal{V}_1) + c(\mathcal{V}_2)$. With $(f_1, f_2, \ldots, f_{n+T}) = \underbrace{(e_l \otimes h, e_l \otimes h, \ldots, e_l \otimes h}_{n}, e_{k_1} \otimes \pi_s(h), \ldots, e_{k_T} \otimes \pi_s(h))$, we obtain

$$
\tau\left(s_{q,m}(e_l \otimes h)^n \prod_{t=1}^{T} s_{q,m}(e_{k_t} \otimes \pi_s(h))^{n_t} \right)
$$

$$
\overset{(2.39)}{=} \sum_{\mathcal{V} \in \mathcal{P}_2(2k)} q^{c(\mathcal{V})} \prod_{(i,j) \in \mathcal{V}} \langle f_i, f_j \rangle_{\ell_m^2(H)}
$$

$$
= \sum_{\mathcal{V}_1, \mathcal{V}_2} q^{c(\mathcal{V}_1)+c(\mathcal{V}_2)} \prod_{(i_1,j_1) \in \mathcal{V}_1} \langle f_{i_1}, f_{j_1} \rangle_{\ell_m^2(H)} \prod_{(i_2,j_2) \in \mathcal{V}_2} \langle f_{i_2}, f_{j_2} \rangle_{\ell_m^2(H)}
$$

$$
= \left(\sum_{\mathcal{V}_1} q^{c(\mathcal{V}_1)} \prod_{(i_1,j_1) \in \mathcal{V}_1} \langle f_{i_1}, f_{j_1} \rangle_{\ell_m^2(H)} \right)\left(\sum_{\mathcal{V}_2} q^{c(\mathcal{V}_2)} \prod_{(i_2,j_2) \in \mathcal{V}_2} \langle f_{i_2}, f_{j_2} \rangle_{\ell_m^2(H)} \right)
$$

$$
\overset{(2.39)}{=} \tau\left(s_{q,m}(e_l \otimes h)^n \right) \tau\left(\prod_{t=1}^{T} s_{q,m}(e_{k_t} \otimes \pi_s(h))^{n_t} \right).
$$

Thus, also in this case (3.13) follows. □

Proof (End of the Proof of Theorem 3.1) Now we are able to apply the noncommutative Rosenthal inequality (2.48) which yields

$$
\|f\|_{L^p(\Gamma_q(H) \rtimes_\alpha G)} \overset{(3.12)}{=} \frac{1}{\sqrt{m}} \left\| \sum_{l=1}^{m} f_l \right\|_{L^p(\Gamma_q(\ell_m^2(H)) \rtimes_{\alpha^m} G)} \tag{3.14}
$$

$$
\overset{(2.48)}{\lesssim} \frac{1}{\sqrt{m}} \left[p\left(\sum_{l=1}^{m} \|f_l\|_{L^p(\Gamma_q(\ell_m^2(H)) \rtimes_{\alpha^m} G)}^p \right)^{\frac{1}{p}} \right. \tag{3.15}
$$

$$
\left. + \sqrt{p} \left\| \left(\sum_{l=1}^{m} \mathbb{E}(f_l^* f_l) \right)^{\frac{1}{2}} \right\|_{L^p(\mathrm{VN}(G))} + \sqrt{p} \left\| \left(\sum_{l=1}^{m} \mathbb{E}(f_l f_l^*) \right)^{\frac{1}{2}} \right\|_{L^p(\mathrm{VN}(G))} \right].
$$

For any integer $1 \leqslant l \leqslant m$, note that

$$
\mathbb{E}(f_l^* f_l) = \mathbb{E}\left(\left(\sum_{s,h} f_{s,h} s_{q,m}(e_l \otimes h) \rtimes \lambda_s \right)^* \left(\sum_{t,k} f_{t,k} s_{q,m}(e_l \otimes k) \rtimes \lambda_t \right) \right)
$$

$$
\overset{(2.57)}{=} \sum_{s,h,t,k} \overline{f_{s,h}} f_{t,k} \mathbb{E}\left(\left(\alpha_{s^{-1}}^m(s_{q,m}(e_l \otimes h)) \right) \rtimes \lambda_{s^{-1}} \right)\left(s_{q,m}(e_l \otimes k) \rtimes \lambda_t \right)
$$

$$
\overset{(2.56)}{=} \sum_{s,h,t,k} \overline{f_{s,h}} f_{t,k} \mathbb{E}\left(\alpha_{s^{-1}}^m(s_{q,m}(e_l \otimes h)) \alpha_{s^{-1}}^m(s_{q,m}(e_l \otimes k)) \rtimes \lambda_{s^{-1}t} \right)
$$

$$= \sum_{s,h,t,k} \overline{f_{s,h}} f_{t,k} \tau(\alpha_{s^{-1}}^m (s_{q,m}(e_l \otimes h)(s_{q,m}(e_l \otimes k)))\lambda_{s^{-1}t}$$

$$= \sum_{s,h,t,k} \overline{f_{s,h}} f_{t,k} \tau(s_{q,m}(e_l \otimes h)s_{q,m}(e_l \otimes k))\lambda_{s^{-1}t}$$

$$\overset{(2.41)}{=} \sum_{i,j,h,s} \overline{f_{s,h}} f_{t,k} \langle h, k \rangle_H \lambda_{s^{-1}t}.$$

and

$$\mathbb{E}(f^*f) = \mathbb{E}\left(\left(\sum_{s,h} f_{s,h} s_q(h) \rtimes \lambda_s\right)^* \left(\sum_{t,k} f_{t,k} s_q(k) \rtimes \lambda_t\right)\right)$$

$$\overset{(2.57)}{=} \sum_{s,h,t,k} \overline{f_{s,h}} f_{t,k} \mathbb{E}\left(\left(\alpha_{s^{-1}}(s_q(h)) \rtimes \lambda_{s^{-1}}\right)\left(s_q(k) \rtimes \lambda_t\right)\right)$$

$$\overset{(2.56)}{=} \sum_{s,h,t,k} \overline{f_{s,h}} f_{t,k} \mathbb{E}\left(\left(\alpha_{s^{-1}}(s_q(h))\alpha_{s^{-1}}(s_q(k)) \rtimes \lambda_{s^{-1}t}\right)\right)$$

$$= \sum_{s,h,t,k} \overline{f_{s,h}} f_{t,k} \tau\left(\alpha_{s^{-1}}(s_q(h)s_q(k))\right)\lambda_{s^{-1}t}$$

$$= \sum_{s,h,t,k} \overline{f_{s,h}} f_{t,k} \tau\left(s_q(h)s_q(k)\right)\lambda_{s^{-1}t} \overset{(2.41)}{=} \sum_{i,j,h,s} \overline{f_{s,h}} f_{t,k} \langle h, k \rangle_H \lambda_{s^{-1}t}$$

and similarly for the row terms. We conclude that

$$\mathbb{E}(f_l^* f_l) = \mathbb{E}(f^*f) \quad \text{and} \quad \mathbb{E}(f_l f_l^*) = \mathbb{E}(ff^*). \tag{3.16}$$

Moreover, for $1 \leqslant l \leqslant m$, using the isometric map $\psi_l: H \to \ell_m^2(H), h \mapsto e_l \otimes h$, we can introduce the second quantization operator $\Gamma_q(\psi_l): \Gamma_q(H) \to \Gamma_q(\ell_m^2(H))$. For any $h \in H$, any $s \in G$, note that

$$\pi_s^m \circ \psi_l(h) = \pi_s^m(e_l \otimes h) = e_l \otimes \pi_s(h) = \psi_l \circ \pi_s(h). \tag{3.17}$$

We deduce that

$$\alpha_s^m \circ \Gamma_q(\psi_l) = \Gamma_q(\pi_s^m) \circ \Gamma_q(\psi_l) = \Gamma_q(\pi_s^m \circ \psi_l) \overset{(3.17)}{=} \Gamma_q(\psi_l \circ \pi_s) = \Gamma_q(\psi_l) \circ \alpha_s.$$

By Lemma 2.6, we obtain a trace preserving unital normal injective $*$-homomorphism $\Gamma_q(\psi_l) \rtimes \mathrm{Id}_{\mathrm{VN}(G)} \colon \Gamma_q(H) \rtimes_\alpha G \to \Gamma_q(\ell^2_m(H)) \rtimes_{\alpha^m} G$. This map induces an isometric map $\Gamma_q^p(\psi_l) \rtimes_\alpha \mathrm{Id}_{L^p(\mathrm{VN}(G))} \colon L^p(\Gamma_q(H) \rtimes_\alpha G) \to L^p(\Gamma_q(\ell^2_m(H)) \rtimes_{\alpha^m} G)$. We have

$$\|f_l\|_{L^p(\Gamma_q(\ell^2_m(H)) \rtimes_{\alpha^m} G)} = \left\| \sum_{s,h} f_{s,h} s_{q,m}(e_l \otimes h) \rtimes \lambda_s \right\|_{L^p(\Gamma_q(\ell^2_m(H)) \rtimes_{\alpha^m} G)}$$

(3.18)

$$\overset{(2.38)}{=} \left\| \sum_{s,h} f_{s,h} \Gamma_q(\psi_l)(s_{q,m}(h)) \rtimes \lambda_s \right\|_{L^p(\Gamma_q(\ell^2_m(H)) \rtimes_{\alpha^m} G)}$$

(3.19)

$$= \left\| \sum_{s,h} f_{s,h} s_q(h) \rtimes \lambda_s \right\|_{L^p(\Gamma_q(H) \rtimes G)} = \|f\|_{L^p(\Gamma_q(H) \rtimes_\alpha G)} .$$

We infer that

$$\|f\|_{L^p(\Gamma_q(H) \rtimes_\alpha G)} \overset{(3.14)(3.18)(3.16)}{\lesssim} \frac{1}{\sqrt{m}} \left[p \left(\sum_{l=1}^m \|f\|^p_{L^p(\Gamma_q(H) \rtimes_\alpha G)} \right)^{\frac{1}{p}} \right.$$

$$+ \sqrt{p} \left\| \left(\sum_{l=1}^m \mathbb{E}(f^* f) \right)^{\frac{1}{2}} \right\|_{L^p(\mathrm{VN}(G))} + \sqrt{p} \left\| \left(\sum_{l=1}^m \mathbb{E}(f f^*) \right)^{\frac{1}{2}} \right\|_{L^p(\mathrm{VN}(G))} \right]$$

$$= \frac{1}{\sqrt{m}} \left[p m^{\frac{1}{p}} \|f\|_{L^p(\Gamma_q(H) \rtimes_\alpha G)} \right.$$

$$+ \sqrt{pm} \left\| (\mathbb{E}(f^* f))^{\frac{1}{2}} \right\|_{L^p(\mathrm{VN}(G))} + \sqrt{pm} \left\| (\mathbb{E}(f f^*))^{\frac{1}{2}} \right\|_{L^p(\mathrm{VN}(G))} \right]$$

$$= p m^{\frac{1}{p} - \frac{1}{2}} \|f\|_{L^p(\Gamma_q(H) \rtimes_\alpha G)} + \sqrt{p} \left\| (\mathbb{E}(f^* f))^{\frac{1}{2}} \right\|_{L^p} + \sqrt{p} \left\| (\mathbb{E}(f f^*))^{\frac{1}{2}} \right\|_{L^p}.$$

Since $p > 2$, passing to the limit when $m \to \infty$ and noting that Rosenthal's inequality comes with an absolute constant not depending on the von Neumann algebras under consideration [125, p. 4303], we finally obtain

$$\|f\|_{L^p(\Gamma_q(H) \rtimes_\alpha G)} \lesssim \sqrt{p} \left[\left\| (\mathbb{E}(f^* f))^{\frac{1}{2}} \right\|_{L^p(\mathrm{VN}(G))} + \left\| (\mathbb{E}(f f^*))^{\frac{1}{2}} \right\|_{L^p(\mathrm{VN}(G))} \right].$$

Using the equivalence $\ell^1_2 \approx \ell^\infty_2$, we obtain the upper estimate of (3.8).

The lower estimate of (3.8) holds with constant 1 from the contractivity of the conditional expectation $\mathbb{E}$ on $L^{\frac{p}{2}}(\Gamma_q(H) \rtimes_\alpha G)$:

$$\max\left\{\left\|\left(\mathbb{E}(f^*f)\right)^{\frac{1}{2}}\right\|_{L^p(VN(G))}, \left\|\left(\mathbb{E}(ff^*)\right)^{\frac{1}{2}}\right\|_{L^p(VN(G))}\right\}$$

$$= \max\left\{\left\|\mathbb{E}(f^*f)\right\|_{L^{\frac{p}{2}}(VN(G))}^{\frac{1}{2}}, \left\|\mathbb{E}(ff^*)\right\|_{L^{\frac{p}{2}}(VN(G))}^{\frac{1}{2}}\right\}$$

$$\leqslant \max\left\{\left\|f^*f\right\|_{L^{\frac{p}{2}}(\Gamma_q(H)\rtimes_\alpha G)}^{\frac{1}{2}}, \left\|ff^*\right\|_{L^{\frac{p}{2}}(\Gamma_q(H)\rtimes_\alpha G)}^{\frac{1}{2}}\right\} = \|f\|_{L^p(\Gamma_q(H)\rtimes_\alpha G)}.$$

1. Now, let us consider the case $1 < p < 2$. We will proceed by duality as follows. Consider the Gaussian projection Q_p from Lemma 3.2 which is a contraction on $L_{cr}^p(\mathbb{E})$. Note that $\tau_\rtimes(f^*g) = \tau_\rtimes(Q_p(f)^*g) = \tau_\rtimes(f^*Q_{p^*}(g))$ for any $g \in L^{p^*}(\Gamma_q(H) \rtimes_\alpha G)$ and the fixed f from the statement of the theorem. Indeed, this can be seen from selfadjointness of P and (3.5). Note also that $L_{cr}^p(\mathbb{E})$ and $L_{cr}^{p^*}(\mathbb{E})$ are dual spaces with respect to each other according to Lemma 2.14, with duality bracket $\langle f, g \rangle = \tau_\rtimes(f^*g)$, i.e. the same as the duality bracket between $L^p(\Gamma_q(H) \rtimes_\alpha G)$ and $L^{p^*}(\Gamma_q(H) \rtimes_\alpha G)$.

Using this in the first two equalities, the contractivity $Q_{p^*}: L_{cr}^{p^*}(\mathbb{E}) \to L_{cr}^{p^*}(\mathbb{E})$ from Lemma 3.2 in the third equality, and the upper estimate of (3.8) together with the density result of Lemma 2.13 in the last inequality, we obtain for any $f \in$ $\text{Gauss}_{q,2}(\mathbb{C}) \rtimes \mathcal{P}_G$

$$\|f\|_{L_{cr}^p(\mathbb{E})} = \sup_{\|g\|_{L_{cr}^{p^*}(\mathbb{E})} \leqslant 1} \left|\tau_\rtimes(f^*g)\right| = \sup_{\|g\|_{L_{cr}^{p^*}(\mathbb{E})} \leqslant 1} \left|\tau_\rtimes(f^*Q_{p^*}(g))\right|$$

$$= \sup_{\substack{\|g\|_{L_{cr}^{p^*}(\mathbb{E})} \leqslant 1 \\ g \in \text{Ran}\, Q_{p^*}}} \left|\tau_\rtimes(f^*g)\right|$$

$$\overset{(3.8)}{\leqslant} \|f\|_{L^p(\Gamma_q(H)\rtimes_\alpha G)} \sup_{\substack{\|g\|_{L_{cr}^{p^*}(\mathbb{E})} \leqslant 1 \\ g \in \text{Ran}\, Q_{p^*}}} \|g\|_{L^{p^*}(\Gamma_q(H)\rtimes_\alpha G)} \lesssim_p \|f\|_{L^p(\Gamma_q(H)\rtimes_\alpha G)}.$$

Note that in the definition of the $L_{cr}^p(\mathbb{E})$-norm, the infimum runs over all $g \in L_c^p(\mathbb{E})$ and $h \in L_r^p(\mathbb{E})$ such that $f = g + h$. However, we can write $f = Q_p(f) = Q_p(g) + Q_p(h)$ and $\|Q_p(g)\|_{L_c^p(\mathbb{E})} \overset{(3.4)}{\leqslant} \|g\|_{L_c^p(\mathbb{E})}$ and $\|Q_p(h)\|_{L_r^p(\mathbb{E})} \overset{(3.4)}{\leqslant} \|h\|_{L_r^p(\mathbb{E})}$. Thus, in the definition of the $L_{cr}^p(\mathbb{E})$-norm of f, we can restrict to $g, h \in \text{Ran}\, Q_p$.

Our next goal is to restrict in the lower estimate of (3.7) to elements g, h of $\text{span}\{s_q(e): e \in H\} \rtimes \mathcal{P}_G$ with $f = g + h$, in the case where f belongs to $\text{span}\{s_q(e): e \in H\} \rtimes \mathcal{P}_G$, under the supplementary assumptions that $L^p(VN(G))$ has CCAP and that $VN(G)$ has QWEP. We consider the approximating net (M_{φ_j})

of Fourier multipliers. Moreover, let $f = g + h$ be a decomposition with $g \in L^p_c(\mathbb{E})$ and $h \in L^p_r(\mathbb{E})$ such that $\|g\|_{L^p_c(\mathbb{E})} + \|h\|_{L^p_r(\mathbb{E})} \leqslant 2\|f\|_{L^p_{cr}(\mathbb{E})}$. In light of this, we can already assume that g, h belong to Ran Q_p. Then we newly decompose for each j

$$f = (\mathrm{Id}_{\Gamma_q(H)} \rtimes M_{\varphi_j})(f) + \left(f - \mathrm{Id}_{\Gamma_q(H)} \rtimes M_{\varphi_j}(f)\right)$$

$$= \underbrace{(\mathrm{Id}_{\Gamma_q(H)} \rtimes M_{\varphi_j})(g) + \left(f - (\mathrm{Id}_{\Gamma_q(H)} \rtimes M_{\varphi_j})(f)\right)}_{\in L^p_c(\mathbb{E})} + \underbrace{(\mathrm{Id}_{\Gamma_q(H)} \rtimes M_{\varphi_j})(h)}_{\in L^p_r(\mathbb{E})}.$$

Note that $(\mathrm{Id}_{\Gamma_q(H)} \rtimes M_{\varphi_j})(g)$ and $(\mathrm{Id}_{\Gamma_q(H)} \rtimes M_{\varphi_j})(h)$ belong to the subspace span $\{s_q(e): e \in H\} \rtimes \mathcal{P}_G$ since φ_j is of finite support. Moreover, also $f - \mathrm{Id}_{\Gamma_q(H)} \rtimes M_{\varphi_j}(f)$ belongs to that space. Now, we will control the norms in this new decomposition. Recall that $f = \sum_{s,e} f_{s,e} s_q(e) \rtimes \lambda_s$ with finite sums. Since $\varphi_j(s)$ converges pointwise to 1, there is j such that

$$\left\| f - \mathrm{Id}_{\Gamma_q(H)} \rtimes M_{\varphi_j}(f) \right\|_{L^p_c(\mathbb{E})} \leqslant \sum_{s,e} |f_{s,e}| \left\| s_q(e) \rtimes \lambda_s - \varphi_j(s) s_q(e) \rtimes \lambda_s \right\|_{L^p_c(\mathbb{E})}$$

$$\leqslant \sum_{s,e} |f_{s,e}| |1 - \varphi_j(s)| \left\| s_q(e) \rtimes \lambda_s \right\|_{L^p_c(\mathbb{E})} \leqslant \varepsilon \leqslant \|f\|_{L^p_{cr}(\mathbb{E})}$$

$$\lesssim \|f\|_{L^p(\Gamma_q(H) \rtimes_\alpha G)}.$$

We turn to the other two parts. With (2.77), we obtain

$$\left\| (\mathrm{Id}_{\Gamma_q(H)} \otimes M_{\varphi_j})(g) \right\|_{L^p_c(\mathbb{E})} + \left\| (\mathrm{Id}_{\Gamma_q(H)} \otimes M_{\varphi_j})(h) \right\|_{L^p_r(\mathbb{E})}$$

$$\lesssim_p \|g\|_{L^p_c(\mathbb{E})} + \|h\|_{L^p_r(\mathbb{E})} \lesssim \|f\|_{L^p_{cr}(\mathbb{E})}$$

$$\lesssim_p \|f\|_{L^p(\Gamma_q(H) \rtimes_\alpha G)}.$$

Together we have shown that also the infimum in (3.7) restricted to elements of the space span $\{s_q(e): e \in H\} \rtimes \mathcal{P}_G$ is also controlled by $\|f\|_{\mathrm{Gauss}_{q,p,\rtimes}(L^p(\mathrm{VN}(G)))}$.

Now, we will prove the remaining estimate, that is, $\|f\|_{\mathrm{Gauss}_{q,p,\rtimes}(L^p(\mathrm{VN}(G)))}$ is controlled by the second expression in (3.7). Since $1 < p < 2$, the function $\mathbb{R}^+ \to \mathbb{R}^+, t \mapsto t^{\frac{p}{2}}$ is operator concave by [37, p. 112]. Using [100, Corollary 2.2] applied with the trace preserving positive map $\mathbb{E}$, we can write

$$\|f\|_{L^p(\Gamma_q(H) \rtimes_\alpha G)} = \left\| |f|^2 \right\|_{L^{\frac{p}{2}}(\Gamma_q(H) \rtimes_\alpha G)}^{\frac{1}{2}} \leqslant \left\| \mathbb{E}(|f|^2) \right\|_{L^{\frac{p}{2}}(\Gamma_q(H) \rtimes_\alpha G)}^{\frac{1}{2}} = \|f\|_{L^p_c(\mathbb{E})}$$

and similarly for the row term. Thus, passing to the infimum over all decompositions $f = g + h$, we obtain $\|f\|_{L^p(\Gamma_q(H) \rtimes_\alpha G)} \leqslant \|f\|_{L^p_{cr}(\mathbb{E})}$, which can be majorised in turn by the infimum of $\|g\|_{L^p_c(\mathbb{E})} + \|h\|_{L^p_r(\mathbb{E})}$, where $f = g + h$ and $g, h \in \mathrm{span}\{s_q(h): h \in H\} \rtimes \mathcal{P}_G$. Hence, we have the last equivalence in the part 1 of the theorem.

The case $p = 2$ is obvious since we have isometrically

$$L^2_{cr}(\mathbb{E}) = L^2(VN(G), L^2(\Gamma_q(H))_{rad,2}) = L^2(\Gamma_q(H)) \otimes_2 L^2(VN(G))$$

by Lemma 2.10 and [128, Remark 2.3 (1)]. □

The remainder of the section is devoted to extend Theorem 3.1 to the case of f being a generic element of $\text{Gauss}_{q,p,\rtimes}(L^p(VN(G)))$. First we have the following lemma.

Lemma 3.4 *Let $-1 \leqslant q \leqslant 1$ and $1 < p < \infty$. Consider again $Q_p = P \rtimes \text{Id}_{L^p(VN(G))}$ the extension of the Gaussian projection from Lemma 3.2. Then Q_p extends to a bounded operator*

$$Q_p : L^p(\Gamma_q(H) \rtimes_\alpha G) \to L^p(\Gamma_q(H) \rtimes_\alpha G). \tag{3.20}$$

Proof First note that the case $G = \{e\}$ is contained in [118, Theorem 3.5], putting there $d = 1$. Note that the closed space spanned by $\{s_q(h) : h \in H\}$ coincides in this case with $\mathcal{G}^1_{p,q}$ there. To see that the projection considered in this source is Q_p, we refer to [118, Proof of Theorem 3.1].

We turn to the case of general discrete group G. Since P is selfadjoint, $(Q_p)^* = Q_{p^*}$. We obtain for $f = \sum_{s \in F} f_s \rtimes \lambda_s$ with $F \subset G$ finite and $f_s \in L^p(\Gamma_q(H))$

$$\|Q_p(f)\|_{\text{Gauss}_{q,p,\rtimes}(L^p(VN(G)))} \overset{\text{Theorem 3.1}}{\approx} \|Q_p(f)\|_{L^p_{cr}(\mathbb{E})}$$

$$= \sup\left\{ |\langle Q_p(f), g \rangle| : \|g\|_{L^{p^*}_{cr}(\mathbb{E})} \leqslant 1 \right\}$$

$$= \sup\left\{ \left|\langle f, Q_{p^*}(g)\rangle_{L^p_{cr}(\mathbb{E}), L^{p^*}_{cr}(\mathbb{E})}\right| : g \right\}$$

$$= \sup\left\{ \left|\langle f, Q_{p^*}(g)\rangle_{L^p(\Gamma_q(H) \rtimes_\alpha G), L^{p^*}}\right| : g \right\}$$

$$\leqslant \|f\|_{L^p(\Gamma_q(H) \rtimes_\alpha G)} \sup\left\{ \|Q_{p^*}(g)\|_{L^{p^*}(\Gamma_q(H) \rtimes_\alpha G)} : g \right\}$$

$$\overset{\text{Theorem 3.1}}{\lesssim} \|f\|_{L^p(\Gamma_q(H) \rtimes_\alpha G)} \sup\left\{ \|Q_{p^*}(g)\|_{L^{p^*}_{cr}(\mathbb{E})} : \|g\|_{L^{p^*}_{cr}(\mathbb{E})} \leqslant 1 \right\}$$

$$\lesssim \|f\|_{L^p(\Gamma_q(H) \rtimes_\alpha G)},$$

where in the last step we used that Q_{p^*} is bounded on $L^{p^*}_{cr}(\mathbb{E})$ according to Lemma 3.2. □

Now we obtain the following extension of Theorem 3.1.

Proposition 3.1 *Let* $-1 \leqslant q \leqslant 1$ *and* $1 < p < \infty$. *Assume that* $\mathrm{L}^p(\mathrm{VN}(G))$ *has* CCAP *and that* $\Gamma_q(H) \rtimes_\alpha G$ *has* QWEP. *Let* f *be an element of* $\mathrm{Ran}\, Q_p = \mathrm{Gauss}_{q,p,\rtimes}(\mathrm{L}^p(\mathrm{VN}(G)))$.[4]

1. Suppose $1 < p < 2$. *Then we have*

$$\|f\|_{\mathrm{Gauss}_{q,p,\rtimes}(\mathrm{L}^p(\mathrm{VN}(G)))} \approx_p \inf_{f=g+h} \left\{ \left\|\left(\mathbb{E}(g^*g)\right)^{\frac{1}{2}}\right\|_{\mathrm{L}^p(\mathrm{VN}(G))}, \left\|\left(\mathbb{E}(hh^*)\right)^{\frac{1}{2}}\right\|_{\mathrm{L}^p} \right\}.$$

$$(3.21)$$

Here the infimum runs over all $g \in \mathrm{L}_c^p(\mathbb{E})$ *and* $h \in \mathrm{L}_r^p(\mathbb{E})$ *such that* $f = g + h$. *Further one can restrict the infimum to all* $g, h \in \mathrm{Ran}\, Q_p$.

2. Suppose $2 \leqslant p < \infty$. *Then we have*

$$\max\left\{ \left\|\left(\mathbb{E}(f^*f)\right)^{\frac{1}{2}}\right\|_{\mathrm{L}^p}, \left\|\left(\mathbb{E}(ff^*)\right)^{\frac{1}{2}}\right\|_{\mathrm{L}^p} \right\} \leqslant \|f\|_{\mathrm{Gauss}_{q,p,\rtimes}(\mathrm{L}^p(\mathrm{VN}(G)))}$$

$$(3.22)$$

$$\lesssim \sqrt{p}\, \max\left\{ \left\|\left(\mathbb{E}(f^*f)\right)^{\frac{1}{2}}\right\|_{\mathrm{L}^p(\mathrm{VN}(G))}, \left\|\left(\mathbb{E}(ff^*)\right)^{\frac{1}{2}}\right\|_{\mathrm{L}^p(\mathrm{VN}(G))} \right\}.$$

Proof We consider the approximating net (M_{φ_j}) of finitely supported Fourier multipliers guaranteed by the CCAP assumption. Observe that $(\mathrm{Id}_{\Gamma_q(H)} \rtimes M_{\varphi_j})$ approximates the identity on $\mathcal{P}_{\rtimes,G}$. Moreover, according to Proposition 2.8, $(\mathrm{Id}_{\Gamma_q(H)} \rtimes M_{\varphi_j})_j$ is a bounded net in $\mathrm{B}(\mathrm{L}^p(\Gamma_q(H) \rtimes_\alpha G))$. We conclude by density of $\mathcal{P}_{\rtimes,G}$ in $\mathrm{L}^p(\Gamma_q(H) \rtimes_\alpha G)$ that $(\mathrm{Id}_{\Gamma_q(H)} \rtimes M_{\varphi_j})_j$ converges in the point norm topology of $\mathrm{L}^p(\Gamma_q(H) \rtimes_\alpha G)$ to the identity. Moreover, replacing in this argument Proposition 2.8 by Lemma 2.12, the same argument yields that $(\mathrm{Id}_{\Gamma_q(H)} \rtimes M_{\varphi_j})_j$ converges to the identity in the point norm topology of $\mathrm{L}_c^p(\mathbb{E})$ and of $\mathrm{L}_r^p(\mathbb{E})$. Then again by the same argument, we infer that $\|f\|_{\mathrm{L}_{cr}^p(\mathbb{E})} = \lim_j \left\|(\mathrm{Id}_{\Gamma_q(H)} \rtimes M_{\varphi_j})(f)\right\|_{\mathrm{L}_{cr}^p(\mathbb{E})}$. Note that for fixed j, $(\mathrm{Id}_{\Gamma_q(H)} \rtimes M_{\varphi_j})(f) = (\mathrm{Id}_{\Gamma_q(H)} \rtimes M_{\varphi_j})(Q_p f) = Q_p(\mathrm{Id}_{\Gamma_q(H)} \rtimes M_{\varphi_j})(f)$ belongs to $\mathrm{span}\{s_q(e) : e \in H\} \rtimes \mathcal{P}_G$. Thus, Theorem 3.1 applies to f replaced by $\mathrm{Id}_{\Gamma_q(H)} \rtimes M_{\varphi_j}(f)$, and therefore,

$$\|f\|_{\mathrm{Gauss}_{q,p,\rtimes}(\mathrm{L}^p(\mathrm{VN}(G)))} = \lim_j \left\|(\mathrm{Id}_{\Gamma_q(H)} \rtimes M_{\varphi_j})(f)\right\|_{\mathrm{Gauss}_{q,p,\rtimes}(\mathrm{L}^p(\mathrm{VN}(G)))}$$

$$\overset{\underset{\text{Theorem 3.1}}{}}{\cong} \lim_j \left\|(\mathrm{Id}_{\Gamma_q(H)} \rtimes M_{\varphi_j})(f)\right\|_{\mathrm{L}_{cr}^p(\mathbb{E})} = \|f\|_{\mathrm{L}_{cr}^p(\mathbb{E})}.$$

---

[4] This equality follows from the fact that $\mathcal{P}_{\rtimes,G}$ is dense in $\mathrm{L}^p(\Gamma_q(H) \rtimes_\alpha G)$ according to Lemma 2.3, so $Q_p(\mathcal{P}_{\rtimes,G}) \subset \mathrm{Gauss}_{q,p,\rtimes}(\mathrm{L}^p(\mathrm{VN}(G)))$ is dense in $\mathrm{Ran}\, Q_p$. The other inclusion $\mathrm{Gauss}_{q,p,\rtimes}(\mathrm{L}^p(\mathrm{VN}(G))) \subset \mathrm{Ran}\, Q_p$ follows from the fact that the span of the $s_q(h) \rtimes x$ with $x \in \mathrm{L}^p(\mathrm{VN}(G))$ is dense in $\mathrm{Gauss}_{q,p,\rtimes}(\mathrm{L}^p(\mathrm{VN}(G)))$ and obviously lies in $\mathrm{Ran}\, Q_p$ which in turn is closed.

Finally, the fact that one can restrict the infimum to all $g, h \in \operatorname{Ran} Q_p$ can be proved in the same way as that in Theorem 3.1.   □

3.2 L^p-Kato's Square Root Problem for Semigroups of Fourier Multipliers

Throughout this section, we consider a discrete group G, and fix a markovian semigroup of Fourier multipliers $(T_t)_{t \geqslant 0}$ acting on $\operatorname{VN}(G)$ with negative generator A and representing objects $b_\psi : G \to H$, $\pi : G \to O(H)$, $\alpha : G \to \operatorname{Aut}(\Gamma_q(H))$, see Proposition 2.3 and (2.83). We also have the noncommutative gradient $\partial_{\psi,q} : \mathcal{P}_G \subset L^p(\operatorname{VN}(G)) \to L^p(\Gamma_q(H) \rtimes_\alpha G)$ from (2.84). The aim of this section is to compare $A^{\frac{1}{2}}(x)$ and $\partial_{\psi,q}(x)$ in the L^p-norm. We shall also extend $\partial_{\psi,q}$ to a closed operator and identify the domain of its closure. To this end, we need some facts from the general theory of strongly continuous semigroups.

The following is a straightforward extension of [25, Lemma 4.2].

Proposition 3.2 *Let $(T_t)_{t \geqslant 0}$ be a strongly continuous bounded semigroup acting on a Banach space X with (negative) generator A. We have*

$$\left\| (\operatorname{Id}_X + A)^{\frac{1}{2}}(x) \right\|_X \approx \|x\|_X + \left\| A^{\frac{1}{2}}(x) \right\|_X, \quad x \in X. \tag{3.23}$$

Proof It is well-known [25, Lemma 2.3] that the function $f_1 : t \mapsto (1+t)^{\frac{1}{2}}(1 + t^{\frac{1}{2}})^{-1}$ is the Laplace transform $\mathcal{L}(\mu_1)$ of some bounded measure μ_1. By [98, Proposition 3.3.2] (note that $f_1 \in H^\infty(\Sigma_{\pi-\varepsilon})$ for any $\varepsilon \in (0, \frac{\pi}{2})$ and f_1 has finite limits $\lim_{z \in \Sigma_{\pi-\varepsilon}, z \to 0} f_1(z)$ and $\lim_{z \in \Sigma_{\pi-\varepsilon}, |z| \to \infty} f_1(z)$) (see also [151, Lemma 2.12]), we have

$$\int_0^\infty T_t \, d\mu_1(t) = \mathcal{L}(\mu_1)(A) = f_1(A) = (\operatorname{Id}_X + A)^{\frac{1}{2}} \left(\operatorname{Id}_X + A^{\frac{1}{2}} \right)^{-1}.$$

Hence for any $x \in \operatorname{dom}(\operatorname{Id}_X + A^{\frac{1}{2}}) = \operatorname{dom} A^{\frac{1}{2}}$, we have

$$(\operatorname{Id}_X + A)^{\frac{1}{2}} x = \int_0^\infty T_t \left(x + A^{\frac{1}{2}} x \right) d\mu_1(t). \tag{3.24}$$

By Haase [98, Remark 3.3.3], we know that $\int_0^\infty T_t \, d\mu_1(t)$ is a bounded operator on X of norm $\leqslant \|\mu_1\|$. Using the triangular inequality in the last inequality, we conclude that

$$\left\| (\mathrm{Id}_X + A)^{\frac{1}{2}} x \right\|_X \overset{(3.24)}{=} \left\| \int_0^\infty T_t \left(x + A^{\frac{1}{2}} x \right) d\mu_1(t) \right\|_X \leqslant \|\mu_1\| \left\| x + A^{\frac{1}{2}} x \right\|_X$$

$$\leqslant \|\mu_1\| \left(\|x\|_X + \left\| A^{\frac{1}{2}} x \right\|_X \right).$$

It is easy to see [25, Lemma 2.3] that the functions $f_2 \colon t \mapsto (1 + t^{\frac{1}{2}})(1 + t)^{-\frac{1}{2}}$ and $f_3 \colon t \mapsto (1 + t)^{-\frac{1}{2}}$ are the Laplace transforms $\mathcal{L}(\mu_2)$ and $\mathcal{L}(\mu_3)$ of some bounded measures μ_2 and μ_3. So we have

$$\int_0^\infty T_t \, d\mu_2(t) = \left(\mathrm{Id}_X + A^{\frac{1}{2}} \right)(\mathrm{Id} + A)^{-\frac{1}{2}} \quad \text{and} \quad \int_0^\infty T_t \, d\mu_3(t) = (\mathrm{Id}_X + A)^{-\frac{1}{2}}.$$

Following the same argument as previously, we obtain

$$\left\| x + A^{\frac{1}{2}} x \right\|_X \leqslant \|\mu_2\| \left\| (\mathrm{Id}_X + A)^{\frac{1}{2}} x \right\|_X \quad \text{and} \quad \|x\|_X \leqslant \|\mu_3\| \left\| (\mathrm{Id}_X + A)^{\frac{1}{2}} x \right\|_X. \tag{3.25}$$

Note that

$$\left\| A^{\frac{1}{2}} x \right\|_X = \left\| -x + x + A^{\frac{1}{2}} x \right\|_X \leqslant \|x\|_X + \left\| x + A^{\frac{1}{2}} x \right\|_X. \tag{3.26}$$

We conclude that

$$\|x\|_X + \left\| A^{\frac{1}{2}} x \right\|_X \overset{(3.26)}{\leqslant} 2\|x\|_X + \left\| x + A^{\frac{1}{2}} x \right\|_X \overset{(3.25)}{\lesssim} \left\| (\mathrm{Id}_X + A)^{\frac{1}{2}} x \right\|_X.$$

$\square$

We start to observe that the estimates in (3.27) below come with a constant independent of the group G and the cocycle (α, H). This is essentially the second proof of the appendix [131, pp. 574–575]. Note that we are unfortunately unable[5] to check the original proof given in [131, p. 544]. Here, $\mathrm{L}^\infty(\Omega) = \Gamma_1(H)$ is the Gaussian space from Sect. 2.2.

---

[5] More precisely, with the notations of [131] we are unable to check that "$H \rtimes \mathrm{Id}_G$ extends to a bounded operator on L_p". A part of the very concise explanation given in [131, p. 544] is "H is G-equivariant". But it seems to be strange. Indeed we have an action $\alpha \colon G \to \mathrm{Aut}(L_\infty(\mathbb{R}^n_{\mathrm{bohr}}))$, $f \mapsto \left[x \mapsto \alpha_g(f)(x) = f(\pi_g(x)) \right]$ for some map $\pi_g \colon \mathbb{R}^n_{\mathrm{bohr}} \to \mathbb{R}^n_{\mathrm{bohr}}$ where $g \in G$ and an induced action α from G on $L_\infty(\mathbb{R}^n_{\mathrm{bohr}} \times \mathbb{R}^n, \nu \times \gamma)$. Now, note that

$$(H(\alpha_g f))(x, y) = \left(\mathrm{p.\,v.} \int_{\mathbb{R}} \beta_t \alpha_g f \, \frac{dt}{t} \right)(x, y) = \left(\mathrm{p.\,v.} \int_{\mathbb{R}} \beta_t (f \circ \pi_g) \, \frac{dt}{t} \right)(x, y)$$

Lemma 3.5 *Suppose* $1 < p < \infty$. *For any* $x \in \operatorname{dom} \mathcal{P}_G$, *we have*

$$\frac{1}{K \max(p, p^*)} \left\| A_p^{\frac{1}{2}}(x) \right\|_{L^p} \leqslant \left\| \partial_{\psi,1,p}(x) \right\|_{L^p(L^\infty(\Omega) \rtimes_\alpha G)}$$

$$\leqslant K \max(p, p^*)^{\frac{3}{2}} \left\| A_p^{\frac{1}{2}}(x) \right\|_{L^p} \tag{3.27}$$

with an absolute constant K *not depending on* G *nor the cocycle* (α, H).

We define the densely defined unbounded operator $\partial^*_{\psi,q} : \mathcal{P}_{\rtimes,G} \subset L^p(\Gamma_q(H) \rtimes_\alpha G) \to L^p(\Gamma_q(H))$ by

$$\partial^*_{\psi,q}(f \rtimes \lambda_s) = \left\langle s_q(b_\psi(s)), f \right\rangle_{L^{p^*}(\Gamma_q(H)), L^p(\Gamma_q(H))} \lambda_s, \quad s \in G, f \in L^p(\Gamma_q(H)). \tag{3.28}$$

The following lemma is left to the reader.

Lemma 3.6 *The operators* $\partial_{\psi,q}$ *and* $\partial^*_{\psi,q}$ *are formal adjoints.*

The next proposition extends (3.27) to the case of q-Gaussians.

Proposition 3.3 *Suppose* $1 < p < \infty$ *and* $-1 \leqslant q \leqslant 1$. *For any* $x \in \mathcal{P}_G$, *we have*

$$\frac{1}{K \max(p, p^*)^{\frac{3}{2}}} \left\| A_p^{\frac{1}{2}}(x) \right\|_{L^p} \leqslant \left\| \partial_{\psi,q,p}(x) \right\|_{L^p(\Gamma_q(H) \rtimes_\alpha G)}$$

$$\leqslant K \max(p, p^*)^2 \left\| A_p^{\frac{1}{2}}(x) \right\|_{L^p} \tag{3.29}$$

with an absolute constant K *not depending on* G *nor the cocycle* (α, b_ψ, H).

Proof Start with the case $2 \leqslant p < \infty$. The case $q = 1$ is covered by Lemma 3.5. Consider now the case $-1 \leqslant q < 1$. Pick some element $x = \sum_s x_s \lambda_s$ of $\mathcal{P}_G$. We recall from Sect. 3.1 that we have a conditional expectation $\mathbb{E} : \Gamma_q(H) \rtimes_\alpha G \to \Gamma_q(H) \rtimes_\alpha G$, $x \rtimes \lambda_s \mapsto \tau_{\Gamma_q(H)}(x) \lambda_s$. In the following calculations, we consider an orthonormal basis (e_k) of H. Therefore, using the orthonormal systems $(W(e_k))$

$$\overline{} = \text{p. v.} \int_{\mathbb{R}} f(\pi_g(x + ty)) \frac{\mathrm{d}t}{t}$$

and

$$\alpha_g(H(f))(x, y) = \alpha_g \left(\text{p. v.} \int_{\mathbb{R}} \beta_t f \, \frac{\mathrm{d}t}{t} \right)(x, y) = \text{p. v.} \int_{\mathbb{R}} f(\pi_g(x) + ty) \frac{\mathrm{d}t}{t}$$

which could be different if π is not trivial.

and $(s_q(e_k))$ in $L^2(\Omega)$ and $L^2(\Gamma_q(H))$ in the third equality, we obtain

$$
\left\| \sum_s x_s W(b_\psi(s)) \rtimes \lambda_s \right\|_{L_c^p(\mathbb{E})} = \left\| \sum_{s,k} x_s \langle e_k, b_\psi(s) \rangle W(e_k) \rtimes \lambda_s \right\|_{L_c^p(\mathbb{E})}
$$

$$
\overset{(2.76)}{=} \left\| \sum_{s,k} x_s \langle e_k, b_\psi(s) \rangle \lambda_s \otimes W(e_k) \right\|_{L^p(\text{VN}(G), L^2(\Omega)_{c,p})}
$$

$$
= \left\| \sum_{s,k} x_s \langle e_k, b_\psi(s) \rangle \lambda_s \otimes s_q(e_k) \right\|_{L^p(\text{VN}(G), L^2(\Gamma_q(H))_{c,p})}
$$

$$
\overset{(2.76)}{=} \left\| \sum_s x_s s_q(b_\psi(s)) \rtimes \lambda_s \right\|_{L_c^p(\mathbb{E})}.
$$

We claim that this equality holds also for the row space instead of the column space. Note that according to Remark 2.2, we have to argue differently. We have with $x_{s,k} \overset{\text{def}}{=} x_s \langle e_k, b_\psi(s) \rangle_H$,

$$
\left\| \sum_s x_s s_q(b_\psi(s)) \rtimes \lambda_s \right\|_{L_r^p(\mathbb{E})}^2 = \left\| \sum_{s,k,t,l} \sum_{t,l} x_{s,k} \overline{x_{t,l}} \mathbb{E}\big((s_q(e_k) \rtimes \lambda_s)(s_q(e_l) \rtimes \lambda_t)^* \big) \right\|_{\frac{p}{2}}
$$

$$
\overset{(2.57)}{=} \left\| \sum_{s,k,t,l} x_{s,k} \overline{x_{t,l}} \mathbb{E}\big((s_q(e_k) \rtimes \lambda_s)(\alpha_{t^{-1}}(s_q(e_l)) \rtimes \lambda_{t^{-1}}) \big) \right\|_{\frac{p}{2}}
$$

$$
\overset{(2.56)}{=} \left\| \sum_{s,k,t,l} x_{s,k} \overline{x_{t,l}} \mathbb{E}\big(s_q(e_k) \alpha_{st^{-1}}(s_q(e_l)) \rtimes \lambda_{st^{-1}} \big) \right\|_{\frac{p}{2}}
$$

$$
= \left\| \sum_{s,k,t,l} x_{s,k} \overline{x_{t,l}} \tau_q\big(s_q(e_k) s_q(\pi_{st^{-1}}(e_l)) \big) \lambda_{st^{-1}} \right\|_{\frac{p}{2}}
$$

$$
\overset{(2.41)}{=} \left\| \sum_{s,k,t,l} x_{s,k} \overline{x_{t,l}} \langle e_k, \pi_{st^{-1}}(e_l) \rangle_H \lambda_{st^{-1}} \right\|_{\frac{p}{2}}.
$$

The point is that this last quantity does not depend on q. We infer that

$$
\left\| \sum_s x_s W(b_\psi(s)) \rtimes \lambda_s \right\|_{L_{cr}^p(\mathbb{E})} = \left\| \sum_s x_s s_q(b_\psi(s)) \rtimes \lambda_s \right\|_{L_{cr}^p(\mathbb{E})}. \tag{3.30}
$$

Then we have

$$\left\|A_p^{\frac{1}{2}}(x)\right\|_{L^p(\text{VN}(G))} \overset{(3.27)}{\leqslant} Kp\left\|\partial_{\psi,1,p}(x)\right\|_{L^p(L^\infty(\Omega)\rtimes_\alpha G)}$$

$$= Kp\left\|\sum_s x_s \text{W}(b_\psi(s)) \rtimes \lambda_s\right\|_p = Kp\left\|\sum_{s,k} x_s \langle e_k, b_\psi(s)\rangle \text{W}(e_k) \rtimes \lambda_s\right\|_p$$

$$\overset{(3.8)}{\leqslant} K'p \cdot p^{\frac{1}{2}}\left\|\sum_{s,k} x_s \langle e_k, b_\psi(s)\rangle \text{W}(e_k) \rtimes \lambda_s\right\|_{L^p_{cr}(\mathbb{E})}$$

$$\overset{(3.30)}{=} K'p^{\frac{3}{2}}\left\|\sum_{s,k} x_s \langle e_k, b_\psi(s)\rangle s_q(e_k) \rtimes \lambda_s\right\|_{L^p_{cr}(\mathbb{E})}$$

$$\overset{(3.8)}{\leqslant} K'p^{\frac{3}{2}}\left\|\sum_{s,k} x_s \langle e_k, b_\psi(s)\rangle s_q(e_k) \rtimes \lambda_s\right\|_{L^p(\Gamma_q(H)\rtimes_\alpha G)}$$

$$= K'p^{\frac{3}{2}}\left\|\partial_{\psi,q,p}(x)\right\|_{L^p(\Gamma_q(H)\rtimes_\alpha G)}.$$

We pass to the converse inequality. We have

$$\left\|A_p^{\frac{1}{2}}(x)\right\|_{L^p(\text{VN}(G))} \overset{(3.27)}{\geqslant} \frac{1}{Kp^{\frac{3}{2}}}\left\|\partial_{\psi,1,p}(x)\right\|_{L^p(L^\infty(\Omega)\rtimes_\alpha G)}$$

$$= \frac{1}{Kp^{\frac{3}{2}}}\left\|\sum_s x_s \text{W}(b_\psi(s)) \rtimes \lambda_s\right\|_p = \frac{1}{Kp^{\frac{3}{2}}}\left\|\sum_{s,k} x_s \langle e_k, b_\psi(s)\rangle \text{W}(e_k) \rtimes \lambda_s\right\|_p$$

$$\overset{(3.8)}{\geqslant} \frac{1}{K'p^{\frac{3}{2}}}\left\|\sum_{s,k} x_s \langle e_k, b_\psi(s)\rangle \text{W}(e_k) \rtimes \lambda_s\right\|_{L^p_{cr}(E)}$$

$$\overset{(3.30)}{=} \frac{1}{K'p^{\frac{3}{2}}}\left\|\sum_{s,k} x_s \langle e_k, b_\psi(s)\rangle s_q(e_k) \rtimes \lambda_s\right\|_{L^p_{cr}(E)}$$

$$\overset{(3.8)}{\geqslant} \frac{1}{K''p^{\frac{3}{2}} \cdot p^{\frac{1}{2}}}\left\|\sum_{s,k} x_s \langle e_k, b_\psi(s)\rangle s_q(e_k) \rtimes \lambda_s\right\|_{L^p(\Gamma_q(H)\rtimes_\alpha G)}$$

$$= \frac{1}{K''p^2}\left\|\partial_{\psi,q,p}(x)\right\|_{L^p(\Gamma_q(H)\rtimes_\alpha G)}.$$

Altogether we have shown (3.29) in the case $2 \leqslant p < \infty$.

We turn to the case $1 < p < 2$. Note that (3.30) still holds in this case. Indeed, for elements $f \in \mathrm{Gauss}_{q,p,\rtimes}(\mathrm{L}^p(\mathrm{VN}(G)))$, the norm $\|f\|_{\mathrm{L}^p_{cr}(\mathbb{E})} = \inf\{\|g\|_{\mathrm{L}^p_c(\mathbb{E})} + \|h\|_{\mathrm{L}^p_r(\mathbb{E})} : f = g + h\}$ remains unchanged if g, h are choosen in $\mathrm{Ran}\, Q_p|_{\mathrm{L}^p_c(\mathbb{E})}$ and $\mathrm{Ran}\, Q_p|_{\mathrm{L}^p_r(\mathbb{E})}$, see Theorem 3.1. But for those elements g, h, the $\mathrm{L}^p_c(\mathbb{E})$ and $\mathrm{L}^p_r(\mathbb{E})$ norms remain unchanged upon replacing classical Gaussian variables $\mathrm{W}(e_k)$ by q-Gaussian variables $s_q(e_k)$, see the beginning of the proof. Then the proof in the case $1 < p < 2$ can be executed as in the case $2 \leqslant p < \infty$, noting that the additional factor $p^{*\frac{1}{2}}$ will appear at another step of the estimate, in accordance with the two parts of Theorem 3.1.   □

Proposition 3.4 *Let G be a discrete group. Suppose $1 < p < \infty$ and $-1 \leqslant q \leqslant 1$.*

1. *The operator $\partial_{\psi,q} \colon \mathcal{P}_G \subset \mathrm{L}^p(\mathrm{VN}(G)) \to \mathrm{L}^p(\Gamma_q(H) \rtimes_\alpha G)$ is closable as a densely defined operator on $\mathrm{L}^p(\mathrm{VN}(G))$ into $\mathrm{L}^p(\Gamma_q(H) \rtimes_\alpha G)$. We denote by $\partial_{\psi,q,p}$ its closure. So $\mathcal{P}_G$ is a core of $\partial_{\psi,q,p}$.*

2. *$\mathcal{P}_G$ is a core of $A_p^{\frac{1}{2}}$.*

3. *We have $\mathrm{dom}\, \partial_{\psi,q,p} = \mathrm{dom}\, A_p^{\frac{1}{2}}$. Moreover, for any $x \in \mathrm{dom}\, A_p^{\frac{1}{2}}$, we have*

$$\left\| A_p^{\frac{1}{2}}(x) \right\|_{\mathrm{L}^p(\mathrm{VN}(G))} \approx_p \left\| \partial_{\psi,q,p}(x) \right\|_{\mathrm{L}^p(\Gamma_q(H) \rtimes_\alpha G)}. \tag{3.31}$$

Finally, for any $x \in \mathrm{dom}\, A_p^{\frac{1}{2}}$, there exists a sequence (x_n) of elements of $\mathcal{P}_G$ such that $x_n \to x$, $A_p^{\frac{1}{2}}(x_n) \to A_p^{\frac{1}{2}}(x)$ and $\partial_{\psi,q,p}(x_n) \to \partial_{\psi,q,p}(x)$.

4. *If $x \in \mathrm{dom}\, \partial_{\psi,q,p}$, we have $x^* \in \mathrm{dom}\, \partial_{\psi,q,p}$ and*

$$(\partial_{\psi,q,p}(x))^* = -\partial_{\psi,q,p}(x^*). \tag{3.32}$$

5. *Let $M_\psi \colon \mathrm{L}^p(\mathrm{VN}(G)) \to \mathrm{L}^p(\mathrm{VN}(G))$ be a finitely supported bounded Fourier multiplier such that the map $\mathrm{Id} \rtimes M_\psi \colon \mathrm{L}^p(\Gamma_q(H) \rtimes_\alpha G) \to \mathrm{L}^p(\Gamma_q(H) \rtimes_\alpha G)$ is a well-defined bounded operator. For any $x \in \mathrm{dom}\, \partial_{\psi,q,p}$, the element $M_\psi(x)$ belongs to $\mathrm{dom}\, \partial_{\psi,q,p}$ and we have*

$$\partial_{\psi,q,p} M_\psi(x) = (\mathrm{Id} \rtimes M_\psi) \partial_{\psi,q,p}(x). \tag{3.33}$$

Proof

1. This is a consequence of [136, Theorem 5.28 p. 168] together with Lemma 3.6.

2. Since $\mathrm{dom}\, A_p$ is a core of $\mathrm{dom}\, A_p^{\frac{1}{2}}$, this is a consequence of a classical argument [180, p.29].

3. Let $x \in \mathrm{dom}\, A_p^{\frac{1}{2}}$. By the point 3, $\mathcal{P}_G$ is dense in $\mathrm{dom}\, A_p^{\frac{1}{2}}$ equipped with the graph norm. Hence we can find a sequence (x_n) of $\mathcal{P}_G$ such that $x_n \to x$ and

$A_p^{\frac{1}{2}}(x_n) \to A_p^{\frac{1}{2}}(x)$. For any integers n, m, we obtain

$$\|x_n - x_m\|_{L^p(\mathrm{VN}(G))} + \left\|\partial_{\psi,q,p}(x_n) - \partial_{\psi,q,p}(x_m)\right\|_{L^p(\Gamma_q(H)\rtimes_\alpha G)}$$

$$\overset{(3.29)}{\lesssim_p} \|x_n - x_m\|_{L^p(\mathrm{VN}(G))} + \left\|A_p^{\frac{1}{2}}(x_n) - A_p^{\frac{1}{2}}(x_m)\right\|_{L^p(\mathrm{VN}(G))}.$$

which shows that (x_n) is a Cauchy sequence in $\mathrm{dom}\,\partial_{\psi,q,p}$. By the closedness of $\partial_{\psi,q,p}$, we infer that this sequence converges to some $x' \in \mathrm{dom}\,\partial_{\psi,q,p}$ equipped with the graph norm. Since $\mathrm{dom}\,\partial_{\psi,q,p}$ is continuously embedded into $L^p(\mathrm{VN}(G))$, we have $x_n \to x'$ in $L^p(\mathrm{VN}(G))$, and therefore $x = x'$ since $x_n \to x$. It follows that $x \in \mathrm{dom}\,\partial_{\psi,q,p}$. This proves the inclusion $\mathrm{dom}\,A_p^{\frac{1}{2}} \subset \mathrm{dom}\,\partial_{\psi,q,p}$. Moreover, for any integer n, we have

$$\left\|\partial_{\psi,q,p}(x_n)\right\|_{L^p(\Gamma_q(H)\rtimes_\alpha G)} \overset{(3.29)}{\lesssim_p} \left\|A_p^{\frac{1}{2}}(x_n)\right\|_{L^p(\mathrm{VN}(G))}.$$

Since $x_n \to x$ in $\mathrm{dom}\,\partial_{\psi,q,p}$ and in $\mathrm{dom}\,A_p^{\frac{1}{2}}$ both equipped with the graph norm, we conclude that

$$\left\|\partial_{\psi,q,p}(x)\right\|_{L^p(\Gamma_q(H)\rtimes_\alpha G)} \lesssim_p \left\|A_p^{\frac{1}{2}}(x)\right\|_{L^p(\mathrm{VN}(G))}.$$

The proof of the reverse inclusion and of the reverse estimate are similar. Indeed, by part 1, $\mathcal{P}_G$ is a dense subspace of $\mathrm{dom}\,\partial_{\psi,q,p}$ equipped with the graph norm.

4. Recall that $b_\psi(e) = 0$. For any $s \in G$, we have

$$\left(\partial_{\psi,q,p}(\lambda_s)\right)^* \overset{(2.84)}{=} \left(s_q(b_\psi(s)) \rtimes \lambda_s\right)^* \overset{(2.57)}{=} \alpha_{s^{-1}}(s_q(b_\psi(s))) \rtimes \lambda_{s^{-1}}$$

$$\overset{(2.33)}{=} s_q(b_\psi(e) - b_\psi(s^{-1})) \rtimes \lambda_{s^{-1}}$$

$$= -s_q(b_\psi(s^{-1})) \rtimes \lambda_{s^{-1}} \overset{(2.84)}{=} -\partial_{\psi,q,p}(\lambda_{s^{-1}}) = -\partial_{\psi,q,p}(\lambda_s^*).$$

Let $x \in \mathrm{dom}\,\partial_{\psi,q,p}$. By the point 1, $\mathcal{P}_G$ is core of $\partial_{\psi,q,p}$. Hence there exists a sequence (x_n) of $\mathcal{P}_G$ such that $x_n \to x$ and $\partial_{\psi,q,p}(x_n) \to \partial_{\psi,q,p}(x)$. We have $x_n^* \to x^*$ and by the first part of the proof $\partial_{\psi,q,p}(x_n^*) = -(\partial_{\psi,q,p}(x_n))^* \to -(\partial_{\psi,q,p}(x))^*$. By (2.7), we conclude that $x^* \in \mathrm{dom}\,\partial_{\psi,q,p}$ and that $\partial_{\psi,q,p}(x^*) = -(\partial_{\psi,q,p}(x))^*$.

5. If $s \in G$, we have

$$(\mathrm{Id} \rtimes M_\psi)\partial_{\psi,q,p}(\lambda_s) \overset{(2.84)}{=} (\mathrm{Id} \rtimes M_\psi)(s_q(b_\psi(s)) \rtimes \lambda_s) = \psi_s s_q(b_\psi(s)) \rtimes \lambda_s$$

$$\overset{(2.84)}{=} \psi_s \partial_{\psi,q,p}(\lambda_s) = \partial_{\psi,q,p} M_\psi(\lambda_s).$$

By linearity, (3.33) is true for elements of $\mathcal{P}_G$. Let $x \in$ dom $\partial_{\psi,q,p}$. There exists a sequence (x_n) of $\mathcal{P}_G$ such that $x_n \to x$ and $\partial_{\psi,q,p}(x_n) \to \partial_{\psi,q,p}(x)$. We have $M_\psi(x_n) \to M_\psi(x)$ and

$$\partial_{\psi,q,p}M_\psi(x_n) = (\mathrm{Id} \rtimes M_\psi)\partial_{\psi,q,p}(x_n) \xrightarrow[n\to+\infty]{} (\mathrm{Id} \rtimes M_\psi)\partial_{\psi,q,p}(x).$$

By (2.1), we deduce (3.33). □

For the property AP in the next proposition, we refer to the preliminary Sect. 2.4.

Proposition 3.5 *Assume* $-1 \leqslant q < 1$. *Let* G *be a discrete group with* AP. *The operator* $\partial_{\psi,q} \colon \mathcal{P}_G \subset \mathrm{VN}(G) \to \Gamma_q(H) \rtimes_\alpha G$ *is weak* closable.*[6] *We denote by* $\partial_{\psi,q,\infty}$ *its weak* closure.*

Proof Suppose that (x_i) is a net of $\mathcal{P}_G$ which converges to 0 for the weak* topology with $x_i = \sum_{s\in G} x_{i,s}\lambda_s$ such that the net $(\partial_{\psi,q}(x_i))$ converges for the weak* topology to some y belonging to $\Gamma_q(H) \rtimes_\alpha G$. Let (M_{φ_j}) be the net of Fourier multipliers approximating the identity from Proposition 2.9. For any j, we have

$$\partial_{\psi,q}(M_{\varphi_j}x_i) = \partial_{\psi,q}\left(M_{\varphi_j}\left(\sum_{s\in G}x_{i,s}\lambda_s\right)\right) = \partial_{\psi,q}\left(\sum_{s\in\mathrm{supp}\,\varphi_j}\varphi_j(s)x_{i,s}\lambda_s\right)$$

$$= \sum_{s\in\mathrm{supp}\,\varphi_j}\varphi_j(s)x_{i,s}\partial_{\psi,q}(\lambda_s) = (\mathrm{Id} \rtimes M_{\varphi_j})\partial_{\psi,q}\left(\sum_{s\in G}x_{i,s}\lambda_s\right)$$

$$= (\mathrm{Id} \rtimes M_{\varphi_j})\partial_{\psi,q}(x_i) \xrightarrow[i]{} (\mathrm{Id} \rtimes M_{\varphi_j})(y).$$

On the other hand, for all $s \in$ supp φ_j we have $x_{i,s} \to 0$ as $i \to \infty$. Hence

$$\partial_{\psi,q}(M_{\varphi_j}x_i) = \partial_{\psi,q}\left(\sum_{s\in\mathrm{supp}(\varphi_j)}\varphi_j(s)x_{i,s}\lambda_s\right)$$

$$\overset{(2.84)}{=} \sum_{s\in\mathrm{supp}(\varphi_j)}\varphi_j(s)x_{i,s}s_q(b_\psi(s)) \rtimes \lambda_s \xrightarrow[i]{} 0.$$

This implies by uniqueness of the limit that $(\mathrm{Id} \rtimes M_{\varphi_j})(y) = 0$ for any j. By the point 3 of Proposition 2.9, we deduce that $y = \mathrm{w}^*\text{-}\lim_j(\mathrm{Id}_{\Gamma_q(H)} \rtimes M_{\varphi_j})(y) = 0$. □

---

[6] That is, if (x_n) is a sequence in $\mathcal{P}_G$ such that $x_n \to 0$ and $\partial_{\psi,q}(x_n) \to y$ for some $y \in \Gamma_q(H) \rtimes_\alpha G$ both for the weak* topology, then $y = 0$.

Remark 3.1 We do not know if the assumption "AP" in Proposition 3.5 is really necessary.

The following generalizes an observation of [131] (in the case $q = 1$).

Lemma 3.7 *Let* $-1 \leqslant q \leqslant 1$. *Suppose* $1 < p < \infty$. *For any* $x \in \mathrm{dom}\, A_p^{\frac{1}{2}}$ *and any* $y \in \mathrm{dom}\, A_{p*}^{\frac{1}{2}}$, *we have*

$$\tau_{\Gamma_q(H) \rtimes_\alpha G}\big(\partial_{\psi,q,p}(x)(\partial_{\psi,q,p*}(y))^*\big) = \tau_G\Big(A_p^{\frac{1}{2}}(x)\big(A_{p*}^{\frac{1}{2}}(y)\big)^*\Big). \tag{3.34}$$

Proof By the fact that $\mathcal{P}_G$ is a core for $\partial_{\psi,q,p}$, $\partial_{\psi,q,p*}$, $A_p^{\frac{1}{2}}$ and A_{p*} and the norm equivalence (3.31), we can assume $x, y \in \mathcal{P}_G$. Consider some elements $x = \sum_{s \in G} x_s \lambda_s$ and $y = \sum_{r \in G} y_r \lambda_r$ of $\mathcal{P}_G$ where both sums are finite. On the one hand, we have

$$\tau_{\Gamma_q(H) \rtimes_\alpha G}\big(\partial_{\psi,q}(x)(\partial_{\psi,q}(y))^*\big) = \sum_{s,r \in G} x_s \overline{y_r} \tau_{\Gamma_q(H) \rtimes_\alpha G}\big(\partial_{\psi,q}(\lambda_s)(\partial_{\psi,q}(\lambda_r))^*\big)$$

$$\overset{(2.84)}{=} \sum_{s,r \in G} x_s \overline{y_r} \tau_{\Gamma_q(H) \rtimes_\alpha G}\big((s_q(b_\psi(s)) \rtimes \lambda_s)(s_q(b_\psi(r)) \rtimes \lambda_r)^*\big)$$

$$\overset{(2.57)}{=} \sum_{s,r \in G} x_s \overline{y_r} \tau_{\Gamma_q(H) \rtimes_\alpha G}\big((s_q(b_\psi(s)) \rtimes \lambda_s)(\alpha_{r-1}(s_q(b_\psi(r))) \rtimes \lambda_{r-1})\big)$$

$$\overset{(2.56)}{=} \sum_{s,r \in G} x_s \overline{y_r} \tau_{\Gamma_q(H) \rtimes_\alpha G}\big((s_q(b_\psi(s))\alpha_{sr-1}(s_q(b_\psi(r)))) \rtimes \lambda_{sr-1}\big)$$

$$= \sum_{s \in G} x_s \overline{y_s} \tau_{\Gamma_q(H)}\big(s_q(b_\psi(s))s_q(b_\psi(s))\big) \overset{(2.41)}{=} \sum_{s \in G} x_s \overline{y_s} \|b_\psi(s)\|_H^2.$$

On the other hand, we have

$$\tau_G\Big(A^{\frac{1}{2}}(x)\big(A^{\frac{1}{2}}(y)\big)^*\Big) = \sum_{s,r \in G} x_s \overline{y_r} \tau_G\Big(A^{\frac{1}{2}}(\lambda_s)\big(A^{\frac{1}{2}}(\lambda_r)\big)^*\Big)$$

$$= \sum_{s,r \in G} x_s \overline{y_r} \|b_\psi(s)\|_H \|b_\psi(r)\|_H \tau_G(\lambda_s \lambda_r^*) = \sum_{s \in G} x_s \overline{y_s} \|b_\psi(s)\|_H^2.$$

$\square$

3.3   Extension of the Carré du Champ Γ for Fourier Multipliers

In this section, we consider again a markovian semigroup $(T_t)_{t \geqslant 0}$ of Fourier multipliers acting on the von Neumann algebra VN(G) where G is a discrete group. We shall extend the carré du champ Γ associated with $(T_t)_{t \geqslant 0}$ to a closed form and identify its domain. It will be easier to consider simultaneously $x \mapsto \Gamma(x, x)$ and $x \mapsto \Gamma(x^*, x^*)$ and the case $2 \leqslant p < \infty$ throughout the section. We will also link the carré du champ with $A_p^{\frac{1}{2}}$ and the gradient $\partial_{\psi, q, p}$. In most of the results of this section, we need approximation properties of $L^p(\text{VN}(G))$ and $L^p(\Gamma_q(H) \rtimes_\alpha G)$. Note that it is stated in [121, Theorem 1.2] that if G is a discrete group with AP such that VN(G) has QWEP, then $L^p(\text{VN}(G))$ has the completely contractive approximation property CCAP for any $1 < p < \infty$. With the following lemma, one can extend the definition of $\Gamma(x, y)$ to a larger domain.

Lemma 3.8 *Suppose* $2 \leqslant p < \infty$. *Let* G *be a discrete group. The forms* $a: \mathcal{P}_G \times \mathcal{P}_G \to L^{\frac{p}{2}}(\text{VN}(G)) \oplus_\infty L^{\frac{p}{2}}(\text{VN}(G))$, $(x, y) \mapsto \Gamma(x, y) \oplus \Gamma(x^*, y^*)$ *and* $\Gamma: \mathcal{P}_G \times \mathcal{P}_G \to L^{\frac{p}{2}}(\text{VN}(G))$, $(x, y) \mapsto \Gamma(x, y)$ *are symmetric, positive and satisfy the Cauchy-Schwarz inequality. The first is in addition closable and the domain of the closure* $\overline{a}$ *is* dom $A_p^{\frac{1}{2}}$.

Proof According to the point 3 of Lemma 2.15, we have $\Gamma(x, y)^* = \Gamma(y, x)$, so a is symmetric. Moreover, again according to Lemma 2.15, a is positive. For any $x, y \in \mathcal{P}_G$, we have

$$\|a(x, y)\|_{L^{\frac{p}{2}}(\text{VN}(G)) \oplus_\infty L^{\frac{p}{2}}(\text{VN}(G))} = \left\| \Gamma(x, y) \oplus \Gamma(x^*, y^*) \right\|_{L^{\frac{p}{2}}(\text{VN}(G)) \oplus_\infty L^{\frac{p}{2}}(\text{VN}(G))}$$

$$= \max \left\{ \|\Gamma(x, y)\|_{\frac{p}{2}} , \|\Gamma(x^*, y^*)\|_{\frac{p}{2}} \right\}$$

$$\overset{(2.81)}{\leqslant} \max \left\{ \|\Gamma(x, x)\|_{\frac{p}{2}}^{\frac{1}{2}} \|\Gamma(y, y)\|_{\frac{p}{2}}^{\frac{1}{2}} , \|\Gamma(x^*, x^*)\|_{\frac{p}{2}}^{\frac{1}{2}} \|\Gamma(y^*, y^*)\|_{\frac{p}{2}}^{\frac{1}{2}} \right\}$$

$$\leqslant \max \left\{ \|\Gamma(x, x)\|_{\frac{p}{2}}^{\frac{1}{2}} , \|\Gamma(x^*, x^*)\|_{\frac{p}{2}}^{\frac{1}{2}} \right\} \max \left\{ \|\Gamma(y, y)\|_{\frac{p}{2}}^{\frac{1}{2}} , \|\Gamma(y^*, y^*)\|_{\frac{p}{2}}^{\frac{1}{2}} \right\}$$

$$= \|a(x, x)\|_{L^{\frac{p}{2}}(\text{VN}(G)) \oplus_\infty L^{\frac{p}{2}}(\text{VN}(G))}^{\frac{1}{2}} \|a(y, y)\|_{L^{\frac{p}{2}}(\text{VN}(G)) \oplus_\infty L^{\frac{p}{2}}(\text{VN}(G))}^{\frac{1}{2}} .$$

So a satisfies the Cauchy-Schwarz inequality. The assertions concerning Γ are similar. Suppose $x_n \xrightarrow{a} 0$ that is $x_n \in \mathcal{P}_G$, $x_n \to 0$ and $a(x_n - x_m, x_n - x_m) \to 0$. For any integer n, m, we have

$$\left\| A_p^{\frac{1}{2}}(x_n - x_m) \right\|_{L^p(\mathrm{VN}(G))}$$

$$\overset{(2.90)}{\lesssim_p} \max\left\{ \left\| \Gamma(x_n - x_m, x - n - x_m)^{\frac{1}{2}} \right\|_{L^p}, \left\| \Gamma((x_n - x_m)^*, (x_n - x_m)^*)^{\frac{1}{2}} \right\|_{L^p} \right\}$$

$$= \left\| a(x_n - x_m, x_n - x_m) \right\|_{L^{\frac{p}{2}}(\mathrm{VN}(G)) \oplus_\infty L^{\frac{p}{2}}(\mathrm{VN}(G))} \to 0.$$

We infer that $\left(A_p^{\frac{1}{2}}(x_n) \right)$ is a Cauchy sequence, hence a convergent sequence. Since $x_n \to 0$, by the closedness of $A_p^{\frac{1}{2}}$, we deduce that $A_p^{\frac{1}{2}}(x_n) \to 0$. Now, we have

$$\left\| a(x_n, x_n) \right\|_{L^{\frac{p}{2}}(\mathrm{VN}(G)) \oplus_\infty L^{\frac{p}{2}}(\mathrm{VN}(G))}$$

$$= \max\left\{ \left\| \Gamma(x_n, x_n)^{\frac{1}{2}} \right\|_{L^p(\mathrm{VN}(G))}, \left\| \Gamma(x_n^*, x_n^*))^{\frac{1}{2}} \right\|_{L^p(\mathrm{VN}(G))} \right\}$$

$$\overset{(2.90)}{\lesssim_p} \left\| A_p^{\frac{1}{2}}(x_n) \right\|_{L^p(\mathrm{VN}(G))} \to 0.$$

By Proposition 2.7, we obtain the closability of the form a.

Let $x \in L^p(\mathrm{VN}(G))$. By (2.52), we have $x \in \operatorname{dom} \overline{a}$ if and only if there exists a sequence (x_n) of $\mathcal{P}_G$ satisfying $x_n \xrightarrow{a} x$, that is satisfying $x_n \to x$ and $\Gamma(x_n - x_m, x_n - x_m) \to 0$, $\Gamma((x_n - x_m)^*, (x_n - x_m)^*) \to 0$ as $n, m \to \infty$. By (2.90), this is equivalent to the existence of a sequence $x_n \in \mathcal{P}_G$, such that $x_n \to x$, $A_p^{\frac{1}{2}}(x_n - x_m) \to 0$ as $n, m \to \infty$. Now recalling that $\mathcal{P}_G$ is a core of the operator $A_p^{\frac{1}{2}}$, we conclude that this is equivalent to $x \in \operatorname{dom} A_p^{\frac{1}{2}}$. $\square$

Remark 3.2 The closability and biggest reasonable domain of the form Γ are unclear.

If $2 \leqslant p < \infty$, and $x, y \in \operatorname{dom} A_p^{\frac{1}{2}}$, then we let $\Gamma(x, y)$ be the first component of $\overline{a}(x, y)$, where $a : \mathcal{P}_G \times \mathcal{P}_G \to L^{\frac{p}{2}}(\mathrm{VN}(G)) \oplus_\infty L^{\frac{p}{2}}(\mathrm{VN}(G))$, $(u, v) \mapsto \Gamma(u, v) \oplus \Gamma(u^*, v^*)$ is the form in Lemma 3.8.

Lemma 3.9 *Suppose* $2 \leqslant p < \infty$. *Let* G *be a discrete group and assume that* $L^p(\mathrm{VN}(G))$ *has CCAP and that* $\mathrm{VN}(G)$ *has QWEP. Let* (φ_j) *be a net of functions* $\varphi_j : G \to \mathbb{C}$ *with finite support such that the net* (M_{φ_j}) *converges to*

$\mathrm{Id}_{\mathrm{L}^p(\mathrm{VN}(G))}$ *in the point-norm topology with* $\sup_j \left\| M_{\varphi_j} \right\|_{\mathrm{cb},\mathrm{L}^p(\mathrm{VN}(G)) \to \mathrm{L}^p(\mathrm{VN}(G))} \leqslant 1.$
If $x \in \mathrm{dom}\, A_p^{\frac{1}{2}}$ *then for any* j

$$\left\| \Gamma\big(M_{\varphi_j}(x), M_{\varphi_j}(x)\big) \right\|_{\mathrm{L}^{\frac{p}{2}}(\mathrm{VN}(G))}^{\frac{1}{2}} \leqslant \left\| \partial_{\psi,q,p}(x) \right\|_{\mathrm{L}^p(\mathrm{VN}(G), \mathrm{L}^2(\Gamma_q(H))_{c,p})}.$$

Proof Let $x \in \mathrm{dom}\, A_p^{\frac{1}{2}} = \mathrm{dom}\, \partial_{\psi,q,p}$. Since $\mathcal{P}_G$ is a core of $\partial_{\psi,q,p}$ by Proposition 3.4, there exists a sequence (x_n) of elements of $\mathcal{P}_G$ such that $x_n \to x$ and $\partial_{\psi,q,p}(x_n) \to \partial_{\psi,q,p}(x)$. Note that $M_{\varphi_j} \colon \mathrm{L}^p(\mathrm{VN}(G)) \to \mathrm{L}^p(\mathrm{VN}(G))$ is a complete contraction. The linear map $M_{\varphi_j} \otimes \mathrm{Id}_{\mathrm{L}^2(\Gamma_q(H))} \colon \mathrm{L}^p(\mathrm{VN}(G), \mathrm{L}^2(\Gamma_q(H))_{c,p}) \to \mathrm{L}^p(\mathrm{VN}(G), \mathrm{L}^2(\Gamma_q(H))_{c,p})$ is also a contraction according to (2.70). We deduce that

$$\partial_{\psi,q,p} M_{\varphi_j}(x_n)$$
$$= \sum_{s \in \mathrm{supp}\, \varphi_j} \varphi_j(s) x_{ns} \partial_{\psi,q,p}(\lambda_s)$$
$$= \big(M_{\varphi_j} \otimes \mathrm{Id}_{\mathrm{L}^2(\Gamma_q(H))}\big) \bigg(\sum_{s \in G} x_{ns} \partial_{\psi,q,p}(\lambda_s) \bigg)$$
$$= \big(M_{\varphi_j} \otimes \mathrm{Id}_{\mathrm{L}^2(\Gamma_q(H))}\big)(\partial_{\psi,q,p} x_n) \xrightarrow[n \to +\infty]{} \big(M_{\varphi_j} \otimes \mathrm{Id}_{\mathrm{L}^2(\Gamma_q(H))}\big)(\partial_{\psi,q,p} x).$$

Since $\partial_{\psi,q,p} M_{\varphi_j}$ is bounded by Kato [136, Problem 5.22], we deduce that $\partial_{\psi,q,p} M_{\varphi_j}(x) = \big(M_{\varphi_j} \otimes \mathrm{Id}_{\mathrm{L}^2(\Omega)}\big)(\partial_{\psi,q,p} x)$. Now, we have

$$\left\| \Gamma(M_{\varphi_j} x, M_{\varphi_j} x) \right\|_{\mathrm{L}^{\frac{p}{2}}}^{\frac{1}{2}} = \left\| \Gamma(M_{\varphi_j} x, M_{\varphi_j} x)^{\frac{1}{2}} \right\|_{\mathrm{L}^p}$$
$$\overset{(2.89)}{=} \left\| \partial_{\psi,q,p} M_{\varphi_j}(x) \right\|_{\mathrm{L}^p(\mathrm{VN}(G), \mathrm{L}^2(\Gamma_q(H))_{c,p})}$$
$$= \left\| \big(M_{\varphi_j} \otimes \mathrm{Id}_{\mathrm{L}^2(\Gamma_q(H))}\big)(\partial_{\psi,q,p} x) \right\|_{\mathrm{L}^p(\mathrm{VN}(G), \mathrm{L}^2(\Gamma_q(H))_{c,p})}$$
$$\leqslant \left\| \partial_{\psi,q,p}(x) \right\|_{\mathrm{L}^p(\mathrm{VN}(G), \mathrm{L}^2(\Gamma_q(H))_{c,p})}.$$

$\square$

Now, we give a very concrete way to approximate the carré du champ for a large class of groups.

Lemma 3.10 *Let* $2 \leqslant p < \infty$. *Assume that* $\mathrm{L}^p(\mathrm{VN}(G))$ *has CCAP and that* $\mathrm{VN}(G)$ *has QWEP. Let* (φ_j) *be a net of functions* $\varphi_j \colon G \to \mathbb{C}$ *with finite support such that the net* (M_{φ_j}) *converges to* $\mathrm{Id}_{\mathrm{L}^p(\mathrm{VN}(G))}$ *in the point-norm topology with*

$\sup_j \|M_{\varphi_j}\|_{\mathrm{cb}, L^p(\mathrm{VN}(G)) \to L^p(\mathrm{VN}(G))} \leqslant 1$. *For any* $x, y \in \mathrm{dom}\, A_p^{\frac{1}{2}}$, *we have in* $L^{\frac{p}{2}}(\mathrm{VN}(G))$

$$\Gamma(x, y) = \lim_j \Gamma\big(M_{\varphi_j}(x), M_{\varphi_j}(y)\big). \tag{3.35}$$

Proof If $x \in \mathrm{dom}\, A_p^{\frac{1}{2}}$, we have for any j, k

$$\big\|\Gamma(M_{\varphi_j}x - M_{\varphi_k}x, M_{\varphi_j}x - M_{\varphi_k}x)\big\|_{L^{\frac{p}{2}}(\mathrm{VN}(G))}^{\frac{1}{2}}$$

$$= \big\|\Gamma(M_{\varphi_j}x - M_{\varphi_k}x, M_{\varphi_j}x - M_{\varphi_k}x)^{\frac{1}{2}}\big\|_{L^p(\mathrm{VN}(G))}$$

$$\overset{(2.90)}{\lesssim_p} \big\|A^{\frac{1}{2}}(M_{\varphi_j}x - M_{\varphi_k}x)\big\|_{L^p(\mathrm{VN}(G))} = \big\|M_{\varphi_j}A_p^{\frac{1}{2}}x - M_{\varphi_k}A_p^{\frac{1}{2}}x\big\|_{L^p(\mathrm{VN}(G))}.$$

The same inequality holds with $\Gamma((M_{\varphi_j}x - M_{\varphi_k}x)^*, (M_{\varphi_j}x - M_{\varphi_k}x)^*)$ on the left hand side. Note that since $A_p^{\frac{1}{2}}(x)$ belongs to $L^p(\mathrm{VN}(G))$, $(M_{\varphi_j}A_p^{\frac{1}{2}}(x))_p$ is a Cauchy net of $L^p(\mathrm{VN}(G))$. Since $M_{\varphi_j}(x) \to x$, we infer that $M_{\varphi_j}(x) \overset{a}{\to} x$, where a is the first form from Lemma 3.8. Then (3.35) is a consequence of Lemma 3.9 and Proposition 2.6. □

Lemma 3.11 *Suppose* $2 \leqslant p < \infty$ *and* $-1 \leqslant q \leqslant 1$.

1. For any $x, y \in \mathrm{dom}\, A_p^{\frac{1}{2}} = \mathrm{dom}\, \partial_{\psi,q,p}$, *we have*

$$\Gamma(x, y) = \big\langle \partial_{\psi,q,p}(x), \partial_{\psi,q,p}(y)\big\rangle_{L^p(\mathrm{VN}(G), L^2(\Gamma_q(H))_{c,p})} \tag{3.36}$$

$$= \mathbb{E}\big((\partial_{\psi,q,p}(x))^* \partial_{\psi,q,p}(y)\big).$$

2. For any $x \in \mathrm{dom}\, A_p^{\frac{1}{2}}$, *we have*

$$\big\|\Gamma(x, x)^{\frac{1}{2}}\big\|_{L^p(\mathrm{VN}(G))} = \big\|\partial_{\psi,q,p}(x)\big\|_{L^p(\mathrm{VN}(G), L^2(\Gamma_q(H))_{c,p})}. \tag{3.37}$$

Proof

1. Consider some elements x, y of $\mathrm{dom}\, A_p^{\frac{1}{2}} = \mathrm{dom}\, \partial_{\psi,q,p}$. According to the point 2 of Proposition 3.4, $\mathcal{P}_G$ is a core of $\mathrm{dom}\, \partial_{\psi,q,p}$. So by (2.3) there exist sequences (x_n) and (y_n) of $\mathcal{P}_G$ such that $x_n \to x$, $y_n \to y$, $\partial_{\psi,q,p}(x_n) \to \partial_{\psi,q,p}(x)$ and $\partial_{\psi,q,p}(y_n) \to \partial_{\psi,q,p}(y)$. By Lemma 2.8, we have, since $p \geqslant 2$, a contractive inclusion

$$L^p(\Gamma_q(H) \rtimes_\alpha G) \hookrightarrow L_c^p(\mathbb{B}) \overset{(2.76)}{=} L^p(\mathrm{VN}(G), L^2(\Gamma_q(H))_{c,p}).$$

We deduce that $\partial_{\psi,q,p}(x_n) \to \partial_{\psi,q,p}(x)$ and $\partial_{\psi,q,p}(y_n) \to \partial_{\psi,q,p}(y)$ in the space $L^p(\mathrm{VN}(G), L^2(\Gamma_q(H))_{c,p})$. We obtain

$$\|\Gamma(x_n - x_m, x_n - x_m)\|_{L^{\frac{p}{2}}(\mathrm{VN}(G))}$$

$$\overset{(2.89)}{=} \left\|\langle\partial_{\psi,q,p}(x_n - x_m), \partial_{\psi,q,p}(x_n - x_m)\rangle_{L^p(\mathrm{VN}(G), L^2(\Gamma_q(H)))_{c,p}}\right\|_{L^{\frac{p}{2}}(\mathrm{VN}(G))}$$

$$\leqslant \left\|\partial_{\psi,q,p}(x_n - x_m)\right\|_{L^p(L^2(\Gamma_q(H))_{c,p})} \left\|\partial_{\psi,q,p}(x_n - x_m)\right\|_{L^p(L^2(\Gamma_q(H))_{c,p})}$$

$$\xrightarrow[n,m\to+\infty]{} 0$$

and similarly for x_n^*, y_n and y_n^*. We deduce that

$$\Gamma(x, y) \overset{(2.53)}{=} \lim_{n\to+\infty} \Gamma(x_n, y_n)$$

$$\overset{(2.88)}{=} \lim_{n\to+\infty} \langle\partial_{\psi,q,p}(x_n), \partial_{\psi,q,p}(y_n)\rangle_{L^p(\mathrm{VN}(G), L^2(\Gamma_q(H))_{c,p})}$$

$$= \langle\partial_{\psi,q,p}(x), \partial_{\psi,q,p}(y)\rangle_{L^p(\mathrm{VN}(G), L^2(\Gamma_q(H))_{c,p})}.$$

2. If $x \in \mathrm{dom}\, A_p^{\frac{1}{2}}$, we have

$$\left\|\partial_{\psi,q,p}(x)\right\|_{L^p(\mathrm{VN}(G), L^2(\Gamma_q(H))_{c,p})}$$

$$\overset{(2.66)}{=} \left\|\langle\partial_{\psi,q,p}(x), \partial_{\psi,q,p}(x)\rangle_{L^p(\mathrm{VN}(G), L^2(\Gamma_q(H))_{c,p})}^{\frac{1}{2}}\right\|_{L^p(\mathrm{VN}(G))}$$

$$\overset{(3.36)}{=} \left\|\Gamma(x, x)^{\frac{1}{2}}\right\|_{L^p(\mathrm{VN}(G))}.$$

$\square$

Now, we can extend Proposition 2.12.

Theorem 3.2 *Suppose that $L^p(\mathrm{VN}(G))$ has CCAP and that $\mathrm{VN}(G)$ has QWEP. Let $2 \leqslant p < \infty$. For any $x \in \mathrm{dom}\, A_p^{\frac{1}{2}}$, we have*

$$\left\|A_p^{\frac{1}{2}}(x)\right\|_{L^p(\mathrm{VN}(G))} \approx_p \max\left\{\left\|\Gamma(x, x)^{\frac{1}{2}}\right\|_{L^p(\mathrm{VN}(G))}, \left\|\Gamma(x^*, x^*)^{\frac{1}{2}}\right\|_{L^p(\mathrm{VN}(G))}\right\}. \tag{3.38}$$

Proof Pick any $-1 \leqslant q \leqslant 1$. For any $x \in \mathrm{dom}\, A_p^{\frac{1}{2}}$, first note that

$$\mathbb{E}\big(\partial_{\psi,q,p}(x)(\partial_{\psi,q,p}(x)^*)\big) \overset{(3.32)}{=} \mathbb{E}\big(\partial_{\psi,q,p}(x^*)^*(\partial_{\psi,q,p}(x^*))\big) \overset{(3.36)}{=} \Gamma(x^*, x^*). \tag{3.39}$$

Using Proposition 3.1, in the second equivalence, we conclude that

$$\left\| A_p^{\frac{1}{2}}(x) \right\|_{L^p(\mathrm{VN}(G))} \overset{(3.31)}{\approx_p} \left\| \partial_{\psi,q,p}(x) \right\|_{L^p(\Gamma_q(H) \rtimes_\alpha G)}$$

$$\approx_p \max \left\{ \left\| \left(\mathbb{E}((\partial_{\psi,q,p}(x))^* \partial_{\psi,q,p}(x)) \right)^{\frac{1}{2}} \right\|_p, \left\| \left(\mathbb{E}(\partial_{\psi,q,p}(x)(\partial_{\psi,q,p}(x)^*)) \right)^{\frac{1}{2}} \right\|_p \right\}$$

$$\overset{(3.36)(3.39)}{=} \max \left\{ \left\| \Gamma(x,x)^{\frac{1}{2}} \right\|_{L^p(\mathrm{VN}(G))}, \left\| \Gamma(x^*,x^*)^{\frac{1}{2}} \right\|_{L^p(\mathrm{VN}(G))} \right\}.$$

$\square$

3.4 L^p-Kato's Square Root Problem for Semigroups of Schur Multipliers

In this section, we shall consider the Kato square root problem for markovian semigroups of Schur multipliers. Thus, we fix for the whole section such a markovian semigroup from Proposition 2.4, with its generator A_p on S_I^p and also the gradient type operator $\partial_{\alpha,q}$ from (2.95). An L^p-variant of Kato's square root problem is then the question whether $A_p^{\frac{1}{2}}(x)$ and $\partial_{\alpha,q}(x)$ are comparable for the L^p norms. The main results in this section, answering affirmatively to this question, are then Theorem 3.3 (case of classical Gaussians) and Proposition 3.10 (case of q-deformed Gaussians), together with Lemma 3.19 on a Khintchine type equivalence in some Gaussian subspace of $L^p(\Gamma_q(H) \overline{\otimes} B(\ell_I^2))$. Finally, the problem of exact description of the domain of the closure of the gradients is investigated in Proposition 3.11. Throughout this section, if H is a Hilbert space, we denote by H_{disc} the abelian group $(H, +)$ equipped with the *discrete* topology. We will use the trace preserving normal unital injective $*$-homomorphism map

$$J: \mathrm{VN}(H_{\mathrm{disc}}) \to L^\infty(\Omega) \overline{\otimes} \mathrm{VN}(H_{\mathrm{disc}}), \quad \lambda_h \mapsto 1 \otimes \lambda_h \qquad (3.40)$$

and the unbounded Fourier multiplier

$$(-\Delta)^{-\frac{1}{2}}: \mathcal{P}_{H_{\mathrm{disc}}} \subset L^p(\mathrm{VN}(H_{\mathrm{disc}})) \to L^p(\mathrm{VN}(H_{\mathrm{disc}}))$$

defined by $(-\Delta)^{-\frac{1}{2}}(\lambda_0) = 0$ and for $h \neq 0$,

$$(-\Delta)^{-\frac{1}{2}}(\lambda_h) \overset{\mathrm{def}}{=} \frac{1}{2\pi \|h\|_H} \lambda_h. \qquad (3.41)$$

The following is inspired by [15].

Proposition 3.6 *Let H be a Hilbert space. There exists a unique weak* continuous group $(U_t)_{t \in \mathbb{R}}$ of *-automorphisms of $\mathrm{L}^\infty(\Omega) \overline{\otimes} \mathrm{VN}(H_{\mathrm{disc}})$ such that*

$$U_t(f \otimes \lambda_h) = \mathrm{e}^{\sqrt{2} \mathrm{i} t\, \mathrm{W}(h)} f \otimes \lambda_h, \qquad t \in \mathbb{R},\ f \in \mathrm{L}^\infty(\Omega),\ h \in H. \tag{3.42}$$

Moreover, each U_t is trace preserving.

Proof For any $t \in \mathbb{R}$, we consider the (continuous) function $u_t \colon H_{\mathrm{disc}} \to \mathrm{U}(\mathrm{L}^\infty(\Omega))$, $h \mapsto \mathrm{e}^{-\sqrt{2}\mathrm{i} t\, \mathrm{W}(h)}$. For any $t \in \mathbb{R}$ and any $h_1, h_2 \in H_{\mathrm{disc}}$, note that

$$u_t(h_1 + h_2) = \mathrm{e}^{-\sqrt{2}\mathrm{i} t\, \mathrm{W}(h_1 + h_2)} = \mathrm{e}^{-\sqrt{2}\mathrm{i} t\, \mathrm{W}(h_1)}\mathrm{e}^{-\sqrt{2}\mathrm{i} t\, \mathrm{W}(h_2)} = u_t(h_1)u_t(h_2).$$

By Arhancet [13, Proposition 2.3] applied with $M = \mathrm{L}^\infty(\Omega)$, $G = H_{\mathrm{disc}}$ and by considering the trivial action α, for any $t \in \mathbb{R}$, we have a unitary $V_t \colon \mathrm{L}^2(H_{\mathrm{disc}}, \mathrm{L}^2(\Omega)) \to \mathrm{L}^2(H_{\mathrm{disc}}, \mathrm{L}^2(\Omega))$, $\xi \mapsto (h \mapsto u_t(-h)(\xi(h)))$ and a *-isomorphism

$$U_t \colon \mathrm{L}^\infty(\Omega) \overline{\otimes} \mathrm{VN}(H_{\mathrm{disc}}) \to \mathrm{L}^\infty(\Omega) \overline{\otimes} \mathrm{VN}(H_{\mathrm{disc}}), \quad x \mapsto V_t x V_t^*$$

satisfying (3.42). The uniqueness is clear by density.

For any $t, t' \in \mathbb{R}$ any $\xi \in \mathrm{L}^2(H_{\mathrm{disc}}, \mathrm{L}^2(\Omega))$, note that almost everywhere

$$\big(V_t V_{t'}(\xi)\big)(h) = u_t(-h)\big((u_{t'}(-h)(\xi(h))\big) = \mathrm{e}^{-\sqrt{2}\mathrm{i} t\, \mathrm{W}(-h)}\mathrm{e}^{-\sqrt{2}\mathrm{i} t'\, \mathrm{W}(-h)}\xi(h)$$

$$= \mathrm{e}^{-\sqrt{2}\mathrm{i}(t+t')\mathrm{W}(-h)}\xi(h) = u_{t+t'}(-h)(\xi(h)) = \big(V_{t+t'}(\xi)\big)(h).$$

We conclude that $V_t V_{t'} = V_{t+t'}$. Moreover, for any $\xi, \eta \in \mathrm{L}^2(H_{\mathrm{disc}}, \mathrm{L}^2(\Omega)) = \mathrm{L}^2(H_{\mathrm{disc}} \times \Omega)$, using dominated convergence theorem, we obtain

$$\big\langle V_t(\xi), \eta \big\rangle_{\mathrm{L}^2(H_{\mathrm{disc}}, \mathrm{L}^2(\Omega))} = \int_{H_{\mathrm{disc}} \times \Omega} \overline{V_t(\xi)(h)(\omega)} \eta(h, \omega)\, \mathrm{d}\mu_{H_{\mathrm{disc}}}(h)\, \mathrm{d}\mu(\omega)$$

$$= \int_{H_{\mathrm{disc}} \times \Omega} \overline{u_t(-h)(\xi(h))(\omega)} \eta(h, \omega)\, \mathrm{d}\mu_{H_{\mathrm{disc}}}(h)\, \mathrm{d}\mu(\omega)$$

$$= \int_{H_{\mathrm{disc}} \times \Omega} \mathrm{e}^{\sqrt{2}\mathrm{i} t\, \mathrm{W}(-h)(\omega)}\overline{\xi(h, \omega)} \eta(h, \omega)\, \mathrm{d}\mu_{H_{\mathrm{disc}}}(h)\, \mathrm{d}\mu(\omega).$$

$$\xrightarrow[t \to 0]{} \int_{H_{\mathrm{disc}} \times \Omega} \overline{\xi(h, \omega)} \eta(h, \omega)\, \mathrm{d}\mu_{H_{\mathrm{disc}}}(s)\, \mathrm{d}\mu(\omega) = \langle \xi, \eta \rangle_{\mathrm{L}^2(H_{\mathrm{disc}}, \mathrm{L}^2(\Omega))}.$$

So $(V_t)_{t \in \mathbb{R}}$ is a weakly continuous group of unitaries hence a strongly continuous group by [212, Lemma 13.4] or [216, p. 239]. By [216, p. 238], we conclude that $(U_t)_{t \in \mathbb{R}}$ is a weak* continuous group of *-automorphisms.

Finally, for any $t \geq 0$, any $f \in L^\infty(\Omega)$ and any $h \in H$, we have

$$\left(\int_\Omega \cdot \otimes \tau_{\mathrm{VN}(H_{\mathrm{disc}})} \right) (U_t(f \otimes \lambda_h)) = \left(\int_\Omega \cdot \otimes \tau_{\mathrm{VN}(H_{\mathrm{disc}})} \right) \left(e^{\sqrt{2} it \mathrm{W}(h)} f \otimes \lambda_h \right)$$

$$= \left(\int_\Omega e^{\sqrt{2} it \mathrm{W}(h)} f \, d\mu \right) \tau_{\mathrm{VN}(H_{\mathrm{disc}})}(\lambda_h) = \left(\int_\Omega e^{\sqrt{2} it \mathrm{W}(h)} f \, d\mu \right) \delta_{0,h}$$

$$= \left(\int_\Omega f \, d\mu \right) \tau_{\mathrm{VN}(H_{\mathrm{disc}})}(\lambda_h) = \left(\int_\Omega \cdot \otimes \tau_{\mathrm{VN}(H_{\mathrm{disc}})} \right) (f \otimes \lambda_h).$$

$$\square$$

We consider the unbounded operator

$$\delta \colon \operatorname{span}\{\lambda_h : h \in H_{\mathrm{disc}}\} \subset L^p(\mathrm{VN}(H_{\mathrm{disc}})) \to L^p(\Omega, L^p(\mathrm{VN}(H_{\mathrm{disc}})))$$

defined by

$$\delta(\lambda_h) \overset{\mathrm{def}}{=} 2\pi i \mathrm{W}(h) \otimes \lambda_h. \tag{3.43}$$

Recall the classical transference principle [36, Theorem 2.8]. Let G be a locally compact abelian group and $G \to B(X)$, $t \mapsto \pi_t$ be a strongly continuous representation of G on a Banach space X such that $c = \sup\{\|\pi_t\| : t \in G\} < \infty$. Let $k \in L^1(G)$ and let $T_k \colon X \to X$ be the operator defined by $T_k(x) = \int_G k(t) \pi_{-t}(x) \, d\mu_G(t)$. Then

$$\|T_k\|_{X \to X} \leq c^2 \|k * \cdot\|_{L^p(G,X) \to L^p(G,X)}. \tag{3.44}$$

If $0 < \varepsilon < R$, we will use the function [109, p. 388]

$$k_{\varepsilon,R}(t) = \frac{1}{\pi t} 1_{\varepsilon < |t| < R}. \tag{3.45}$$

Recall that we say that a function $f \in L^1_{\mathrm{loc}}(\mathbb{R}^*, X)$ admits a Cauchy principal value if the limit $\lim_{\varepsilon \to 0^+} \left(\int_{-1}^{-\varepsilon} f(t) \, dt + \int_\varepsilon^{\frac{1}{\varepsilon}} f(t) \, dt \right)$ exists and we let

$$\mathrm{p.\,v.} \int_\mathbb{R} f(t) \, dt \overset{\mathrm{def}}{=} \lim_{\varepsilon \to 0^+} \left(\int_{-\frac{1}{\varepsilon}}^{-\varepsilon} f(t) \, dt + \int_\varepsilon^{\frac{1}{\varepsilon}} f(t) \, dt \right).$$

Proposition 3.7 *Suppose* $1 < p < \infty$. *For any Hilbert space* H, *the map*

$$\mathcal{H}_{\text{disc}}\colon L^p(\Omega, L^p(\text{VN}(H_{\text{disc}}))) \longrightarrow L^p(\Omega, L^p(\text{VN}(H_{\text{disc}})))$$

$$z \longmapsto \text{p. v.} \frac{1}{\pi} \int_{\mathbb{R}} U_t(z) \, \frac{dt}{t} \, . \tag{3.46}$$

is well-defined and completely bounded.

Proof For any $\varepsilon > 0$ large enough, using the fact that $L^p(\Omega, L^p(\text{VN}(H_{\text{disc}}), S_I^p))$ is UMD and [18, p. 485], [99, Theorem 13.5] in the last inequality, we have by transference

$$\left\| \frac{1}{\pi} \int_{\varepsilon < |t| < \frac{1}{\varepsilon}} (U_t \otimes \text{Id}_{S_I^p}) \, \frac{dt}{t} \right\|_{L^p(\Omega, L^p(\text{VN}(H_{\text{disc}})), S_I^p) \to L^p(\Omega, L^p(\text{VN}(H_{\text{disc}})), S_I^p))}$$

$$\overset{(3.45)}{=} \left\| \int_{\mathbb{R}} k_{\varepsilon, \frac{1}{\varepsilon}}(t)(U_t \otimes \text{Id}_{S_I^p}) \, dt \right\|_{L^p(\Omega, L^p(\text{VN}(H_{\text{disc}})), S_I^p) \to L^p(\Omega, L^p(\text{VN}(H_{\text{disc}})), S_I^p))}$$

$$\overset{(3.44)}{\leqslant} \left\| (k_{\varepsilon, \frac{1}{\varepsilon}} * \cdot) \otimes \text{Id}_{L^p(\Omega, L^p(\text{VN}(H_{\text{disc}})), S_I^p))} \right\|_{L^p(\mathbb{R} \times \Omega, L^p(\text{VN}(H_{\text{disc}})), S_I^p)) \to L^p}$$

$$\lesssim_p 1.$$

If $z = f \otimes \lambda_h$, we have

$$\int_{-\frac{1}{\varepsilon}}^{-\varepsilon} U_t(z) \, \frac{dt}{t} + \int_{\varepsilon}^{\frac{1}{\varepsilon}} U_t(z) \, \frac{dt}{t} = \int_{\varepsilon}^{\frac{1}{\varepsilon}} (U_t(z) - U_{-t}(z)) \, \frac{dt}{t}$$

$$\overset{(3.42)}{=} 2\mathrm{i} \int_{\varepsilon}^{\frac{1}{\varepsilon}} \sin\left(\sqrt{2}W(h)t\right)(f \otimes \lambda_h) \, \frac{dt}{t}$$

$$= 2\mathrm{i}\left(\int_{\varepsilon}^{\frac{1}{\varepsilon}} \sin\left(\sqrt{2}W(h)t\right) \frac{dt}{t} \right)(f \otimes \lambda_h)$$

which admits a limit when $\varepsilon \to 0$. We have the existence of the principal value by linearity and density. □

The following is a variation of Pisier formula. Here $Qf = \sum_k \left(\int_{\Omega} f\gamma_k \right) \cdot \gamma_k$ is the Gaussian projection where (γ_k) is a family of independent standard Gaussian variables given by $\gamma_k = W(e_k)$, here e_k is running through an orthonormal basis of H. Note that if H were to be non-separable, then for any fixed $f \in L^2(\Omega)$, only countably many $\int_\Omega f \gamma_k$ would be non-zero. Then Q is independent of the particular choice of the basis (e_k).[7]

We have the following fundamental formula.

Lemma 3.12 *Suppose* $1 < p < \infty$. *If* $h \neq 0$, *we have in* $L^p(\Omega)$

$$Q\left(\text{p. v.} \frac{1}{\pi} \int_{\mathbb{R}} e^{i\sqrt{2}t W(h)} \frac{dt}{t}\right) = i\sqrt{\frac{2}{\pi}} \frac{W(h)}{\|h\|_H}. \qquad (3.47)$$

Proof Recall that for any $\alpha \in \mathbb{R}$ we have $\int_0^{+\infty} \frac{\sin(\alpha t)}{t} dt = \text{sign}(\alpha)\frac{\pi}{2}$. Consequently

$$Q\left(\text{p. v.} \frac{1}{\pi} \int_{\mathbb{R}} e^{i\sqrt{2}t W(h)} \frac{dt}{t}\right) = \frac{2i}{\pi} Q\left(\int_0^{+\infty} \sin\left(\sqrt{2}W(h)t\right) \frac{dt}{t}\right)$$

$$= iQ\left(\text{sign} \circ W(h)\right).$$

The Gaussian projection Q is independent of the choice of the family (γ_k) of $L^2(\Omega)$ consisting of independent standard Gaussian variables stemming from the process $W(e)$, $e \in H$. Choosing $\gamma_1 \overset{\text{def}}{=} \frac{W(h)}{\|W(h)\|_{L^2(\Omega)}} \overset{(2.44)}{=} \frac{W(h)}{\|h\|_H}$, for any $k \geqslant 2$ the random variables γ_k and $W(h)$ are independent. Thus γ_k and $\text{sign} \circ W(h)$ are also independent. So $\int_\Omega \text{sign} \circ W(h) \cdot \gamma_k = \int_\Omega \text{sign} \circ W(h) \int_\Omega \gamma_k = 0$. Using [72, p. 100] or [110, (E.2)] in the last equality, we infer that

$$Q(\text{sign} \circ W(h)) = \left(\int_\Omega \text{sign} \circ W(h) \cdot \gamma_1\right) \gamma_1 = \left(\int_\Omega \text{sign} \circ \gamma_1 \cdot \gamma_1\right) \gamma_1$$

$$= \left(\int_\Omega |\gamma_1|\right) \gamma_1 = \sqrt{\frac{2}{\pi}} \frac{W(h)}{\|h\|_H}.$$

We conclude that (3.47) is true.   □

---

[7] If (e_k) and (f_j) are two such orthonormal bases of H, then $f_j = \sum_k u_{jk} e_k$, where the sum is over at most countably many k and the coefficients u_{jk} form an orthogonal matrix, $\sum_j u_{jk} u_{jl} = \delta_{kl}$. We have

$$Q_{(f_j)}(f) = \sum_j \langle W(f_j), f\rangle W(f_j) = \sum_j \left\langle \sum_k u_{jk} W(e_k), f\right\rangle \sum_l u_{jl} W(e_l)$$

$$= \sum_{k,l} \sum_j u_{jk} u_{jl} \langle W(e_k), f\rangle W(e_l)$$

$$= \sum_{k,l} \delta_{kl} \langle W(e_k), f\rangle W(e_l) = \sum_l \langle W(e_k), f\rangle W(e_k) = Q_{(e_k)}(f).$$

Recall that J is given in (3.40) and $(-\Delta)^{-\frac{1}{2}}$ is defined in (3.41) at the beginning of this section.

Proposition 3.8 *Let* $1 < p < \infty$. *For any* $h \in H$, *we have*

$$\sqrt{\frac{2}{\pi}}\delta(-\Delta)^{-\frac{1}{2}}(\lambda_h) = \left(Q \otimes \mathrm{Id}_{L^p(\mathrm{VN}(H_{\mathrm{disc}}))}\right)\mathcal{H}_{\mathrm{disc}}J(\lambda_h). \qquad (3.48)$$

Proof If $h = 0$, then the left hand side of (3.48) equals 0 by definition of $(-\Delta)^{-\frac{1}{2}}(\lambda_0)$, and the right hand side equals 0 since $J(\lambda_0) = 1$ and $\mathcal{H}_{\mathrm{disc}}(1) = 0$. For any non-zero $h \in H$, we have

$$\sqrt{\frac{2}{\pi}}\delta(-\Delta)^{-\frac{1}{2}}(\lambda_h) \overset{(3.41)}{=} \frac{1}{\sqrt{2\pi^3}\,\|h\|_H}\delta(\lambda_h) \overset{(3.43)}{=} \frac{\sqrt{2}i}{\sqrt{\pi}\,\|h\|_H}W(h) \otimes \lambda_h.$$

Consequently, we have

$$\left(Q \otimes \mathrm{Id}_{L^p(\mathrm{VN}(H_{\mathrm{disc}}))}\right)\mathcal{H}_{\mathrm{disc}}J(\lambda_h) \overset{(3.40)}{=} \left(Q \otimes \mathrm{Id}_{L^p(\mathrm{VN}(H_{\mathrm{disc}}))}\right)\mathcal{H}_{\mathrm{disc}}(1 \otimes \lambda_h)$$

$$\overset{(3.46)}{=} \left(Q \otimes \mathrm{Id}_{L^p(\mathrm{VN}(H_{\mathrm{disc}}))}\right)\left(\mathrm{p.\,v.}\,\frac{1}{\pi}\int_{\mathbb{R}}e^{i\sqrt{2}tW(h)}\,\frac{dt}{t} \otimes \lambda_h\right)$$

$$= Q\left(\mathrm{p.\,v.}\,\frac{1}{\pi}\int_{\mathbb{R}}e^{i\sqrt{2}tW(h)}\,\frac{dt}{t}\right) \otimes \lambda_h \overset{(3.47)}{=} \frac{i\sqrt{2}}{\sqrt{\pi}\,\|h\|_H}W(h) \otimes \lambda_h.$$

$\square$

Lemma 3.13 *If* $1 < p < \infty$, *the operator*

$$\delta(-\Delta)^{-\frac{1}{2}} \otimes \mathrm{Id}_{S_I^p} : \mathcal{P}_{H_{\mathrm{disc}}} \otimes S_I^p \subset L^p\left(\mathrm{VN}(H_{\mathrm{disc}}), S_I^p\right) \to L^p\left(\Omega, L^p(\mathrm{VN}(H_{\mathrm{disc}}), S_I^p)\right) \qquad (3.49)$$

extends to a well-defined and bounded mapping.

Proof Note that the map $J : L^p(\mathrm{VN}(H_{\mathrm{disc}})) \to L^p(\Omega, L^p(\mathrm{VN}(H_{\mathrm{disc}})))$ is completely bounded. By Proposition 3.7, the operator

$$\mathcal{H}_{\mathrm{disc}} \otimes \mathrm{Id}_{S_I^p} : L^p(\Omega, L^p(\mathrm{VN}(H_{\mathrm{disc}}), S_I^p)) \to L^p(\Omega, L^p(\mathrm{VN}(H_{\mathrm{disc}}), S_I^p))$$

is well-defined and bounded. Moreover, since the Banach space $L^p(\mathrm{VN}(H_{\mathrm{disc}}), S_I^p)$ is K-convex, the map $Q \otimes \mathrm{Id}_{L^p(\mathrm{VN}(H_{\mathrm{disc}}), S_I^p)}$ induces a bounded operator on the Banach space $L^p(\Omega, L^p(\mathrm{VN}(H_{\mathrm{disc}}), S_I^p))$. Using the equality (3.48), we obtain the result by composition. $\square$

For the remainder of this section, we consider and fix a markovian semigroup $(T_t)_{t \geqslant 0}$ of Schur multipliers on $B(\ell_I^2)$ from Proposition 2.4. Recall that the real Hilbert space H and the family $(\alpha_i)_{i \in I}$ in H is defined in (2.36).

Note that $u = \mathrm{diag}(\lambda_{\alpha_i} : i \in I)$ is a unitary of the von Neumann algebra $M_I(\mathrm{VN}(H_{\mathrm{disc}})) = \mathrm{VN}(H_{\mathrm{disc}}) \overline{\otimes} B(\ell_I^2)$. Hence we have a trace preserving normal unital injective $*$-homomorphism $\pi \colon B(\ell_I^2) \to \mathrm{VN}(H_{\mathrm{disc}}) \overline{\otimes} B(\ell_I^2)$, $x \mapsto u(1 \otimes x)u^*$. For any $i, j \in I$, note that

$$\pi(e_{ij}) = \lambda_{\alpha_i - \alpha_j} \otimes e_{ij}. \tag{3.50}$$

Moreover, if $1 \leqslant p \leqslant \infty$ and if $y \in L^p(\Omega, S_I^p)$, we have

$$\left\| (\mathrm{Id}_{L^p(\Omega)} \otimes \pi_p)(y) \right\|_{L^p(\Omega, L^p(\mathrm{VN}(H_{\mathrm{disc}}), S_I^p))} = \|y\|_{L^p(\Omega, S_I^p)}. \tag{3.51}$$

The next two lemmas prepare our first main result in this section, which is Theorem 3.3. In the following lemma, we consider the operator $A_p^{-\frac{1}{2}}$ on $M_{I, \mathrm{fin}} \subset S_I^p$, given as the Schur multiplier defined by $A_p^{-\frac{1}{2}}(e_{ij}) = a_{ij}^{-\frac{1}{2}} e_{ij}$. Note that this makes sense when $a_{ij} \neq 0$, so for $e_{ij} \in \mathrm{Ran}\, A_p$, and we interpret $A_p^{-\frac{1}{2}} x$ to be 0 if $x \in \mathrm{Ker}\, A_p$. Note that by (2.16), we have a direct sum decomposition $S_I^p = \overline{\mathrm{Ran}\, A_p} \oplus \mathrm{Ker}\, A_p$. Then the statement of (3.52) below includes that the right hand side vanishes for $x \in \mathrm{Ker}\, A_p$, and both sides are bounded operators on $\mathrm{Ran}(A_p)$, and thus on $\overline{\mathrm{Ran}\, A_p}$.

Lemma 3.14 (Intertwining Formula) *Suppose* $1 \leqslant p < \infty$. *We have*

$$\mathrm{i}\big(\mathrm{Id}_{L^p(\Omega)} \otimes \pi_p\big)\partial_{\alpha,1} A_p^{-\frac{1}{2}} = \big(\delta(-\Delta)^{-\frac{1}{2}} \otimes \mathrm{Id}_{S_I^p}\big)\pi_p. \tag{3.52}$$

Proof For any $i, j \in I$ such that $\alpha_i \neq \alpha_j$, we have

$$\mathrm{i}\big(\mathrm{Id}_{L^p(\Omega)} \otimes \pi_p\big)\partial_{\alpha,1} A_p^{-\frac{1}{2}}(e_{ij}) = \frac{\mathrm{i}}{\|\alpha_i - \alpha_j\|_H}\big(\mathrm{Id}_{L^p(\Omega)} \otimes \pi_p\big)\partial_{\alpha,1}(e_{ij})$$

$$\overset{(2.95)}{=} \frac{\mathrm{i}}{\|\alpha_i - \alpha_j\|_H}\big(\mathrm{Id}_{L^p(\Omega)} \otimes \pi_p\big)\big(W(\alpha_i - \alpha_j) \otimes e_{ij}\big)$$

$$= \frac{\mathrm{i}}{\|\alpha_i - \alpha_j\|_H} W(\alpha_i - \alpha_j) \otimes \pi_p(e_{ij})$$

$$\overset{(3.50)}{=} \frac{\mathrm{i}}{\|\alpha_i - \alpha_j\|_H} W(\alpha_i - \alpha_j) \otimes \lambda_{\alpha_i - \alpha_j} \otimes e_{ij}$$

and

$$\left(\delta(-\Delta)^{-\frac{1}{2}} \otimes \mathrm{Id}_{S_I^p}\right)\pi_p(e_{ij}) \overset{(3.50)}{=} \left(\delta(-\Delta)^{-\frac{1}{2}} \otimes \mathrm{Id}_{S_I^p}\right)(\lambda_{\alpha_i - \alpha_j} \otimes e_{ij})$$

$$= \delta(-\Delta)^{-\frac{1}{2}}(\lambda_{\alpha_i - \alpha_j}) \otimes e_{ij}$$

$$\overset{(3.41)}{=} \frac{1}{2\pi\|\alpha_i - \alpha_j\|_H}\delta(\lambda_{\alpha_i - \alpha_j}) \otimes e_{ij}$$

$$\overset{(3.43)}{=} \frac{i}{\|\alpha_i - \alpha_j\|_H}\left(\mathrm{W}(\alpha_i - \alpha_j) \otimes \lambda_{\alpha_i - \alpha_j}\right) \otimes e_{ij}.$$

We conclude by linearity.   □

For future use, we state the next lemma equally for the case of q-deformed algebras.

Lemma 3.15 *For* $-1 \leqslant q \leqslant 1$ *and any* $x, y \in \mathrm{M}_{I,\mathrm{fin}}$, *we have*

$$(\tau_{\Gamma_q(H)} \otimes \mathrm{Tr}_{\mathrm{B}(\ell_I^2)})\left((\partial_{\alpha,q}(x))^*\partial_{\alpha,q}(y)\right) = \mathrm{Tr}_{\mathrm{B}(\ell_I^2)}\left((A^{\frac{1}{2}}(x))^*A^{\frac{1}{2}}(y)\right). \qquad (3.53)$$

Proof For any $i, j, k, l \in I$, we have

$$\mathrm{Tr}\left((A^{\frac{1}{2}}(e_{ij}))^*A^{\frac{1}{2}}(e_{kl})\right) = \|\alpha_i - \alpha_j\|_H \|\alpha_k - \alpha_l\|_H \, \mathrm{Tr}(e_{ij}^*e_{kl})$$

$$= \delta_{i=k}\delta_{j=l}\|\alpha_i - \alpha_j\|_H^2.$$

Moreover, we have

$$(\tau \otimes \mathrm{Tr})\left((\partial_{\alpha,q}(e_{ij}))^*\partial_{\alpha,q}(e_{kl})\right)$$

$$\overset{(2.95)}{=} (\tau \otimes \mathrm{Tr})\left((s_q(\alpha_i - \alpha_j) \otimes e_{ij})^*(s_q(\alpha_k - \alpha_l) \otimes e_{kl})\right)$$

$$= \tau(s_q(\alpha_i - \alpha_j)s_q(\alpha_k - \alpha_l)) \, \mathrm{Tr}(e_{ij}^*e_{kl})$$

$$\overset{(2.41)(2.44)}{=} \delta_{i=k}\delta_{j=l}\|\alpha_i - \alpha_j\|_H^2.$$

We conclude by linearity.   □

Suppose $1 \leqslant p < \infty$. We denote by $A_p^{\frac{1}{2}} \colon \mathrm{dom}\, A_p^{\frac{1}{2}} \subset S_I^p \to S_I^p$ the square root of the sectorial operator $A_p \colon \mathrm{dom}\, A_p \subset S_I^p \to S_I^p$. It is again a Schur multiplier associated with the symbol $[\|\alpha_i - \alpha_j\|]$.

Theorem 3.3 *Suppose* $1 < p < \infty$. *For any* $x \in \mathrm{M}_{I,\mathrm{fin}}$, *we have*

$$\left\| A_p^{\frac{1}{2}}(x) \right\|_{S_I^p} \approx_p \left\| \partial_{\alpha,1}(x) \right\|_{L^p(\Omega,S_I^p)}. \tag{3.54}$$

Proof We write in short $\partial_{\alpha,1} = \partial_\alpha$ in this proof. Using Lemma 3.14, for any $x \in \mathrm{M}_{I,\mathrm{fin}}$, we obtain the inequality

$$\left\| \partial_\alpha(x) \right\|_{L^p(\Omega,S_I^p)}$$

$$= \left\| \partial_\alpha A_p^{-\frac{1}{2}} A_p^{\frac{1}{2}}(x) \right\|_{L^p(\Omega,S_I^p)}$$

$$\overset{(3.51)}{=} \left\| \left(\mathrm{Id}_{L^p(\Omega)} \otimes \pi_p \right) \partial_\alpha A_p^{-\frac{1}{2}} A_p^{\frac{1}{2}}(x) \right\|_{L^p(\Omega,L^p(\mathrm{VN}(H_{\mathrm{disc}}),S_I^p))}$$

$$\overset{(3.52)}{=} \left\| \left(\delta(-\Delta)^{-\frac{1}{2}} \otimes \mathrm{Id}_{S_I^p} \right) \pi_p A_p^{\frac{1}{2}}(x) \right\|_{L^p(\Omega,L^p(\mathrm{VN}(H_{\mathrm{disc}}),S_I^p))}$$

$$\leqslant \left\| \left(\delta(-\Delta)^{-\frac{1}{2}} \otimes \mathrm{Id}_{S_I^p} \right) \right\|_{L^p(\mathrm{VN}(H_{\mathrm{disc}}),S_I^p) \to L^p(\Omega,L^p(\mathrm{VN}(H_{\mathrm{disc}}),S_I^p))} \|\pi_p\| \left\| A_p^{\frac{1}{2}}(x) \right\|_{S_I^p}$$

$$\overset{(3.49)}{\lesssim_p} \left\| A_p^{\frac{1}{2}}(x) \right\|_{S_I^p}.$$

The reverse estimate follows with essentially the same constant from a duality argument. Indeed, if we fix x to be an element of $\mathrm{M}_{I,\mathrm{fin}} \subset S_I^p$, by the Hahn-Banach theorem, there exists an element y of $S_I^{p^*}$ with $\|y\|_{S_I^{p^*}} = 1$ satisfying

$$\left\| A_p^{\frac{1}{2}}(x) \right\|_{S_I^p} = \mathrm{Tr}\left(y^* A_p^{\frac{1}{2}}(x) \right). \tag{3.55}$$

We can suppose that $y \in \mathrm{M}_{I,\mathrm{fin}}$. Note that $A^{-\frac{1}{2}}$ is well-defined on e_{ij} only if $\|\alpha_i - \alpha_j\| \neq 0$. We consider on the set I the equivalence relation $i \cong j \overset{\mathrm{def}}{\Longleftrightarrow} \alpha_i = \alpha_j$. Then we can write $I = \bigcup_{k \in K} J_k$ partitioned into pairwise disjoint subsets J_k according to this equivalence relation. Consider the subalgebra $M = \overline{\bigoplus_{k \in K} \mathrm{B}(\ell_{J_k}^2)}^{\mathrm{w}^*}$ of $\mathrm{B}(\ell_I^2)$ and the conditional expectation $\mathbb{E}_I \colon S_I^p \to L^p(M, \mathrm{Tr}\,|M)$. We obtain an element $y_0 \overset{\mathrm{def}}{=} y - \mathbb{E}_I(y)$ with

$$\|y_0\|_{S_I^{p^*}} = \|y - \mathbb{E}_I(y)\|_{S_I^{p^*}} \leqslant \|y\|_{S_I^{p^*}} + \|\mathbb{E}_I(y)\|_{S_I^{p^*}} \leqslant 2.$$

Note that $y_{0ij} = 0$ if $\alpha_i = \alpha_j$, hence $A_p^{-\frac{1}{2}}(y_0)$ is well-defined. For any $i \in I$, we have

$$\left[(\mathbb{E}_I(y))^* A_p^{\frac{1}{2}}(x)\right]_{ii} = \sum_{k \in I}(\mathbb{E}_I(y)^*)_{ik} \, \|\alpha_k - \alpha_i\| \, x_{ki}$$

$$= \sum_{j=1}^{|K|} \sum_{l \in J_j}(\mathbb{E}_I(y)^*)_{il} \, \|\alpha_l - \alpha_i\| \, x_{li}$$

$$= \sum_{j=1}^{|K|} \sum_{l \in J_j} \delta_{i \in J_j} y_{li} \, \|\alpha_l - \alpha_i\| \, x_{li} = \sum_{j=1}^{|K|} \sum_{l \in J_j} \delta_{i \in J_j} y_{li} \cdot 0 \cdot x_{li} = 0.$$

Hence, $\mathrm{Tr}\left((\mathbb{E}_I(y))^* A_p^{\frac{1}{2}}(x)\right) = 0$. Using the first part of the proof in the last estimate, we conclude that

$$\left\|A_p^{\frac{1}{2}}(x)\right\|_{S_I^p} \overset{(3.55)}{=} \mathrm{Tr}\left(y^* A_p^{\frac{1}{2}}(x)\right) = \mathrm{Tr}\left(y^* A_p^{\frac{1}{2}}(x)\right) - \mathrm{Tr}\left((\mathbb{E}_I(y))^* A_p^{\frac{1}{2}}(x)\right)$$

$$= \mathrm{Tr}\left(y_0^* A_p^{\frac{1}{2}}(x)\right) = \mathrm{Tr}\left((A_{p^*}^{\frac{1}{2}} A_{p^*}^{-\frac{1}{2}}(y_0))^* A_p^{\frac{1}{2}}(x)\right)$$

$$\overset{(3.53)}{=} \tau_{L^\infty(\Omega)} \otimes \mathrm{Tr}_{B(\ell_I^2)}\left((\partial_\alpha A_{p^*}^{-\frac{1}{2}}(y_0))^* \partial_\alpha(x)\right)$$

$$\leqslant \left\|\partial_\alpha A_{p^*}^{-\frac{1}{2}}(y_0)\right\|_{L^{p^*}(\Omega, S_I^{p^*})} \|\partial_\alpha(x)\|_{L^p(\Omega, S_I^p)} \lesssim_p \|\partial_\alpha(x)\|_{L^p(\Omega, S_I^p)} .$$

$\square$

Remark 3.3 Keeping track of the constants in the two-sided estimate of Theorem 3.3, we obtain

$$\frac{1}{K \max(p, p^*)^{\frac{3}{2}}} \|\partial_{\alpha,1}(x)\|_{L^p(\Omega, S_I^p)} \leqslant \left\|A_p^{\frac{1}{2}}(x)\right\|_{S_I^p} \leqslant K \max(p, p^*)^{\frac{3}{2}}$$

$$\|\partial_{\alpha,1}(x)\|_{L^p(\Omega, S_I^p)} \tag{3.56}$$

where K is an absolute constant. Indeed, for the upper estimate, looking into the proof of Theorem 3.3, we have a control by

$$\left\|\delta(-\Delta)^{-\frac{1}{2}} \otimes \mathrm{Id}_{S_I^p}\right\|_{L^p(\mathrm{VN}(H_{\mathrm{disc}}), S_I^p) \to L^p(\Omega, L^p(\mathrm{VN}(H_{\mathrm{disc}}), S^p))} \|\pi_p\|_{p \to p}$$

$$\overset{\text{Proposition 3.8}}{=} \sqrt{\frac{\pi}{2}} \left\|Q \otimes \mathrm{Id}_{L^p(\mathrm{VN}(H_{\mathrm{disc}}))} \mathcal{H}_{\mathrm{disc}} J\right\|_{\mathrm{cb}} \cdot 1$$

$$\lesssim \left\|Q \otimes \mathrm{Id}_{L^p(\mathrm{VN}(H_{\mathrm{disc}}), S^p)}\right\| \|\mathcal{H}_{\mathrm{disc}}\|_{\mathrm{cb}} \|J\|_{\mathrm{cb}} .$$

Let us estimate the three factors. For the first, we recall that Q is the Gaussian projection $Q: L^p(\Omega) \to L^p(\Omega)$, $f \mapsto \sum_k \langle f, W(e_k)\rangle W(e_k)$. Its norm is controlled by $C\sqrt{\max(p, p^*)}$ according to Lemma 3.16 below. Furthermore, going into the proof of Proposition 3.7, we see that the norm $\|\mathcal{H}_{\text{disc}}\|_{\text{cb}}$ is no more than the truncated Hilbert transform norm

$$\sup_{\varepsilon \in (0,1)} \left\| k_{\varepsilon, \frac{1}{\varepsilon}} * \cdot \right\|_{B(L^p(\mathbb{R}, L^p(\Omega, L^p(\text{VN}(H_{\text{disc}})), S^p)))}.$$

The last expression in turn is controlled by an absolute constant times the norm of the Hilbert transform $\|H\|_{B(L^p(\mathbb{R}, L^p(\Omega, L^p(\text{VN}(H_{\text{disc}})), S^p)))}$ according to [94, p. 222-223]. This in turn is controlled by $\frac{8}{\pi}\max(p, p^*)$ [109, Proposition 5.4.2] and [179, Theorem 4.3]. Finally, J is a complete contraction. In all, we obtain

$$\left\| Q \otimes \text{Id}_{L^p(\text{VN}(H_{\text{disc}}), S^p)} \right\| \|\mathcal{H}_{\text{disc}}\|_{\text{cb}} \|J\|_{\text{cb}} \leqslant K\sqrt{\max(p, p^*)} \cdot \max(p, p^*).$$

The lower estimate in (3.56) follows the same way, since in the proof of Theorem 3.3, it was obtained by duality.

Lemma 3.16 *Let $1 < p < \infty$ and X be a σ-finite measure space. We have a control of the Gaussian projection*

$$\left\| Q \otimes \text{Id}_{L^p(X, S^p)} \right\|_{L^p(\Omega, L^p(X, S^p)) \to L^p(\Omega, L^p(X, S^p))} \leqslant C\sqrt{\max(p, p^*)}, \qquad (3.57)$$

where C is some absolute constant.

Proof First some remarks are in order. Note that the Gaussian projection was defined as $Q(f) \overset{\text{def}}{=} \sum_{k \in K} \langle f, \gamma_k \rangle \gamma_k$, where $(\gamma_k)_{k \in K}$ is an orthonormal family of standard Gaussians. If K is at most countable, then the main part of the proof below applies directly. If K happened to be larger, then consider for a subset K_0 of K the modified Gaussian projection $Q_{K_0} f = \sum_{k \in K_0} \langle f, \gamma_k \rangle \gamma_k$. Note that for fixed $f \in D \overset{\text{def}}{=} L^p(\Omega, L^p(X, S^p)) \cap (L^2(\Omega) \otimes L^p(X, S^p))$, the sum over K_0 has at most countably many non-zero entries. This also implies by density of D that

$$\|Q\|_{B(L^p(\Omega, L^p(X, S^p)))} = \sup\left\{ \frac{\|Q(f)\|_p}{\|f\|_p} : f \in D\backslash\{0\} \right\}$$

$$= \sup\left\{ \frac{\|Q_{K_0}(f)\|_p}{\|f\|_p} : f \in D\backslash\{0\}, \ K_0 \subset K \text{ countable} \right\}$$

$$= \sup\left\{ \|Q_{K_0}\|_{L^p(\Omega, L^p(X, S^p)) \to L^p(\Omega, L^p(X, S^p))} : K_0 \subset K \text{ countable} \right\}.$$

Then the proof below first gives an estimate for $\|Q_{K_0}\|$, which is uniform in K_0, and consequently gives an estimate for $Q_K = Q$.

Now we proceed to the proof. We recall constants of noncommutative Khintchine inequalities. Namely, for $2 \leqslant p < \infty$ and a sequence of independent standard Gaussians (γ_k) over Ω, we have[8]

$$\left\| \sum_k \gamma_k \otimes x_k \right\|_{L^p(\Omega, L^p(X, S^p))} \leqslant C\sqrt{p} \left\| \sum_k x_k \otimes e_k \right\|_{L^p(X, S^p, H_{\mathrm{rad}, p})}, \tag{3.58}$$

with an absolute constant C, where the latter space was defined in Sect. 2.5 and (e_k) is any orthonormal sequence in H. Moreover, we claim that for $1 \leqslant p \leqslant 2$ we have

$$\left\| \sum_k \gamma_k \otimes x_k \right\|_{L^p(\Omega, L^p(X, S^p))} \leqslant C \left\| \sum_k x_k \otimes e_k \right\|_{L^p(X, S^p(H_{\mathrm{rad}, p}))} \tag{3.59}$$

with some universal constant C. Indeed, this follows since (3.59) holds with constant C_1 (resp. C_2) for $p = 1$ (resp. $p = 2$) according to noncommutative

---

[8] Indeed, according to [189, p. 193], [190, p. 271], (3.58) holds for S^p instead of $L^p(X, S^p)$. Then (3.58) follows with $L^p(X, S^p)$ by the following Fubini argument. For any finite family (x_k) of $L^p(X, S^p)$, we have

$$\left\| \sum_k \gamma_k \otimes x_k \right\|_{L^p(\Omega, L^p(X, S^p))}^p = \int_X \left\| \sum_k \gamma_k \otimes x_k(t) \right\|_{L^p(\Omega, S^p)}^p dt$$

$$\leqslant C\sqrt{p}^p \int_X \left\| \sum_k x_k(t) \otimes e_k \right\|_{S^p(H_{\mathrm{rad}, p})}^p dt$$

$$= C\sqrt{p}^p \int_X \max \left\{ \left\| \left(\sum_k |x_k(t)|^2 \right)^{\frac{1}{2}} \right\|_{S^p}, \left\| \left(\sum_k |x_k(t)^*|^2 \right)^{\frac{1}{2}} \right\|_{S^p} \right\}^p dt$$

$$\leqslant C\sqrt{p}^p \int_X \left\| \left(\sum_k |x_k(t)|^2 \right)^{\frac{1}{2}} \right\|_{S^p}^p + \left\| \left(\sum_k |x_k(t)^*|^2 \right)^{\frac{1}{2}} \right\|_{S^p}^p dt$$

$$= C\sqrt{p}^p \left(\left\| \left(\sum_k |x_k|^2 \right)^{\frac{1}{2}} \right\|_{L^p(X, S^p)}^p + \left\| \left(\sum_k |x_k^*|^2 \right)^{\frac{1}{2}} \right\|_{L^p(X, S^p)}^p \right)$$

$$\leqslant 2C\sqrt{p}^p \max \left\{ \left\| \left(\sum_k |x_k|^2 \right)^{\frac{1}{2}} \right\|_{L^p(X, S^p)}, \left\| \left(\sum_k |x_k^*|^2 \right)^{\frac{1}{2}} \right\|_{L^p(X, S^p)} \right\}^p$$

$$= 2C\sqrt{p}^p \left\| \sum_k x_k \otimes e_k \right\|_{L^p(X, S^p(H_{\mathrm{rad}, p}))}^p .$$

Khintchine inequalities. Moreover, both $L^p(X, S^p, H_{\mathrm{rad},p})$ and $L^p(\Omega, L^p(X, S^p))$ form a complex interpolation scale, so that we have $C_p \leqslant C_1^\theta C_2^{1-\theta} \leqslant C$ with the correct interpolation parameter $\theta \in [0, 1]$.

Now the end of the argument for (3.57) is as follows. Write $(Q \otimes \mathrm{Id})(f) = \sum_k \gamma_k \otimes x_k$ with $x_k = \int_\Omega f \gamma_k$. Then for $1 < p \leqslant 2$, by duality $(L^p(X, S^p, H_{\mathrm{rad},p}))^* \overset{(2.69)}{=} L^{p^*}(X, S^{p^*}, \overline{H}_{\mathrm{rad},p^*})$,

$$\|(Q \otimes \mathrm{Id})(f)\|_p = \left\|\sum_k \gamma_k \otimes x_k\right\|_{L^p(\Omega, L^p(X, S^p))} \overset{(3.59)}{\leqslant} C \left\|\sum_k x_k \otimes e_k\right\|_{L^p(X, S^p(H_{\mathrm{rad},p}))}$$

$$\overset{(2.69)}{=} C \sup\left\{\left|\sum_k \langle x_k, y_k \rangle\right| : \left\|\sum_k y_k \otimes e_k\right\|_{L^{p^*}(X, S^{p^*}(H_{\mathrm{rad},p^*}))} \leqslant 1\right\}$$

$$= C \sup\left\{\left|\int_\Omega \sum_k \langle x_k, y_k \rangle \gamma_k^2\right| : \left\|\sum_k y_k \otimes e_k\right\|_{L^{p^*}(X, S^{p^*}(H_{\mathrm{rad},p^*}))} \leqslant 1\right\}$$

$$= C \sup\left\{\left|\int_\Omega \langle (Q \otimes \mathrm{Id})(f), (Q \otimes \mathrm{Id})(y) \rangle\right| : y \in L^{p^*}, \left\|\sum_k \left(\int_\Omega y \gamma_k\right) \otimes e_k\right\| \leqslant 1\right\}$$

$$= C \sup\left\{\left|\int_\Omega f (Q \otimes \mathrm{Id})(y)\right| : y \in L^{p^*}, \left\|\sum_k \left(\int_\Omega y \gamma_k\right) \otimes e_k\right\| \leqslant 1\right\}$$

$$\leqslant C \|f\|_p \sup\left\{\|(Q \otimes \mathrm{Id})(y)\|_{L^{p^*}(\Omega, L^{p^*}(S^{p^*}))} : y \in L^{p^*},\right.$$

$$\left.\left\|\sum_k \left(\int_\Omega y \gamma_k\right) \otimes e_k\right\| \leqslant 1\right\} \overset{(3.58)}{\leqslant} C' \sqrt{p^*} \|f\|_{L^p(\Omega, L^p(X, S^p))}.$$

The case $2 \leqslant p < \infty$ follows by duality, since Q is selfadjoint on $L^2(\Omega)$. $\qquad\square$

Using the Hodge-Dirac operator $\mathscr{D}_{\alpha,1,p}$ of Proposition 5.18 and some results of this book, we can give a second proof of Theorem 3.3 with a better constant at right. We also need the normal unital injective $*$-homomorphism map $J : \mathrm{B}(\ell_I^2) \to L^\infty(\Omega)\overline{\otimes}\mathrm{B}(\ell_I^2)$ defined by

$$J(x) = 1 \otimes x, \quad x \in \mathrm{B}(\ell_I^2) \tag{3.60}$$

with associated conditional expectation $\mathbb{E} : L^\infty(\Omega)\overline{\otimes}\mathrm{B}(\ell_I^2) \to \mathrm{B}(\ell_I^2)$.

Proposition 3.9 *Suppose* $1 < p < \infty$. *For any Hilbert space H, the map* $\mathscr{U}_p : L^p(\Omega, S_I^p) \to L^p(\Omega, S_I^p)$, $z \mapsto \mathrm{p.v.} \frac{1}{\pi} \int_\mathbb{R} e^{\mathrm{i}t \mathscr{D}_{\alpha,1,p}}(z) \frac{\mathrm{d}t}{t}$ *is well-defined and bounded with*

$$\|\mathscr{U}_p\|_{L^p(\Omega, S_I^p) \to L^p(\Omega, S_I^p)} \lesssim \max(p, p^*). \tag{3.61}$$

Proof For any $\varepsilon > 0$ large enough, using the fact that the Banach space $L^p(\Omega, S_I^p)$ is UMD and [99, Theorem 13.5] combined with [193, Corollary 4.5] (see also [109, page 484]) in the last inequality, we have by transference

$$\left\| \frac{1}{\pi} \int_{\varepsilon < |t| < \frac{1}{\varepsilon}} e^{it \mathscr{D}_{\alpha,1,p}} \frac{dt}{t} \right\|_{L^p(\Omega, S_I^p) \to L^p(\Omega, S_I^p)}$$

$$\overset{(3.45)}{=} \left\| \int_{\mathbb{R}} k_{\varepsilon, \frac{1}{\varepsilon}}(t) e^{it \mathscr{D}_{\alpha,1,p}} \, dt \right\|_{L^p(\Omega, S_I^p) \to L^p(\Omega, S_I^p)}$$

$$\overset{(3.44)}{\leqslant} \left\| (k_{\varepsilon, \frac{1}{\varepsilon}} * \cdot) \otimes \mathrm{Id}_{L^p(\Omega, S_I^p)} \right\|_{L^p(\mathbb{R} \times \Omega, S_I^p) \to L^p(\mathbb{R} \times \Omega, S_I^p)} \lesssim \max(p, p^*).$$

If $z = f \otimes e_{ij}$, we have

$$\int_{-\frac{1}{\varepsilon}}^{-\varepsilon} e^{it \mathscr{D}_{\alpha,1,p}}(z) \frac{dt}{t} + \int_{\varepsilon}^{\frac{1}{\varepsilon}} e^{it \mathscr{D}_{\alpha,1,p}}(z) \frac{dt}{t} = \int_{\varepsilon}^{\frac{1}{\varepsilon}} \left(e^{it \mathscr{D}_{\alpha,1,p}}(z) - e^{-it \mathscr{D}_{\alpha,1,p}}(z) \right) \frac{dt}{t}$$

$$\overset{(5.98)}{=} 2i \int_{\varepsilon}^{\frac{1}{\varepsilon}} \sin(W(\alpha_i - \alpha_j)t)(f \otimes e_{ij}) \frac{dt}{t}$$

$$= 2i \left(\int_{\varepsilon}^{\frac{1}{\varepsilon}} \sin(W(\alpha_i - \alpha_j)t) \frac{dt}{t} \right) (f \otimes e_{ij})$$

which admits a limit when $\varepsilon \to 0$. The existence of the principal value follows by linearity and density. $\qquad\qquad\square$

We define the operator $R_p \overset{\mathrm{def}}{=} \partial_{\alpha,1,p} A_p^{-\frac{1}{2}} \colon \mathrm{Ran}\, A_p^{\frac{1}{2}} \to L^p(\Omega, S_I^p)$ where $\partial_{\alpha,1,p}$ is defined in Proposition 3.11. Note that if $i, j \in I$ with $\|\alpha_i - \alpha_j\|_H \neq 0$ we have

$$R_p(e_{ij}) = \frac{1}{\|\alpha_i - \alpha_j\|_H} W(\alpha_i - \alpha_j) \otimes e_{ij}. \tag{3.62}$$

Lemma 3.17 *Let* $1 < p < \infty$*. On* $\mathrm{Ran}\, A_p$*, we have*

$$R_p = \frac{-i}{\sqrt{2\pi}} (Q \otimes \mathrm{Id}_{S_I^p}) \circ \underbrace{\left(\mathrm{p.\,v.} \, \frac{1}{\pi} \int_{\mathbb{R}} e^{it \mathscr{D}_{\alpha,1,p}} \frac{dt}{t} \right)}_{\mathcal{U}_p} \circ J. \tag{3.63}$$

Proof For any $i, j \in I$ with $\|\alpha_i - \alpha_j\|_H \neq 0$, we have using Proposition 5.18 in the third equality

$$\frac{-i}{\sqrt{2\pi}}(Q \otimes \mathrm{Id}_{S_I^p})\left(\text{p. v. } \frac{1}{\pi}\int_{\mathbb{R}} e^{it\mathscr{D}_{\alpha,1,p}} \frac{dt}{t}\right) \circ J(e_{ij})$$

$$= \frac{-i}{\sqrt{2\pi}}(Q \otimes \mathrm{Id}_{S^p})\left(\text{p. v. } \frac{1}{\pi}\int_{\mathbb{R}} e^{it\mathscr{D}_{\alpha,1,p}} \circ J(e_{ij}) \frac{dt}{t}\right)$$

$$\overset{(3.60)}{=} \frac{-i}{\sqrt{2\pi}}(Q \otimes \mathrm{Id}_{S^p})\left(\text{p. v. } \frac{1}{\pi}\int_{\mathbb{R}} e^{it\mathscr{D}_{\alpha,1,p}}(1 \otimes e_{ij}) \frac{dt}{t}\right)$$

$$\overset{(5.98)}{=} \frac{-i}{\sqrt{2\pi}}(Q \otimes \mathrm{Id}_{S^p})\left(\text{p. v. } \frac{1}{\pi}\int_{\mathbb{R}} \left(e^{itW(\alpha_i - \alpha_j)} \otimes e_{ij}\right) \frac{dt}{t}\right)$$

$$= \frac{-i}{\sqrt{2\pi}}(Q \otimes \mathrm{Id}_{S^p})\left(\left(\text{p. v. } \frac{1}{\pi}\int_{\mathbb{R}} e^{itW(\alpha_i - \alpha_j)} \frac{dt}{t}\right) \otimes e_{ij}\right)$$

$$= \frac{-i}{\sqrt{2\pi}}Q\left(\text{p. v. } \frac{1}{\pi}\int_{\mathbb{R}} e^{itW(\alpha_i - \alpha_j)} \frac{dt}{t}\right) \otimes e_{ij}$$

$$\overset{(3.47)}{=} \frac{1}{\pi\|\alpha_i - \alpha_j\|_H} W(\alpha_i - \alpha_j) \otimes e_{ij} \overset{(3.62)}{=} \frac{1}{\pi} R_p(e_{ij}).$$

$\square$

Recall that $\mathrm{Ran}\, A_p^{\frac{1}{2}} \overset{(2.21)}{=} \overline{\mathrm{Ran}\, A_p}$.

Theorem 3.4 *Suppose* $1 < p < \infty$. *We have*

$$\|R_p\|_{\overline{\mathrm{Ran}\, A_p} \to L^p(\Omega, S_I^p)} \lesssim \max(p, p^*)^{\frac{3}{2}} \tag{3.64}$$

where $1 = \frac{1}{p} + \frac{1}{p^*}$ *and if* $x \in M_{I,\mathrm{fin}}$

$$\frac{1}{\max(p, p^*)^{\frac{3}{2}}}\|\partial_{\alpha,1}(x)\|_{L^p(\Omega, S_I^p)} \lesssim \|A_p^{\frac{1}{2}}(x)\|_{S_I^p} \lesssim \max(p, p^*)\|\partial_{\alpha,1}(x)\|_{L^p(\Omega, S_I^p)}. \tag{3.65}$$

Proof If $x \in \mathrm{Ran}\, A_p^{\frac{1}{2}}$, we have

$$\|R_p(x)\|_{L^p(\Omega, B(\ell_I^2))} \overset{(3.63)}{\lesssim} \|(Q \otimes \mathrm{Id}_{S_I^p}) \circ \mathcal{U}_p \circ J(x)\|_{L^p(\Omega, S_I^p)}$$

$$\leqslant \|Q \otimes \mathrm{Id}_{S_I^p}\| \|\mathcal{U}_p\| \|J(x)\|_{L^p(\Omega, S_I^p)}$$

$$\overset{(3.57)}{\lesssim} \sqrt{\max(p, p^*)} \|\mathcal{U}_p\| \|x\|_{S_I^p} \overset{(3.61)}{\lesssim} \max(p, p^*)^{\frac{3}{2}} \|x\|_{S_I^p}$$

If $x \in M_{I,\mathrm{fin}}$, we have

$$
\big\|\partial_{\alpha,1}(x)\big\|_{L^p(\Omega,S_I^p)} = \big\|\partial_{\alpha,1}A_p^{-\frac{1}{2}}A_p^{\frac{1}{2}}(x)\big\|_{L^p(\Omega,S_I^p)} = \big\|R_p A_p^{\frac{1}{2}}(x)\big\|_{L^p(\Omega,S_I^p)}
$$

$$
\overset{(3.64)}{\lesssim} \max(p,p^*)^{\frac{3}{2}}\big\|A_p^{\frac{1}{2}}(x)\big\|_{S_I^p}.
$$

Moreover, if $y \in \operatorname{Ran} A_p^{\frac{1}{2}}$ we have

$$
y \overset{(4.50)}{=} A_p^{-\frac{1}{2}}(\partial_{\alpha,1,p^*})^*\partial_{\alpha,1,p}A_p^{-\frac{1}{2}}y = (\partial_{\alpha,1,p^*}A_{p^*}^{-\frac{1}{2}})^*\partial_{\alpha,1,p}A_p^{-\frac{1}{2}}y \tag{3.66}
$$

$$
= (R_{p^*})^* R_p(y) \overset{(3.63)}{\approx} J_{p^*}^* \mathcal{U}_{p^*}^*(Q_{p^*}\otimes \operatorname{Id}_{S^{p^*}})^*(Q_p \otimes \operatorname{Id}_{S_I^p})\mathcal{U}_p J_p y
$$

$$
= \mathbb{E}_p\mathcal{U}_{p^*}^*(Q_{p^*}^* \otimes \operatorname{Id}_{S_I^p})(Q_p \otimes \operatorname{Id}_{S_I^p})\mathcal{U}_p J_p y = \mathbb{E}_p\mathcal{U}_p(Q_p \otimes \operatorname{Id}_{S_I^p})\mathcal{U}_p J_p y
$$

$$
\overset{(3.63)}{=} \mathbb{E}_p\mathcal{U}_p R_p y.
$$

Hence

$$
\|y\|_{S_I^p} \overset{(3.66)}{\approx} \big\|\mathbb{E}\mathcal{U}_{p^*}R_p(y)\big\|_{S_I^p} \leqslant \|\mathbb{E}\|\,\|\mathcal{U}_{p^*}\|\,\|R_p(y)\|_{S_I^p} \tag{3.67}
$$

$$
\overset{(3.61)}{\lesssim} \max(p,p^*)\big\|R_p(y)\big\|_{S_I^p}.
$$

Replacing y by $A_p^{\frac{1}{2}}(x)$ with $x \in M_{I,\mathrm{fin}}$ we finally obtain

$$
\big\|A_p^{\frac{1}{2}}(x)\big\|_{S_I^p} \overset{(3.67)}{\lesssim} \max(p,p^*)\big\|R_p\big(A_p^{\frac{1}{2}}(x)\big)\big\|_{L^p(\Omega,S_I^p)}
$$

$$
= \max(p,p^*)\big\|\partial_{\alpha,1}A_p^{-\frac{1}{2}}\big(A_p^{\frac{1}{2}}(x)\big)\big\| = \max(p,p^*)\big\|\partial_{\alpha,1}(x)\big\|_{L^p(\Omega,S_I^p)}.
$$

$\square$

The q-Gaussian gradients equally satisfy the equivalence with $A_p^{\frac{1}{2}}$. To this end, the following Lemma 3.19 (which is probably folklore) will be useful. We denote here the canonical conditional expectation $\mathbb{E}\colon L^p(\Gamma_q(H)\overline{\otimes}B(\ell_I^2)) \to S_I^p$. We let

$$
\mathrm{Gauss}_{q,p}(S_I^p) \overset{\mathrm{def}}{=} \overline{\operatorname{span}\big\{s_q(h)\otimes x : h \in H, x \in S_I^p\big\}} \tag{3.68}
$$

where the closure is taken in $L^p(\Gamma_q(H)\overline{\otimes}B(\ell_I^2))$ (for the weak* topology if $p = \infty$ and $-1 \leqslant q < 1$). If $(e_k)_{k\in K}$ is an orthonormal basis of the Hilbert space H, note that $\mathrm{Gauss}_{q,p}(S_I^p)$ is also the closure of the span of the $s_q(e_k)\otimes x$'s with $x \in S_I^p$, closure which we denote temporarily by G'. Indeed, it is trivial that $G' \subset$

Gauss$_{q,p}(S_I^p)$. For the reverse inclusion, it suffices by linearity and density to show that $s_q(h) \otimes x$ belongs to G' for any $h \in H$ and any $x \in S_I^p$. Let $\varepsilon > 0$ be fixed. Then there exists some finite subset F of K and $\alpha_k \in \mathbb{C}$ for $k \in F$ such that $\left\| h - \sum_{k \in F} \alpha_k e_k \right\|_H < \varepsilon$. We have $\sum_{k \in F} \alpha_k s_q(e_k) \otimes x \in G'$ and

$$
\left\| s_q(h) \otimes x - \sum_{k \in F} \alpha_k s_q(e_k) \otimes x \right\|_{L^p(\Gamma_q(H) \overline{\otimes} B(\ell_I^2))} = \left\| s_q\left(h - \sum_{k \in F} \alpha_k e_k\right) \otimes x \right\|_{L^p}
$$

$$
= \left\| s_q\left(h - \sum_{k \in F} \alpha_k e_k\right) \right\|_{L^p(\Gamma_q(H))} \|x\|_{S_I^p}
$$

$$
\cong \left\| s_q\left(h - \sum_{k \in F} \alpha_k e_k\right) \right\|_{L^2(\Gamma_q(H))} \|x\|_{S_I^p} \leqslant \varepsilon \|x\|_{S_I^p}.
$$

We conclude Gauss$_{q,p}(S_I^p) \subset G'$ by closedness of G'.

If $2 \leqslant p < \infty$, recall that we have a contractive injective inclusion

$$
\mathrm{Gauss}_{q,p}(S_I^p) \subset L^p(\Gamma_q(H) \overline{\otimes} B(\ell_I^2)) \subset L^p_{cr}(\mathbb{E}) \tag{3.69}
$$

and that if $1 < p < 2$ we have an inclusion

$$
\mathrm{Gauss}_{q,2}(\mathbb{C}) \otimes S_I^p \subset L^2(\Gamma_q(H)) \otimes S_I^p \subset S_I^p(L^2(\Gamma_q(H))_{\mathrm{rad},p}) \overset{(2.74)}{=} L^p_{cr}(\mathbb{E}). \tag{3.70}
$$

We introduce the orthogonal projection $P \colon L^2(\Gamma_q(H)) \to L^2(\Gamma_q(H))$ onto the closed span Gauss$_{q,2}(\mathbb{C})$ of the $s_q(e_k)$. We let $Q_p \overset{\mathrm{def}}{=} \mathrm{Id}_{S_I^p} \otimes P$ initially defined on $M_{I,\mathrm{fin}} \otimes L^2(\Gamma_q(H))$. We recall some properties of Q_p.

Lemma 3.18 *Let $1 < p < \infty$ and $-1 \leqslant q \leqslant 1$. Then Q_p induces well-defined contractions*

$$
Q_p \colon L^p_c(\mathbb{E}) \to L^p_c(\mathbb{E}) \quad and \quad Q_p \colon L^p_r(\mathbb{E}) \to L^p_r(\mathbb{E}) \tag{3.71}
$$

Thus, according to Lemma 2.11, Q_p also extends to a contraction $Q_p \colon L^p_{cr}(\mathbb{E}) \to L^p_{cr}(\mathbb{E})$.

Proof Since P is a contraction, by (2.71) and (2.74), the column part of (3.71) follows. Then by Remark 2.2, also the row part of (3.71) follows. □

Note that the noncommutative Khintchine inequalities can be rewritten under the following form.

Lemma 3.19 *Consider* $-1 \leqslant q \leqslant 1$.

1. *Suppose* $1 < p < 2$. *For any element* $f = \sum_{i,j,h} f_{i,j,h} s_q(h) \otimes e_{ij}$ *of* $\operatorname{span}\{s_q(h) : h \in H\} \otimes M_{I,\mathrm{fin}}$, *we have*

$$\|f\|_{\mathrm{Gauss}_{q,p}(S_I^p)} \approx_p \|f\|_{S_I^p(\mathrm{L}^2(\Gamma_q(H))_{\mathrm{rad},p})} \tag{3.72}$$

$$\approx_p \inf_{f = g + h} \left\{ \left\| \left(\mathbb{E}(g^* g) \right)^{\frac{1}{2}} \right\|_{S_I^p}, \left\| \left(\mathbb{E}(hh^*) \right)^{\frac{1}{2}} \right\|_{S_I^p} \right\}$$

where the infimum can equally be taken over all $g, h \in \operatorname{span}\{s_q(e) : e \in H\} \otimes M_{I,\mathrm{fin}}$.

2. *Suppose* $2 \leqslant p < \infty$. *For any element* $f = \sum_{i,j,h} f_{i,j,h} s_q(h) \otimes e_{ij}$ *of* $\mathrm{Gauss}_{q,p}(S_I^p)$ *with* $f_{i,j,h} \in \mathbb{C}$

$$\max \left\{ \left\| \left(\mathbb{E}(f^* f) \right)^{\frac{1}{2}} \right\|_{S_I^p}, \left\| \left(\mathbb{E}(ff^*) \right)^{\frac{1}{2}} \right\|_{S_I^p} \right\} \leqslant \|f\|_{\mathrm{Gauss}_{q,p}(S_I^p)}$$

$$\lesssim \sqrt{p} \underbrace{\max \left\{ \left\| \left(\mathbb{E}(f^* f) \right)^{\frac{1}{2}} \right\|_{S_I^p}, \left\| \left(\mathbb{E}(ff^*) \right)^{\frac{1}{2}} \right\|_{S_I^p} \right\}}_{\|f\|_{L_{cr}^p(\mathbb{E})}}. \tag{3.73}$$

Proof 2. Suppose $2 < p < \infty$. We begin by proving the upper estimate of (3.73). Fix $m \geqslant 1$. We have[9] an isometric embedding $J_m \colon \mathrm{H} \to \ell_m^2(H)$ defined by

$$J_m(h) \overset{\mathrm{def}}{=} \frac{1}{\sqrt{m}} \left(\sum_{l=1}^m e_l \right) \otimes h. \tag{3.74}$$

Hence we can consider the operator $\Gamma_p(J_m) \colon \mathrm{L}^p(\Gamma_q(H)) \to \mathrm{L}^p(\Gamma_q(\ell_m^2(H)))$ of second quantization (2.38) which is an isometric completely positive map. By tensorizing with the identity $\mathrm{Id}_{S_I^p}$, we obtain an isometric map $\pi_I \overset{\mathrm{def}}{=} \Gamma_p(J_m) \otimes \mathrm{Id}_{S_I^p} \colon \mathrm{L}^p(\Gamma_q(H) \overline{\otimes} \mathrm{B}(\ell_I^2)) \to \mathrm{L}^p(\Gamma_q(\ell_m^2(H)) \overline{\otimes} \mathrm{B}(\ell_I^2))$. For any finite sum

$$f = \sum_{i,j,h} f_{i,j,h} s_q(h) \otimes e_{ij}$$

---

[9] If $h \in H$, we have

$$\left\| \frac{1}{\sqrt{m}} \left(\sum_{l=1}^m e_l \right) \otimes h \right\|_{\ell_m^2(H)} = \frac{1}{\sqrt{m}} \left\| \sum_{l=1}^m e_l \right\|_{\ell_m^2} \|h\|_H = \|h\|_H.$$

of $L^p(\Gamma_q(H)\overline{\otimes}B(\ell_I^2))$ with $f_{i,j,h} \in \mathbb{C}$, we obtain

$$\|f\|_{L^p(\Gamma_q(H)\overline{\otimes}B(\ell_I^2))} \tag{3.75}$$

$$= \left\|\sum_{i,j,h} f_{i,j,h} s_q(h) \otimes e_{ij}\right\|_{L^p(\Gamma_q(H)\overline{\otimes}B(\ell_I^2))}$$

$$= \left\|\pi_I\left(\sum_{i,j,h} f_{i,j,h} s_q(h) \otimes e_{ij}\right)\right\|_{L^p(\Gamma_q(\ell_m^2(H))\overline{\otimes}B(\ell_I^2))}$$

$$= \left\|\sum_{i,j,h} f_{i,j,h} \Gamma_p(J_m)(s_q(h)) \otimes e_{ij}\right\|_{L^p(\Gamma_q(\ell_m^2(H))\overline{\otimes}B(\ell_I^2))}$$

$$\overset{(2.38)}{=} \left\|\sum_{i,j,h} f_{i,j,h} s_{q,m}(J_m(h)) \otimes e_{ij}\right\|_{L^p(\Gamma_q(\ell_m^2(H))\overline{\otimes}B(\ell_I^2))}$$

$$\overset{(3.74)}{=} \left\|\sum_{i,j,h} f_{i,j,h} s_{q,m}\left(\frac{1}{\sqrt{m}}\left(\sum_{l=1}^m e_l\right) \otimes h\right) \otimes e_{ij}\right\|_{L^p(\Gamma_q(\ell_m^2(H))\overline{\otimes}B(\ell_I^2))}$$

$$= \frac{1}{\sqrt{m}}\left\|\sum_{i,j,h,l} f_{i,j,h} s_{q,m}(e_l \otimes h) \otimes e_{ij}\right\|_{L^p(\Gamma_q(\ell_m^2(H))\overline{\otimes}B(\ell_I^2))}.$$

For any $1 \leqslant l \leqslant m$, consider the element $f_l \overset{\text{def}}{=} \sum_{i,j,h} f_{i,j,h} s_{q,m}(e_l \otimes h) \otimes e_{ij}$ and the conditional expectation $\mathbb{E}\colon L^p(\Gamma_q(\ell_m^2(H))\overline{\otimes}B(\ell_I^2)) \to S_I^p$. For any $1 \leqslant l \leqslant m$, note that

$$\mathbb{E}(f_l) = \mathbb{E}\left(\sum_{i,j,h} f_{i,j,h} s_{q,m}(e_l \otimes h) \otimes e_{ij}\right) = \sum_{i,j,h} f_{i,j,h}\mathbb{E}\big(s_{q,m}(e_l \otimes h) \otimes e_{ij}\big)$$

$$= \sum_{i,j,h} f_{i,j,h}\tau(s_{q,m}(e_l \otimes h))e_{ij} \overset{(2.40)}{=} 0.$$

We deduce that the random variables f_l are mean-zero. Now, we prove in addition that the f_l are independent over $B(\ell_I^2)$, so that we will be able to apply the noncommutative Rosenthal inequality (2.48) to them. The rest of the proof follows page 113. $\qquad\square$

Lemma 3.20 *The f_l defined here previously are independent over $B(\ell_I^2)$. More precisely, the von Neumann algebras generated by f_l are independent with respect to the von Neumann algebra $B(\ell_I^2)$.*

Proof Due to normality of $\mathbb{E}$, it suffices to prove

$$\mathbb{E}(xy) = \mathbb{E}(x)\mathbb{E}(y) \tag{3.76}$$

for x (resp. y) belonging to a weak* dense subset of $\mathrm{W}^*(f_l)$ (resp. of $\mathrm{W}^*((f_k)_{k \neq l})$).
Moreover, by bilinearity in x, y of both sides of (3.76) and selfadjointness of
$s_{q,m}(e_l \otimes h)$, it suffices to take $x = s_{q,m}(e_l \otimes h)^n \otimes e_{ij}$ for some $n \in \mathbb{N}$
and $y = \prod_{t=1}^T s_{q,m}(e_{k_t} \otimes h)^{n_t} \otimes e_{rs}$ for some $T \in \mathbb{N}$, $k_t \neq l$ and $n_t \in \mathbb{N}$.
We have $\mathbb{E}(x)\mathbb{E}(y) = \tau(s_{q,m}(e_l \otimes h)^n)\tau\left(\prod_{t=1}^T s_{q,m}(e_{k_t} \otimes h)^{n_t}\right) \otimes e_{ij}e_{rs}$ and
$\mathbb{E}(xy) = \tau\left(s_{q,m}(e_l \otimes h)^n \prod_{t=1}^T s_{q,m}(e_{k_t} \otimes h)^{n_t}\right) \otimes e_{ij}e_{rs}$. We shall now apply the
Wick formulae (2.39) and (2.40) to the preceding trace term.

Note that if $n + \sum_{t=1}^T n_t$ is odd, then according to the Wick formula (2.40) we
have $\mathbb{E}(xy) = 0$. On the other hand, then either n or $\sum_{t=1}^T n_t$ is odd, so according to
the Wick formula (2.40), either $\mathbb{E}(x) = 0$ or $\mathbb{E}(y) = 0$. Thus, (3.76) follows in this
case.

Now suppose that $2k \overset{\mathrm{def}}{=} n + \sum_{t=1}^T n_t$ is even. Consider a 2-partition $\mathcal{V} \in \mathcal{P}_2(2k)$.
If both n and $\sum_{t=1}^T n_t$ are odd, then we must have some $(i, j) \in \mathcal{V}$ such that one
term $\langle f_i, f_j \rangle_{\ell_m^2(H)}$ in the Wick formula (2.39) equals $\langle e_l \otimes h, e_{k_t} \otimes h \rangle = 0$, since $k_t \neq$
l. Thus, $\mathbb{E}(xy) = 0$, and since n is odd, also $\mathbb{E}(x) = 0$, and therefore, (3.76) follows.
If both n and $\sum_{t=1}^T n_t$ are even, then in the Wick formula (2.39), we only need to
consider those 2-partitions $\mathcal{V}$ without a mixed term as the previous $\langle f_i, f_j \rangle = 0$.
Such a $\mathcal{V}$ is clearly the disjoint union $\mathcal{V} = \mathcal{V}_1 \cup \mathcal{V}_2$ of 2-partitions corresponding to
n and to $\sum_{t=1}^T n_t$. Moreover, we have for the number of crossings, $c(\mathcal{V}) = c(\mathcal{V}_1) +$
$c(\mathcal{V}_2)$. With $(f_1, f_2, \ldots, f_{n+T}) = (\underbrace{e_l \otimes h, e_l \otimes h, \ldots, e_l \otimes h}_{n}, e_{k_1} \otimes h, \ldots, e_{k_T} \otimes$
$h)$, we obtain

$$\tau\left(s_{q,m}(e_l \otimes h)^n \prod_{t=1}^T s_{q,m}(e_{k_t} \otimes h)^{n_t}\right) \overset{(2.39)}{=} \sum_{\mathcal{V} \in \mathcal{P}_2(2k)} q^{c(\mathcal{V})} \prod_{(i,j) \in \mathcal{V}} \langle f_i, f_j \rangle_{\ell_m^2(H)}$$

$$= \sum_{\mathcal{V}_1, \mathcal{V}_2} q^{c(\mathcal{V}_1) + c(\mathcal{V}_2)} \prod_{(i_1, j_1) \in \mathcal{V}_1} \langle f_{i_1}, f_{j_1} \rangle_{\ell_m^2(H)} \prod_{(i_2, j_2) \in \mathcal{V}_2} \langle f_{i_2}, f_{j_2} \rangle_{\ell_m^2(H)}$$

$$= \left(\sum_{\mathcal{V}_1} q^{c(\mathcal{V}_1)} \prod_{(i_1, j_1) \in \mathcal{V}_1} \langle f_{i_1}, f_{j_1} \rangle_{\ell_m^2(H)}\right)\left(\sum_{\mathcal{V}_2} q^{c(\mathcal{V}_2)} \prod_{(i_2, j_2) \in \mathcal{V}_2} \langle f_{i_2}, f_{j_2} \rangle_{\ell_m^2(H)}\right)$$

$$\overset{(2.39)}{=} \tau\left(s_{q,m}(e_l \otimes h)^n\right) \tau\left(\prod_{t=1}^T s_{q,m}(e_{k_t} \otimes h)^{n_t}\right).$$

Thus, also in this case (3.76) follows. $\square$

Proof (End of Proof of Lemma 3.19) Now we are able to apply the noncommutative Rosenthal inequality (2.48) which yields

$$\|f\|_{L^p(\Gamma_q(H)\overline{\otimes}B(\ell_I^2))} \overset{(3.75)}{=} \frac{1}{\sqrt{m}}\left\|\sum_{l=1}^m f_l\right\|_{L^p(\Gamma_q(\ell_m^2(H))\overline{\otimes}B(\ell_I^2))} \tag{3.77}$$

$$\overset{(2.48)}{\lesssim} \frac{1}{\sqrt{m}}\left[p\left(\sum_{l=1}^m \|f_l\|^p_{L^p(\Gamma_q(\ell_m^2(H))\overline{\otimes}B(\ell_I^2))}\right)^{\frac{1}{p}}\right. \tag{3.78}$$

$$\left.+\sqrt{p}\left\|\left(\sum_{l=1}^m \mathbb{E}(f_l^* f_l)\right)^{\frac{1}{2}}\right\|_{S_I^p} + \sqrt{p}\left\|\left(\sum_{l=1}^m \mathbb{E}(f_l f_l^*)\right)^{\frac{1}{2}}\right\|_{S_I^p}\right].$$

For any integer $1 \leqslant l \leqslant m$, note that

$$\mathbb{E}(f_l^* f_l) = \mathbb{E}\left(\left(\sum_{i,j,h} f_{i,j,h}s_{q,m}(e_l \otimes h)\otimes e_{ij}\right)^*\left(\sum_{r,s,k} f_{r,s,k}s_{q,m}(e_l \otimes k)\otimes e_{rs}\right)\right)$$

$$= \sum_{i,j,h,r,s,k} \overline{f_{i,j,h}}\,f_{r,s,k}\left(\tau(s_{q,m}(e_l \otimes h)s_{q,m}(e_l \otimes k))\right)e_{ji}e_{rs}$$

$$\overset{(2.41)}{=} \sum_{i,j,h,s,k} \overline{f_{i,j,h}}\,f_{i,s,k}\langle h,k\rangle_H e_{js}.$$

and

$$\mathbb{E}(f^* f) = \mathbb{E}\left(\left(\sum_{i,j,h} f_{i,j,h}s_q(h)\otimes e_{ij}\right)^*\left(\sum_{r,s,k} f_{r,s,k}s_q(k)\otimes e_{rs}\right)\right)$$

$$= \sum_{i,j,h,r,s,k}\left(\overline{f_{i,j,h}}\,f_{r,s,k}\tau\left(s_q(h)s_q(k)\right)\right)e_{ji}e_{rs}$$

$$\overset{(2.41)}{=} \sum_{i,j,h,s,k} \overline{f_{i,j,h}}\,f_{i,s,k}\langle h,k\rangle_H e_{js}$$

and similarly for the row terms. We conclude that

$$\mathbb{E}(f_l^* f_l) = \mathbb{E}(f^* f) \quad \text{and} \quad \mathbb{E}(f_l f_l^*) = \mathbb{E}(f f^*). \tag{3.79}$$

Moreover, for $1 \leqslant l \leqslant m$, using the isometric map $\psi_l \colon H \to \ell_m^2(H)$, $h \mapsto e_l \otimes h$, we can introduce the second quantization operator $\Gamma_p(\psi_l) \colon L^p(\Gamma_q(\ell_m^2(H))) \to L^p(\Gamma_q(\ell_m^2(H)))$. We have

$$\|f_l\|_{L^p(\Gamma_q(\ell_m^2(H))\overline{\otimes}B(\ell_I^2))} \tag{3.80}$$

$$= \left\|\sum_{i,j,h} f_{i,j,h} s_{q,m}(e_l \otimes h) \otimes e_{ij}\right\|_{L^p(\Gamma_q(\ell_m^2(H))\overline{\otimes}B(\ell_I^2))}$$

$$\overset{(2.38)}{=} \left\|\sum_{i,j,h} f_{i,j,h} \Gamma_p(\psi_l)(s_{q,m}(h)) \otimes e_{ij}\right\|_{L^p(\Gamma_q(\ell_m^2(H))\overline{\otimes}B(\ell_I^2))}$$

$$= \left\|\sum_{i,j,h} f_{i,j,h} s_q(h) \otimes e_{ij}\right\|_{L^p(\Gamma_q(H)\overline{\otimes}B(\ell_I^2))} = \|f\|_{L^p(\Gamma_q(H)\overline{\otimes}B(\ell_I^2))}.$$

We infer that

$$\|f\|_{L^p(\Gamma_q(H)\overline{\otimes}B(\ell_I^2))} \overset{(3.77)(3.80)(3.79)}{\lesssim} \frac{1}{\sqrt{m}}\left[p\left(\sum_{l=1}^m \|f\|_{L^p(\Gamma_q(H)\overline{\otimes}B(\ell_I^2))}^p\right)^{\frac{1}{p}}\right.$$

$$\left. + \sqrt{p}\left\|\left(\sum_{l=1}^m \mathbb{E}(f^*f)\right)^{\frac{1}{2}}\right\|_p + \sqrt{p}\left\|\left(\sum_{l=1}^m \mathbb{E}(ff^*)\right)^{\frac{1}{2}}\right\|_p\right]$$

$$= \frac{1}{\sqrt{m}}\left[pm^{\frac{1}{p}}\|f\|_{L^p} + \sqrt{pm}\left\|(\mathbb{E}(f^*f))^{\frac{1}{2}}\right\|_{S_I^p} + \sqrt{pm}\left\|(\mathbb{E}(ff^*))^{\frac{1}{2}}\right\|_{S_I^p}\right]$$

$$= pm^{\frac{1}{p}-\frac{1}{2}}\|f\|_{L^p} + \sqrt{p}\left\|(\mathbb{E}(f^*f))^{\frac{1}{2}}\right\|_{S_I^p} + \sqrt{p}\left\|(\mathbb{E}(ff^*))^{\frac{1}{2}}\right\|_{S_I^p}.$$

Since $p > 2$, passing to the limit when $m \to \infty$, we finally obtain

$$\|f\|_{L^p(\Gamma_q(H)\overline{\otimes}B(\ell_I^2))} \lesssim \sqrt{p}\left[\left\|(\mathbb{E}(f^*f))^{\frac{1}{2}}\right\|_{S_I^p} + \left\|(\mathbb{E}(ff^*))^{\frac{1}{2}}\right\|_{S_I^p}\right].$$

Using the equivalence $\ell_2^1 \approx \ell_2^\infty$, we obtain the upper estimate of (3.73).

The lower estimate of (3.73) holds with constant 1 from the contractivity of the conditional expectation $\mathbb{E}$ on $L^{\frac{p}{2}}(\Gamma_q(H)\overline{\otimes}B(\ell_I^2))$:

$$\max\left\{\left\|(\mathbb{E}(f^*f))^{\frac{1}{2}}\right\|_{S_I^p}, \left\|(\mathbb{E}(ff^*))^{\frac{1}{2}}\right\|_{S_I^p}\right\} = \max\left\{\left\|\mathbb{E}(f^*f)\right\|_{S_I^{\frac{p}{2}}}^{\frac{1}{2}}, \left\|\mathbb{E}(ff^*)\right\|_{S_I^{\frac{p}{2}}}^{\frac{1}{2}}\right\}$$

$$\leqslant \max\left\{\|f^*f\|_{L^{\frac{p}{2}}(\Gamma_q(H)\overline{\otimes}B(\ell_I^2))}^{\frac{1}{2}}, \|ff^*\|_{L^{\frac{p}{2}}(\Gamma_q(H)\overline{\otimes}B(\ell_I^2))}^{\frac{1}{2}}\right\}$$

$$= \|f\|_{L^p(\Gamma_q(H)\overline{\otimes}B(\ell_I^2))}.$$

1. Let us now consider the case $1 < p < 2$. We will proceed by duality as follows. Recall the Gaussian projection Q_p from Lemma 3.18.

Using in the second equality that $Q_p^* = Q_{p^*}$ and that Q_{p^*} extends to a contraction on $S_I^{p^*}(\mathrm{L}^2(\Gamma_q(H))_{\mathrm{rad},p^*})$ according to Lemma 3.18, and using the upper estimate of (3.73) and the density of $\mathrm{span}\{s_q(h) : h \in H\} \otimes \mathrm{M}_{I,\mathrm{fin}}$ in $S_I^{p^*}(\mathrm{L}^2(\Gamma_q(H))_{\mathrm{rad},p^*})$ in the last inequality, we obtain for any $f = Q_p(f) \in \mathrm{Gauss}_{q,2}(\mathbb{C}) \otimes S_I^p$

$$
\|f\|_{S_I^p(\mathrm{L}^2(\Gamma_q(H))_{\mathrm{rad},p})} \overset{(2.69)}{=} \sup_{\|g\|_{S_I^{p^*}(\mathrm{L}^2(\Gamma_q(H))_{\mathrm{rad},p^*})} \leqslant 1} |\langle f, g \rangle|
$$

$$
= \sup_{\substack{\|g\|_{S_I^{p^*}(\mathrm{L}^2(\Gamma_q(H))_{\mathrm{rad},p^*})} \leqslant 1 \\ g \in \mathrm{Ran}\, Q_{p^*}}} |\langle f, g \rangle|
$$

$$
\leqslant \|f\|_{\mathrm{L}^p(\Gamma_q(H)\overline{\otimes}\mathrm{B}(\ell_I^2))} \sup_{\substack{\|g\|_{S_I^{p^*}(\mathrm{L}^2(\Gamma_q(H))_{\mathrm{rad},p^*})} \leqslant 1 \\ g \in \mathrm{Ran}\, Q_{p^*}}} \|g\|_{\mathrm{L}^{p^*}(\Gamma_q(H)\overline{\otimes}\mathrm{B}(\ell_I^2))}
$$

$$
\overset{(3.72)}{\lesssim_p} \|f\|_{\mathrm{L}^p(\Gamma_q(H)\overline{\otimes}\mathrm{B}(\ell_I^2))}.
$$

Note that in the definition of the norm of $S_I^p(\mathrm{L}^2(\Gamma_q(H))_{\mathrm{rad},p})$, the infimum runs over g, h belonging to $\mathrm{L}_c^p(\mathbb{E})$ and $\mathrm{L}_r^p(\mathbb{E})$. Our next goal is to restrict to those g, h belonging to $\mathrm{span}\{s_q(e) : e \in H\} \otimes \mathrm{M}_{I,\mathrm{fin}}$. To this end, consider a decomposition $f = g + h$ with $g \in \mathrm{L}_c^p(\mathbb{E})$ and $h \in \mathrm{L}_r^p(\mathbb{E})$ such that

$$
\|g\|_{\mathrm{L}_c^p(\mathbb{E})} + \|h\|_{\mathrm{L}_r^p(\mathbb{E})} \leqslant 2 \|f\|_{S_I^p(\mathrm{L}^2(\Gamma_q(H))_{\mathrm{rad},p})}.
$$

Then for some large enough J and with as before Q_p the Gaussian projection, we have

$$
f = Q_p(\mathrm{Id}_{\Gamma_q(H)} \otimes \mathcal{T}_J)(f) = Q_p(\mathrm{Id}_{\Gamma_q(H)} \otimes \mathcal{T}_J)(g) + Q_p(\mathrm{Id}_{\Gamma_q(H)} \otimes \mathcal{T}_J)(h)
$$

and $Q_p(\mathrm{Id}_{\Gamma_q(H)} \otimes \mathcal{T}_J)(g), Q_p(\mathrm{Id}_{\Gamma_q(H)} \otimes \mathcal{T}_J)(h)$ belong to $\mathrm{span}\{s_q(e) : e \in H\} \otimes \mathrm{M}_{I,\mathrm{fin}}$. We claim that we have

$$
\left\|Q_p(\mathrm{Id}_{\Gamma_q(H)} \otimes \mathcal{T}_J)(g)\right\|_{\mathrm{L}_c^p(\mathbb{E})} \leqslant \|g\|_{\mathrm{L}_c^p(\mathbb{E})}
$$

and

$$
\left\|Q_p(\mathrm{Id}_{\Gamma_q(H)} \otimes \mathcal{T}_J)(h)\right\|_{\mathrm{L}_r^p(\mathbb{E})} \leqslant \|h\|_{\mathrm{L}_r^p(\mathbb{E})}.
$$

Indeed, first note that $Q_p \colon L_c^p(\mathbb{E}) \to L_c^p(\mathbb{E})$ and $Q_p \colon L_r^p(\mathbb{E}) \to L_r^p(\mathbb{E}))$ are contractive according to (3.71). Moreover, since $\mathcal{T}_J \colon S_I^p \to S_I^p$ is a complete contraction, according to (2.70), the linear map $\mathcal{T}_J \otimes \mathrm{Id}_{L^2(\Gamma_q(H))} \colon S_I^p(L^2(\Gamma_q(H))_{c,p}) \to S_I^p(L^2(\Gamma_q(H))_{c,p})$ is also a contraction. So we have

$$
\left\| Q_p\big(\mathrm{Id}_{\Gamma_q(H)} \otimes \mathcal{T}_J\big)(g) \right\|_{L_c^p(\mathbb{E})} \overset{(2.74)}{=} \left\| Q_p\big(\mathcal{T}_J \otimes \mathrm{Id}_{L^2(\Gamma_q(H))}\big)(g) \right\|_{S_I^p(L^2(\Gamma_q(H))_{c,p})}
$$

$$
\overset{(2.70)}{\leqslant} \|g\|_{S_I^p(L^2(\Gamma_q(H))_{c,p})} \overset{(2.74)}{=} \|g\|_{L_c^p(\mathbb{E})} .
$$

The row estimate is similar. Together we have shown that the third expression in (3.72) is controlled by $\|f\|_{\mathrm{Gauss}_{q,p}(S_I^p)}$.

Now, we will prove the remaining estimate, that is, $\|f\|_{\mathrm{Gauss}_{q,p}(S_I^p)}$ is controlled by the second expression in (3.72). Since $1 < p < 2$, the function $\mathbb{R}^+ \to \mathbb{R}^+$, $t \mapsto t^{\frac{p}{2}}$ is operator concave by [37, p. 112]. Using [100, Corollary 2.2] applied with the trace preserving positive map $\mathbb{E}$, we can write

$$
\|f\|_{L^p(\Gamma_q(H)\overline{\otimes}B(\ell_I^2))} = \left\| |f|^2 \right\|_{L^{\frac{p}{2}}(\Gamma_q(H)\overline{\otimes}B(\ell_I^2))}^{\frac{1}{2}}
$$

$$
\leqslant \left\| \mathbb{E}(|f|^2) \right\|_{L^{\frac{p}{2}}(\Gamma_q(H)\overline{\otimes}B(\ell_I^2))}^{\frac{1}{2}} = \|f\|_{L_c^p(\mathbb{E})}
$$

and similarly for the row term. Thus, passing to the infimum over all decompositions $f = g + h$, we obtain $\|f\|_{L^p(\Gamma_q(H)\overline{\otimes}B(\ell_I^2))} \leqslant \|f\|_{S_I^p(L^2(\Gamma_q(H))_{\mathrm{rad},p})}$, which can be majorised in turn by the infimum of $\|g\|_{L_c^p(\mathbb{E})} + \|h\|_{L_r^p(\mathbb{E})}$, where $f = g + h$ and $g, h \in \mathrm{span}\{s_q(e) : e \in H\} \otimes M_{I,\mathrm{fin}}$. Hence, we have the last equivalence in the part 1 of the theorem.

The case $p = 2$ is obvious since we have isometrically

$$
L_{cr}^2(\mathbb{E}) = S_I^2(L^2(\Gamma_q(H))_{\mathrm{rad},2}) = L^2(\Gamma_q(H)) \otimes_2 S_I^2
$$

by Lemma 2.9 and [128, Remark 2.3 (1)]. □

We can equally extend Lemma 3.19 to the case that $f \in L^p(\Gamma_q(H)\overline{\otimes}B(\ell_I^2))$ belongs to $\mathrm{Ran}\, Q_p$, where we recall that $Q_p = \mathrm{Id}_{S_I^p} \otimes P \colon L^p(\Gamma_q(H)\overline{\otimes}B(\ell_I^2)) \to L^p(\Gamma_q(H)\overline{\otimes}B(\ell_I^2))$ is the Gaussian projection from Lemma 3.18.

Lemma 3.21 *Let* $-1 \leqslant q \leqslant 1$. *Suppose* $1 < p < \infty$. *Then the linear map* $Q_p \colon L^p(\Gamma_q(H)\overline{\otimes}B(\ell_I^2)) \to L^p(\Gamma_q(H)\overline{\otimes}B(\ell_I^2))$ *is completely bounded.*

Proof Since P is selfadjoint, we have $Q_p^* = Q_{p^*}$. We obtain for $f \in \Gamma_q(H)\overline{\otimes}M_{I,\text{fin}}$,

$$\|Q_p(f)\|_{\text{Gauss}_{q,p}(S_I^p)} \overset{\text{Lemma 3.19}}{\lesssim} \|Q_p(f)\|_{L_{cr}^p(\mathbb{E})}$$

$$= \sup\left\{ |\langle Q_p(f), g\rangle| : \|g\|_{L_{cr}^{p^*}(\mathbb{E})} \leqslant 1 \right\}$$

$$= \sup\left\{ |\langle f, Q_{p^*}(g)\rangle_{L_{cr}^p(\mathbb{E}), L_{cr}^{p^*}(\mathbb{E})}| : g \right\}$$

$$= \sup\left\{ |\langle f, Q_{p^*}(g)\rangle_{L^p(\Gamma_q(H)\overline{\otimes}B(\ell_I^2)), L^{p^*}}| : g \right\}$$

$$\leqslant \|f\|_{L^p(\Gamma_q(H)\overline{\otimes}B(\ell_I^2))} \sup\left\{ \|Q_{p^*}(g)\|_{L^{p^*}(\Gamma_q(H)\overline{\otimes}B(\ell_I^2))} : g \right\}$$

$$\overset{\text{Lemma 3.19}}{\approx} \|f\|_{L^p(\Gamma_q(H)\overline{\otimes}B(\ell_I^2))} \sup\left\{ \|Q_{p^*}(g)\|_{L_{cr}^{p^*}(\mathbb{E})} : \|g\|_{L_{cr}^{p^*}(\mathbb{E})} \leqslant 1 \right\}$$

$$\leqslant \|f\|_{L^p(\Gamma_q(H)\overline{\otimes}B(\ell_I^2))},$$

where in the last step we used that Q_{p^*} is contractive on $L_{cr}^{p^*}(\mathbb{E})$ according to Lemma 3.18. This shows that Q_p is bounded. Then the fact that Q_p is completely bounded follows from a standard matrix amplification argument since we can replace I by $I \times \{1, \ldots, N\}$. □

Lemma 3.22 *Consider* $-1 \leqslant q \leqslant 1$ *and let* $f \in \text{Ran } Q_p = \text{Gauss}_{q,p}(S_I^p)$.

1. Suppose $1 < p < 2$. *We have*

$$\|f\|_{\text{Gauss}_{q,p}(S_I^p)} \approx_p \|f\|_{S_I^p(L^2(\Gamma_q(H))_{\text{rad},p})}$$

$$\approx_p \inf_{f=g+h} \left\{ \|(\mathbb{E}(g^*g))^{\frac{1}{2}}\|_{S_I^p}, \|(\mathbb{E}(hh^*))^{\frac{1}{2}}\|_{S_I^p} \right\}$$

where the infimum can be taken over all $g, h \in \text{Ran } Q_p$.
2. Suppose $2 \leqslant p < \infty$. *We have*

$$\max\left\{ \|(\mathbb{E}(f^*f))^{\frac{1}{2}}\|_{S_I^p}, \|(\mathbb{E}(ff^*))^{\frac{1}{2}}\|_{S_I^p} \right\} \leqslant \|f\|_{\text{Gauss}_{q,p}(S_I^p)}$$

$$\lesssim \sqrt{p} \max\left\{ \|(\mathbb{E}(f^*f))^{\frac{1}{2}}\|_{S_I^p}, \|(\mathbb{E}(ff^*))^{\frac{1}{2}}\|_{S_I^p} \right\}.$$

(3.81)

Proof Observe that $\text{Id}_{\Gamma_q(H)} \otimes \mathcal{T}_J$ approximates the identity on $\Gamma_q(H)\overline{\otimes}M_{I,\text{fin}}$. Moreover, according to Lemma 5.3, the net $(\text{Id}_{\Gamma_q(H)} \otimes \mathcal{T}_J)$ is bounded in the Banach space $B(L^p(\Gamma_q(H)\overline{\otimes}B(\ell_I^2)))$. We conclude by density of $\Gamma_q(H)\overline{\otimes}M_{I,\text{fin}}$ in $L^p(\Gamma_q(H)\overline{\otimes}B(\ell_I^2))$ that the net $(\text{Id}_{\Gamma_q(H)} \otimes \mathcal{T}_J)$ converges in the point norm topology of $L^p(\Gamma_q(H)\overline{\otimes}B(\ell_I^2))$ to the identity. Moreover, replacing in this argument boundedness in $B(L^p(\Gamma_q(H)\overline{\otimes}B(\ell_I^2)))$ by boundedness in $B(L_c^p(\mathbb{E}))$ (resp. $B(L_r^p(\mathbb{E}))$, $B(L_{cr}^p(\mathbb{E}))$) according to (2.70) and (2.74) (resp. Remark 2.2), we obtain that the net $(\mathrm{Id}_{\Gamma_q(H)} \otimes \mathcal{T}_J)$ converges in the point norm topology of $L_c^p(\mathbb{E})$ (resp. $L_r^p(\mathbb{E})$, $L_{cr}^p(\mathbb{E})$) to the identity. Note that for fixed J, $(\mathrm{Id}_{\Gamma_q(H)} \otimes \mathcal{T}_J)(f) = (\mathrm{Id}_{\Gamma_q(H)} \otimes \mathcal{T}_J)(Q_p f) = Q_p(\mathrm{Id}_{\Gamma_q(H)} \otimes \mathcal{T}_J)(f)$ belongs to $\mathrm{span}\{s_q(e) : e \in H\} \otimes M_{I,\mathrm{fin}}$. Thus, Lemma 3.19 applies to f replaced by $(\mathrm{Id}_{\Gamma_q(H)} \otimes \mathcal{T}_J)(f)$ and therefore,

$$\|f\|_{\mathrm{Gauss}_{q,p}(S_I^p)} = \lim_J \left\|(\mathrm{Id}_{\Gamma_q(H)} \otimes \mathcal{T}_J)(f)\right\|_{\mathrm{Gauss}_{q,p}(S_I^p)}$$

$$\overset{\mathrm{Lemma\ 3.19}}{\cong} \lim_J \left\|(\mathrm{Id}_{\Gamma_q(H)} \otimes \mathcal{T}_J)(f)\right\|_{L_{cr}^p(\mathbb{E})} = \|f\|_{L_{cr}^p(\mathbb{E})}.$$

Finally, the fact that one can restrict the infimum to all $g, h \in \mathrm{Ran}\, Q_p$ can be proved in the same way as that in Lemma 3.19. $\square$

Now we can state the Kato square root problem for the case of the gradient taking values in a q-deformed algebra.

Proposition 3.10 *Suppose* $-1 \leqslant q \leqslant 1$ *and* $1 < p < \infty$. *For any* $x \in M_{I,\mathrm{fin}}$, *we have*

$$\left\|A_p^{\frac{1}{2}}(x)\right\|_{S_I^p} \approx_p \left\|\partial_{\alpha,q}(x)\right\|_{L^p(\Gamma_q(H)\overline{\otimes}B(\ell_I^2))}. \tag{3.82}$$

Proof Note that if $q = 1$, then this is a consequence of Theorem 3.3. We show the remaining case $-1 \leqslant q < 1$ by reducing it to the case $q = 1$. Let $x = \sum_{i,j} x_{ij}e_{ij} \in M_{I,\mathrm{fin}}$. Choose an orthonormal basis (e_k) of $\mathrm{span}\{\alpha_i\}$, where we consider only those indices i appearing in the previous double sum describing x. Thus, $\alpha_i - \alpha_j = \sum_k \langle \alpha_i - \alpha_j, e_k\rangle_H e_k$, where the sum is finite. Then

$$\left\|A_p^{\frac{1}{2}}(x)\right\|_{S_I^p} \overset{(3.54)}{\approx_p} \left\|\partial_{\alpha,1}(x)\right\|_{L^p(\Omega,S_I^p)} = \left\|\partial_{\alpha,1}(x)\right\|_{\mathrm{Gauss}_{1,p}(S_I^p)}$$

$$\overset{(2.95)}{=} \left\|\sum_{i,j} x_{ij} W(\alpha_i - \alpha_j) \otimes e_{ij}\right\|_{\mathrm{Gauss}_{1,p}(S_I^p)}$$

$$= \left\|\sum_{i,j} x_{ij} W\left(\sum_k \langle e_k, \alpha_i - \alpha_j\rangle_H e_k\right) \otimes e_{ij}\right\|_{\mathrm{Gauss}_{1,p}(S_I^p)}$$

$$= \left\|\sum_{i,j} x_{ij} \sum_k \langle e_k, \alpha_i - \alpha_j\rangle_H W(e_k) \otimes e_{ij}\right\|_{\mathrm{Gauss}_{1,p}(S_I^p)}$$

$$= \left\| \sum_{k} W(e_k) \otimes \left(\sum_{i,j} x_{ij} \langle e_k, \alpha_i - \alpha_j \rangle_H e_{ij} \right) \right\|_{\mathrm{Gauss}_{1,p}(S_I^p)}$$

$$\overset{[128,\,(2.21),\,(2.22)\,\mathrm{p.}\,12]}{\approx} \left\| \sum_{k} \left(\sum_{i,j} x_{ij} \langle e_k, \alpha_i - \alpha_j \rangle_H e_{ij} \right) \otimes e_k \right\|_{S_I^p(H_{rc})}$$

$$\overset{\mathrm{Lemma}\,3.19}{\approx} \left\| \sum_{k} s_q(e_k) \otimes \left(\sum_{i,j} x_{ij} \langle e_k, \alpha_i - \alpha_j \rangle_H e_{ij} \right) \right\|_{\mathrm{Gauss}_{q,p}(S_I^p)}$$

$$= \left\| \sum_{i,j} x_{ij} s_q(\alpha_i - \alpha_j) \otimes e_{ij} \right\|_{\mathrm{Gauss}_{q,p}(S_I^p)} \overset{(2.95)}{=} \| \partial_{\alpha,q}(x) \|_{\mathrm{Gauss}_{q,p}(S_I^p)} .$$

□

Remark 3.4 Assume $-1 \leqslant q \leqslant 1$. In Proposition 3.10, we obtain again constants depending on p as in (3.56) of a slightly different form, that is, for some absolute constant $K > 0$, we have for all $x \in M_{I,\mathrm{fin}}$,

$$\frac{1}{Kp^2} \| \partial_{\alpha,q}(x) \|_{L^p(\Gamma_q(H) \overline{\otimes} B(\ell_I^2))} \leqslant \| A_p^{\frac{1}{2}}(x) \|_{S_I^p} \leqslant Kp^{\frac{3}{2}} \| \partial_{\alpha,q}(x) \|_{L^p} \tag{3.83}$$

$(2 \leqslant p < \infty)$ and

$$\frac{1}{K(p^*)^2} \| \partial_{\alpha,q}(x) \|_{L^p(\Gamma_q(H) \overline{\otimes} B(\ell_I^2))} \leqslant \| A_p^{\frac{1}{2}}(x) \|_{S_I^p} \leqslant K(p^*)^{\frac{3}{2}} \| \partial_{\alpha,q}(x) \|_{L^p} \tag{3.84}$$

$(1 < p \leqslant 2)$.

Let us prove this and start with the case $2 \leqslant p < \infty$. Note that $W(e_k)$ and $s_q(e_k)$ are orthonormal systems in their spaces $L^2(\Omega)$ and $L^2(\Gamma_q(H))$ respectively. Thus, we have isometrically, for $x_k \in M_{I,\mathrm{fin}}$,

$$\left\| \sum_{k} x_k \otimes W(e_k) \right\|_{S_I^p(L^2(\Omega)_{\mathrm{rad},p})} = \left\| \sum_{k} x_k \otimes s_q(e_k) \right\|_{S_I^p(L^2(\Gamma_q(H))_{\mathrm{rad},p})} . \tag{3.85}$$

Thus, for the upper estimate in (3.83), we have

$$\left\| A_p^{\frac{1}{2}}(x) \right\|_{S_I^p} \overset{(3.65)}{\leqslant} Kp \left\| \partial_{\alpha,1}(x) \right\|_{\mathrm{Gauss}_{1,p}(S_I^p)}$$

$$= Kp \left\| \sum_k \mathrm{W}(e_k) \otimes \left(\sum_{i,j} x_{ij} \langle e_k, \alpha_i - \alpha_j \rangle_H e_{ij} \right) \right\|_{\mathrm{Gauss}_{1,p}(S_I^p)}$$

$$\overset{\mathrm{Lemma\ 3.19}}{\leqslant} K'p \cdot p^{\frac{1}{2}} \left\| \sum_k \left(\sum_{i,j} x_{ij} \langle e_k, \alpha_i - \alpha_j \rangle_H e_{ij} \right) \otimes \mathrm{W}(e_k) \right\|_{S_I^p(\mathrm{L}^2(\Omega)_{\mathrm{rad},p})}$$

$$\overset{(3.85)}{=} K'p^{\frac{3}{2}} \left\| \sum_k \left(\sum_{i,j} x_{ij} \langle e_k, \alpha_i - \alpha_j \rangle_H e_{ij} \right) \otimes s_q(e_k) \right\|_{S_I^p(\mathrm{L}^2(\Gamma_q(H))_{\mathrm{rad},p})}$$

$$\overset{\mathrm{Lemma\ 3.19}}{\leqslant} K'p^{\frac{3}{2}} \left\| \sum_k s_q(e_k) \otimes \left(\sum_{i,j} x_{ij} \langle e_k, \alpha_i - \alpha_j \rangle_H e_{ij} \right) \right\|_{\mathrm{Gauss}_{q,p}(S_I^p)}$$

$$= K'p^{\frac{3}{2}} \left\| \partial_{\alpha,q}(x) \right\|_{\mathrm{Gauss}_{q,p}(S_I^p)}.$$

In the other direction, we have, again for $2 \leqslant p < \infty$,

$$\left\| A_p^{\frac{1}{2}}(x) \right\|_{S_I^p} \overset{(3.56)}{\geqslant} \frac{1}{Kp^{\frac{3}{2}}} \left\| \partial_{\alpha,1}(x) \right\|_{\mathrm{Gauss}_{1,p}(S_I^p)}$$

$$= \frac{1}{Kp^{\frac{3}{2}}} \left\| \sum_k \mathrm{W}(e_k) \otimes \left(\sum_{i,j} x_{ij} \langle e_k, \alpha_i - \alpha_j \rangle_H e_{ij} \right) \right\|_{\mathrm{Gauss}_{1,p}(S_I^p)}$$

$$\overset{\mathrm{Lemma\ 3.19}}{\geqslant} \frac{1}{Kp^{\frac{3}{2}}} \left\| \sum_k \left(\sum_{i,j} x_{ij} \langle e_k, \alpha_i - \alpha_j \rangle_H e_{ij} \right) \otimes \mathrm{W}(e_k) \right\|_{S_I^p(\mathrm{L}^2(\Omega)_{\mathrm{rad},p})}$$

$$\overset{(3.85)}{=} \frac{1}{Kp^{\frac{3}{2}}} \left\| \sum_k \left(\sum_{i,j} x_{ij} \langle e_k, \alpha_i - \alpha_j \rangle_H e_{ij} \right) \otimes s_q(e_k) \right\|_{S_I^p(\mathrm{L}^2(\Gamma_q(H))_{\mathrm{rad},p})}$$

$$\overset{\mathrm{Lemma\ 3.19}}{\geqslant} \frac{1}{K'p^{\frac{3}{2}} \cdot p^{\frac{1}{2}}} \left\| \sum_k s_q(e_k) \otimes \left(\sum_{i,j} x_{ij} \langle e_k, \alpha_i - \alpha_j \rangle_H e_{ij} \right) \right\|_{\mathrm{Gauss}_{q,p}(S_I^p)}$$

$$= \frac{1}{K'p^2} \left\| \partial_{\alpha,q}(x) \right\|_{\mathrm{Gauss}_{q,p}(S_I^p)}.$$

We turn to the case $1 < p \leqslant 2$. Note that (3.85) equally holds in this case. Then we argue in the same way as in the case $p \geqslant 2$, noting that the additional factor $p*^{\frac{1}{2}}$ appears at another step of the estimate, in accordance with Lemma 3.19.

In the remainder of this section, we shall extend the gradient $\partial_{\alpha,q}$ to a closed (weak* closed if $p = \infty$) operator $S_I^p \to \mathrm{L}^p(\Gamma_q(H) \overline{\otimes} \mathrm{B}(\ell_I^2))$, and in particular identify its domain in terms of the generator of the markovian semigroup. All this will be achieved in Proposition 3.11.

We define the densely defined unbounded operator

$$\partial_{\alpha,q}^* : L^p(\Gamma_q(H)) \otimes M_{I,\mathrm{fin}} \subset L^p(\Gamma_q(H))\overline{\otimes}B(\ell_I^2)) \to S_I^p$$

by

$$\partial_{\alpha,q}^*(f \otimes e_{ij}) = \langle s_q(\alpha_i - \alpha_j), f\rangle_{L^{p^*},L^p} e_{ij} = \tau(s_q(\alpha_i - \alpha_j)f)e_{ij} \qquad (3.86)$$

where $i, j \in I$ and $f \in L^p(\Gamma_q(H))$.

Lemma 3.23 *The operators $\partial_{\alpha,q}$ and $\partial_{\alpha,q}^*$ are formal adjoints (with respect to duality brackets $\langle x, y\rangle = \mathrm{Tr}(x^*y)$ and $\langle x, y\rangle = \tau \otimes \mathrm{Tr}(x^*y)$).*

Proof For any $i, j, k, l \in I$ and any $f \in L^p(\Gamma_q(H))$, we have

$$\langle \partial_{\alpha,q}(e_{kl}), f \otimes e_{ij}\rangle_{L^{p^*}(\Gamma_q(H)\overline{\otimes}B(\ell_I^2)),L^p(\Gamma_q(H)\overline{\otimes}B(\ell_I^2))} \qquad (3.87)$$

$$\overset{(2.95)}{=} \langle s_q(\alpha_k - \alpha_l) \otimes e_{kl}, f \otimes e_{ij}\rangle_{L^{p^*}(\Gamma_q(H)\overline{\otimes}B(\ell_I^2)),L^p(\Gamma_q(H)\overline{\otimes}B(\ell_I^2))}$$

$$= \delta_{ki}\delta_{lj}\langle s_q(\alpha_i - \alpha_j), f\rangle_{L^{p^*}(\Gamma_q(H)),L^p(\Gamma_q(H))}$$

$$= \left\langle e_{kl}, \langle s_q(\alpha_i - \alpha_j), f\rangle_{L^{p^*}(\Gamma_q(H)),L^p(\Gamma_q(H))}e_{ij}\right\rangle_{S_I^{p^*},S_I^p}$$

$$= \left\langle e_{kl}, \partial_{\alpha,q}^*(f \otimes e_{ij})\right\rangle_{S_I^{p^*},S_I^p}.$$

$\square$

Proposition 3.11 *Suppose $1 < p < \infty$ and $-1 \leqslant q \leqslant 1$.*

1. *The operator $\partial_{\alpha,q} : M_{I,\mathrm{fin}} \subset S_I^p \to L^p(\Gamma_q(H)\overline{\otimes}B(\ell_I^2))$ is closable as a densely defined operator on S_I^p into $L^p(\Gamma_q(H)\overline{\otimes}B(\ell_I^2))$. We denote by $\partial_{\alpha,q,p}$ its closure. So $M_{I,\mathrm{fin}}$ is a core of $\partial_{\alpha,q,p}$.*

2. *$M_{I,\mathrm{fin}}$ is a core of $\mathrm{dom}\, A_p^{\frac{1}{2}}$.*

3. *We have $\mathrm{dom}\, \partial_{\alpha,q,p} = \mathrm{dom}\, A_p^{\frac{1}{2}}$. Moreover, for any $x \in \mathrm{dom}\, A_p^{\frac{1}{2}}$, we have*

$$\left\|A_p^{\frac{1}{2}}(x)\right\|_{S_I^p} \approx_p \left\|\partial_{\alpha,q,p}(x)\right\|_{\mathrm{Gauss}_{q,p}(S_I^p)}. \qquad (3.88)$$

Finally, for any $x \in \mathrm{dom}\, A_p^{\frac{1}{2}}$ there exists a sequence (x_n) of elements of $M_{I,\mathrm{fin}}$ such that $x_n \to x$, $A_p^{\frac{1}{2}}(x_n) \to A_p^{\frac{1}{2}}(x)$ and $\partial_{\alpha,q,p}(x_n) \to \partial_{\alpha,q,p}(x)$.

4. *If $x \in \operatorname{dom} \partial_{\alpha,q,p}$, we have $x^* \in \operatorname{dom} \partial_{\alpha,q,p}$ and*

$$\left(\partial_{\alpha,q,p}(x)\right)^* = -\partial_{\alpha,q,p}(x^*). \tag{3.89}$$

5. *Suppose that $-1 \leqslant q < 1$. Then the operator $\partial_{\alpha,q} \colon M_{I,\mathrm{fin}} \subset \mathrm{B}(\ell_I^2) \to \Gamma_q(H)\overline{\otimes}\mathrm{B}(\ell_I^2)$ is weak* closable. We denote by $\partial_{\alpha,q,\infty}$ its weak* closure.*
6. *Let $M_B \colon S_I^p \to S_I^p$ be a finitely supported bounded Schur multiplier such that the map $\mathrm{Id} \otimes M_B \colon L^p(\Gamma_q(H)\overline{\otimes}\mathrm{B}(\ell_I^2)) \to L^p(\Gamma_q(H)\overline{\otimes}\mathrm{B}(\ell_I^2))$ is a well-defined bounded operator. For any $x \in \operatorname{dom}\partial_{\alpha,q,p}$, the element $M_B(x)$ belongs to $\operatorname{dom}\partial_{\alpha,q,p}$ and we have*

$$\partial_{\alpha,q,p}M_B(x) = (\mathrm{Id} \otimes M_B)\partial_{\alpha,q,p}(x). \tag{3.90}$$

Proof The proofs of the first three points and of the last point are identical to the proof of Proposition 3.4 replacing $\mathcal{P}_G$ by $M_{I,\mathrm{fin}}$, $L^p(\mathrm{VN}(G))$ by S_I^p and $L^p(\Gamma_q(H) \rtimes_\alpha G)$ by $L^p(\Gamma_q(H) \otimes \mathrm{B}(\ell_I^2))$.

4. For any $i, j \in I$, we have

$$\left(\partial_{\alpha,q,p}(e_{ij})\right)^* \overset{(2.95)}{=} \left(s_q(\alpha_i - \alpha_j) \otimes e_{ij}\right)^* = s_q(\alpha_i - \alpha_j) \otimes e_{ji}$$

$$= -s_q(\alpha_j - \alpha_i) \otimes e_{ji} \overset{(2.95)}{=} -\partial_{\alpha,q,p}(e_{ji}) = -\partial_{\alpha,q,p}(e_{ij}^*).$$

Let $x \in \operatorname{dom}\partial_{\alpha,q,p}$. By the point 1, $M_{I,\mathrm{fin}}$ is core of $\partial_{\alpha,q,p}$. Hence there exists a sequence (x_n) of $M_{I,\mathrm{fin}}$ such that $x_n \to x$ and $\partial_{\alpha,q,p}(x_n) \to \partial_{\alpha,q,p}(x)$. We have $x_n^* \to x^*$ and by the first part of the proof $\partial_{\alpha,q,p}(x_n^*) = -(\partial_{\alpha,q,p}(x_n))^* \to -(\partial_{\alpha,q,p}(x))^*$. By (2.7), we conclude that $x^* \in \operatorname{dom}\partial_{\alpha,q,p}$ and that $\partial_{\alpha,q,p}(x^*) = -(\partial_{\alpha,q,p}(x))^*$.

5. Note that $\left(\mathrm{Id}_{\Gamma_q(H)} \otimes \mathcal{T}_J\right)$ converges for the point weak* topology to $\mathrm{Id}_{\Gamma_q(H)\overline{\otimes}\mathrm{B}(\ell_I^2)}$. Suppose that (x_k) is a net of $M_{I,\mathrm{fin}}$ with $x_k = \sum_{i,j\in I} x_{k,i,j}e_{ij}$ which converges to 0 for the weak* topology such that the net $(\partial_{\alpha,q}(x_k))$ converges for the weak* topology to some y belonging to $\Gamma_q(H)\overline{\otimes}\mathrm{B}(\ell_I^2)$. For any finite subset J of I, we have

$$\partial_{\alpha,q}\mathcal{T}_J(x_k) = \partial_{\alpha,q}\left(\sum_{i,j\in J} x_{k,i,j}e_{ij}\right) = \sum_{i,j\in J} x_{k,i,j}\partial_{\alpha,q}(e_{ij})$$

$$= \left(\mathrm{Id}_{\Gamma_q(H)} \otimes \mathcal{T}_J\right)\partial_{\alpha,q}\left(\sum_{i,j\in I} x_{k,i,j}e_{ij}\right) = \left(\mathrm{Id}_{\Gamma_q(H)} \otimes \mathcal{T}_J\right)\partial_{\alpha,q}(x_k)$$

$$\underset{k}{\to} \left(\mathrm{Id}_{\Gamma_q(H)} \otimes \mathcal{T}_J\right)(y)$$

for the weak* topology. On the other hand, for any $i, j \in I$ we have $x_{k,ij} \to 0$ as $k \to \infty$. Hence for any finite subset J of I

$$\partial_{\alpha,q} \mathcal{T}_J(x_k) = \partial_{\alpha,q} \left(\sum_{i,j \in J} x_{k,i,j} e_{ij} \right) \overset{(2.95)}{=} \sum_{i,j \in J} x_{k,i,j} s_q(\alpha_i - \alpha_j) \otimes e_{ij} \underset{k}{\to} 0.$$

This implies by uniqueness of the $\Gamma_q(H) \overline{\otimes} B(\ell_I^2)$-limit that $\left(\mathrm{Id}_{\Gamma_q(H)} \otimes \mathcal{T}_J \right)(y) = 0$. We deduce that $y = \mathrm{w}^*\text{-}\lim_J \left(\mathrm{Id}_{\Gamma_q(H)} \otimes \mathcal{T}_J \right)(y) = 0$. □

Remark 3.5 It would be more natural to replace the definition (2.95) by the formula

$$\partial_{\alpha,q}(e_{ij}) \overset{\mathrm{def}}{=} 2\pi \mathrm{i} s_q(\alpha_i - \alpha_j) \otimes e_{ij}.$$

With this new definition, the derivation is symmetric, i.e. $\partial_{\alpha,q}(x^*) = \partial_{\alpha,q}(x)^*$. We refer to [225] for more information on weak* closed derivations on von Neumann algebras.

Finally, note that the extension of Lemma 3.15 which can be proved with the point 3 of Proposition 3.11.

Lemma 3.24 *Suppose* $1 < p < \infty$ *and* $-1 \leqslant q \leqslant 1$. *For any* $x \in \mathrm{dom}\, A_p^{\frac{1}{2}}$ *and any* $y \in \mathrm{dom}\, A_{p*}^{\frac{1}{2}}$, *we have*

$$(\tau_{\Gamma_q(H)} \otimes \mathrm{Tr}_{B(\ell_I^2)})\left((\partial_{\alpha,q,p}(x))^* \partial_{\alpha,q,p*}(y) \right) = \mathrm{Tr}_{B(\ell_I^2)}\left((A_p^{\frac{1}{2}}(x))^* A_{p*}^{\frac{1}{2}}(y) \right). \quad (3.91)$$

3.5 Meyer's Problem for Semigroups of Schur Multipliers

In this section, we again fix a markovian semigroup of Schur multipliers, and thus we have some Hilbert space H and a family $(\alpha_i)_{i \in I}$ in H representing the semigroup in the sense of Proposition 2.4. We investigate the so-called Meyer's problem which consists in expressing the L^p norm of the gradient form (3.88) in terms of the carré du champ Γ. Here our results split into the cases $1 < p < 2$ and $2 \leqslant p < \infty$, and are more satisfactory in the second one. The first main result will be Theorem 3.5. In the second part of this section, we shall have a look at Riesz transforms (3.99) associated with the semigroup, carrying a direction vector in the Hilbert space H. Then these directional Riesz transforms have some sort of square functions expressing the S_I^p norm, see Theorem 3.6.

In the case of finite sized matrices, we are already in the position to state the essence of Meyer's problem for semigroups of Schur multipliers. Namely, the following is an immediate consequence of Lemma 3.19, Proposition 3.10 and (2.97). We recall that we have a canonical conditional expectation $\mathbb{E} \colon L^p(\Gamma_q(H)\overline{\otimes}\mathrm{B}(\ell_I^2)) \to S_I^p$.

Corollary 3.1 *1. Suppose $1 < p < 2$. Let $-1 \leqslant q \leqslant 1$. For any $x \in \mathrm{M}_{I,\mathrm{fin}}$ we have*

$$\left\| A_p^{\frac{1}{2}}(x) \right\|_{S_I^p} \approx_p \inf_{\partial_{\alpha,q,p}(x)=y+z} \|y\|_{L_r^p(\mathbb{E})} + \|z\|_{L_c^p(\mathbb{E})}$$

where the infimum is taken over all $y \in \mathrm{Gauss}_{q,2}(\mathbb{C}) \otimes \mathrm{M}_{I,\mathrm{fin}}$ and all $z \in \mathrm{Gauss}_{q,2}(\mathbb{C}) \otimes \mathrm{M}_{I,\mathrm{fin}}$.
2. Suppose $2 \leqslant p < \infty$. For any $x \in \mathrm{M}_{I,\mathrm{fin}}$, we have

$$\left\| A_p^{\frac{1}{2}}(x) \right\|_{S_I^p} \approx_p \max\left\{ \left\|\Gamma(x,x)^{\frac{1}{2}}\right\|_{S_I^p}, \left\|\Gamma(x^*,x^*)^{\frac{1}{2}}\right\|_{S_I^p} \right\}. \tag{3.92}$$

Now, we will extend the definition of $\Gamma(x,y)$ to a larger domain.

Lemma 3.25 *Suppose $2 \leqslant p < \infty$. The forms $a \colon \mathrm{M}_{I,\mathrm{fin}} \times \mathrm{M}_{I,\mathrm{fin}} \to S_I^{\frac{p}{2}} \oplus_\infty S_I^{\frac{p}{2}}$,
$(x,y) \mapsto \Gamma(x,y) \oplus \Gamma(x^*,y^*)$ and $\Gamma \colon \mathrm{M}_{I,\mathrm{fin}} \times \mathrm{M}_{I,\mathrm{fin}} \to S_I^{\frac{p}{2}}$, $(x,y) \mapsto \Gamma(x,y)$
are symmetric, positive and satisfy the Cauchy-Schwarz inequality. The first is in
addition closable and the domain of the closure $\overline{a}$ is $\mathrm{dom}\, A_p^{\frac{1}{2}}$.*

Proof According to the point 3 of Lemma 2.17, we have $\Gamma(x,y)^* = \Gamma(y,x)$, so a is symmetric. Moreover, again according to Lemma 2.17, a is positive. For any $x, y \in \mathrm{M}_{I,\mathrm{fin}}$, we have

$$\|a(x,y)\|_{S_I^{\frac{p}{2}} \oplus_\infty S_I^{\frac{p}{2}}} = \left\|\Gamma(x,y) \oplus \Gamma(x^*,y^*)\right\|_{S_I^{\frac{p}{2}} \oplus_\infty S_I^{\frac{p}{2}}}$$

$$= \max\left\{ \|\Gamma(x,y)\|_{\frac{p}{2}}, \|\Gamma(x^*,y^*)\|_{\frac{p}{2}} \right\}$$

$$\overset{(2.94)}{\leqslant} \max\left\{ \|\Gamma(x,x)\|_{\frac{p}{2}}^{\frac{1}{2}} \|\Gamma(y,y)\|_{\frac{p}{2}}^{\frac{1}{2}}, \|\Gamma(x^*,x^*)\|_{\frac{p}{2}}^{\frac{1}{2}} \|\Gamma(y^*,y^*)\|_{\frac{p}{2}}^{\frac{1}{2}} \right\}$$

$$\leqslant \max\left\{ \|\Gamma(x,x)\|_{\frac{p}{2}}^{\frac{1}{2}}, \|\Gamma(x^*,x^*)\|_{\frac{p}{2}}^{\frac{1}{2}} \right\} \max\left\{ \|\Gamma(y,y)\|_{\frac{p}{2}}^{\frac{1}{2}}, \|\Gamma(y^*,y^*)\|_{\frac{p}{2}}^{\frac{1}{2}} \right\}$$

$$= \|a(x,x)\|_{S_I^{\frac{p}{2}} \oplus_\infty S_I^{\frac{p}{2}}}^{\frac{1}{2}} \|a(y,y)\|_{S_I^{\frac{p}{2}} \oplus_\infty S_I^{\frac{p}{2}}}^{\frac{1}{2}}.$$

So a satisfies the Cauchy-Schwarz inequality. The assertions concerning Γ are similar. Suppose $x_n \xrightarrow{a} 0$ that is $x_n \in M_{I,\mathrm{fin}}$, $x_n \to 0$ and $a(x_n - x_m, x_n - x_m) \to 0$. For any integer n, m, we have

$$\left\| A_p^{\frac{1}{2}}(x_n - x_m) \right\|_{S_I^p} \overset{(3.92)}{\lesssim_p}$$

$$\max\left\{ \left\| \Gamma(x_n - x_m, x_n - x_m)^{\frac{1}{2}} \right\|_{S_I^p}, \left\| \Gamma((x_n - x_m)^*, (x_n - x_m)^*)^{\frac{1}{2}} \right\|_{S_I^p} \right\}$$

$$= \| a(x_n - x_m, x_n - x_m) \|_{S_I^{\frac{p}{2}} \oplus_\infty S_I^{\frac{p}{2}}} \to 0.$$

We infer that $\left(A_p^{\frac{1}{2}}(x_n) \right)$ is a Cauchy sequence, hence a convergent sequence. Since $x_n \to 0$, by the closedness of $A_p^{\frac{1}{2}}$, we deduce that $A_p^{\frac{1}{2}}(x_n) \to 0$. Now, we have

$$\| a(x_n, x_n) \|_{S_I^{\frac{p}{2}} \oplus_\infty S_I^{\frac{p}{2}}}$$

$$= \max\left\{ \left\| \Gamma(x_n, x_n)^{\frac{1}{2}} \right\|_p, \left\| \Gamma(x_n^*, x_n^*))^{\frac{1}{2}} \right\|_p \right\} \overset{(3.92)}{\lesssim_p} \left\| A_p^{\frac{1}{2}}(x_n) \right\|_p \to 0.$$

By Proposition 2.7, we obtain the closability of the form a.

Let $x \in S_I^p$. By (2.52), we have $x \in \mathrm{dom}\,\overline{a}$ if and only if there exists a sequence (x_n) of $M_{I,\mathrm{fin}}$ satisfying $x_n \xrightarrow{a} x$, that is satisfying $x_n \to x$ and $\Gamma(x_n - x_m, x_n - x_m) \to 0$, $\Gamma((x_n - x_m)^*, (x_n - x_m)^*) \to 0$ as $n, m \to \infty$. By the equivalence (3.92), this is equivalent to the existence of a sequence $x_n \in M_{I,\mathrm{fin}}$, such that $x_n \to x$, $A_p^{\frac{1}{2}}(x_n - x_m) \to 0$ as $n, m \to \infty$. Now recalling that $M_{I,\mathrm{fin}}$ is a core of the operator $A_p^{\frac{1}{2}}$, we conclude that this is equivalent to $x \in \mathrm{dom}\,A_p^{\frac{1}{2}}$. $\quad\square$

Remark 3.6 The closability of Γ and the biggest reasonable domain of the form Γ are unclear.

If $2 \leqslant p < \infty$, and $x, y \in \mathrm{dom}\,A_p^{\frac{1}{2}}$, then we let $\Gamma(x, y)$ be the first component of $\overline{a}(x, y)$, where $a \colon M_{I,\mathrm{fin}} \times M_{I,\mathrm{fin}} \to S_I^{\frac{p}{2}} \oplus_\infty S_I^{\frac{p}{2}}$, $(u, v) \mapsto \Gamma(u, v) \oplus \Gamma(u^*, v^*)$ is the form in Lemma 3.25.

Lemma 3.26 *Suppose $-1 \leqslant q \leqslant 1$ and $2 \leqslant p < \infty$.*[10]

1. For any $x, y \in \mathrm{dom}\,A_p^{\frac{1}{2}} = \mathrm{dom}\,\partial_{\alpha,q,p}$, we have

$$\Gamma(x, y) = \langle \partial_{\alpha,q,p}(x), \partial_{\alpha,q,p}(y) \rangle_{S_I^p(L^2(\Omega)_{c,p})} = \mathbb{E}\big((\partial_{\alpha,q,p}(x))^* \partial_{\alpha,q,p}(y) \big). \tag{3.93}$$

---

[10] In the proof, we recall that $L^p(\Gamma_q(H) \overline{\otimes} B(\ell_I^2)) \subset S_I^p(L^2(\Gamma_q(H))_{c,p})$.

2. *For any* $x \in \mathrm{dom}\, A_p^{\frac{1}{2}} = \mathrm{dom}\, \partial_{\alpha,q,p}$, *we have*

$$\left\| \Gamma(x,x)^{\frac{1}{2}} \right\|_{S_I^p} = \left\| \partial_{\alpha,q,p}(x) \right\|_{S_I^p(\mathrm{L}^2(\Gamma_q(H))_{c,p})}. \tag{3.94}$$

Proof Consider some elements $x, y \in \mathrm{dom}\, A_p^{\frac{1}{2}} = \mathrm{dom}\, \partial_{\alpha,q,p}$. According to Proposition 3.11, $\mathrm{M}_{I,\mathrm{fin}}$ is a core of $\mathrm{dom}\, \partial_{\alpha,q,p}$. So by (2.3), there exist sequences (x_n) and (y_n) of $\mathrm{M}_{I,\mathrm{fin}}$ such that $x_n \to x$, $y_n \to y$, $\partial_{\alpha,q,p}(x_n) \to \partial_{\alpha,q,p}(x)$ and $\partial_{\alpha,q,p}(y_n) \to \partial_{\alpha,q,p}(y)$. Since $p \geqslant 2$, we have a contractive inclusion $\mathrm{L}^p(\Gamma_q(H)\overline{\otimes}\mathrm{B}(\ell_I^2)) \hookrightarrow \mathrm{L}_c^p(\mathbb{E}) \overset{(2.74)}{=} S_I^p(\mathrm{L}^2(\Gamma_q(H))_{c,p})$. We deduce that $\partial_{\alpha,q,p}(x_n) \to \partial_{\alpha,q,p}(x)$ and $\partial_{\alpha,q,p}(y_n) \to \partial_{\alpha,q,p}(y)$ in the space $S_I^p(\mathrm{L}^2(\Gamma_q(H))_{c,p})$. Note that

$$\left\| \Gamma(x_n - x_m, x_n - x_m) \right\|_{S_I^{\frac{p}{2}}}$$

$$\overset{(2.97)}{=} \left\| \left\langle \partial_{\alpha,q,p}(x_n - x_m), \partial_{\alpha,q,p}(x_n - x_m) \right\rangle_{S_I^p(\mathrm{L}^2(\Gamma_q(H))_{c,p})} \right\|_{S_I^{\frac{p}{2}}}$$

$$\overset{(2.65)}{\leqslant} \left\| \partial_{\alpha,q,p}(x_n - x_m) \right\|_{S_I^p(\mathrm{L}^2(\Gamma_q(H))_{c,p})} \left\| \partial_{\alpha,q,p}(x_n - x_m) \right\|_{S_I^p(\mathrm{L}^2(\Gamma_q(H))_{c,p})}$$

$$\xrightarrow[n,m\to+\infty]{} 0$$

and similarly for y_n. Hence $x_n \overset{\Gamma}{\to} x$ and $y_n \overset{\Gamma}{\to} y$. We deduce that

$$\Gamma(x,y) \overset{(2.53)}{=} \lim_{n\to+\infty} \Gamma(x_n, y_n) \overset{(2.97)}{=} \lim_{n\to+\infty} \left\langle \partial_{\alpha,q,p}(x_n), \partial_{\alpha,q,p}(y_n) \right\rangle_{S_I^p(\mathrm{L}^2(\Gamma_q(H))_{c,p})}$$

$$= \left\langle \partial_{\alpha,q,p}(x), \partial_{\alpha,q,p}(y) \right\rangle_{S_I^p(\mathrm{L}^2(\Gamma_q(H))_{c,p})}.$$

2. If $x \in \mathrm{dom}\, A_p^{\frac{1}{2}}$, we have

$$\left\| \partial_{\alpha,q,p}(x) \right\|_{S_I^p(\mathrm{L}^2(\Gamma_q(H))_{c,p})} \overset{(2.66)}{=} \left\| \left\langle \partial_{\alpha,q,p}(x), \partial_{\alpha,q,p}(x) \right\rangle_{S_I^p(\mathrm{L}^2(\Gamma_q(H))_{c,p})}^{\frac{1}{2}} \right\|_{S_I^p}$$

$$\overset{(3.93)}{=} \left\| \Gamma(x,x)^{\frac{1}{2}} \right\|_{S_I^p}.$$

$\square$

The following is our first main theorem of this section.

Theorem 3.5 *Suppose* $2 \leqslant p < \infty$. *For any* $x \in \mathrm{dom}\, A_p^{\frac{1}{2}}$, *we have*

$$\left\| A_p^{\frac{1}{2}}(x) \right\|_{S_I^p} \approx_p \max\left\{ \left\| \Gamma(x,x)^{\frac{1}{2}} \right\|_{S_I^p}, \left\| \Gamma(x^*, x^*)^{\frac{1}{2}} \right\|_{S_I^p} \right\}. \tag{3.95}$$

Proof Pick any $-1 \leqslant q \leqslant 1$. For any $x \in \operatorname{dom} A_p^{\frac{1}{2}}$, first note that

$$\mathbb{E}\big(\partial_{\alpha,q,p}(x)(\partial_{\alpha,q,p}(x))^*\big) \overset{(3.89)}{=} \mathbb{E}\big((\partial_{\alpha,q,p}(x^*))^* \partial_{\alpha,q,p}(x^*)\big) \overset{(3.93)}{=} \Gamma(x^*, x^*).$$

$$(3.96)$$

We conclude that

$$\big\| A_p^{\frac{1}{2}}(x) \big\|_{S_I^p} \overset{(3.88)}{\approx_p} \big\| \partial_{\alpha,q,p}(x) \big\|_{\mathrm{Gauss}_{q,p}(S_I^p)}$$

$$\overset{(3.81)}{\approx_p} \max \left\{ \left\| \big(\mathbb{E}((\partial_{\alpha,q,p}(x))^* \partial_{\alpha,q,p}(x))\big)^{\frac{1}{2}} \right\|_p, \right.$$

$$\left. \left\| \big(\mathbb{E}(\partial_{\alpha,q,p}(x)(\partial_{\alpha,q,p}(x)^*))\big)^{\frac{1}{2}} \right\|_p \right\}$$

$$\overset{(3.93)(3.96)}{=} \max \left\{ \big\| \Gamma(x,x)^{\frac{1}{2}} \big\|_{S_I^p}, \big\| \Gamma(x^*,x^*)^{\frac{1}{2}} \big\|_{S_I^p} \right\}.$$

$$\square$$

In Lemma 3.25, we have shown that the carré du champ of Γ is a closable form. In fact, one can give a concrete way how to reach all of its domain by approximation with $M_{I,\mathrm{fin}}$ matrices. This is the content of the next two lemmas. Recall the truncations $\mathcal{T}_J$ from Definition 2.3.

Lemma 3.27 *Suppose* $-1 \leqslant q \leqslant 1$ *and* $2 \leqslant p < \infty$. *Let* J *be a finite subset of* I. *If* $x \in \operatorname{dom} A_p^{\frac{1}{2}}$, *then*

$$\big\| \Gamma\big(\mathcal{T}_J(x), \mathcal{T}_J(x)\big) \big\|_{S_I^{\frac{p}{2}}}^{\frac{1}{2}} \leqslant \big\| \partial_{\alpha,q,p}(x) \big\|_{S_I^p(\mathrm{L}^2(\Gamma_q(H))_{c,p})}.$$

Proof Let $x \in \operatorname{dom} A_p^{\frac{1}{2}} = \operatorname{dom} \partial_{\alpha,q,p}$. Since $M_{I,\mathrm{fin}}$ is a core of $\partial_{\alpha,q,p}$ by Proposition 3.11, there exists a sequence (x_n) of $M_{I,\mathrm{fin}}$ such that $x_n \to x$ and $\partial_{\alpha,q,p}(x_n) \to \partial_{\alpha,q,p}(x)$. Note that since $\mathcal{T}_J \colon S_I^p \to S_I^p$ is a complete contraction, the linear map $\mathcal{T}_J \otimes \operatorname{Id}_{\mathrm{L}^2(\Gamma_q(H))} \colon S_I^p(\mathrm{L}^2(\Gamma_q(H))_{c,p}) \to S_I^p(\mathrm{L}^2(\Gamma_q(H))_{c,p})$ is also a contraction according to (2.70). We deduce that

$$\partial_{\alpha,q,p} \mathcal{T}_J(x_n) = \sum_{i,j \in J} x_{nij} \partial_{\alpha,q,p}(e_{ij}) = \big(\operatorname{Id}_{\mathrm{L}^2(\Gamma_q(H))} \otimes \mathcal{T}_J\big)\left(\sum_{i,j \in I} x_{nij} \partial_{\alpha,q,p}(e_{ij})\right)$$

$$= \big(\operatorname{Id}_{\mathrm{L}^2(\Gamma_q(H))} \otimes \mathcal{T}_J\big)(\partial_{\alpha,q,p} x_n) \xrightarrow[n \to +\infty]{} \big(\operatorname{Id}_{\mathrm{L}^2(\Gamma_q(H))} \otimes \mathcal{T}_J\big)(\partial_{\alpha,q,p} x).$$

Since $\partial_{\alpha,q,p} \mathcal{T}_J$ is bounded by [136, Problem 5.22], we deduce that $\partial_{\alpha,q,p}\mathcal{T}_J(x) = \big(\mathrm{Id}_{\mathrm{L}^2(\Gamma_q(H))} \otimes \mathcal{T}_J\big)(\partial_{\alpha,q,p}x)$. Now, we have

$$\big\|\Gamma(\mathcal{T}_J x, \mathcal{T}_J x)\big\|^{\frac{1}{2}}_{S_I^{\frac{p}{2}}} = \big\|\Gamma(\mathcal{T}_J x, \mathcal{T}_J x)^{\frac{1}{2}}\big\|_{S_I^p} \overset{(2.98)}{=} \big\|\partial_{\alpha,q,p}\mathcal{T}_J(x)\big\|_{S_I^p(\mathrm{L}^2(\Gamma_q(H))_{c,p})}$$

$$= \big\|(\mathrm{Id}_{\mathrm{L}^2(\Gamma_q(H))} \otimes \mathcal{T}_J)(\partial_{\alpha,q,p}x)\big\|_{S_I^p(\mathrm{L}^2(\Gamma_q(H))_{c,p})}$$

$$\overset{(2.70)}{\leqslant} \big\|\partial_{\alpha,q,p}(x)\big\|_{S_I^p(\mathrm{L}^2(\Gamma_q(H))_{c,p})}.$$

$\square$

Now, we give a very concrete way to approximate the carré du champ.

Lemma 3.28 *Let* $2 \leqslant p < \infty$. *For any* $x, y \in \mathrm{dom}\, A_p^{\frac{1}{2}}$, *we have in* $S_I^{\frac{p}{2}}$

$$\Gamma(x, y) = \lim_J \Gamma\big(\mathcal{T}_J(x), \mathcal{T}_J(y)\big). \tag{3.97}$$

Proof If $x \in \mathrm{dom}\, A_p^{\frac{1}{2}}$, we have for any finite subsets J, K of I

$$\big\|\Gamma(\mathcal{T}_J x - \mathcal{T}_K x, \mathcal{T}_J x - \mathcal{T}_K x)\big\|^{\frac{1}{2}}_{S_I^{\frac{p}{2}}} = \big\|\Gamma(\mathcal{T}_J x - \mathcal{T}_K x, \mathcal{T}_J x - \mathcal{T}_K x)^{\frac{1}{2}}\big\|_{S_I^p}$$

$$\overset{(3.95)}{\lesssim_p} \big\|A_p^{\frac{1}{2}}(\mathcal{T}_J x - \mathcal{T}_K x)\big\|_{S_I^p} = \big\|\mathcal{T}_J A_p^{\frac{1}{2}} x - \mathcal{T}_K A_p^{\frac{1}{2}} x\big\|_{S_I^p}.$$

Note that since $A_p^{\frac{1}{2}}(x)$ belongs to S_I^p, $(\mathcal{T}_J A_p^{\frac{1}{2}}(x))_p$ is a Cauchy net of S_I^p. Since $\mathcal{T}_J(x) \to x$ in S_I^p, we infer that $\mathcal{T}_J(x) \overset{\Gamma}{\to} x$. Then (3.97) is a consequence of Lemma 3.27 and Proposition 2.6. $\square$

In the second half of this section, we shall show that directional Riesz transforms associated with markovian semigroups of Schur multipliers decompose the S_I^p norm. Since $A_p(e_{ij}) = \|\alpha_i - \alpha_j\|_H^2 e_{ij}$ for any $i, j \in I$, we have

$$\overline{\mathrm{Ran}\, A_p} = \big\{x \in S_I^p : x_{ij} = 0 \text{ for all } i, j \text{ with } \alpha_i = \alpha_j\big\}. \tag{3.98}$$

Note that $(T_t)_{t \geqslant 0}$ is strongly continuous on S_I^∞ and that it is not difficult to show that (3.98) also holds for $\overline{\mathrm{Ran}\, A_\infty}$, where A_∞ denotes the generator of $(T_t)_{t \geqslant 0}$ on S_I^∞. If $h \in H$, we define the h-directional Riesz transform $R_{\alpha,h}$ defined on $\mathrm{M}_{I,\mathrm{fin}}$ by

$$R_{\alpha,h}(e_{ij}) \overset{\mathrm{def}}{=} \frac{\langle \alpha_i - \alpha_j, h\rangle_H}{\|\alpha_i - \alpha_j\|_H} e_{ij} \text{ if } i, j \text{ satisfy } \alpha_i \neq \alpha_j \tag{3.99}$$

and $R_{\alpha,h}(e_{ij}) = 0$ if it is not the case. If $(e_k)_{k \in K}$ is an orthonormal basis of the Hilbert space H, we let

$$R_{\alpha,k} \stackrel{\text{def}}{=} R_{\alpha,e_k}. \tag{3.100}$$

Suppose $1 < p < \infty$. Consider the contractive linear map

$$U : L^2(\Gamma_q(H)) \to \ell_K^2, g \mapsto \sum_k \langle s_q(e_k), g \rangle_{L^2(\Gamma_q(H))} e_k.$$

By (2.71), we have a bounded map $u \stackrel{\text{def}}{=} \mathrm{Id}_{S_I^p} \otimes U : S_I^p(L^2(\Gamma_q(H))_{c,p}) \to S_I^p(\ell_{K,c,p}^2) \subset S_I^p(S_K^p)$. For any $g \in L^2(\Gamma_q(H))$ and any $i, j \in I$ we have

$$u(e_{ij} \otimes g) = e_{ij} \otimes \sum_k \langle s_q(e_k), g \rangle_{L^2(\Gamma_q(H))} e_{k1} = \sum_k \langle s_q(e_k), g \rangle_{L^2(\Gamma_q(H))} e_{ij} \otimes e_{k1}. \tag{3.101}$$

With the linear form $\langle s_q(e_k), \cdot \rangle : L^2(\Gamma_q(H)) \to \mathbb{C}$ and (2.71), we can introduce the map

$$u_k \stackrel{\text{def}}{=} \mathrm{Id}_{S_I^p} \otimes \langle s_q(e_k), \cdot \rangle : S_I^p(L^2(\Gamma_q(H))_{c,p}) \to S_I^p.$$

For any $g \in L^2(\Gamma_q(H))$, any $i, j \in I$ and any $k \in K$, we have

$$u_k(e_{ij} \otimes g) = \langle s_q(e_k), g \rangle_{L^2(\Gamma_q(H))} e_{ij}. \tag{3.102}$$

For any $k \in I$ and any $f \in L^2(\Gamma_q(H)) \otimes S_I^p$, we have

$$u(f) = \sum_k u_k(f) \otimes e_{k1}. \tag{3.103}$$

In the following proposition, recall again that we have a canonical conditional expectation $\mathbb{E} : L^p(\Gamma_q(H) \overline{\otimes} B(\ell_I^2)) \to S_I^p$.

Proposition 3.12 *Let* $-1 \leqslant q \leqslant 1$.

1. Suppose $2 \leqslant p < \infty$. *For any* $f, h \in \mathrm{Gauss}_{q,p}(S_I^p)$ *we have*

$$\mathbb{E}(f^*h) = (u(f))^*u(h) \tag{3.104}$$

where we identify $(u(f))^*u(h)$ *as an element of* $S_I^{\frac{p}{2}}$ *(with its unique non-zero entry).*

2. *Suppose $1 < p < 2$. For any $f, h \in \mathrm{Gauss}_{q,2}(\mathbb{C}) \otimes S_I^p$ we have*

$$\langle f, h \rangle_{\mathrm{L}_c^p(\mathbb{E})} = (u(f))^* u(h) \tag{3.105}$$

where we identify $(u(f))^* u(h)$ as an element of $S_I^{\frac{p}{2}}$ (with its unique non-zero entry).

3. *Suppose $1 < p < \infty$. If $x \in M_{I,\mathrm{fin}} \cap \mathrm{Ran}\, A$, we have*[11]

$$u_k \partial_{\alpha,q,p} A_p^{-\frac{1}{2}}(x) = R_{\alpha,k}(x). \tag{3.106}$$

4. *Suppose $1 < p < \infty$. If $h \in S_I^p(\mathrm{L}^2(\Gamma_q(H))_{c,p})$, for any $k \in K$ we have*

$$\left(u_k(h^*) \right)^* = u_k(h). \tag{3.107}$$

Proof 1 and 2. For any $i, j, r, s \in I$ and any $g, w \in \mathrm{Gauss}_{q,2}(\mathbb{C})$ (this assumption is used in a crucial way in the last equality), we have

$$(u(e_{ij} \otimes g))^* u(e_{rs} \otimes w)$$

$$\overset{(3.101)}{=} \left(\sum_k \langle s_q(e_k), g \rangle_{\mathrm{L}^2} e_{ij} \otimes e_{k1} \right)^* \left(\sum_l \langle s_q(e_l), w \rangle_{\mathrm{L}^2} e_{rs} \otimes e_{l1} \right)$$

$$= \left(\sum_k \overline{\langle s_q(e_k), g \rangle}_{\mathrm{L}^2(\Gamma_q(H))} e_{ij}^* \otimes e_{1k} \right) \left(\sum_l \langle s_q(e_l), w \rangle_{\mathrm{L}^2(\Gamma_q(H))} e_{rs} \otimes e_{l1} \right)$$

$$= \sum_{k,l} \left(\overline{\langle s_q(e_k), g \rangle}_{\mathrm{L}^2(\Gamma_q(H))} \langle s_q(e_l), w \rangle_{\mathrm{L}^2(\Gamma_q(H))} e_{ij}^* e_{rs} \otimes e_{1k} e_{l1} \right)$$

$$= \sum_k \overline{\langle s_q(e_k), g \rangle}_{\mathrm{L}^2(\Gamma_q(H))} \langle s_q(e_k), w \rangle_{\mathrm{L}^2(\Gamma_q(H))} e_{ij}^* e_{rs} \otimes e_{11}$$

$$= \sum_k \langle g, w \rangle_{\mathrm{L}^2(\Gamma_q(H))} e_{ij}^* e_{rs} \otimes e_{11}.$$

On the other hand, we have

$$\langle e_{ij} \otimes g, e_{rs} \otimes w \rangle_{S_I^p(\mathrm{L}^2(\Gamma_q(H))_{c,p})} \overset{(2.67)}{=} \langle g, w \rangle_{\mathrm{L}^2(\Gamma_q(H))} e_{ij}^* e_{rs}.$$

We conclude by bilinearity and density.

---

[11] Recall that if $2 \leqslant p < \infty$, we have $\mathrm{L}^p(\Gamma_q(H) \overline{\otimes} B(\ell_I^2)) \subset S_I^p(\mathrm{L}^2(\Gamma_q(H))_{c,p})$.

3. For any $i, j \in I$ such that $\alpha_i \neq \alpha_j$, we have

$$u_k \partial_{\alpha,q,p} A_p^{-\frac{1}{2}}(e_{ij}) = u_k \partial_{\alpha,q,p}\left(\frac{1}{\|\alpha_i - \alpha_j\|_H} e_{ij}\right) = \frac{1}{\|\alpha_i - \alpha_j\|_H} u_k \partial_{\alpha,q,p}(e_{ij})$$

$$\overset{(2.95)}{=} \frac{1}{\|\alpha_i - \alpha_j\|_H} u_k\left(e_{ij} \otimes s_q(\alpha_i - \alpha_j)\right)$$

$$\overset{(3.102)}{=} \frac{1}{\|\alpha_i - \alpha_j\|_H} \langle s_q(e_k), s_q(\alpha_i - \alpha_j)\rangle_{L^2(\Gamma_q(H))} e_{ij}$$

$$\overset{(2.41)}{=} \frac{\langle e_k, \alpha_i - \alpha_j\rangle_H}{\|\alpha_i - \alpha_j\|_H} e_{ij} \overset{(3.99)(3.100)}{=} R_{\alpha,k}(e_{ij}).$$

4. Since $s_q(e_k)$ is selfadjoint, we have for any $i, j \in I$ and any $g \in L^2(\Gamma_q(H))$

$$\left[u_k((e_{ij} \otimes g)^*)\right]^* = \left[u_k(e_{ji} \otimes g^*)\right]^* \overset{(3.102)}{=} \left(\langle s_q(e_k), g^*\rangle_{L^2(\Gamma_q(H))} e_{ji}\right)^*$$

$$= \overline{\langle s_q(e_k), g^*\rangle}_{L^2(\Gamma_q(H))} e_{ij} = \langle g^*, s_q(e_k)\rangle_{L^2(\Gamma_q(H))} e_{ij} = \tau(g s_q(e_k)) e_{ij}$$

$$= \tau(s_q(e_k) g) e_{ij} = \langle s_q(e_k), g\rangle_{L^2(\Gamma_q(H))} e_{ij} \overset{(3.102)}{=} u_k(e_{ij} \otimes g).$$

We conclude by linearity and density. $\qquad\square$

Lemma 3.29 *Suppose* $1 < p \leqslant 2$. *The restriction of* $u \colon S_I^p(L^2(\Gamma_q(H))_{c,p}) \to S_I^p(S_K^p)$ *on* $S_I^p(\mathrm{Gauss}_{q,2}(\mathbb{C})_{c,p})$ *induces an isometric map and the range of this restriction is* $S_I^p(\ell_{K,c,p}^2)$.

Proof For any $f \in S_I^p \otimes \mathrm{Gauss}_{q,2}(\mathbb{C})$, we have

$$\|f\|_{L_c^p(\mathbb{E})} \overset{(2.66)}{=} \left\|\langle f, f\rangle_{L_c^p(\mathbb{E})}\right\|_{S_I^{\frac{p}{2}}}^{\frac{1}{2}} \overset{(3.105)}{=} \left\|(u(f))^* u(f)\right\|_{S_I^{\frac{p}{2}}}^{\frac{1}{2}} = \|u(f)\|_{S_I^p(S_K^p)}.$$

and $u(e_{ij} \otimes s_q(e_l)) \overset{(3.101)}{=} e_{ij} \otimes e_l$ for any $i, j \in I$ and any $k \in K$. Alternatively, note that the linear map $\mathrm{Gauss}_{q,2}(\mathbb{C}) \to \ell_K^2$, $g \mapsto \sum_{k \in K} \langle s_q(e_k), g\rangle_{L^2(\Gamma_q(H))} e_k$ is a surjective isometry. So the associated map $S_I^p(\mathrm{Gauss}_{q,2}(\mathbb{C})_{c,p}) \to S_I^p(\ell_{K,c,p}^2)$ is also a surjective isometry. $\qquad\square$

Theorem 3.6 *Suppose* $1 < p < \infty$.

1. If $1 < p \leqslant 2$ *and if* $x \in M_{I,\mathrm{fin}} \cap \mathrm{Ran}\, A$ *we have*

$$\|x\|_{S_I^p} \approx_p \inf_{R_{\alpha,k}(x)=a_k+b_k} \left\|\left(\sum_{k \in K} |a_k|^2\right)^{\frac{1}{2}}\right\|_{S_I^p} + \left\|\left(\sum_{k \in K} |b_k^*|^2\right)^{\frac{1}{2}}\right\|_{S_I^p} \qquad (3.108)$$

where the infimum is taken over all $(a_k), (b_k) \in S_I^p(\ell_{K,c}^2)$.

2. *If $2 \leqslant p < \infty$ and if $x \in M_{I,\text{fin}} \cap \text{Ran } A$, we have*

$$\|x\|_{S_I^p} \approx_p \max\left\{ \left\|\left(\sum_{k \in K} |R_{\alpha,k}(x)|^2\right)^{\frac{1}{2}}\right\|_{S_I^p}, \left\|\left(\sum_{k \in K} |(R_{\alpha,k}(x))^*|^2\right)^{\frac{1}{2}}\right\|_{S_I^p}\right\}.$$

Proof Here we use some fixed $-1 \leqslant q \leqslant 1$ and we drop the index in the notation $\partial_{\alpha,(q,)p}$.

1. Suppose $1 < p < 2$. For any $x \in M_{I,\text{fin}} \cap \text{Ran } A$, we have

$$\|x\|_{S_I^p} = \left\|A_p^{\frac{1}{2}} A_p^{-\frac{1}{2}}(x)\right\|_{S_I^p} \overset{(3.88)}{\approx_p} \left\|\partial_{\alpha,p} A_p^{-\frac{1}{2}}(x)\right\|_{L^p(\Gamma_q(H)\overline{\otimes}B(\ell_I^2))}$$

$$\overset{(3.72)}{\approx_p} \inf_{\partial_{\alpha,p} A_p^{-\frac{1}{2}}(x)=g+h} \|g\|_{L_c^p(\mathbb{E})} + \|h\|_{L_r^p(\mathbb{E})}$$

where the infimum is taken over all $g, h \in \text{Gauss}_2(\mathbb{C}) \otimes M_{I,\text{fin}}$. By Proposition 3.12, we infer that

$$\langle g, g\rangle_{L_c^p(\mathbb{E})}^{\frac{1}{2}} \overset{(3.105)}{=} \left(u(g)^* u(g)\right)^{\frac{1}{2}} \overset{(3.103)}{=} \left(\left(\sum_k u_k(g) \otimes e_{k1}\right)^*\left(\sum_l u_l(g) \otimes e_{l1}\right)\right)^{\frac{1}{2}}$$

$$= \left(\sum_{k,l} (u_k(g))^* u_l(g) \otimes e_{1k}e_{l1}\right)^{\frac{1}{2}} = \left(\sum_k (u_k(g))^* u_k(g)\right)^{\frac{1}{2}}.$$

Similarly, we have

$$\langle h^*, h^*\rangle_{L_c^p(\mathbb{E})}^{\frac{1}{2}} \overset{(3.105)}{=} \left((u(h^*))^* u(h^*)\right)^{\frac{1}{2}}$$

$$\overset{(3.103)}{=} \left(\left(\sum_k u_k(h^*) \otimes e_{k1}\right)^*\left(\sum_l u_l(h^*) \otimes e_{l1}\right)\right)^{\frac{1}{2}}$$

$$\overset{(3.107)}{=} \left(\left(\sum_k u_k(h) \otimes e_{1k}\right)\left(\sum_l u_l(h^*) \otimes e_{l1}\right)\right)^{\frac{1}{2}}$$

$$= \left(\sum_k u_k(h)u_k(h^*)\right)^{\frac{1}{2}}.$$

By Lemma 3.29, the restriction of the map $u \colon S_I^p(L^2(\Gamma_q(H))_{c,p}) \to S_I^p(S_I^p)$ on $S_I^p(\text{Gauss}_{q,2}(\mathbb{C})_{c,p})$ is injective. So we have $\partial_{\alpha,p} A_p^{-\frac{1}{2}}(x) = g + h$ if and only if $u\partial_{\alpha,p} A_p^{-\frac{1}{2}}(x) = u(g) + u(h)$. By (3.106) and (3.103), this is equivalent to

$R_{\alpha,k}(x) = u_k(g) + u_k(h)$ for any k. Hence this computation gives that $\|x\|_{S_I^p}$ is comparable to the norm

$$\inf_{R_{\alpha,k}(x)=u_k(g)+u_k(h)} \left\| \left(\sum_{k\in K} (u_k(g))^* u_k(g) \right)^{\frac{1}{2}} \right\|_{S_I^p} + \left\| \left(\sum_{k\in K} (u_k(h) u_k(h^*) \right)^{\frac{1}{2}} \right\|_{S_I^p}.$$
(3.109)

Therefore, using again Lemma 3.29, this infimum is finally equal to the infimum (3.108).

2. Now, suppose $2 \leqslant p < \infty$. For any $x \in M_{I,\mathrm{fin}} \cap \operatorname{Ran} A$, we have

$$\|x\|_{S_I^p} = \left\| A_p^{\frac{1}{2}} A_p^{-\frac{1}{2}}(x) \right\|_{S_I^p} \overset{(3.88)}{\approx_p} \left\| \partial_{\alpha,p} A_p^{-\frac{1}{2}}(x) \right\|_{\mathrm{Gauss}_{q,p}(S_I^p)}$$

$$\overset{(3.73)}{\approx_p} \left\| \partial_{\alpha,p} A_p^{-\frac{1}{2}}(x) \right\|_{L_{rc}^p(\mathbb{E})}$$

$$= \max\left\{ \left\| \left(\mathbb{E}\left((\partial_{\alpha,p} A_p^{-\frac{1}{2}}(x))^* \partial_{\alpha,p} A_p^{-\frac{1}{2}}(x) \right) \right)^{\frac{1}{2}} \right\|_p , \right.$$

$$\left. \left\| \left(\mathbb{E}\left(\partial_{\alpha} A_p^{-\frac{1}{2}}(x) (\partial_{\alpha,p} A_p^{-\frac{1}{2}}(x))^* \right) \right)^{\frac{1}{2}} \right\|_p \right\}.$$

We obtain

$$\mathbb{E}\left((\partial_{\alpha,p} A_p^{-\frac{1}{2}} x)^* \partial_{\alpha,p} A_p^{-\frac{1}{2}} x \right) \overset{(3.104)}{=} \left[u(\partial_{\alpha,p} A_p^{-\frac{1}{2}}(x)) \right]^* u(\partial_{\alpha,p} A_p^{-\frac{1}{2}}(x))$$

$$\overset{(3.106)}{=} \left(\sum_{k\in K} R_{\alpha,k}(x) \otimes e_{k1} \right)^* \left(\sum_{l\in K} R_{\alpha,l}(x) \otimes e_{l1} \right)$$

$$= \left(\sum_{k\in K} (R_{\alpha,k}(x))^* \otimes e_{1k} \right) \left(\sum_{l\in K} R_{\alpha,l}(x) \otimes e_{l1} \right)$$

$$= \sum_{k\in K} (R_{\alpha,k}(x))^* R_{\alpha,k}(x) = \sum_{k\in K} |R_{\alpha,k}(x)|^2.$$

Similarly, using (3.107) in addition, we obtain

$$\mathbb{E}\left(\partial_{\alpha,p} A_p^{-\frac{1}{2}}(x) (\partial_{\alpha,p} A_p^{-\frac{1}{2}} x)^* \right) \overset{(3.104)}{=} \left(u((\partial_{\alpha,p} A_p^{-\frac{1}{2}} x)^*) \right)^* u((\partial_{\alpha,p} A_p^{-\frac{1}{2}} x)^*)$$

$$\overset{(3.103)}{=} \left(\sum_{k} u_k((\partial_{\alpha,p} A_p^{-\frac{1}{2}} x)^*) \otimes e_{k1} \right)^* \left(\sum_{l} u_l((\partial_{\alpha,p} A_p^{-\frac{1}{2}} x)^*) \otimes e_{l1} \right)$$

$$\overset{(3.107)}{=} \left(\sum_{k} (u_k(\partial_{\alpha,p} A_p^{-\frac{1}{2}} x))^* \otimes e_{k1} \right)^* \left(\sum_{l} (u_l(\partial_{\alpha,p} A_p^{-\frac{1}{2}} x))^* \otimes e_{l1} \right)$$

$$= \left(\sum_k u_k \big(\partial_{\alpha,p} A_p^{-\frac{1}{2}} x \big) \otimes e_{1k} \right) \left(\sum_l \big(u_l \big(\partial_{\alpha,p} A_p^{-\frac{1}{2}} x \big) \big)^* \otimes e_{l1} \right)$$

$$\overset{(3.106)}{=} \left(\sum_k R_{\alpha,k}(x) \otimes e_{1k} \right) \left(\sum_l (R_{\alpha,l}(x))^* \otimes e_{l1} \right) = \sum_{k \in K} \big| (R_{\alpha,k}(x))^* \big|^2.$$

The proof is complete.

$\square$

Chapter 4
Boundedness of H$^\infty$ Functional Calculus of Hodge-Dirac Operators

Abstract In this chapter, we introduce Hodge-Dirac operators associated with markovian semigroups of Fourier multipliers and we show that these operators admit a bounded H$^\infty$ functional calculus on a bisector. We also provide Hodge decompositions. We equally show a similar result for Hodge-Dirac operators associated with markovian semigroups of Schur multipliers. Particular attention is paid to the domain of the Hodge-Dirac operators and different choices are discussed in the course of the chapter. We also prove independence of the bounds of the H$^\infty$ functional calculus, on the group G or index set I, and on the markovian semigroup of multipliers.

4.1 Boundedness of Functional Calculus of Hodge-Dirac Operators for Fourier Multipliers

In this section, we consider a discrete group G and a semigroup $(T_t)_{t \geqslant 0}$ of Markov Fourier multipliers satisfying Proposition 2.3. If $1 \leqslant p < \infty$, we denote by A_p the (negative) infinitesimal generator on L^p(VN(G)). By Junge et al. [128, (5.2)], we have $(A_p)^* = A_{p^*}$ if $1 < p < \infty$. If $-1 \leqslant q \leqslant 1$, recall that by Proposition 3.4, we have a closed operator

$$\partial_{\psi,q,p} \colon \operatorname{dom} \partial_{\psi,q,p} \subset \mathrm{L}^p(\mathrm{VN}(G)) \to \mathrm{L}^p(\Gamma_q(H) \rtimes_\alpha G), \quad \lambda_s \mapsto s_q(b_\psi(s)) \rtimes \lambda_s$$

and a closed operator $(\partial_{\psi,q,p^*})^* \colon \operatorname{dom}(\partial_{\psi,q,p^*})^* \subset \mathrm{L}^p(\Gamma_q(H) \rtimes G) \to$ L^p(VN(G)). Now, we will define a Hodge-Dirac operator $D_{\psi,q,p}$ in (4.16), from $\partial_{\psi,q,p}$ and its adjoint. Then the main topic of this section will be to show that $D_{\psi,q,p}$ is R-bisectorial (Theorem 4.2) and has a bounded H$^\infty$ functional calculus on L^p(VN(G)) $\oplus \overline{\operatorname{Ran} \partial_{\psi,q,p}}$ (Theorem 4.3). By Remark 4.1, this extends the Kato square root equivalence from (3.31).

Note that we use Proposition 2.8, so we need some assumption on the group G. First, with exception of Proposition 4.4, we assume that $\Gamma_q(H) \rtimes_\alpha G$ has QWEP (e.g. G is amenable or is a free group and $q = \pm 1$ by an adaptation of

© The Author(s), under exclusive license to Springer Nature Switzerland AG 2022
C. Arhancet, C. Kriegler, *Riesz Transforms, Hodge-Dirac Operators and Functional Calculus for Multipliers*, Lecture Notes in Mathematics 2304, https://doi.org/10.1007/978-3-030-99011-4_4

[10, Proposition 4.8]). Second, since we need approximating compactly supported Fourier multipliers, from Lemma 4.4 on, we also assume that G is weakly amenable.

We start with the following intuitive formula which show that $\partial_{\psi,q,p}$ can be seen as a "gradient" for A_p in the spirit of the link between the classical Laplacian and the classical gradient.

Proposition 4.1 *Suppose $1 < p < \infty$ and $-1 \leqslant q \leqslant 1$. As unbounded operators, we have*

$$A_p = (\partial_{\psi,q,p^*})^* \partial_{\psi,q,p}. \tag{4.1}$$

Proof By Lemma 3.6 and [136, p. 167], $\partial_{\psi,q,p}(\mathcal{P}_G)$ is a subspace of $\mathrm{dom}(\partial_{\psi,q,p^*})^*$. For any $s \in G$, we have

$$(\partial_{\psi,q,p^*})^* \partial_{\psi,q,p}(\lambda_s) \overset{(2.84)}{=} (\partial_{\psi,q,p^*})^* \big(s_q(b_\psi(s)) \rtimes \lambda_s \big) \tag{4.2}$$

$$\overset{(3.28)}{=} \tau\big(s_q(b_\psi(s)) s_q(b_\psi(s)) \big) \lambda_s \overset{(2.41)}{=} \big\| b_\psi(s) \big\|_H^2 \, \lambda_s = A_p(\lambda_s).$$

Hence for any $x, y \in \mathcal{P}_G$, by linearity we have

$$\big\langle A_p^{\frac{1}{2}}(x), A_{p^*}^{\frac{1}{2}}(y) \big\rangle_{\mathrm{L}^p(\mathrm{VN}(G)), \mathrm{L}^{p^*}(\mathrm{VN}(G))} = \big\langle A_p(x), y \big\rangle_{\mathrm{L}^p(\mathrm{VN}(G)), \mathrm{L}^{p^*}(\mathrm{VN}(G))}$$

$$\overset{(4.2)}{=} \big\langle (\partial_{\psi,q,p^*})^* \partial_{\psi,q,p}(x), y \big\rangle_{\mathrm{L}^p(\mathrm{VN}(G)), \mathrm{L}^{p^*}(\mathrm{VN}(G))}$$

$$\overset{(2.5)}{=} \big\langle \partial_{\psi,q,p}(x), \partial_{\psi,q,p^*}(y) \big\rangle_{\mathrm{L}^p, \mathrm{L}^{p^*}}.$$

Using the part 3 of Proposition 3.4, it is not difficult to see that this identity extends to elements $x \in \mathrm{dom}\, A_p$. For any $x \in \mathrm{dom}\, A_p$ and any $y \in \mathcal{P}_G$, we obtain

$$\big\langle A_p(x), y \big\rangle_{\mathrm{L}^p(\mathrm{VN}(G)), \mathrm{L}^{p^*}(\mathrm{VN}(G))} = \big\langle \partial_{\psi,q,p}(x), \partial_{\psi,q,p^*}(y) \big\rangle_{\mathrm{L}^p, \mathrm{L}^{p^*}}.$$

Recall that $\mathcal{P}_G$ is a core of ∂_{ψ,q,p^*} by the part 1 of Proposition 3.4. So using (2.3), it is easy to check that this identity remains true for elements y of $\mathrm{dom}\, \partial_{\psi,q,p^*}$. By (2.4), this implies that $\partial_{\psi,q,p}(x) \in \mathrm{dom}(\partial_{\psi,q,p^*})^*$ and that $(\partial_{\psi,q,p^*})^* \partial_{\psi,q,p}(x) = A_p(x)$. We conclude that $A_p \subset (\partial_{\psi,q,p^*})^* \partial_{\psi,q,p}$.

To prove the other inclusion we consider some $x \in \mathrm{dom}\, \partial_{\psi,q,p}$ such that $\partial_{\psi,q,p}(x)$ belongs to $\mathrm{dom}(\partial_{\psi,q,p^*})^*$. By Kato [136, Theorem 5.29 p. 168], we have $(\partial_{\psi,q,p^*})^{**} = \partial_{\psi,q,p^*}$. We infer that $(\partial_{\psi,q,p})^* \partial_{\psi,q,p^*} \overset{(2.10)}{\subset} \big((\partial_{\psi,q,p^*})^* \partial_{\psi,q,p} \big)^* \overset{(2.9)}{\subset} A_p^*$. For any $y \in \mathcal{P}_G$, using $\partial_{\psi,q,p}(x) \in \mathrm{dom}(\partial_{\psi,q,p^*})^*$ in the last equality, we deduce that

$$\langle A_p^*(y), x \rangle_{L^{p^*}(\text{VN}(G)), L^p(\text{VN}(G))} = \langle (\partial_{\psi,q,p})^* \partial_{\psi,q,p^*}(y), x \rangle_{L^{p^*}(\text{VN}(G)), L^p(\text{VN}(G))}$$

$$\overset{(2.5)}{=} \langle \partial_{\psi,q,p^*}(y), \partial_{\psi,q,p}(x) \rangle_{L^{p^*}, L^p}$$

$$\overset{(2.5)}{=} \langle y, (\partial_{\psi,q,p^*})^* \partial_{\psi,q,p}(x) \rangle_{L^{p^*}, L^p}.$$

Since $\mathcal{P}_G$ is a core for $A_p^* = A_{p^*}$ by definition, this implies [136, Problem 5.24 p. 168] that $x \in \text{dom } A_p^{**} = A_p$ and that $A_p(x) = (\partial_{\psi,q,p^*})^* \partial_{\psi,q,p}(x)$. □

Now, we show how the noncommutative gradient $\partial_{\psi,q,p}$ commutes with the semigroup and the resolvents of its generator.

Lemma 4.1 *Let G be a discrete group such that $\Gamma_q(H) \rtimes_\alpha G$ has* QWEP. *Suppose $1 < p < \infty$ and $-1 \leqslant q \leqslant 1$. If $x \in \text{dom } \partial_{\psi,q,p}$ and $t \geqslant 0$, then $T_{t,p}(x)$ belongs to $\text{dom } \partial_{\psi,q,p}$ and we have*

$$\left(\text{Id}_{L^p(\Gamma_q(H))} \rtimes T_{t,p} \right) \partial_{\psi,q,p}(x) = \partial_{\psi,q,p} T_{t,p}(x). \tag{4.3}$$

Proof For any $s \in G$, we have

$$\left(\text{Id}_{L^p(\Gamma_q(H))} \rtimes T_{t,p} \right) \partial_{\psi,q,p}(\lambda_s) \overset{(2.84)}{=} \left(\text{Id}_{L^p(\Gamma_q(H))} \rtimes T_{t,p} \right) \left(s_q(b_\psi(s)) \rtimes \lambda_s \right)$$

$$= e^{-t \|b_\psi(s)\|^2} s_q(b_\psi(s)) \rtimes \lambda_s$$

$$\overset{(2.84)}{=} e^{-t \|b_\psi(s)\|^2} \partial_{\psi,q,p}(\lambda_s)$$

$$= \partial_{\psi,q,p} \left(e^{-t \|b_\psi(s)\|^2} \lambda_s \right) = \partial_{\psi,q,p} T_{t,p}(\lambda_s).$$

So by linearity the equality (4.3) is true for elements of $\mathcal{P}_G$. Now consider some $x \in \text{dom } \partial_{\psi,q,p}$. By Kato [136, p. 166], since $\partial_{\psi,q,p}$ is the closure of $\partial_{\psi,q} \colon \mathcal{P}_G \subset L^p(\text{VN}(G)) \to L^p(\Gamma_q(H) \rtimes_\alpha G)$, there exists a sequence (x_n) of elements of $\mathcal{P}_G$ converging to x in $L^p(\text{VN}(G))$ such that the sequence $(\partial_{\psi,q,p}(x_n))$ converges to $\partial_{\psi,q,p}(x)$. The complete boundedness of $T_{t,p} \colon L^p(\text{VN}(G)) \to L^p(\text{VN}(G))$ implies by Proposition 2.8 that we have a (completely) bounded linear operator $\text{Id}_{L^p(\Gamma_q(H))} \rtimes T_{t,p} \colon L^p(\Gamma_q(H) \rtimes_\alpha G) \to L^p(\Gamma_q(H) \rtimes_\alpha G)$. We infer that in $L^p(\text{VN}(G))$ and $L^p(\Gamma_q(H) \rtimes_\alpha G)$ we have $T_{t,p}(x_n) \xrightarrow[n \to +\infty]{} T_{t,p}(x)$ and

$$\left(\text{Id}_{L^p(\Gamma_q(H))} \rtimes T_{t,p} \right) \partial_{\psi,q,p}(x_n) \xrightarrow[n \to +\infty]{} \left(\text{Id}_{L^p(\Gamma_q(H))} \rtimes T_{t,p} \right) \partial_{\psi,q,p}(x).$$

For any integer n, we have $\left(\text{Id}_{L^p(\Gamma_q(H))} \rtimes T_{t,p} \right) \partial_{\psi,q,p}(x_n) = \partial_{\psi,q,p} T_{t,p}(x_n)$ by the first part of the proof. Since the left-hand side converges, we obtain that the sequence $(\partial_{\psi,q,p} T_{t,p}(x_n))$ converges to $(\text{Id}_{L^p(\Gamma_q(H))} \rtimes T_{t,p}) \partial_{\psi,q,p}(x)$ in $L^p(\Gamma_q(H) \rtimes_\alpha G)$.

Since each $T_{t,p}(x_n)$ belongs to dom $\partial_{\psi,q,p}$, the closedness of $\partial_{\psi,q,p}$ and (2.1) shows that $T_{t,p}(x)$ belongs to dom $\partial_{\psi,q,p}$ and that $\partial_{\psi,q,p}T_{t,p}(x) = (\mathrm{Id}_{L^p(\Gamma_q(H))} \rtimes T_{t,p})\partial_{\psi,q,p}(x)$. □

Proposition 4.2 *Let G be a discrete group such that $\Gamma_q(H) \rtimes_\alpha G$ has QWEP. Suppose $1 < p < \infty$ and $-1 \leqslant q \leqslant 1$. For any $s \geqslant 0$ and any $x \in \mathrm{dom}\,\partial_{\psi,q,p}$, we have $(\mathrm{Id} + sA_p)^{-1}(x) \in \mathrm{dom}\,\partial_{\psi,q,p}$ and*

$$\left(\mathrm{Id} \rtimes (\mathrm{Id} + sA_p)^{-1}\right)\partial_{\psi,q,p}(x) = \partial_{\psi,q,p}\left(\mathrm{Id} + sA_p\right)^{-1}(x). \tag{4.4}$$

Proof Note that for any $s > 0$ and any $x \in L^p(\mathrm{VN}(G))$ (resp. $x \in \mathrm{dom}\,\partial_{\psi,q,p}$) the continuous functions $\mathbb{R}^+ \to L^p(\mathrm{VN}(G))$, $t \mapsto e^{-s^{-1}t}T_{t,p}(x)$ and $\mathbb{R}^+ \to L^p(\Gamma_q(H) \rtimes_\alpha G)$, $t \mapsto e^{-s^{-1}t}(\mathrm{Id} \rtimes T_{t,p})\partial_{\psi,q,p}(x)$ are Bochner integrable. If $t > 0$ and if $x \in \mathrm{dom}\,\partial_{\psi,q,p}$, taking Laplace transforms on both sides of (4.3) and using [110, Theorem 1.2.4] and the closedness of $\partial_{\psi,q,p}$ in the penultimate equality, we obtain that $\int_0^\infty e^{-ts^{-1}}T_{t,p}(x)\,dt$ belongs to dom $\partial_{\psi,q,p}$ and that

$$\left(\mathrm{Id} \rtimes (s^{-1}\mathrm{Id} + A_p)^{-1}\right)\partial_{\psi,q,p}(x) = -\left(-s^{-1}\mathrm{Id} - (\mathrm{Id} \rtimes A_p)\right)^{-1}\partial_{\psi,q,p}(x)$$

$$\overset{(2.18)}{=} \int_0^\infty e^{-s^{-1}t}(\mathrm{Id} \rtimes T_{s,p})\partial_{\psi,q,p}(x)\,dt$$

$$\overset{(4.3)}{=} \int_0^\infty e^{-s^{-1}t}\partial_{\psi,q,p}T_{t,p}(x)\,dt$$

$$= \partial_{\psi,q,p}\left(\int_0^\infty e^{-s^{-1}t}T_{t,p}(x)\,dt\right)$$

$$\overset{(2.18)}{=} \partial_{\psi,q,p}\left(s^{-1}\mathrm{Id} + A_p\right)^{-1}(x).$$

We deduce the desired identity by multiplying by s^{-1}. □

Now, we prove a result which gives some R-boundedness.

Proposition 4.3 *Suppose $1 < p < \infty$ and $-1 \leqslant q \leqslant 1$. Let G be a discrete group. The family*

$$\left\{t\partial_{\psi,q,p}(\mathrm{Id} + t^2A_p)^{-1} : t > 0\right\} \tag{4.5}$$

of operators of $\mathrm{B}(L^p(\mathrm{VN}(G)), L^p(\Gamma_q(H) \rtimes_\alpha G))$ is R-bounded.

Proof Note that the operator $\partial_{\psi,q,p}A_p^{-\frac{1}{2}} : \overline{\mathrm{Ran}\,A_p} \to L^p(\Gamma_q(H) \rtimes_\alpha G)$ is bounded by (3.31). Suppose $t > 0$. A standard functional calculus argument gives

$$t\partial_{\psi,q,p}(\mathrm{Id} + t^2A_p)^{-1} = \partial_{\psi,q,p}A_p^{-\frac{1}{2}}\left((t^2A_p)^{\frac{1}{2}}(\mathrm{Id} + t^2A_p)^{-1}\right). \tag{4.6}$$

By Arhancet [13, 14] or [12], note that A_p has a bounded $H^\infty(\Sigma_\theta)$ functional calculus for some $0 < \theta < \frac{\pi}{2}$. Moreover, the Banach space $L^p(\mathrm{VN}(G))$ is UMD by Pisier and Xu [191, Corollary 7.7], hence has the triangular contraction property (Δ) by Hytönen et al. [110, Theorem 7.5.9]. We deduce by Hytönen et al. [110, Theorem 10.3.4 (2)] that the operator A_p is R-sectorial. By Hytönen et al. [110, Example 10.3.5] applied with $\alpha = \frac{1}{2}$ and $\beta = 1$, we infer that the set

$$\left\{ (t^2 A_p)^{\frac{1}{2}} (\mathrm{Id} + t^2 A_p)^{-1} : t > 0 \right\}$$

of operators of $B(L^p(\mathrm{VN}(G)))$ is R-bounded. Recalling that a singleton is R-bounded by Hytönen et al. [110, Example 8.1.7], we obtain by composition [110, Proposition 8.1.19 (3)] that the set

$$\left\{ \partial_{\psi,q,p} A_p^{-\frac{1}{2}} \left((t^2 A_p)^{\frac{1}{2}} (\mathrm{Id} + t^2 A_p)^{-1} \right) : t > 0 \right\}$$

of operators of $B(L^p(\mathrm{VN}(G)), L^p(\Gamma_q(H) \rtimes_\alpha G))$ is R-bounded. Hence with (4.6) we conclude that the subset (4.5) is R-bounded. $\qquad\square$

Our Hodge-Dirac operator in (4.16) below will be constructed out of $\partial_{\psi,q,p}$ and the unbounded operator $(\partial_{\psi,q,p^*})^* | \overline{\mathrm{Ran}\, \partial_{\psi,q,p}}$. Note that the latter is by definition an unbounded operator on the Banach space $\overline{\mathrm{Ran}\, \partial_{\psi,q,p}}$ with values in $L^p(\mathrm{VN}(G))$ having domain $\mathrm{dom}(\partial_{\psi,q,p^*})^* \cap \overline{\mathrm{Ran}\, \partial_{\psi,q,p}}$.

Lemma 4.2 *Let G be a discrete group. Suppose $1 < p < \infty$ and $-1 \leqslant q \leqslant 1$. The operator $(\partial_{\psi,q,p^*})^* | \overline{\mathrm{Ran}\, \partial_{\psi,q,p}}$ is densely defined and is closed. More precisely, the subspace $\partial_{\psi,q,p}(\mathcal{P}_G)$ of $\mathrm{dom}(\partial_{\psi,q,p^*})^*$ is dense in $\overline{\mathrm{Ran}\, \partial_{\psi,q,p}}$.*

Proof Let $y \in \overline{\mathrm{Ran}\, \partial_{\psi,q,p}}$. Let $\varepsilon > 0$. There exists $x \in \mathrm{dom}\, \partial_{\psi,q,p}$ such that $\| y - \partial_{\psi,q,p}(x) \| < \varepsilon$. By Proposition 3.4, there exist $x_{\mathrm{fin}} \in \mathcal{P}_G$ such that $\| x - x_{\mathrm{fin}} \|_{L^p(\mathrm{VN}(G))} < \varepsilon$ and

$$\left\| \partial_{\psi,q,p}(x) - \partial_{\psi,q,p}(x_{\mathrm{fin}}) \right\|_{L^p(\Gamma_q(H) \rtimes_\alpha G)} < \varepsilon.$$

We deduce that $\| y - \partial_{\psi,q,p}(x_{\mathrm{fin}}) \|_{L^p(\Gamma_q(H) \rtimes_\alpha G)} < 2\varepsilon$. By Proposition 4.1, $\partial_{\psi,q,p}(\mathcal{P}_G)$ is a subspace of $\mathrm{dom}(\partial_{\psi,q,p^*})^*$. So $\partial_{\psi,q,p}(x_{\mathrm{fin}})$ belongs to $\mathrm{dom}(\partial_{\psi,q,p^*})^*$.

Since $(\partial_{\psi,q,p^*})^*$ is closed, the assertion on the closedness is (really) obvious. $\qquad\square$

According to Lemma 4.1, $\mathrm{Id} \rtimes T_t$ leaves $\mathrm{Ran}\, \partial_{\psi,q,p}$ invariant for any $t \geqslant 0$, so by continuity of $\mathrm{Id} \rtimes T_t$ also leaves $\overline{\mathrm{Ran}\, \partial_{\psi,q,p}}$ invariant. By Engel and Nagel [85, pp. 60-61], we can consider the generator

$$B_p \overset{\text{def}}{=} (\mathrm{Id}_{L^p(\Gamma_q(H))} \rtimes A_p) | \overline{\mathrm{Ran}\, \partial_{\psi,q,p}}. \tag{4.7}$$

of the restriction of $(\mathrm{Id} \rtimes T_t)_{t \geqslant 0}$ on $\overline{\mathrm{Ran}\, \partial_{\psi,q,p}}$.

Lemma 4.3 *Let G be a discrete group such that $\Gamma_q(H) \rtimes_\alpha G$ has QWEP. Suppose $1 < p < \infty$ and $-1 \leqslant q \leqslant 1$. The operator B_p is injective and sectorial on $\overline{\text{Ran}\, \partial_{\psi,q,p}}$.*

Proof The operator B_p is sectorial of type $\frac{\pi}{2}$ by e.g. [128, p. 25].

For the injectivity, we note that once B_p is known to be sectorial on a reflexive space, we have the projection (2.17) onto the kernel of B_p essentially given by the strong limit

$$P = \lim_{\lambda \to 0^+} \lambda(\lambda + \text{Id} \rtimes A_p)^{-1}|_{\overline{\text{Ran}\, \partial_{\psi,q,p}}}.$$

It is easy to check that $P(y) = 0$ for any $y \in \partial_{\psi,q,p}(\mathscr{P}_G)$. Indeed, by linearity, we can assume that $y = \partial_{\psi,q}(\lambda_s) \overset{(2.84)}{=} s_q(b_\psi(s)) \rtimes \lambda_s$ for some $s \in G$ such that in addition $b_\psi(s) \neq 0$. We claim that

$$(\lambda + \text{Id} \rtimes A_p)^{-1}|_{\overline{\text{Ran}\, \partial_{\psi,q,p}}}\left(s_q(b_\psi(s)) \rtimes \lambda_s\right) = \frac{1}{\lambda + \|b_\psi(s)\|^2} s_q(b_\psi(s)) \rtimes \lambda_s. \quad (4.8)$$

To see (4.8), it suffices to calculate

$$(\lambda + \text{Id} \rtimes A_p)\left(\frac{1}{\lambda + \|b_\psi(s)\|^2} s_q(b_\psi(s)) \rtimes \lambda_s\right)$$

$$= \frac{1}{\lambda + \|b_\psi(s)\|^2}\left(\lambda s_q(b_\psi(s)) \rtimes \lambda_s + \|b_\psi(s)\|^2 s_q(b_\psi(s)) \rtimes \lambda_s\right)$$

$$= s_q(b_\psi(s)) \rtimes \lambda_s.$$

Now $P(y) = 0$ follows from $\frac{\lambda}{\lambda + \|b_\psi(s)\|^2} \to \frac{0}{\|b_\psi(s)\|^2} = 0$ when $\lambda \to 0^+$ since $\|b_\psi(s)\|^2 \neq 0$. Since $\mathscr{P}_G$ is a core for $\partial_{\psi,q,p}$, it is not difficult to see that $P(y) = 0$ also for $y \in \text{Ran}\, \partial_{\psi,q,p}$. Finally by continuity of P, we deduce $P = 0$. Thus, B_p is injective. $\qquad\qquad\square$

Recall that if G is a weakly amenable discrete group, then by the considerations at the end of Sect. 2.4, there exists a net $(\text{Id}_{L^p(\Gamma_q(H))} \rtimes M_{\varphi_j})$ of finitely supported crossed product Fourier multipliers converging strongly to the identity. We also recall from (2.58) that $\mathscr{P}_{p,\rtimes,G}$ is the span of the $x \rtimes \lambda_s$'s, where $x \in L^p(\text{VN}(G))$ and $s \in G$.

We will use the following result.

Lemma 4.4 *Let G be a weakly amenable discrete group such that $\Gamma_q(H) \rtimes_\alpha G$ has QWEP. If $z \in \mathscr{P}_{p,\rtimes,G} \cap \overline{\text{Ran}\, \partial_{\psi,q,p}}$ then $z \in \partial_{\psi,q,p}(\mathscr{P}_G)$.*

Proof Let (z_n) be a sequence in $\text{dom}\, \partial_{\psi,q,p}$ such that $\partial_{\psi,q,p}(z_n) \to z$. Let M_φ be a compactly supported completely bounded Fourier multiplier such that $(\text{Id} \rtimes$

$M_\varphi)(z) = z$. Then

$$\left\| z - \partial_{\psi,q,p} M_\varphi(z_n) \right\|_{L^p} \overset{(3.33)}{=} \left\| z - (\mathrm{Id} \rtimes M_\varphi) \partial_{\psi,q,p}(z_n) \right\|_{L^p}$$

$$\leqslant \left\| z - (\mathrm{Id} \rtimes M_\varphi)(z) \right\|_{L^p} + \left\| (\mathrm{Id} \rtimes M_\varphi)(z - \partial_{\psi,q,p}(z_n)) \right\|_{L^p}$$

$$\lesssim \left\| z - (\mathrm{Id} \rtimes M_\varphi)(z) \right\|_{L^p} + \left\| z - \partial_{\psi,q,p}(z_n) \right\|_{L^p} \xrightarrow[n \to +\infty]{} 0.$$

Therefore, the sequence $(\partial_{\psi,q,p} M_\varphi(z_n))$ is convergent to z for $n \to \infty$. Write $M_\varphi(z_n) = \sum_{s \in F} z_{n,s} \lambda_s$ with some $z_{n,s} \in \mathbb{C}$ and a finite subset F of G. Since for any $s \in G$, the Fourier multiplier M_s associated with the symbol $t \mapsto \delta_s(t)$ is completely bounded, $(\mathrm{Id} \rtimes M_s) \partial_{\psi,q,p} M_\varphi(z_n) = z_{n,s} s_q(b_\psi(s)) \rtimes \lambda_s$ is convergent for $n \to \infty$. Thus, either $b_\psi(s) = 0$ or $(z_{n,s})$ is a convergent scalar sequence with limit, say, $z^s \in \mathbb{C}$. We infer that

$$\partial_{\psi,q,p} M_\varphi(z_n) = \sum_{s \in F} z_{n,s} s_q(b_\psi(s)) \rtimes \lambda_s$$

$$\xrightarrow[n \to +\infty]{} \sum_{s \in F} z^s s_q(b_\psi(s)) \rtimes \lambda_s \overset{(2.84)}{=} \partial_{\psi,q,p}\left(\sum_{s \in F} z^s \lambda_s \right).$$

We conclude that $z = \partial_{\psi,q,p}(\sum_{s \in F} z^s \lambda_s)$ belongs to $\partial_{\psi,q,p}(\mathcal{P}_G)$. $\qquad \square$

Proposition 4.4 *Let G be a discrete group such that $\Gamma_q(H) \rtimes_\alpha G$ has QWEP. In the first three points below, assume in addition that G is weakly amenable, so that the approximating Fourier multipliers M_{φ_j} exist. Suppose $1 < p < \infty$ and $-1 \leqslant q \leqslant 1$.*

1. $\mathcal{P}_{p,\rtimes,G}$ is a core of $(\partial_{\psi,q,p^*})^*$. Furthermore, if $y \in \mathrm{dom}(\partial_{\psi,q,p^*})^*$, then $(\mathrm{Id} \rtimes M_{\varphi_j})(y)$ belongs to $\mathrm{dom}(\partial_{\psi,q,p^*})^*$ for any j and

$$(\partial_{\psi,q,p^*})^*(\mathrm{Id} \rtimes M_{\varphi_j})(y) = M_{\varphi_j}(\partial_{\psi,q,p^*})^*(y). \tag{4.9}$$

2. $\partial_{\psi,q,p}(\mathcal{P}_G)$ is a core of $(\partial_{\psi,q,p^*})^*|\overline{\mathrm{Ran}\, \partial_{\psi,q,p}}$.
3. $\partial_{\psi,q,p}(\mathcal{P}_G)$ is equally a core of $\partial_{\psi,q,p}(\partial_{\psi,q,p^*})^*|\overline{\mathrm{Ran}\, \partial_{\psi,q,p}}$.
4. $\partial_{\psi,q,p}(\mathcal{P}_G)$ is equally a core of B_p.

Proof

1. Let $y \in \mathrm{dom}(\partial_{\psi,q,p^*})^*$. Then $(\mathrm{Id}_{L^p(\Gamma_q(H))} \rtimes M_{\varphi_j})(y)$ belongs to $\mathcal{P}_{p,\rtimes,G}$. It remains to show that $(\mathrm{Id}_{L^p(\Gamma_q(H))} \rtimes M_{\varphi_j})(y)$ converges to y in the graph norm. Recall that $(\mathrm{Id}_{L^p(\Gamma_q(H))} \rtimes M_{\varphi_j})(y)$ converges to y in $L^p(\Gamma_q(H) \rtimes_\alpha G)$ according to the assumptions. For any $s \in G$, we have

$$\left\langle (\partial_{\psi,q,p^*})^*(\mathrm{Id} \rtimes M_{\varphi_j})(y), \lambda_s \right\rangle$$

$$\overset{(2.5)}{=} \left\langle (\mathrm{Id} \rtimes M_{\varphi_j})(y), \partial_{\psi,q,p^*}(\lambda_s) \right\rangle$$

$$= \big\langle y, (\mathrm{Id} \rtimes M_{\overline{\varphi_j}})(\partial_{\psi,q,p^*}(\lambda_s)) \big\rangle \overset{(3.33)}{=} \big\langle y, \partial_{\psi,q,p^*} M_{\overline{\varphi_j}}(\lambda_s) \big\rangle$$

$$= \big\langle (\partial_{\psi,q,p^*})^*(y), M_{\overline{\varphi_j}}(\lambda_s) \big\rangle = \big\langle M_{\varphi_j}(\partial_{\psi,q,p^*})^*(y), \lambda_s \big\rangle.$$

By linearity and density, we deduce the relation (4.9), i.e.

$$(\partial_{\psi,q,p^*})^*(\mathrm{Id} \rtimes M_{\varphi_j})(y) = M_{\varphi_j}(\partial_{\psi,q,p^*})^*(y)$$

which converges to $(\partial_{\psi,q,p^*})^*(y)$ in $\mathrm{L}^p(\mathrm{VN}(G))$.

2. Let $y \in \mathrm{dom}(\partial_{\psi,q,p^*})^* | \overline{\mathrm{Ran}\, \partial_{\psi,q,p}}$ and $\varepsilon > 0$. By the proof of the first point, we already know that for some j large enough, there is $z \overset{\mathrm{def}}{=} (\mathrm{Id} \rtimes M_{\varphi_j})(y)$ belonging to $\mathcal{P}_{p,\rtimes,G}$ such that

$$\|y - z\|_{\mathrm{L}^p} < \varepsilon \quad \text{and} \quad \big\|(\partial_{\psi,q,p^*})^*(y) - (\partial_{\psi,q,p^*})^*(z)\big\|_{\mathrm{L}^p} < \varepsilon.$$

We claim that z belongs to $\overline{\mathrm{Ran}\, \partial_{\psi,q,p}}$. Indeed, for $\delta > 0$, since $y \in \overline{\mathrm{Ran}\, \partial_{\psi,q,p}}$, there is some $a \in \mathrm{dom}\, \partial_{\psi,q,p}$ such that $\big\|y - \partial_{\psi,q,p}(a)\big\|_{\mathrm{L}^p} < \delta$. But then we have

$$\big\|z - (\mathrm{Id} \rtimes M_{\varphi_j})\partial_{\psi,q,p}(a)\big\|_{\mathrm{L}^p} = \big\|(\mathrm{Id} \rtimes M_{\varphi_j})(y - \partial_{\psi,q,p}(a))\big\|_{\mathrm{L}^p}$$

$$\lesssim \big\|y - \partial_{\psi,q,p}(a)\big\|_{\mathrm{L}^p} < \delta$$

and $(\mathrm{Id} \rtimes M_{\varphi_j})\partial_{\psi,q,p}(a) \overset{(3.33)}{=} \partial_{\psi,q,p} M_{\varphi_j}(a)$ belongs again to $\mathrm{Ran}\, \partial_{\psi,q,p}$. Letting $\delta \to 0$, it follows that z belongs to $\overline{\mathrm{Ran}\, \partial_{\psi,q,p}}$.

By Lemma 4.4, we conclude that z belongs to $\partial_{\psi,q,p}(\mathcal{P}_G)$.

3. Take $y \in \mathrm{dom}\, \partial_{\psi,q,p}(\partial_{\psi,q,p^*})^* | \overline{\mathrm{Ran}\, \partial_{\psi,q,p}}$, that is, $y \in \mathrm{dom}(\partial_{\psi,q,p^*})^* \cap \overline{\mathrm{Ran}\, \partial_{\psi,q,p}}$ such that $(\partial_{\psi,q,p^*})^*(y) \in \mathrm{dom}\, \partial_{\psi,q,p}$. We have $(\mathrm{Id} \rtimes M_{\varphi_j})(y) \to y$ and $(\mathrm{Id} \rtimes M_{\varphi_j})(y) \in \partial_{\psi,q,p}(\mathcal{P}_G)$ by the proof of part 2. Moreover, we have

$$\partial_{\psi,q,p}(\partial_{\psi,q,p^*})^*(\mathrm{Id} \rtimes M_{\varphi_j})(y)$$

$$\overset{(4.9)}{=} \partial_{\psi,q,p} M_{\varphi_j}(\partial_{\psi,q,p^*})^*(y)$$

$$\overset{(3.33)}{=} (\mathrm{Id} \rtimes M_{\varphi_j})\partial_{\psi,q,p}(\partial_{\psi,q,p^*})^*(y) \underset{j}{\longrightarrow} \partial_{\psi,q,p}(\partial_{\psi,q,p^*})^*(y).$$

4. Note that $\partial_{\psi,q,p}(\mathcal{P}_G)$ is a dense subspace of $\overline{\mathrm{Ran}\, \partial_{\psi,q,p}}$ which is clearly a subspace of $\mathrm{dom}\, B_p$ and invariant under each operator $(\mathrm{Id} \rtimes T_t) | \overline{\mathrm{Ran}\, \partial_{\psi,q,p}}$ by Lemma 4.1. By Lemma 2.2, we deduce that $\partial_{\psi,q,p}(\mathcal{P}_G)$ is a core of $(\mathrm{Id} \rtimes A_p) | \overline{\mathrm{Ran}\, \partial_{\psi,q,p}} \overset{(4.7)}{=} B_p$.

$\square$

Proposition 4.5 *Let G be a weakly amenable discrete group such that $\Gamma_q(H) \rtimes_\alpha G$ has QWEP. Suppose $1 < p < \infty$ and $-1 \leqslant q \leqslant 1$.*

1. *For any $s > 0$, the operator $(\mathrm{Id} + s A_p)^{-1} (\partial_{\psi,q,p^*})^*$ induces a bounded operator on $\overline{\mathrm{Ran}\,\partial_{\psi,q,p}}$.*

2. *For any $s \geqslant 0$ and any $y \in \overline{\mathrm{Ran}\,\partial_{\psi,q,p}} \cap \mathrm{dom}(\partial_{\psi,q,p^*})^*$, the element $\big(\mathrm{Id} \rtimes (\mathrm{Id} + s A_p)^{-1}\big)(y)$ belongs to $\mathrm{dom}(\partial_{\psi,q,p^*})^*$ and*

$$\big(\mathrm{Id} + s A_p\big)^{-1}(\partial_{\psi,q,p^*})^*(y) = (\partial_{\psi,q,p^*})^*\big(\mathrm{Id} \rtimes (\mathrm{Id} + s A_p)^{-1}\big)(y). \qquad (4.10)$$

3. *For any $t \geqslant 0$ and any $y \in \overline{\mathrm{Ran}\,\partial_{\psi,q,p}} \cap \mathrm{dom}(\partial_{\psi,q,p^*})^*$, the element $(\mathrm{Id} \rtimes T_t)(y)$ belongs to $\mathrm{dom}(\partial_{\psi,q,p^*})^*$ and*

$$T_t(\partial_{\psi,q,p^*})^*(y) = (\partial_{\psi,q,p^*})^*\big(\mathrm{Id} \rtimes T_t\big)(y).$$

Proof

1. Note that $(\mathrm{Id} + s A_p)^{-1}(\partial_{\psi,q,p^*})^* \overset{(2.10)}{\subset} \big(\partial_{\psi,q,p^*}(\mathrm{Id} + s A_{p^*})^{-1}\big)^*$. By Proposition 4.3, the operator $\big(\partial_{\psi,q,p^*}(\mathrm{Id} + s A_{p^*})^{-1}\big)^*$ is bounded. By Lemma 4.2, the subspace $\partial_{\psi,q,p}(\mathcal{P}_G)$ of $\mathrm{dom}(\partial_{\psi,q,p^*})^*$ is dense in $\overline{\mathrm{Ran}\,\partial_{\psi,q,p}}$. Now, the conclusion is immediate.

2. By Proposition 4.1, for any $x \in \mathrm{dom}\,A_p$ we have $x \in \mathrm{dom}\,\partial_{\psi,q,p}$ and $\partial_{\psi,q,p}(x) \in \mathrm{dom}(\partial_{\psi,q,p^*})^*$. Moreover, for all $t > 0$ we have

$$T_t(\partial_{\psi,q,p^*})^* \partial_{\psi,q,p}(x) \overset{(4.1)}{=} T_t A_p(x) \overset{(2.13)}{=} A_p T_t(x) \overset{(4.1)}{=} (\partial_{\psi,q,p^*})^* \partial_{\psi,q,p} T_t(x)$$

$$\overset{(4.3)}{=} (\partial_{\psi,q,p^*})^*(\mathrm{Id} \rtimes T_t)\partial_{\psi,q,p}(x).$$

By taking Laplace transforms with (2.18) and using the closedness of $(\partial_{\psi,q,p^*})^*$, we deduce that the element $(\mathrm{Id} + s\,\mathrm{Id} \rtimes A_p)^{-1} \partial_{\psi,q,p}(x)$ belongs to $\mathrm{dom}(\partial_{\psi,q,p^*})^*$ for any $s \geqslant 0$ and that

$$(\mathrm{Id} + s A_p)^{-1}(\partial_{\psi,q,p^*})^* \partial_{\psi,q,p}(x) = (\partial_{\psi,q,p^*})^*(\mathrm{Id} \rtimes (\mathrm{Id} + s A_p))^{-1} \partial_{\psi,q,p}(x). \qquad (4.11)$$

Let $y \in \overline{\mathrm{Ran}\,\partial_{\psi,q,p}} \cap \mathrm{dom}(\partial_{\psi,q,p^*})^*$. Then according to Lemma 4.2, there exists a sequence (x_n) of $\mathcal{P}_G$ such that $\partial_{\psi,q,p}(x_n) \to y$. We have $(\mathrm{Id} \rtimes (\mathrm{Id} + s A_p))^{-1} \partial_{\psi,q,p}(x_n) \to (\mathrm{Id} \rtimes (\mathrm{Id} + s A_p))^{-1}(y)$. Since each x_n belongs to $\mathrm{dom}\,A_p$, using the first point in the passage to the limit we deduce that

$$(\partial_{\psi,q,p^*})^*(\mathrm{Id} \rtimes (\mathrm{Id} + s A_p))^{-1} \partial_{\psi,q,p}(x_n) \overset{(4.11)}{=} (\mathrm{Id} + s A_p)^{-1}(\partial_{\psi,q,p^*})^* \partial_{\psi,q,p}(x_n)$$

$$\xrightarrow[n \to +\infty]{} (\mathrm{Id} + s A_p)^{-1}(\partial_{\psi,q,p^*})^*(y).$$

Since $(\partial_{\psi,q,p^*})^*$ is closed, we infer by (2.1) that $(\mathrm{Id} \rtimes (\mathrm{Id} + sA_p))^{-1}(y)$ belongs to $\mathrm{dom}(\partial_{\psi,q,p^*})^*$ and that

$$(\partial_{\psi,q,p^*})^*(\mathrm{Id} \rtimes (\mathrm{Id} + sA_p))^{-1}(y) = (\mathrm{Id} + sA_p)^{-1}(\partial_{\psi,q,p^*})^*(y). \qquad (4.12)$$

Thus (4.10) follows.

3. Let $y \in \overline{\mathrm{Ran}\, \partial_{\psi,q,p}} \cap \mathrm{dom}(\partial_{\psi,q,p^*})^*$. If $t \geqslant 0$, note that

$$\left(\mathrm{Id} \rtimes \left(\mathrm{Id} + \frac{t}{n}A_p\right)\right)^{-n}(y) \xrightarrow[n\to+\infty]{(2.12)} (\mathrm{Id} \rtimes T_t)(y).$$

Repeating the commutation relation (4.12) together with the observation that $(\mathrm{Id} \rtimes (\mathrm{Id} + sA_p))^{-1}$ maps $\partial_{\psi,q,p}(\mathcal{P}_G)$ into itself hence by continuity $\overline{\mathrm{Ran}\, \partial_{\psi,q,p}}$ into itself yields[1] for any integer $n \geqslant 1$ and any $t \geqslant 0$

$$(\partial_{\psi,q,p^*})^*\left(\mathrm{Id} \rtimes \left(\mathrm{Id} + \frac{t}{n}A_p\right)\right)^{-n}(y) = \left(\mathrm{Id} + \frac{t}{n}A_p\right)^{-n}(\partial_{\psi,q,p^*})^*(y)$$

$$\xrightarrow[n\to+\infty]{(2.12)} T_t(\partial_{\psi,q,p^*})^*(y).$$

Then by the closedness of $(\partial_{\psi,q,p^*})^*$, we deduce that $(\mathrm{Id} \rtimes T_t)(y)$ belongs to $\mathrm{dom}(\partial_{\psi,q,p^*})^*$ and that

$$(\partial_{\psi,q,p^*})^*(\mathrm{Id} \rtimes T_t)(y) = T_t(\partial_{\psi,q,p^*})^*(y).$$

Thus the third point follows.

$\qquad\qquad\qquad\qquad\qquad\qquad\qquad\qquad\qquad\qquad\qquad\qquad\qquad\qquad\qquad\square$

Proposition 4.4 enables us to identify B_p in terms of $\partial_{\psi,q,p}$ and its adjoint.

Proposition 4.6 *Let G be a weakly amenable discrete group such that $\Gamma_q(H) \rtimes_\alpha G$ has QWEP. Let $1 < p < \infty$ and $-1 \leqslant q \leqslant 1$. As unbounded operators, we have*

$$B_p = \partial_{\psi,q,p}(\partial_{\psi,q,p^*})^* | \overline{\mathrm{Ran}\, \partial_{\psi,q,p}}. \qquad (4.13)$$

Proof For any $s \in G$, we have

$$\partial_{\psi,q,p}(\partial_{\psi,q,p^*})^* \partial_{\psi,q,p}(\lambda_s) \overset{(4.1)}{=} \partial_{\psi,q,p} A_p(\lambda_s) = \left\| b_\psi(s) \right\|_H^2 \partial_{\psi,q,p}(\lambda_s)$$

$$\overset{(2.84)}{=} \left\| b_\psi(s) \right\|_H^2 s_q(b_\psi(s)) \rtimes \lambda_s$$

---

[1] Note that we replace s by $\frac{t}{n}$.

$$= (\mathrm{Id} \rtimes A_p)\big(s_q(b_\psi(s)) \rtimes \lambda_s\big)$$

$$\overset{(2.84)}{=} (\mathrm{Id} \rtimes A_p)\big(\partial_{\psi,q,p}(\lambda_s)\big).$$

We deduce that the operators $\partial_{\psi,q,p}(\partial_{\psi,q,p^*})^*|\overline{\mathrm{Ran}\,\partial_{\psi,q,p}}$ and $(\mathrm{Id}_{L^p(\Gamma_q(H))} \rtimes A_p)$ coincide on $\partial_{\psi,q,p}(\mathcal{P}_G)$. By Proposition 4.4, $\partial_{\psi,q,p}(\mathcal{P}_G)$ is a core for each operator. We conclude that they are equal. $\square$

In the proof of Theorem 4.1, we shall use the following folklore lemma, see e.g. [151, Proposition 2.13] relying essentially on [110, Proposition 10.7.2].

Lemma 4.5 *Let* $(T_t)_{t \geqslant 0}$ *be a bounded strongly continuous semigroup on a Banach space X. Let* $-A$ *be its infinitesimal generator. For any* $\frac{\pi}{2} < \theta < \pi$ *the following are equivalent*

1. *A admits a bounded* $\mathrm{H}^\infty(\Sigma_\theta)$ *functional calculus.*
2. *There exists a constant* $C > 0$ *such that for any* $b \in \mathrm{L}^1(\mathbb{R}^+)$ *whose Laplace transform* $\mathscr{L}(b)$ *belongs to* $\mathrm{H}_0^\infty(\Sigma_\theta)$, *we have*

$$\left\| \int_0^{+\infty} b(t) T_t \, dt \right\|_{X \to X} \leqslant C \, \|\mathscr{L}(b)\|_{\mathrm{H}^\infty(\Sigma_\theta)}. \tag{4.14}$$

Recall that the preceding integral is defined in the strong operator topology sense.

Note that the QWEP assumption of the following result is satisfied if $G = \mathbb{F}_n$ is a free group and $q = -1$.

Theorem 4.1 *Let G be a discrete group. Suppose* $1 < p < \infty$ *and* $-1 \leqslant q \leqslant 1$. *Let G be a weakly amenable discrete group such that* $\Gamma_q(H) \rtimes_\alpha G$ *has QWEP. The operators* A_p *and* $\mathrm{Id} \rtimes A_p$ *have a bounded* $\mathrm{H}^\infty(\Sigma_\theta)$ *functional calculus of angle* θ *for any* $\theta > \pi|\frac{1}{p} - \frac{1}{2}|$.

Proof For the QWEP property in case of G being a free group, we refer to [10, Proposition 4.8]. According to Proposition 2.8, $\mathrm{Id} \rtimes T_t$ extends to a (completely) contractive operator on $\mathrm{L}^p(\Gamma_q(H) \rtimes_\alpha G)$. Moreover, since $\mathcal{P}_{\rtimes,G}$ is dense in $\mathrm{L}^p(\Gamma_q(H) \rtimes_\alpha G)$, the property of $(T_t)_{t \geqslant 0}$ being a strongly continuous semigroup carries over to $(\mathrm{Id} \rtimes T_t)_{t \geqslant 0}$. According to [13, Theorem 4.1], [14] or [86, Corollary 4] (see also [12]), A_p has a (completely) bounded $\mathrm{H}^\infty(\Sigma_\theta)$ functional calculus on $\mathrm{L}^p(\mathrm{VN}(G))$ for any $\frac{\pi}{2} < \theta < \pi$. By Lemma 4.5, for any $b \in \mathrm{L}^1(\mathbb{R}_+)$ such that $\mathscr{L}(b)$ belongs to $\mathrm{H}_0^\infty(\Sigma_\theta)$ we have

$$\left\| \int_0^\infty b(t) T_t \, dt \right\|_{\mathrm{cb}, \mathrm{L}^p(\mathrm{VN}(G)) \to \mathrm{L}^p(\mathrm{VN}(G))} \lesssim \|\mathscr{L}(b)\|_{\mathrm{H}^\infty(\Sigma_\theta)}. \tag{4.15}$$

Note that for any $\lambda_s \in \mathcal{P}_G$, we have

$$\left(\int_0^\infty b(t)T_t \, \mathrm{d}t\right)(\lambda_s) = \int_0^\infty b(t)T_t(\lambda_s) \, \mathrm{d}t = \int_0^\infty b(t)\mathrm{e}^{-t\|b_\psi(s)\|^2}\lambda_s \, \mathrm{d}t$$

$$= \left(\int_0^\infty b(t)\mathrm{e}^{-t\|b_\psi(s)\|^2} \, \mathrm{d}t\right)(\lambda_s).$$

So the map $\int_0^\infty b(t)T_t \, \mathrm{d}t$ is also a Fourier multiplier. Thus, by Proposition 2.8

$$\left\|\int_0^\infty b(t)\mathrm{Id} \rtimes T_t(x) \, \mathrm{d}t\right\|_{\mathrm{L}^p(\Gamma_q(H)\rtimes_\alpha G)\to \mathrm{L}^p(\Gamma_q(H)\rtimes_\alpha G)}$$

$$= \left\|\mathrm{Id} \rtimes \left(\int_0^\infty b(t)T_t(x) \, \mathrm{d}t\right)\right\|$$

$$\overset{(2.62)}{\leqslant} \left\|\int_0^\infty b(t)T_t(x) \, \mathrm{d}t\right\|_{\mathrm{cb},\mathrm{L}^p(\mathrm{VN}(G))\to \mathrm{L}^p(\mathrm{VN}(G))} \overset{(4.15)}{\lesssim} \|\mathscr{L}(b)\|_{\mathrm{H}^\infty(\Sigma_\theta)}.$$

By Lemma 4.5, $\mathrm{Id} \rtimes A_p$ admits a bounded $\mathrm{H}^\infty(\Sigma_\theta)$ functional calculus for any $\theta > \frac{\pi}{2}$. Now, we reduce the angle and conclude with [128, Proposition 5.8]. $\qquad\square$

Suppose $1 < p < \infty$. We introduce the unbounded operator

$$D_{\psi,q,p} \overset{\mathrm{def}}{=} \begin{bmatrix} 0 & (\partial_{\psi,q,p^*})^* \\ \partial_{\psi,q,p} & 0 \end{bmatrix} \tag{4.16}$$

on the Banach space $\mathrm{L}^p(\mathrm{VN}(G)) \oplus_p \overline{\mathrm{Ran}\,\partial_{\psi,q,p}}$ defined by

$$D_{\psi,q,p}(x, y) \overset{\mathrm{def}}{=} \big((\partial_{\psi,q,p^*})^*(y), \partial_{\psi,q,p}(x)\big) \tag{4.17}$$

for $x \in \mathrm{dom}\,\partial_{\psi,q,p}$, $y \in \mathrm{dom}(\partial_{\psi,q,p^*})^* \cap \overline{\mathrm{Ran}\,\partial_{\psi,q,p}}$. We call it the Hodge-Dirac operator of the semigroup. This operator is a closed operator and can be seen as a differential square root of the generator of the semigroup $(T_{t,p})_{t\geqslant 0}$ since we have Proposition 4.7.

Theorem 4.2 *Let G be a weakly amenable discrete group such that $\Gamma_q(H) \rtimes_\alpha G$ has QWEP. Suppose $1 < p < \infty$ and $-1 \leqslant q \leqslant 1$. The Hodge-Dirac operator $D_{\psi,q,p}$ is R-bisectorial on $\mathrm{L}^p(\mathrm{VN}(G)) \oplus_p \overline{\mathrm{Ran}\,\partial_{\psi,q,p}}$.*

Proof We will start by showing that the set $\{\mathrm{i}t : t \in \mathbb{R}, t \neq 0\}$ is contained in the resolvent set of $D_{\psi,q,p}$. We will do this by showing that $\mathrm{Id} - \mathrm{i}tD_{\psi,q,p}$ has a two-sided bounded inverse $(\mathrm{Id} - it D_{\psi,q,p})^{-1}$ given by

$$\begin{bmatrix} (\mathrm{Id} + t^2 A_p)^{-1} & it(\mathrm{Id} + t^2 A_p)^{-1}(\partial_{\psi,q,p^*})^* \\ it\partial_{\psi,q,p}(\mathrm{Id} + t^2 A_p)^{-1} & \mathrm{Id} \rtimes (\mathrm{Id} + t^2 A_p)^{-1} \end{bmatrix} \tag{4.18}$$

acting on $L^p(\mathrm{VN}(G)) \oplus_p \overline{\mathrm{Ran}\,\partial_{\psi,q,p}}$. By Proposition 4.3 and since the operators A_p and $\mathrm{Id}_{L^p} \rtimes A_p$ satisfy the property (2.15) of R-sectoriality (see Theorem 4.1), the four entries are bounded. It only remains to check that this matrix defines a two-sided inverse of $\mathrm{Id} - it D_{\psi,q,p}$. We have the following equalities of operators acting on $\mathrm{dom}\, D_{\psi,q,p}$.

$$\begin{bmatrix} (\mathrm{Id} + t^2 A_p)^{-1} & it(\mathrm{Id}_{L^p} + t^2 A_p)^{-1}(\partial_{\psi,q,p^*})^* \\ it\partial_{\psi,q,p}(\mathrm{Id} + t^2 A_p)^{-1} & \mathrm{Id} \rtimes (\mathrm{Id} + t^2 A_p)^{-1} \end{bmatrix}(\mathrm{Id} - it D_{\psi,q,p})$$

$$\overset{(4.16)}{=} \begin{bmatrix} (\mathrm{Id} + t^2 A_p)^{-1} & it(\mathrm{Id} + t^2 A_p)^{-1}(\partial_{\psi,q,p^*})^* \\ it\partial_{\psi,q,p}(\mathrm{Id} + t^2 A_p)^{-1} & \mathrm{Id} \rtimes (\mathrm{Id} + t^2 A_p)^{-1} \end{bmatrix}$$

$$\times \begin{bmatrix} \mathrm{Id}_{L^p} & -it(\partial_{\psi,q,p^*})^* \\ -it\partial_{\psi,q,p} & \mathrm{Id}_{\overline{\mathrm{Ran}\,\partial_{\psi,q,p}}} \end{bmatrix}$$

$$= \begin{bmatrix} (\mathrm{Id} + t^2 A_p)^{-1} + t^2(\mathrm{Id} + t^2 A_p)^{-1}(\partial_{\psi,q,p^*})^* \partial_{\psi,q,p} \\ it\partial_{\psi,q,p}(\mathrm{Id} + t^2 A_p)^{-1} - it(\mathrm{Id} \rtimes (\mathrm{Id} + t^2 A_p)^{-1})\partial_{\psi,q,p} \end{bmatrix}$$

$$\begin{bmatrix} -it(\mathrm{Id} + t^2 A_p)^{-1}(\partial_{\psi,q,p^*})^* + it(\mathrm{Id} + t^2 A_p)^{-1}(\partial_{\psi,q,p^*})^* \\ t^2\partial_{\psi,q,p}(\mathrm{Id} + t^2 A_p)^{-1}(\partial_{\psi,q,p^*})^* + \mathrm{Id} \rtimes (\mathrm{Id} + t^2 A_p)^{-1} \end{bmatrix}$$

$$\overset{(4.1)(4.4)(4.10)}{=} \begin{bmatrix} (\mathrm{Id} + t^2 A_p)^{-1} + t^2(\mathrm{Id} + t^2 A_p)^{-1} A_p \\ it\partial_{\psi,q,p}(\mathrm{Id} + t^2 A_p)^{-1} - it\partial_{\psi,q,p}(\mathrm{Id} + t^2 A_p)^{-1} \end{bmatrix}$$

$$\begin{bmatrix} 0 \\ (t^2\partial_{\psi,q,p}(\partial_{\psi,q,p^*})^* + \mathrm{Id})(\mathrm{Id} \rtimes (\mathrm{Id} + t^2 A_p)^{-1}) \end{bmatrix}$$

$$\overset{(4.13)(4.7)}{=} \begin{bmatrix} \mathrm{Id}_{L^p(\mathrm{VN}(G))} & 0 \\ 0 & \mathrm{Id}_{\overline{\mathrm{Ran}\,\partial_{\psi,q,p}}} \end{bmatrix}$$

and similarly

$$(\mathrm{Id} - it D_{\psi,p}) \begin{bmatrix} (\mathrm{Id} + t^2 A_p)^{-1} & it(\mathrm{Id} + t^2 A_p)^{-1}(\partial_{\psi,q,p^*})^* \\ it\partial_{\psi,q,p}(\mathrm{Id} + t^2 A_p)^{-1} & \mathrm{Id} \rtimes (\mathrm{Id} + t^2 A_p)^{-1} \end{bmatrix}$$

$$= \begin{bmatrix} \mathrm{Id} & -it(\partial_{\psi,q,p^*})^* \\ -it\partial_{\psi,q,p} & \mathrm{Id}_{\overline{\mathrm{Ran}\,\partial_{\psi,q,p}}} \end{bmatrix}$$

$$\times \begin{bmatrix} (\mathrm{Id} + t^2 A_p)^{-1} & it(\mathrm{Id} + t^2 A_p)^{-1}(\partial_{\psi,q,p^*})^* \\ it\partial_{\psi,q,p}(\mathrm{Id} + t^2 A_p)^{-1} & \mathrm{Id} \rtimes (\mathrm{Id} + t^2 A_p)^{-1} \end{bmatrix}$$

$$
= \begin{bmatrix} (\mathrm{Id} + t^2 A_p)^{-1} + t^2 (\partial_{\psi,q,p*})^* \partial_{\psi,q,p} (\mathrm{Id} + t^2 A_p)^{-1} \\ -it \partial_{\psi,q,p} (\mathrm{Id} + t^2 A_p)^{-1} + it \partial_{\psi,q,p} (\mathrm{Id} + t^2 A_p)^{-1} \end{bmatrix}
$$

$$
\begin{bmatrix} it (\mathrm{Id} + t^2 A_p)^{-1} (\partial_{\psi,q,p*})^* - it (\partial_{\psi,q,p*})^* \big(\mathrm{Id} \rtimes (\mathrm{Id} + t^2 A_p)^{-1} \big) \\ t^2 \partial_{\psi,q,p} (\mathrm{Id} + t^2 A_p)^{-1} (\partial_{\psi,q,p*})^* + \mathrm{Id} \rtimes (\mathrm{Id} + t^2 A_p)^{-1} \end{bmatrix}
$$

$$
= \begin{bmatrix} \mathrm{Id}_{L^p} & 0 \\ 0 & \mathrm{Id}_{\overline{\mathrm{Ran}\, \partial_{\psi,q,p}}} \end{bmatrix}.
$$

It remains to show that the set $\{it (it - D_{\psi,q,p})^{-1} : t \neq 0\} = \{(\mathrm{Id} - it D_{\psi,q,p})^{-1} : t \neq 0\}$ is R-bounded. For this, observe that the diagonal entries of (4.18) are R-bounded by the R-sectoriality of A_p and $\mathrm{Id}_{L^p} \rtimes A_p$. The R-boundedness of the other entries follows from the R-gradient bounds of Proposition 4.3. Since a set of operator matrices is R-bounded precisely when each entry is R-bounded, we conclude that (2.19) is satisfied, i.e. that $D_{\psi,q,p}$ is R-bisectorial.   □

Proposition 4.7 *Let G be a weakly amenable discrete group such that $\Gamma_q(H) \rtimes_\alpha G$ has QWEP. Suppose $1 < p < \infty$ and $-1 \leqslant q \leqslant 1$. As densely defined closed operators on $\mathrm{L}^p(\mathrm{VN}(G)) \oplus_p \overline{\mathrm{Ran}\, \partial_{\psi,q,p}}$, we have*

$$
D_{\psi,q,p}^2 = \begin{bmatrix} A_p & 0 \\ 0 & (\mathrm{Id}_{\mathrm{L}^p(\Gamma_q(H))} \rtimes A_p)|\overline{\mathrm{Ran}\, \partial_{\psi,q,p}} \end{bmatrix}. \tag{4.19}
$$

Proof By Proposition 4.6, we have

$$
C_p \overset{\mathrm{def}}{=} \begin{bmatrix} A_p & 0 \\ 0 & (\mathrm{Id}_{\mathrm{L}^p(\Gamma_q(H))} \rtimes A_p)|\overline{\mathrm{Ran}\, \partial_{\psi,q,p}} \end{bmatrix}
$$

$$
\overset{(4.1)(4.13)}{=} \begin{bmatrix} (\partial_{\psi,q,p*})^* \partial_{\psi,q,p} & 0 \\ 0 & \partial_{\psi,q,p} (\partial_{\psi,q,p*})^* |\overline{\mathrm{Ran}\, \partial_{\psi,q,p}} \end{bmatrix}
$$

$$
= \begin{bmatrix} 0 & (\partial_{\psi,q,p*})^* |\overline{\mathrm{Ran}\, \partial_{\psi,q,p}} \\ \partial_{\psi,q,p} & 0 \end{bmatrix}^2 \overset{(4.16)}{=} D_{\psi,q,p}^2.
$$

          □

Now, we can state the following main result of this section.

Theorem 4.3 *Suppose $1 < p < \infty$ and $-1 \leqslant q \leqslant 1$. Let G be a weakly amenable discrete group such that $\Gamma_q(H) \rtimes_\alpha G$ has QWEP. The Hodge-Dirac operator $D_{\psi,q,p}$ is R-bisectorial on $\mathrm{L}^p(\mathrm{VN}(G)) \oplus_p \overline{\mathrm{Ran}\, \partial_{\psi,q,p}}$ and admits a bounded $\mathrm{H}^\infty(\Sigma_\omega^\pm)$ functional calculus on a bisector.*

Proof By Theorem 4.1, the operator $D_{\psi,q,p}^2 \overset{(4.19)}{=} \begin{bmatrix} A_p & 0 \\ 0 & B_p \end{bmatrix}$ has a bounded H^∞ functional calculus of some angle $2\omega < \frac{\pi}{2}$. Since $D_{\psi,q,p}$ is R-bisectorial by

Theorem 4.2, we deduce by Proposition 2.1 that the operator $D_{\psi,q,p}$ has a bounded $H^\infty(\Sigma_\omega^\pm)$ functional calculus on a bisector. $\qquad\square$

Remark 4.1 The boundedness of the H^∞ functional calculus of the operator $D_{\psi,q,p}$ implies the boundedness of the Riesz transforms and this result may be thought of as a strengthening of the equivalence (3.31). Indeed, consider the function sgn $\in$ $H^\infty(\Sigma_\omega^\pm)$ defined by $\text{sgn}(z) \overset{\text{def}}{=} 1_{\Sigma_\omega^+}(z) - 1_{\Sigma_\omega^-}(z)$. By Theorem 4.3, the operator $D_{\psi,q,p}$ has a bounded $H^\infty(\Sigma_\omega^\pm)$ functional calculus on $L^p(\text{VN}(G)) \oplus_p \overline{\text{Ran}\,\partial_{\psi,q,p}}$. Hence the operator $\text{sgn}(D_{\psi,q,p})$ is bounded. This implies that

$$|D_{\psi,q,p}| = \text{sgn}(D_{\psi,q,p})D_{\psi,q,p} \quad \text{and} \quad D_{\psi,q,p} = \text{sgn}(D_{\psi,q,p})|D_{\psi,q,p}|. \quad (4.20)$$

For any element ξ of dom $D_{\psi,q,p} = \text{dom}\,|D_{\psi,q,p}|$, we deduce that

$$\left\| D_{\psi,q,p}(\xi) \right\|_{L^p(\text{VN}(G))\oplus_p L^p(\Gamma_q(H)\rtimes_\alpha G)}$$

$$\overset{(4.20)}{=} \left\| \text{sgn}(D_{\psi,q,p})|D_{\psi,q,p}|(\xi) \right\|_{L^p(\text{VN}(G))\oplus_p L^p(\Gamma_q(H)\rtimes_\alpha G)}$$

$$\lesssim_p \left\| |D_{\psi,q,p}|(\xi) \right\|_{L^p(\text{VN}(G))\oplus_p L^p(\Gamma_q(H)\rtimes_\alpha G)}$$

and

$$\left\| |D_{\psi,q,p}|(\xi) \right\|_{L^p(\text{VN}(G))\oplus_p L^p(\Gamma_q(H)\rtimes_\alpha G)}$$

$$\overset{(4.20)}{=} \left\| \text{sgn}(D_{\psi,q,p})D_{\psi,q,p}(\xi) \right\|_{L^p(\text{VN}(G))\oplus_p L^p(\Gamma_q(H)\rtimes_\alpha G)}$$

$$\lesssim_p \left\| D_{\psi,q,p}(\xi) \right\|_{L^p(\text{VN}(G))\oplus_p L^p(\Gamma_q(H)\rtimes_\alpha G)}.$$

Recall that on $L^p(\text{VN}(G)) \oplus_p \overline{\text{Ran}\,\partial_{\psi,q,p}}$, we have

$$|D_{\psi,q,p}| \overset{(4.19)}{=} \begin{bmatrix} A_p^{\frac{1}{2}} & 0 \\ 0 & \text{Id}_{L^p(\Gamma_q(H))} \rtimes A_p^{\frac{1}{2}}|\overline{\text{Ran}\,\partial_{\psi,q,p}} \end{bmatrix}. \quad (4.21)$$

By restricting to elements of the form $(x, 0)$ with $x \in \text{dom}\,A_p^{\frac{1}{2}}$, we obtain the desired result.

Remark 4.2 In a similar way to Remark 4.1, we also obtain for any element (x, y) of dom $\partial_{\psi,q,p} \oplus \text{dom}(\partial_{\psi,q,p^*})^*|\overline{\text{Ran}\,\partial_{\psi,q,p}}$ that

$$\left\| |D_{\psi,q,p}|(x, y) \right\|_p \overset{(4.21)}{\cong} \left\| A_p^{\frac{1}{2}}(x) \right\|_p + \left\| (\text{Id}_{L^p(\Gamma_q(H))} \rtimes A_p^{\frac{1}{2}})(y) \right\|_p$$

$$\cong \left\| \partial_{\psi,q,p}(x) \right\|_p + \left\| (\partial_{\psi,q,p^*})^*(y) \right\|_p \overset{(4.16)}{\cong} \left\| D_{\psi,q,p}(x, y) \right\|_p.$$

Similarly, we have

$$\left\| \left(\mathrm{Id}_{L^p(\Gamma_q(H))} \rtimes A_p^{\frac{1}{2}} \right)(y) \right\|_p \cong \left\| (\partial_{\psi,q,p^*})^*(y) \right\|_p, \quad y \in \mathrm{dom}(\partial_{\psi,q,p^*})^* \cap \overline{\mathrm{Ran}\, \partial_{\psi,q,p}}. \tag{4.22}$$

Proposition 4.8 *Let G be a weakly amenable discrete group such that $\Gamma_q(H) \rtimes_\alpha G$ has QWEP. Suppose $1 < p < \infty$ and $-1 \leqslant q \leqslant 1$. We have $\overline{\mathrm{Ran}\, A_p} = \overline{\mathrm{Ran}(\partial_{\psi,q,p^*})^*}$, $\overline{\mathrm{Ran}\, B_p} = \overline{\mathrm{Ran}\, \partial_{\psi,q,p}}$, $\mathrm{Ker}\, A_p = \mathrm{Ker}\, \partial_{\psi,q,p}$, $\mathrm{Ker}\, B_p = \mathrm{Ker}(\partial_{\psi,q,p^*})^* = \{0\}$ and*

$$L^p(\mathrm{VN}(G)) = \overline{\mathrm{Ran}(\partial_{\psi,q,p^*})^*} \oplus \mathrm{Ker}\, \partial_{\psi,q,p}. \tag{4.23}$$

Here, by $(\partial_{\psi,q,p^*})^*$ we understand its restriction to $\overline{\mathrm{Ran}\, \partial_{\psi,q,p}}$. However, we have $\mathrm{Ran}(\partial_{\psi,q,p^*})^* = \mathrm{Ran}(\partial_{\psi,q,p^*})^* | \overline{\mathrm{Ran}\, \partial_{\psi,q,p}}$ (see Corollary 4.1).

Proof By (2.20), we have $\overline{\mathrm{Ran}\, D_{\psi,q,p}^2} = \overline{\mathrm{Ran}\, D_{\psi,q,p}}$ and $\mathrm{Ker}\, D_{\psi,q,p}^2 = \mathrm{Ker}\, D_{\psi,q,p}$. It is not difficult to prove the first four equalities using (4.19) and (4.17). The last one is a consequence of the definition of A_p and of [110, p. 361]. $\qquad \square$

4.2 Extension to Full Hodge-Dirac Operator and Hodge Decomposition

We keep the standing assumptions of the preceding section and thus we have a markovian semigroup $(T_t)_{t \geqslant 0}$ of Fourier multipliers with generator A_p, the noncommutative gradient $\partial_{\psi,q,p}$ and its adjoint $(\partial_{\psi,q,p^*})^*$, together with the Hodge-Dirac operator $D_{\psi,q,p}$. We shall now extend the operator $D_{\psi,q,p}$ to a densely defined bisectorial operator $\mathcal{D}_{\psi,q,p}$ on $L^p(\mathrm{VN}(G)) \oplus L^p(\Gamma_q(H) \rtimes_\alpha G)$ which will also be bisectorial and will have an $\mathrm{H}^\infty(\Sigma_\omega^\pm)$ functional calculus on a bisector. The key will be Corollary 4.1 below. We let

$$\mathcal{D}_{\psi,q,p} \overset{\mathrm{def}}{=} \begin{bmatrix} 0 & (\partial_{\psi,q,p^*})^* \\ \partial_{\psi,q,p} & 0 \end{bmatrix} \tag{4.24}$$

along the decomposition $L^p(\mathrm{VN}(G)) \oplus L^p(\Gamma_q(H) \rtimes_\alpha G)$, with natural domains for $\partial_{\psi,q,p}$ and $(\partial_{\psi,q,p^*})^*$. Again, except in Lemma 4.6, we need in this section approximation properties of G, see before Lemma 4.7.

Consider the sectorial operator $A_p^{\frac{1}{2}}$ on $L^p(\mathrm{VN}(G))$. According to (2.16), we have the *topological* direct sum decomposition $L^p(\mathrm{VN}(G)) = \overline{\mathrm{Ran}\, A_p^{\frac{1}{2}}} \oplus \mathrm{Ker}\, A_p^{\frac{1}{2}}$. We define the operator $R_p \overset{\mathrm{def}}{=} \partial_{\psi,q,p} A_p^{-\frac{1}{2}} : \overline{\mathrm{Ran}\, A_p^{\frac{1}{2}}} \to L^p(\Gamma_q(H) \rtimes_\alpha G)$.

According to the point 3 of Proposition 3.4, R_p is bounded on $\operatorname{Ran} A_p^{\frac{1}{2}}$, so extends to a bounded operator on $\overline{\operatorname{Ran} A_p^{\frac{1}{2}}} \overset{(2.21)}{=} \overline{\operatorname{Ran} A_p}$. We extend it to a bounded operator $R_p \colon \mathrm{L}^p(\mathrm{VN}(G)) \to \mathrm{L}^p(\Gamma_q(H) \rtimes_\alpha G)$, called Riesz transform, by putting $R_p | \operatorname{Ker} A_p^{\frac{1}{2}} = 0$ along the previous decomposition of $\mathrm{L}^p(\mathrm{VN}(G))$. We equally let $R_{p^*}^* \overset{\mathrm{def}}{=} (R_{p^*})^*$.

Lemma 4.6 *Let* $-1 \leqslant q \leqslant 1$ *and* $1 < p < \infty$. *Then we have the decomposition*

$$\mathrm{L}^p(\Gamma_q(H) \rtimes_\alpha G) = \overline{\operatorname{Ran} \partial_{\psi,q,p}} + \operatorname{Ker}(\partial_{\psi,q,p^*})^*. \qquad (4.25)$$

Proof Let $y \in \mathrm{L}^p(\Gamma_q(H) \rtimes_\alpha G)$ be arbitrary. We claim that $y = R_p R_{p^*}^*(y) + (\mathrm{Id} - R_p R_{p^*}^*)(y)$ is the needed decomposition for (4.25). Note that R_p maps $\operatorname{Ran} A_p^{\frac{1}{2}}$ into $\overline{\operatorname{Ran} \partial_{\psi,q,p}}$, so by boundedness, R_p maps $\overline{\operatorname{Ran} A_p^{\frac{1}{2}}}$ to $\overline{\operatorname{Ran} \partial_{\psi,q,p}}$. Thus, we indeed have $R_p R_{p^*}^*(y) \in \overline{\operatorname{Ran} \partial_{\psi,q,p}}$. Next we claim that for any $z \in \mathrm{L}^p(\mathrm{VN}(G))$ and any $x \in \operatorname{dom} \partial_{\psi,q,p^*}$, we have

$$\big\langle R_p(z), \partial_{\psi,q,p^*}(x) \big\rangle_{\mathrm{L}^p(\Gamma_q(H) \rtimes_\alpha G), \mathrm{L}^{p^*}(\Gamma_q(H) \rtimes_\alpha G)} = \big\langle z, A_{p^*}^{\frac{1}{2}}(x) \big\rangle_{\mathrm{L}^p, \mathrm{L}^{p^*}}. \qquad (4.26)$$

According to the decomposition $\mathrm{L}^p(\mathrm{VN}(G)) = \overline{\operatorname{Ran} A_p^{\frac{1}{2}}} \oplus \operatorname{Ker} A_p^{\frac{1}{2}}$, we can write $z = \lim_{n \to +\infty} A_p^{\frac{1}{2}}(z_n) + z_0$ with $z_n \in \operatorname{dom} A_p^{\frac{1}{2}}$ and $z_0 \in \operatorname{Ker} A_p^{\frac{1}{2}}$. Then using Lemma 3.7 in the third equality, we have

$$\big\langle R_p(z), \partial_{\psi,q,p^*}(x) \big\rangle = \lim_{n \to +\infty} \big\langle R_p\big(A_p^{\frac{1}{2}}(z_n) + z_0\big), \partial_{\psi,q,p^*}(x) \big\rangle$$

$$= \lim_{n \to +\infty} \big\langle \partial_{\psi,q,p}(z_n), \partial_{\psi,q,p^*}(x) \big\rangle = \lim_{n \to +\infty} \big\langle A_p^{\frac{1}{2}}(z_n), A_{p^*}^{\frac{1}{2}}(x) \big\rangle$$

$$= \big\langle z - z_0, A_{p^*}^{\frac{1}{2}}(x) \big\rangle = \big\langle z, A_{p^*}^{\frac{1}{2}}(x) \big\rangle - \big\langle z_0, A_{p^*}^{\frac{1}{2}}(x) \big\rangle = \big\langle z, A_{p^*}^{\frac{1}{2}}(x) \big\rangle.$$

Thus, (4.26) is proved. Now, for any $x \in \operatorname{dom} \partial_{\psi,q,p^*}$, we have

$$\big\langle (\mathrm{Id} - R_p R_{p^*}^*)(y), \partial_{\psi,q,p^*}(x) \big\rangle = \big\langle y, \partial_{\psi,q,p^*}(x) \big\rangle - \big\langle R_p R_{p^*}^*(y), \partial_{\psi,q,p^*}(x) \big\rangle$$

$$\overset{(4.26)}{=} \big\langle y, \partial_{\psi,q,p^*}(x) \big\rangle - \big\langle R_{p^*}^*(y), A_{p^*}^{\frac{1}{2}}(x) \big\rangle$$

$$= \big\langle y, \partial_{\psi,q,p^*}(x) \big\rangle - \big\langle y, R_{p^*} A_{p^*}^{\frac{1}{2}}(x) \big\rangle$$

$$= \big\langle y, \partial_{\psi,q,p^*}(x) \big\rangle - \big\langle y, \partial_{\psi,q,p^*} A_{p^*}^{-\frac{1}{2}} A_{p^*}^{\frac{1}{2}}(x) \big\rangle = 0.$$

By (2.8), we conclude that $\big(\mathrm{Id} - R_p R_{p^*}^*\big)(y)$ belongs to $\operatorname{Ker}(\partial_{\psi,q,p^*})^*$. $\qquad \square$

From now on, we suppose the discrete group G to be weakly amenable such that $\Gamma_q(H) \rtimes_\alpha G$ has QWEP (e.g. if G is amenable). In the proof of Proposition 4.9 below, we shall need some information on the Wiener-Ito chaos decomposition for q-Gaussians. This is collected in the following lemma.

Lemma 4.7 *Let* $-1 \leqslant q \leqslant 1$ *and* $1 < p < \infty$. *Let* G *be a weakly amenable discrete group such that* $\Gamma_q(H) \rtimes_\alpha G$ *has* QWEP. *Consider an approximating net* (M_{φ_j}) *of finitely supported Fourier multipliers.*

1. *There exists a completely bounded projection* $\mathcal{P} \colon L^p(\Gamma_q(H)) \to L^p(\Gamma_q(H))$ *onto the closed space spanned by* $\{s_q(h) : h \in H\}$. *Moreover, the projections are compatible for different values of p. The mapping* $\mathcal{P} \rtimes \mathrm{Id}_{L^p(\mathrm{VN}(G))}$ *extends to a bounded operator on* $L^p(\Gamma_q(H) \rtimes_\alpha G)$.
2. *For any j and any* $y \in L^p(\Gamma_q(H) \rtimes_\alpha G)$, *the element* $(\mathcal{P} \rtimes M_{\varphi_j})(y)$ *can be written as* $\sum_{s \in \mathrm{supp}\,\varphi_j} s_q(h_s) \rtimes \lambda_s$ *for some* $h_s \in H$.
3. *Denoting temporarily by* $\mathcal{P}_p$ *and* $M_{\varphi_j,p}$ *the operator* $\mathcal{P}$ *and* M_{φ_j} *on the p-level, the identity mapping on* $\mathrm{Gauss}_{q,p}(\mathbb{C}) \rtimes \mathrm{span}\{\lambda_s : s \in \mathrm{supp}\,\varphi_j\}$ *extends to an isomorphism*

$$J_{p,2,j} \colon \mathrm{Ran}(\mathcal{P}_p \rtimes M_{\varphi_j,p}) \to \mathrm{Ran}(\mathcal{P}_2 \rtimes M_{\varphi_j,2}) \tag{4.27}$$

where $\mathrm{Ran}(\mathcal{P}_p \rtimes M_{\varphi_j,p}) \subset L^p(\Gamma_q(H) \rtimes_\alpha G)$.

Proof

1. This is contained in [118, Theorem 3.5], putting there $d = 1$. Note that the closed space spanned by $\{s_q(h), h \in H\}$ coincides in this case with $G^1_{p,q}$ there. For the fact that the projections are compatible for different values of p, we refer to [118, Proof of Theorem 3.1]. See also the mapping Q_p from Lemma 3.4.
2. Once we know that $\mathcal{P} \rtimes M_{\varphi_j} = (\mathcal{P} \rtimes \mathrm{Id}) \circ (\mathrm{Id} \rtimes M_{\varphi_j})$ is bounded according to point 1. and Proposition 2.8, this is easy and left to the reader.
3. We have

$$\left\| \sum_{s \in \mathrm{supp}\,\varphi_j} s_q(h_s) \rtimes \lambda_s \right\|_{L^p(\Gamma_q(H) \rtimes G)} \leqslant \sum_{s \in \mathrm{supp}\,\varphi_j} \left\| s_q(h_s) \right\|_{L^p(\Gamma_q(H))}$$

$$\lesssim_{q,p} \sum_{s \in \mathrm{supp}\,\varphi_j} \|h_s\|_H$$

(the same estimate on the L^2-level). Note that for any $s_0 \in \mathrm{supp}\,\varphi_j$ fixed, we have a completely contractive projection of $L^p(\Gamma_q(H) \rtimes_\alpha G)$ onto $\mathrm{span}\{x \rtimes \lambda_{s_0} : x \in L^p(\Gamma_q(H))\}$. So we have

$$\left\| \sum_{s \in \mathrm{supp}\,\varphi_j} s_q(h_s) \rtimes \lambda_s \right\|_{L^p(\Gamma_q(H) \rtimes_\alpha G)} \gtrsim \left\| s_q(h_{s_0}) \right\|_{L^p(\Gamma_q(H))} \gtrsim_{q,p} \left\| h_{s_0} \right\|_H .$$

It follows that

$$
\left\| \sum_{s \in \mathrm{supp}\, \varphi_j} s_q(h_s) \rtimes \lambda_s \right\|_{L^p(\Gamma_q(H) \rtimes_\alpha G)} \cong \sum_{s \in \mathrm{supp}\, \varphi_j} \| h_s \|_H \cong \left\| \sum_{s \in \mathrm{supp}\, \varphi_j} s_q(h_s) \rtimes \lambda_s \right\|_{L^2}.
$$

Thus, (4.27) follows.

$\square$

Proposition 4.9 *Let* $-1 \leqslant q \leqslant 1$ *and* $1 < p < \infty$. *Let* G *be a weakly amenable discrete group such that* $\Gamma_q(H) \rtimes_\alpha G$ *is QWEP. Then the subspaces from Lemma 4.6 have trivial intersection, i.e.* $\overline{\mathrm{Ran}\, \partial_{\psi,q,p}} \cap \mathrm{Ker}(\partial_{\psi,q,p*})^* = \{0\}$.

Proof We begin with the case $p = 2$. According to Theorem 5.4, the unbounded operator $\mathcal{D}_{\psi,q,2}$ is selfadjoint on $L^2(\mathrm{VN}(G)) \oplus L^2(\Gamma_q(H) \rtimes_\alpha G)$. We thus have the orthogonal sum $\overline{\mathrm{Ran}\, \mathcal{D}_{\psi,q,2}} \oplus \mathrm{Ker}\, \mathcal{D}_{\psi,q,2} = L^2(\mathrm{VN}(G)) \oplus L^2(\Gamma_q(H) \rtimes_\alpha G)$. Considering vectors in the second component, that is, in $L^2(\Gamma_q(H) \rtimes_\alpha G)$, we deduce that $\overline{\mathrm{Ran}\, \partial_{\psi,q,2}}$ and $\mathrm{Ker}(\partial_{\psi,q,2})^*$ are orthogonal, hence have trivial intersection.

We turn to the case $1 < p < \infty$. Consider an approximating net (M_{φ_j}). According to Lemma 4.7 point 1. and Proposition 2.8, we have for any j a completely bounded mapping $\mathcal{P} \rtimes M_{\varphi_j} = (\mathcal{P} \rtimes \mathrm{Id}) \circ (\mathrm{Id} \rtimes M_{\varphi_j}) : L^p(\Gamma_q(H) \rtimes_\alpha G) \to L^p(\Gamma_q(H) \rtimes_\alpha G)$. We claim that

$$
\text{the subspace } \overline{\mathrm{Ran}\, \partial_{\psi,q,p}} \text{ is invariant under } \mathrm{Id} \rtimes M_{\varphi_j} \tag{4.28}
$$

$$
\text{the subspace } \mathrm{Ker}(\partial_{\psi,q,p*})^* \text{ is invariant under } \mathrm{Id} \rtimes M_{\varphi_j} \tag{4.29}
$$

$$
\text{the restriction of } \mathcal{P} \rtimes \mathrm{Id} \text{ on } \overline{\mathrm{Ran}\, \partial_{\psi,q,p}} \text{ is the identity mapping.} \tag{4.30}
$$

For (4.28), for any $s \in G$, note that

$$
(\mathrm{Id} \rtimes M_{\varphi_j}) \partial_{\psi,q,p}(\lambda_s) \overset{(2.84)}{=} (\mathrm{Id} \rtimes M_{\varphi_j})\big(s_q(b_\psi(s)) \rtimes \lambda_s \big) = \varphi_j(s) s_q(b_\psi(s)) \rtimes \lambda_s.
$$

This element belongs to $\overline{\mathrm{Ran}\, \partial_{\psi,q,p}}$. By linearity and since $\mathcal{P}_G$ is a core for $\partial_{\psi,q,p}$ according to Proposition 3.4, we deduce that $\mathrm{Id} \rtimes M_{\varphi_j}$ maps $\mathrm{Ran}\, \partial_{\psi,q,p}$ into $\overline{\mathrm{Ran}\, \partial_{\psi,q,p}}$. Now (4.28) follows from the continuity of $\mathrm{Id} \rtimes M_{\varphi_j}$.

For (4.29), note that if $x \in \mathrm{dom}\, \partial_{\psi,q,p*}$ and $f \in \mathrm{Ker}(\partial_{\psi,q,p*})^*$, then

$$
\big\langle (\mathrm{Id} \rtimes M_{\varphi_j})(f), \partial_{\psi,q,p*}(x) \big\rangle = \big\langle f, (\mathrm{Id} \rtimes M_{\overline{\varphi_j}}) \partial_{\psi,q,p*}(x) \big\rangle
$$

$$
= \big\langle f, \partial_{\psi,q,p*} M_{\overline{\varphi_j}}(x) \big\rangle = \big\langle (\partial_{\psi,q,p*})^*(f), M_{\overline{\varphi_j}}(x) \big\rangle = 0.
$$

By (2.8), we conclude that $(\mathrm{Id} \rtimes M_{\varphi_j})(f)$ belongs to $\mathrm{Ker}(\partial_{\psi,q,p*})^*$ and (4.29) follows. For (4.30), for any $s \in G$ we have

$$
(\mathcal{P} \rtimes \mathrm{Id})(\partial_{\psi,q,p}(\lambda_s)) \overset{(2.84)}{=} (\mathcal{P} \rtimes \mathrm{Id})\big(s_q(b_\psi(s)) \rtimes \lambda_s \big) = s_q(b_\psi(s)) \rtimes \lambda_s \in \mathrm{Ran}\, \partial_{\psi,q,p}.
$$

Now use in a similar manner as before linearity, the fact that $\mathcal{P}_G$ is a core of $\partial_{\psi,q,p}$ and the continuity of $\mathcal{P} \rtimes \mathrm{Id}$.

Now, let $z \in \overline{\mathrm{Ran}\, \partial_{\psi,q,p}} \cap \mathrm{Ker}(\partial_{\psi,q,p^*})^*$. Then according to (4.28)–(4.30), we infer that $(\mathcal{P} \rtimes M_{\varphi_j})(z)$ belongs[2] again to $\overline{\mathrm{Ran}\, \partial_{\psi,q,p}} \cap \mathrm{Ker}(\partial_{\psi,q,p^*})^*$. We claim that $J_{p,2,j}(\mathcal{P} \rtimes M_{\varphi_j})(z)$ belongs to $\overline{\mathrm{Ran}\, \partial_{\psi,q,2}} \cap \mathrm{Ker}(\partial_{\psi,q,2})^*$, where the mapping $J_{p,2,j}$ was defined in Lemma 4.7. First $(\mathcal{P} \rtimes M_{\varphi_j})(z)$ belongs to $\overline{\mathrm{Ran}\, \partial_{\psi,q,p}}$, so that there exists a sequence (x_n) in dom $\partial_{\psi,q,p}$ such that

$$(\mathcal{P} \rtimes M_{\varphi_j})(z) = \lim_{n \to \infty} \partial_{\psi,q,p}(x_n). \tag{4.31}$$

Then we have in $\mathrm{L}^p(\Gamma_q(H) \rtimes_\alpha G)$, with $\psi_j = 1_{\mathrm{supp}\,\varphi_j}$, which is of finite support and thus induces a completely bounded multiplier M_{ψ_j},[3]

$$(\mathcal{P} \rtimes M_{\varphi_j})(z) = (\mathrm{Id} \rtimes M_{\psi_j}) \cdot (\mathcal{P} \rtimes M_{\varphi_j})(z) \overset{(4.31)}{=} \lim_{n \to \infty} (\mathrm{Id} \rtimes M_{\psi_j}) \partial_{\psi,q,p}(x_n)$$

$$= \lim_{n \to \infty} \partial_{\psi,q,p} M_{\psi_j}(x_n) = \lim_{n \to \infty} \sum_{s \in \mathrm{supp}\,\varphi_j} x_{n,s} \partial_{\psi,q,p}(\lambda_s).$$

Since $J_{p,2,j}$ is an isomorphism, this limit also holds in $\mathrm{L}^2(\Gamma_q(H) \rtimes_\alpha G)$ and the element $J_{p,2,j}(\mathcal{P} \rtimes M_{\varphi_j})(z) = \lim_{n \to \infty} \sum_{s \in \mathrm{supp}\,\varphi_j} x_{n,s} \partial_{\psi,q,2}(\lambda_s)$ belongs to $\overline{\mathrm{Ran}\, \partial_{\psi,q,2}}$. Furthermore, for some family (h_s) of elements of H, we have

$$(\mathcal{P} \rtimes M_{\varphi_j})(z) = \sum_{s \in \mathrm{supp}\,\varphi_j} s_q(h_s) \rtimes \lambda_s. \tag{4.32}$$

Then using that $(\mathcal{P} \rtimes M_{\varphi_j})(z)$ belongs again to $\mathrm{Ker}(\partial_{\psi,q,p^*})^*$ in the last equality, we obtain

$$(\partial_{\psi,q,2})^* J_{p,2,j}(\mathcal{P} \rtimes M_{\varphi_j})(z) \overset{(4.32)}{=} (\partial_{\psi,q,2})^* \left(\sum_{s \in \mathrm{supp}\,\varphi_j} s_q(h_s) \rtimes \lambda_s \right)$$

$$= (\partial_{\psi,q,p^*})^* \left(\sum_{s \in \mathrm{supp}\,\varphi_j} s_q(h_s) \rtimes \lambda_s \right)$$

$$= (\partial_{\psi,q,p^*})^* (\mathcal{P} \rtimes M_{\varphi_j})(z) = 0.$$

We have shown that $J_{p,2,j}(\mathcal{P} \rtimes M_{\varphi_j})(z)$ belongs to $\overline{\mathrm{Ran}\, \partial_{\psi,q,2}} \cap \mathrm{Ker}(\partial_{\psi,q,2})^*$.

---

[2] Note that $(\mathcal{P} \rtimes \mathrm{Id})(z) = z$.

[3] If $x \in \mathrm{dom}\, \partial_{\psi,q,p}$, it is really easy to check that $(\mathrm{Id} \rtimes M_{\psi_j}) \partial_{\psi,q,p}(x) = \partial_{\psi,q,p} M_{\psi_j}(x)$.

According to the beginning of the proof, the last intersection is trivial. It follows that $J_{p,2,j}(\mathcal{P} \rtimes M_{\varphi_j})(z) = 0$. Since $J_{p,2,j}$ is an isomorphism, we infer that $(\mathcal{P} \rtimes M_{\varphi_j})(z) = 0$ for any j. Since G is weakly amenable and $\Gamma_q(H) \rtimes_\alpha G$ has QWEP, the net $(\mathrm{Id} \rtimes M_{\varphi_j})$ converges to $\mathrm{Id}_{L^p(\Gamma_q(H) \rtimes_\alpha G)}$ for the point norm topology of $L^p(\Gamma_q(H) \rtimes_\alpha G)$. We deduce that $(\mathcal{P} \rtimes \mathrm{Id})(z) = 0$. But we had seen in (4.30) that $(\mathcal{P} \rtimes \mathrm{Id})(z) = z$, so that $z = 0$ and we are done.   □

Combining Lemma 4.6 and Proposition 4.9, we can now deduce the following corollary. The QWEP assumption is satisfied if G is amenable or if G is a free group and $q = \pm 1$.

Corollary 4.1 *Let $-1 \leqslant q \leqslant 1$ and $1 < p < \infty$. Let G be a weakly amenable discrete group such that $\Gamma_q(H) \rtimes_\alpha G$ is QWEP. Then we have a topological direct sum decomposition*

$$L^p(\Gamma_q(H) \rtimes_\alpha G) = \overline{\mathrm{Ran}\, \partial_{\psi,q,p}} \oplus \mathrm{Ker}(\partial_{\psi,q,p^*})^*. \tag{4.33}$$

where the associated first bounded projection is $R_p R_{p^*}^*$. In particular, we have

$$\mathrm{Ran}(\partial_{\psi,q,p^*})^* = \mathrm{Ran}(\partial_{\psi,q,p^*})^* | \overline{\mathrm{Ran}\, \partial_{\psi,q,p}}.$$

Proof According to Lemma 4.6, the preceding subspaces add up to $L^p(\Gamma_q(H) \rtimes_\alpha G)$, and according to Proposition 4.9, the sum is direct. By Kadison and Ringrose [134, Theorem 1.8.7], we conclude that the decomposition is topological. In the course of the proof of Lemma 4.6, we have seen that for any $y \in L^p(\Gamma_q(H) \rtimes_\alpha G)$, we have the suitable decomposition $y = R_p R_{p^*}^*(y) + (\mathrm{Id} - R_p R_{p^*}^*)(y)$. So the associated first bounded projection is $R_p R_{p^*}^*$.   □

Theorem 4.4 *Let $-1 \leqslant q \leqslant 1$ and $1 < p < \infty$. Let G be a weakly amenable discrete group such that $\Gamma_q(H) \rtimes_\alpha G$ is QWEP. Consider the operator $\mathcal{D}_{\psi,q,p}$ from (4.24). Then $\mathcal{D}_{\psi,q,p}$ is bisectorial and has a bounded $\mathrm{H}^\infty(\Sigma_\omega^\pm)$ functional calculus.*

Proof According to Corollary 4.1, the space $L^p(\mathrm{VN}(G)) \oplus L^p(\Gamma_q(H) \rtimes_\alpha G)$ admits the topological direct sum decomposition $L^p(\mathrm{VN}(G)) \oplus L^p(\Gamma_q(H) \rtimes_\alpha G) \overset{(4.33)}{=} L^p(\mathrm{VN}(G)) \oplus \overline{\mathrm{Ran}\, \partial_{\psi,q,p}} \oplus \mathrm{Ker}(\partial_{\psi,q,p^*})^*$ into a sum of three subspaces. Along this decomposition, we can write[4]

$$\mathcal{D}_{\psi,q,p} \overset{(4.24)}{=} \begin{bmatrix} 0 & (\partial_{\psi,q,p^*})^* & 0 \\ \partial_{\psi,q,p} & 0 & 0 \\ 0 & 0 & 0 \end{bmatrix} \overset{(4.16)}{=} \begin{bmatrix} \mathcal{D}_{\psi,q,p} & 0 \\ 0 & 0 \end{bmatrix}.$$

---

[4] Here the notation $(\partial_{\psi,q,p^*})^*$ is used for the restriction of $(\partial_{\psi,q,p^*})^*$ on the subspace $\overline{\mathrm{Ran}\, \partial_{\psi,q,p}}$.

According to Theorem 4.3, the operator $D_{\psi,q,p}$ is bisectorial and does have a bounded H$^\infty(\Sigma_\omega^\pm)$ functional calculus. So we conclude the same thing for $\mathcal{D}_{\psi,q,p}$. See also Proposition 4.10 below.   □

Theorem 4.5 (Hodge Decomposition) *Suppose* $1 < p < \infty$ *and* $-1 \leqslant q \leqslant 1$. *Let* G *be a weakly amenable discrete group such that* $\Gamma_q(H) \rtimes_\alpha G$ *is QWEP. If we identify* $\overline{\mathrm{Ran}\,\partial_{\psi,q,p}}$ *and* $\overline{\mathrm{Ran}(\partial_{\psi,q,p^*})}^*$ *as the closed subspaces* $\{0\} \oplus \overline{\mathrm{Ran}\,\partial_{\psi,q,p}}$ *and* $\overline{\mathrm{Ran}(\partial_{\psi,q,p^*})}^* \oplus \{0\}$ *of* $L^p(\mathrm{VN}(G)) \oplus L^p(\Gamma_q(H) \rtimes_\alpha G)$, *we have*

$$L^p(\mathrm{VN}(G)) \oplus L^p(\Gamma_q(H) \rtimes_\alpha G) = \overline{\mathrm{Ran}\,\partial_{\psi,q,p}} \oplus \overline{\mathrm{Ran}(\partial_{\psi,q,p^*})}^* \oplus \mathrm{Ker}\,\mathcal{D}_{\psi,q,p}. \tag{4.34}$$

Proof From the definition (4.24), it is obvious that $\mathrm{Ker}\,\mathcal{D}_{\psi,q,p} = (\mathrm{Ker}\,\partial_{\psi,q,p} \oplus \{0\}) \oplus (\{0\} \oplus \mathrm{Ker}(\partial_{\psi,q,p^*})^*)$. We deduce that

$$L^p(\mathrm{VN}(G)) \oplus L^p(\Gamma_q(H) \rtimes_\alpha G)$$

$$\overset{(4.23)(4.33)}{=} \left(\overline{\mathrm{Ran}(\partial_{\psi,q,p^*})}^* \oplus \mathrm{Ker}\,\partial_{\psi,q,p}\right) \oplus \left(\mathrm{Ker}(\partial_{\psi,q,p^*})^* \oplus \overline{\mathrm{Ran}\,\partial_{\psi,q,p}}\right)$$

$$= (\{0\} \oplus \overline{\mathrm{Ran}\,\partial_{\psi,q,p}}) \oplus (\overline{\mathrm{Ran}(\partial_{\psi,q,p^*})}^* \oplus \{0\}) \oplus (\mathrm{Ker}\,\partial_{\psi,q,p} \oplus \{0\})$$

$$\oplus (\{0\} \oplus \mathrm{Ker}(\partial_{\psi,q,p^*})^*)$$

$$= (\{0\} \oplus \overline{\mathrm{Ran}\,\partial_{\psi,q,p}}) \oplus (\overline{\mathrm{Ran}(\partial_{\psi,q,p^*})}^* \oplus \{0\}) \oplus \mathrm{Ker}\,\mathcal{D}_{\psi,q,p}.$$

□

Remark 4.3 An inspection in all the steps of the proof of Theorem 4.4 shows that the angle of the H$^\infty(\Sigma_\omega^\pm)$ calculus can be chosen $\omega > \frac{\pi}{2}|\frac{1}{p} - \frac{1}{2}|$ and that the norm of the calculus is bounded by a constant K_ω not depending on G nor the cocycle (b_ψ, H), in particular it is independent of the dimension of H.

Proof First note that since A_p has a (sectorial) completely bounded H$^\infty(\Sigma_{2\omega})$ calculus with angle $2\omega > \pi|\frac{1}{p} - \frac{1}{2}|$ by e.g. [13, Theorem 4.1], by the representation (4.19) together with the fact that the spectral multipliers of A_p are Fourier multipliers and with Proposition 2.8, $\mathcal{D}^2_{\psi,q,p}$ also has a sectorial H$^\infty(\Sigma_{2\omega})$ calculus. According to [110, Proof of Theorem 10.6.7, Theorem 10.4.4 (1) and (3), Proof of Theorem 10.4.9], the operator $\mathcal{D}_{\psi,q,p}$ has then an H$^\infty(\Sigma_\omega^\pm)$ bisectorial calculus to the angle $\omega > \frac{\pi}{2}|\frac{1}{p} - \frac{1}{2}|$ with a norm control

$$\|f(\mathcal{D}_{\psi,q,p})\| \leqslant K_\omega \left(M^\infty_{2\omega,\mathcal{D}^2_{\psi,q,\omega}}\right)^2 \left(M^R_{\omega,\mathcal{D}_{\psi,q,\omega}}\right)^2 \|f\|_{\infty,\omega}, \tag{4.35}$$

where K_ω is a constant only depending on ω (and not on G nor the cocycle (b_ψ, H)). Here $M^\infty_{2\omega, \mathcal{D}^2_{\psi,q,p}}$ is the $H^\infty(\Sigma_{2\omega})$ calculus norm of $\mathcal{D}^2_{\psi,q,p}$ and

$$M^R_{\omega, \mathcal{D}_{\psi,q,p}} = R\left(\left\{\lambda(\lambda - \mathcal{D}_{\psi,q,p})^{-1} : \lambda \in \mathbb{C} \backslash \{0\}, \left||\arg(\lambda)| - \frac{\pi}{2}\right| < \frac{\pi}{2} - \omega\right\}\right). \tag{4.36}$$

Thus it remains to show that both $M^\infty_{2\omega, \mathcal{D}^2_{\psi,q,p}}$ and $M^R_{\omega, \mathcal{D}_{\psi,q,p}}$ can be chosen independently of the Hilbert space H and the cocycle b_ψ. Let us start with $M^\infty_{2\omega, \mathcal{D}^2_{\psi,q,p}}$. It is controlled according to the previous reasoning and (4.19), by $M^\infty_{2\omega, \mathrm{Id}_{S^p} \otimes A_p}$, that is, the completely bounded H^∞ calculus norm of A_p. Moreover, an application of [128, Proposition 5.8] shows that it suffices to consider only $2\omega > \frac{\pi}{2}$. According to [13, Theorem 4.1], we have a certain decomposition of the semigroup $(T_{t,p})_{t \geq 0}$ generated by A_p, given by

$$\mathrm{Id}_{S^p} \otimes T_{t,p} = (\mathrm{Id}_{S^p} \otimes \mathbb{E}_p)(\mathrm{Id}_{S^p} \otimes U_{t,p})(\mathrm{Id}_{S^p} \otimes J_p).$$

Here, $\mathrm{Id}_{S^p} \otimes \mathbb{E}_p$ and $\mathrm{Id}_{S^p} \otimes J_p$ are contractions and $\mathrm{Id}_{S^p} \otimes U_{t,p}$ is a group of isometries. An inspection of the proof of Lemma 4.5 shows that the constant in the second condition there is a bound of the H^∞ calculus in the first condition there, so that it suffices to show that the generator of $\mathrm{Id}_{S^p} \otimes U_{t,p}$ has a bounded $H^\infty(\Sigma_{2\omega})$ calculus with a norm controlled by a constant depending only on $2\omega > \frac{\pi}{2}$. By Hytönen et al. [110, Proof of Theorem 10.7.10], the norm of the calculus of $U_{t,p}$ is controlled by $c_{2\omega}\beta^2_{p,X}h_{p,X}$, $c_{2\omega}$ denoting a constant depending only on 2ω, $\beta_{p,X}$ denoting the UMD constant of $X = S^p(L^p(M))$ and $h_{p,X}$ denoting the Hilbert transform norm, i.e. on the space $L^p(\mathbb{R}, S^p(L^p(M)))$.

These are controlled by a constant depending only on p but not on M. Indeed, using [109, Corollary 5.2.11], it suffices to control the UMD constant of $S^p(L^p(M))$. According to [193, Corollary 4.5, Theorem 4.3], this constant is controlled by some universal bound times $p^2/(p-1)$. So we have the desired control of $M^\infty_{2\omega, \mathrm{Id}_{S^p} \otimes A_p}$, and thus of $M^\infty_{2\omega, \mathcal{D}^2_{\psi,q,p}}$.

We turn to the control of $M^R_{\omega, \mathcal{D}_{\psi,q,p}}$ from (4.36). According to (4.18) extended to complex times z belonging to some bisector $\Sigma^\pm_\sigma$ with $\sigma = \frac{\pi}{2} - \omega$, it suffices to control the following R-bounds

$$R\left(\left\{(\mathrm{Id} + z^2 A_p)^{-1} : z \in \Sigma^\pm_\sigma\right\}\right) \tag{4.37}$$

$$R\left(\left\{z(\mathrm{Id} + z^2 A_p)^{-1}(\partial_{\psi,q,p^*})^* : z \in \Sigma^\pm_\sigma\right\}\right) \tag{4.38}$$

$$R\left(\left\{z\partial_{\psi,q,p}(\mathrm{Id} + z^2 A_p)^{-1} : z \in \Sigma^\pm_\sigma\right\}\right) \tag{4.39}$$

$$R\left(\left\{\mathrm{Id} \rtimes (\mathrm{Id} + z^2 A_p)^{-1} : z \in \Sigma^\pm_\sigma\right\}\right). \tag{4.40}$$

Indeed, an operator matrix family is R-bounded if and only if all operator entries in the matrix are R-bounded. According to [110, Theorem 10.3.4 (1)], (4.37) is R-bounded since we can write $(\mathrm{Id} + z^2 A_p)^{-1} = \mathrm{Id} - f(z^2 A_p)$ with $f(\lambda) = \lambda(1+\lambda)^{-1}$. Moreover, by the same reference, its R-bound is controlled by $M_{2\omega-\varepsilon,A_p}^\infty$, which in turn, by the preceding argument of dilation can be controlled independently of the Hilbert space H and the cocycle b_ψ. The same argument shows that also (4.40) is R-bounded. Since the operator family in (4.38) consists of the family of the adjoints in (4.39) and R-boundedness is preserved under adjoints, it suffices to prove that (4.39) is R-bounded. To this end, we decompose for $z \in \Sigma_\omega$ the positive part of the bisector (similarly if z belongs to the negative part of the bisector)

$$z\partial_{\psi,q,p}(\mathrm{Id} + z^2 A_p)^{-1} = \left[\partial_{\psi,q,p} A_p^{-\frac{1}{2}}\right] \left(z^2 A_p\right)^{\frac{1}{2}} (\mathrm{Id} + z^2 A_p)^{-1}$$

$$= \left[\partial_{\psi,q,p} A_p^{-\frac{1}{2}}\right] f(z^2 A_p).$$

with $f(\lambda) = \sqrt{\lambda}(1 + \lambda)^{-1}$. Again [110, Theorem 10.3.4 (1)] shows that the term $f(z^2 A_p)$ is R-bounded with R-bound controlled by some constant independent of the cocycle. Finally, we are left to show that the Riesz transform is bounded by a constant independent of the cocycle, that is,

$$\left\|\partial_{\psi,q,p}(x)\right\|_{\mathrm{L}^p(\Gamma_q(H)\rtimes_\alpha G)} \leqslant C\left\|A_p^{\frac{1}{2}}(x)\right\|_{\mathrm{L}^p(\mathrm{VN}(G))}. \tag{4.41}$$

For this in turn we refer to Proposition 3.3. $\qquad\square$

4.3 Hodge-Dirac Operator on $\mathbf{L}^p(\mathrm{VN}(G)) \oplus \Omega_{\psi,q,p}$

We keep the standing assumptions of the two preceding sections and thus have a markovian semigroup $(T_t)_{t\geqslant 0}$ of Fourier multipliers with generator A_p, the non-commutative gradient $\partial_{\psi,q,p}$ and its adjoint $(\partial_{\psi,q,p^*})^*$. We also fix the parameters $1 < p < \infty$ and $-1 \leqslant q \leqslant 1$, and assume that the discrete group G is weakly amenable such that $\Gamma_q(H) \rtimes_\alpha G$ has QWEP, so that the main results from the preceding section are valid. For the rest of this section we consider the Hodge-Dirac operator

$$\mathcal{D}_{\psi,q,p} \overset{\mathrm{def}}{=} \begin{bmatrix} 0 & (\partial_{\psi,q,p^*})^* & 0 \\ \partial_{\psi,q,p} & 0 & 0 \\ 0 & 0 & 0 \end{bmatrix}$$

on the bigger space $L^p(\mathrm{VN}(G)) \oplus \overline{\mathrm{Ran}\, \partial_{\psi,q,p}} \oplus \mathrm{Ker}(\partial_{\psi,q,p*})^*$, with domain

$$\mathrm{dom}\, \mathcal{D}_{\psi,q,p} \overset{\text{def}}{=} \mathrm{dom}\, \partial_{\psi,q,p} \oplus \big(\mathrm{dom}(\partial_{\psi,q,p*})^* \cap \overline{\mathrm{Ran}\, \partial_{\psi,q,p}}\big) \oplus \mathrm{Ker}(\partial_{\psi,q,p*})^*.$$

In the following, we consider the bounded operator

$$T \overset{\text{def}}{=} \begin{bmatrix} (\mathrm{Id} + t^2 A_p)^{-1} & it(\mathrm{Id} + t^2 A_p)^{-1}(\partial_{\psi,q,p*})^* & 0 \\ it\partial_{\psi,q,p}(\mathrm{Id} + t^2 A_p)^{-1} & \mathrm{Id} \rtimes (\mathrm{Id} + t^2 A_p)^{-1} & 0 \\ 0 & 0 & \mathrm{Id}_{\mathrm{Ker}(\partial_{\psi,q,p*})^*} \end{bmatrix} \tag{4.42}$$

on the space $L^p(\mathrm{VN}(G)) \oplus \overline{\mathrm{Ran}\, \partial_{\psi,q,p}} \oplus \mathrm{Ker}(\partial_{\psi,q,p*})^*$. Here, we interpret the operators $it\partial_{\psi,q,p}(\mathrm{Id} + t^2 A_p)^{-1}$ and $it(\mathrm{Id} + t^2 A_p)^{-1}(\partial_{\psi,q,p*})^*$ as the bounded extensions $L^p(\mathrm{VN}(G)) \to \overline{\mathrm{Ran}\, \partial_{\psi,q,p}} \subset L^p(\Gamma_q(H) \rtimes_\alpha G)$ resp. $\overline{\mathrm{Ran}\, \partial_{\psi,q,p}} \subset L^p(\Gamma_q(H) \rtimes_\alpha G) \to L^p(\mathrm{VN}(G))$ guaranteed by Proposition 4.3. Then this proposition and Theorem 4.5 yield that T is a bounded operator on $L^p(\mathrm{VN}(G)) \oplus L^p(\Gamma_q(H) \rtimes_\alpha G)$.

Proposition 4.10 *Let* $1 < p < \infty$, $-1 \leqslant q \leqslant 1$ *and* G *be a weakly amenable discrete group such that* $\Gamma_q(H) \rtimes_\alpha G$ *has QWEP. We have*

$$T(\mathrm{Id} - it\mathcal{D}_{\psi,q,p}) = \mathrm{Id}_{\mathrm{dom}\, \mathcal{D}_{\psi,q,p}} \tag{4.43}$$

and

$$(\mathrm{Id} - it\mathcal{D}_{\psi,q,p})T = \mathrm{Id}_{L^p(\mathrm{VN}(G)) \oplus L^p(\Gamma_q(H) \rtimes_\alpha G)}. \tag{4.44}$$

Proof If $R \overset{\text{def}}{=} (\mathrm{Id} + t^2 A_p)^{-1}$, a straightforward calculation shows that

$T(\mathrm{Id} - it\mathcal{D}_{\psi,q,p})$

$$= \begin{bmatrix} R(\mathrm{Id} + t^2(\partial_{\psi,q,p*})^*\partial_{\psi,q,p}) & 0 & 0 \\ it\partial_{\psi,q,p}R - it\mathrm{Id} \rtimes R\partial_{\psi,q,p} & t^2\partial_{\psi,q,p}R(\partial_{\psi,q,p*})^* + \mathrm{Id} \rtimes R & 0 \\ 0 & 0 & \mathrm{Id}_{\mathrm{Ker}(\partial_{\psi,q,p*})^*} \end{bmatrix}$$

$$\overset{\text{def}}{=} \begin{bmatrix} (I) & 0 & 0 \\ (II) & (III) & 0 \\ 0 & 0 & \mathrm{Id}_{\mathrm{Ker}(\partial_{\psi,q,p*})^*} \end{bmatrix}.$$

We check the expressions $(I), (II), (III)$. For (I), note that according to Proposition 4.1, we have $(I) = (\mathrm{Id} + t^2 A_p)^{-1} + t^2(\mathrm{Id} + t^2 A_p)^{-1}(\partial_{\psi,q,p*})^*\partial_{\psi,q,p} = \mathrm{Id}_{\mathrm{dom}\, \partial_{\psi,q,p}}$, since we recall that $(\mathrm{Id} + t^2 A_p)^{-1}(\partial_{\psi,q,p*})^*$ is interpreted as a bounded operator $\overline{\mathrm{Ran}\, \partial_{\psi,q,p*}} \to L^p(\mathrm{VN}(G))$. Then $(II) = 0$ on $\mathrm{dom}\, \partial_{\psi,q,p}$ according to Proposition 4.2. Finally, we note that on the one hand, it is easy to check with Proposition 4.2 that $(III) = \text{Id}$ on $\mathcal{P}_{\rtimes,G}$. On the other hand, $\mathcal{P}_{\rtimes,G}$ is a core of $(\partial_{\psi,q,p*})^*$ according to Proposition 4.4 and $\partial_{\psi,q,p}(\text{Id} + t^2 A_p)^{-1}$ is bounded, the two of which imply easily that $(III) = \text{Id}_{\overline{\text{Ran}\,\partial_{\psi,q,p}} \cap \text{dom}(\partial_{\psi,q,p*})^*}$. Altogether, we have shown (4.43). We turn to (4.44). Again a straightforward calculation shows that

$$(\text{Id} - it\mathcal{D}_{\psi,q,p})T$$

$$= \begin{bmatrix} (\text{Id} + t^2(\partial_{\psi,q,p*})^*\partial_{\psi,q,p})R & it R(\partial_{\psi,q,p*})^* - it(\partial_{\psi,q,p*})^*\text{Id} \rtimes R & 0 \\ 0 & t^2 \partial_{\psi,q,p} R(\partial_{\psi,q,p*})^* + \text{Id} \rtimes R & 0 \\ 0 & 0 & \text{Id}_{\text{Ker}(\partial_{\psi,q,p*})^*} \end{bmatrix}$$

$$\overset{\text{def}}{=} \begin{bmatrix} (I) & (II) & 0 \\ 0 & (III) & 0 \\ 0 & 0 & \text{Id}_{\text{Ker}(\partial_{\psi,q,p*})^*} \end{bmatrix}.$$

As for (4.43), one shows that $(I) = \text{Id}_{L^p(\text{VN}(G))}$. With Proposition 4.5, one shows that $(II) = 0$ on $\text{dom}(\partial_{\psi,q,p*})^*$. But (II) is closed, since the product AB of a closed operator A and a bounded operator B is closed, so since $\text{dom}(\partial_{\psi,q,p*})^* \cap \overline{\text{Ran}\,\partial_{\psi,q,p}}$ is dense in $\overline{\text{Ran}\,\partial_{\psi,q,p}}$, $(II) = 0$ on $\overline{\text{Ran}\,\partial_{\psi,q,p}}$. Finally, we have already shown that $(III) = \text{Id}$ on $\text{dom}(\partial_{\psi,q,p*})^*$, and as before, (III) is closed. We infer that $(III) = \text{Id}_{\overline{\text{Ran}\,\partial_{\psi,q,p}}}$. $\qquad \square$

We recall from Theorem 4.4 that $\mathcal{D}_{\psi,q,p}$ has a bounded H$^\infty(\Sigma_\omega^\pm)$ functional calculus on a bisector on the space $L^p(\text{VN}(G)) \oplus L^p(\Gamma_q(H) \rtimes_\alpha G)$. Next we show that an appropriate restriction of $\mathcal{D}_{\psi,q,p}$ to the space $L^p(\text{VN}(G)) \oplus \Omega_{\psi,q,p}$, where

$$\Omega_{\psi,q,p} \overset{\text{def}}{=} \overline{\text{span}}^{L^p} \left\{ s_q(\xi) \rtimes \lambda_s : \xi \in H_\psi, s \in G \right\}$$

and $H_\psi \overset{\text{def}}{=} \overline{\text{span}}\{b_\psi(s) : s \in G\}$, is still bisectorial and admits an H$^\infty(\Sigma_\omega^\pm)$ functional calculus on a bisector.

Lemma 4.8 *Assume* $-1 \leqslant q \leqslant 1$ *and* $1 < p < \infty$. *Let* G *be a weakly amenable discrete group such that* $\Gamma_q(H) \rtimes_\alpha G$ *has QWEP. There is a bounded projection* $W : L^p(\text{VN}(G)) \oplus L^p(\Gamma_q(H) \rtimes_\alpha G) \to L^p(\text{VN}(G)) \oplus L^p(\Gamma_q(H) \rtimes_\alpha G)$ *on* $\Omega_{\psi,q,p}$ *such that* $L^p(\text{VN}(G)) \subset \text{Ker}\, W$.

Proof We remind the reader that the bounded projection $\mathcal{P} : L^p(\Gamma_q(H)) \to L^p(\Gamma_q(H))$ from the part 1 of Lemma 4.7 is given in the following way. If $(e_k)_{k \geqslant 1}$ denotes an orthonormal basis of H and for a multi-index $\underline{i} = (i_1, i_2, \ldots, i_n)$ of length $|\underline{i}| = n \in \mathbb{N}$ we let $e_{\underline{i}} \overset{\text{def}}{=} e_{i_1} \otimes e_{i_2} \otimes \cdots \otimes e_{i_n} \in \mathcal{F}_q(H)$, then the Wick word $w(e_{\underline{i}}) \in \Gamma_q(H)$ is determined by $w(e_{\underline{i}})\Omega = e_{\underline{i}}$, where Ω denotes as usual the vacuum vector. Then a careful inspection of [118], in particular Theorem 3.5 there, shows that we have $\mathcal{P}(\sum_{\underline{i}} \alpha_{\underline{i}} e_{\underline{i}}) = \sum_{|\underline{i}|=1} \alpha_{\underline{i}} w(e_{\underline{i}})$ for any finite sum and

$\alpha_{\underline{i}} \in \mathbb{C}$. Here, $w(e_{\underline{i}}) = s_q(e_{\underline{i}})$ in case $|\underline{i}| = 1$. According to Lemma 3.4 or 4.7, $\mathcal{P} \rtimes \text{Id}_{L^p(\text{VN}(G))} \colon L^p(\Gamma_q(H) \rtimes_\alpha G) \to L^p(\Gamma_q(H) \rtimes_\alpha G)$ is a bounded mapping.

Now note that if $Q \colon H \to H$ denotes the orthogonal projection onto the closed subspace H_ψ of H, then by Bożejko et al. [44, Theorem 2.11] there exists a trace preserving conditional expectation $\mathbb{E} \colon L^p(\Gamma_q(H)) \to L^p(\Gamma_q(H))$ such that $\mathbb{E}(w(e_{\underline{i}})) = w(\mathcal{F}_q(Q)e_{\underline{i}})$. Note that by (2.33), for any $s \in G$, each π_s induces an operator $\pi_s \colon H_\psi \to H_\psi$, and thus by orthogonality of $\pi_{s^{-1}}$ also $\pi_s \colon H_\psi^\perp \to H_\psi^\perp$.

Thus each π_s and Q commute, whence $\alpha_s \overset{(2.83)}{=} \Gamma_q(\pi_s)$ and $\mathbb{E} = \Gamma_q(Q)$ commute. We can use Lemma 2.6 and deduce that $\mathbb{E}$ extends to a normal complete contraction $\mathbb{E} \rtimes \text{Id}_{\text{VN}(G)} \colon \Gamma_q(H) \rtimes_\alpha G \to \Gamma_q(H) \rtimes_\alpha G$. Again by Lemma 2.6, we infer that $\mathbb{E} \rtimes \text{Id}_{\text{VN}(G)}$ is a conditional expectation which is trace preserving, so extends to a contraction on $L^p(\Gamma_q(H) \rtimes_\alpha G)$ for $1 \leqslant p \leqslant \infty$.

We claim that $\mathcal{P}\mathbb{E} \rtimes \text{Id}_{L^p(\text{VN}(G))}$ is a projection. To this end, it suffices[5] to check that $\mathcal{P}$ and $\mathbb{E}$ commute. We assume that the orthonormal basis $(e_k)_{k \in \mathbb{N}}$ is chosen in such a way that $(e_k)_{k \in \mathbb{N}_\psi}$ is an orthonormal basis of H_ψ for some $\mathbb{N}_\psi \subset \mathbb{N}$. From the foregoing, for a Wick word $w(e_{\underline{i}})$ we have

$$\mathcal{P}\mathbb{E}w(e_{\underline{i}}) = \mathcal{P}w(\mathcal{F}_q(Q)e_{\underline{i}}) = \delta_{|\underline{i}|=1}w(\mathcal{F}_q(Q)e_{\underline{i}}) = \delta_{|\underline{i}|=1}\delta_{\underline{i} \in \mathbb{N}_\psi}w(e_{\underline{i}})$$

and

$$\mathbb{E}\mathcal{P}w(e_{\underline{i}}) = \mathbb{E}\delta_{|\underline{i}|=1}w(e_{\underline{i}}) = \delta_{|\underline{i}|=1}w(\mathcal{F}_q(Q)e_{\underline{i}}) = \delta_{|\underline{i}|=1}\delta_{\underline{i} \in \mathbb{N}_\psi}w(e_{\underline{i}}).$$

Thus $\mathcal{P}$ and $\mathbb{E}$ commute on a total set, so commute on all of $L^p(\Gamma_q(H))$. It suffices to consider the projection $W' \overset{\text{def}}{=} \mathcal{P}\mathbb{E} \rtimes \text{Id}_{L^p(\text{VN}(G))}$ and to finally extend it to $W \colon L^p(\text{VN}(G)) \oplus L^p(\Gamma_q(H) \rtimes_\alpha G) \to \Omega_\psi$ by setting $W(x,y) \overset{\text{def}}{=} 0 \oplus W'(y)$ and observe by a standard density and continuity argument that $\text{Ran } W = \Omega_{\psi,q,p}$. $\square$

The proof of the following elementary lemma is left to the reader.

Lemma 4.9 *Let X be a Banach space and let $Q_1, Q_2, Q_3, Q_4 \colon X \to X$ be bounded projections. Assume that $Q_1 + Q_2 + Q_3 + Q_4 = \text{Id}_X$ and that $Q_i Q_j = 0$ for $i < j$. Then we have a direct sum decomposition*

$$X = \text{Ran}(Q_1) \oplus \text{Ran}(Q_2) \oplus \text{Ran}(Q_3) \oplus \text{Ran}(Q_4).$$

Lemma 4.10 *Let $1 < p < \infty$ and $-1 \leqslant q \leqslant 1$. Assume that the discrete group G is weakly amenable and that $\Gamma_q(H) \rtimes_\alpha G$ has QWEP. The subspace $L^p(\text{VN}(G)) \oplus \Omega_{\psi,q,p}$ is invariant under the resolvents T from (4.42) of the Hodge-Dirac operator $\mathcal{D}_{\psi,q,p}$.*

---

[5] Recall that the product of two commuting projections on a Banach space is a projection.

Proof With respect to the decomposition projections from (4.33), we can decompose the identity $\mathrm{Id}_{L^p(\mathrm{VN}(G))\oplus L^p(\Gamma_q(H)\rtimes_\alpha G)}$ as a sum

$$\mathrm{Id}_{L^p(\mathrm{VN}(G))\oplus L^p(\Gamma_q(H)\rtimes_\alpha G)} = P_1 \qquad\qquad \oplus P_2 \oplus P_3$$

$$L^p(\mathrm{VN}(G)) \oplus L^p(\Gamma_q(H)\rtimes_\alpha G) = L^p(\mathrm{VN}(G)) \quad \oplus\overline{\mathrm{Ran}\,\partial_{\psi,q,p}} \oplus \overline{\mathrm{Ran}(\partial_{\psi,q,p^*})^*}.$$

Then we claim that

$$\mathrm{Id}_{L^p(\mathrm{VN}(G))\oplus L^p(\Gamma_q(H)\rtimes_\alpha G)} = P_1 \qquad\qquad \oplus P_2 W \oplus P_3 W \oplus V$$

$$L^p(\mathrm{VN}(G)) \oplus L^p(\Gamma_q(H)\rtimes_\alpha G) = L^p(\mathrm{VN}(G)) \quad \underbrace{\oplus\overline{\mathrm{Ran}\,\partial_{\psi,q,p}} \oplus X_\psi \oplus Y_\psi}_{=\Omega_{\psi,q,p}}$$

for some subspaces X_ψ and Y_ψ, coming with projections P_1, $P_2 W$, $P_3 W$ and $V \overset{\mathrm{def}}{=} (P_2 + P_3)(\mathrm{Id} - W)$. To this end, we apply the auxiliary Lemma 4.9. Note first that $P_1 + P_2 W + P_3 W + V = P_1 + (P_2 + P_3)W + (P_2 + P_3)(\mathrm{Id} - W) = P_1 + (P_2 + P_3) =$ Id. Then $P_2 W$, $P_3 W$ and V are projections. Indeed, $P_2 W P_2 W = P_2 P_2 W$, since $\mathrm{Ran}\, P_2 W \subset \overline{\mathrm{Ran}\,\partial_{\psi,q,p}} \subset \Omega_{\psi,q,p} = \mathrm{Ran}\, W$, and $P_2 P_2 W = P_2 W$, since P_2 is a projection. Moreover, $P_3 W P_3 W = P_3 W (\mathrm{Id} - P_1 - P_2)W = P_3 W - P_3 W P_1 W - P_3 W P_2 W = P_3 W - P_3 \cdot 0 \cdot W - P_3 P_2 W = P_3 W - 0 - 0 \cdot W = P_3 W$. Thus, $P_2 W$ and $P_3 W$ are projections. Moreover, $V = (P_2 + P_3)(\mathrm{Id} - W) = (P_1 + P_2 + P_3)(\mathrm{Id} - W) - P_1 = \mathrm{Id} - P_1 - W$, and $V^2 = (\mathrm{Id} - P_1 - W)^2 = \mathrm{Id} + P_1 + W - 2P_1 - 2W + P_1 W + W P_1 = \mathrm{Id} - P_1 - W + 0 + 0 = V$. Thus, also V is a projection. Now we check that some products of the four projections vanish as needed to apply Lemma 4.9. We choose the order $(Q_1, Q_2, Q_3, Q_4) = (P_1, P_3 W, P_2 W, V)$. First note that this is clear if one of the factors is P_1. Then $P_3 W P_2 W = 0$ since $\mathrm{Ran}(W P_2 W) = \mathrm{Ran}(P_2 W) \subset \overline{\mathrm{Ran}\,\partial_{\psi,q,p}} \subset \mathrm{Ker}\, P_3$. Moreover, $P_3 W V = P_3 W (P_2 + P_3)(\mathrm{Id} - W) = P_3 W (\mathrm{Id} - W) = 0$ and also $P_2 W V = P_2 W (P_2 + P_3)(\mathrm{Id} - W) = 0$. We have shown the claim and thus have a direct sum decomposition of the space into four closed subspaces.

Now write the resolvent

$$T = \begin{bmatrix} A & B & 0 \\ C & D & 0 \\ 0 & 0 & \mathrm{Id}_{\mathrm{Ker}(\partial_{\psi,q,p^*})^*} \end{bmatrix} \tag{4.45}$$

along the Hodge decomposition $\mathrm{Id} = P_1 + P_2 + P_3$. If $x \in L^p(\mathrm{VN}(G)) \oplus L^p(\Gamma_q(H)\rtimes_\alpha G)$, then x belongs to $L^p(\mathrm{VN}(G)) \oplus \Omega_{\psi,q,p}$ if and only if $V(x) = 0$. For such an x, we have

$$T(x) = T(P_1 x + P_2 W x + P_3 W x)$$

$$= P_1 A P_1 x + P_2 W C P_1 x + P_3 W C P_1 x + V C P_1 x + P_1 B P_2 W x$$

$$+ P_2 W D P_2 W x + P_3 W D P_2 W x + V D P_2 W x + P_3 W x.$$

The summands starting with P_1 and P_2 lie in $L^p(\mathrm{VN}(G))$ and $\overline{\mathrm{Ran}\,\partial_{\psi,q,p}} \subset \Omega_{\psi,q,p}$. The remaining summands are $P_3 W C P_1 x = P_3 C P_1 x = 0$ since $C P_1 x \in \overline{\mathrm{Ran}\,\partial_{\psi,q,p}}$; $V C P_1 x = 0$ since $C P_1 x \in \overline{\mathrm{Ran}\,\partial_{\psi,q,p}} \subset \Omega_{\psi,q,p}$; $P_3 W D P_2 W x = P_3 D P_2 W x = 0$; $V D P_2 W x = 0$ since $D P_2 W x \in \overline{\mathrm{Ran}\,\partial_{\psi,q,p}} \subset \Omega_{\psi,q,p}$. Finally, $P_3 W x \in X_\psi \subset \Omega_{\psi,q,p}$. We conclude that $T(x) \in L^p(\mathrm{VN}(G)) \oplus \Omega_{\psi,q,p}$. $\square$

Theorem 4.6 *Let $1 < p < \infty$ and $-1 \leqslant q \leqslant 1$. Assume that the discrete group G is weakly amenable and that $\Gamma_q(H) \rtimes_\alpha G$ has QWEP. Consider the part $\mathcal{D}'_{\psi,q,p}$ of the Hodge-Dirac operator $\mathcal{D}_{\psi,q,p}$: $\mathrm{dom}\,\mathcal{D}_{\psi,q,p} \subset L^p(\mathrm{VN}(G)) \oplus L^p(\Gamma_q(H) \rtimes_\alpha G) \to L^p(\mathrm{VN}(G)) \oplus L^p(\Gamma_q(H) \rtimes_\alpha G)$ on the closed subspace $L^p(\mathrm{VN}(G)) \oplus \Omega_{\psi,q,p}$. Then $\mathcal{D}'_{\psi,q,p}$ is bisectorial and has a bounded $\mathrm{H}^\infty(\Sigma_\omega^\pm)$ functional calculus on a bisector.*

Proof We want to apply [83, Proposition 3.2.15]. To this end, note that we have proved previously that $\mathcal{D}_{\psi,q,p}$ is bisectorial on $X = L^p(\mathrm{VN}(G)) \oplus L^p(\Gamma_q(H) \rtimes_\alpha G)$. Moreover, for any $t \in \mathbb{R}$, $(\mathrm{Id} - it\mathcal{D}_{\psi,q,p})^{-1}$ leaves invariant $Y = L^p(\mathrm{VN}(G)) \oplus \Omega_{\psi,q,p}$ according to Lemma 4.10. Note that then the same holds for t belonging to any bisector $\Sigma^\pm$ to which $\mathcal{D}_{\psi,q,p}$ is bisectorial. Indeed, $z \mapsto (\mathrm{Id} - iz\mathcal{D}_{\psi,q,p})^{-1}$ is an analytic function on the subset of $\mathbb{C}$ where it is defined. Then also $P_i(\mathrm{Id} - iz\mathcal{D}_{\psi,q,p})^{-1}P_j$ is analytic for $i = 1,2,3$. Therefore by the uniqueness theorem of analytic functions, $(\mathrm{Id} - iz\mathcal{D}_{\psi,q,p})^{-1}$ has the same form as (4.45) at least for z belonging to such a bisector $\Sigma^\pm$. But then the proof of Lemma 4.10 goes through for such z in place of t. Now the theorem follows from an application of [83, Proposition 3.2.15] together with Proposition 4.10. $\square$

4.4 Bimodule $\Omega_{\psi,q,p,c}$

In this short section, we continue to consider a markovian semigroup $(T_t)_{t \geqslant 0}$ of Fourier multipliers as in Proposition 2.3. This time, we do not need approximation properties on the discrete group G. We shall clarify and generalize some results of [131, pp. 585-586]. We need the following notion of bimodule which is different from the notion of [123, Definition 5.4] and is inspired by the well-known theory of Hilbert bimodules. Recall that the notion of right L^p-M-module is defined in Sect. 2.5.

Definition 4.1 Let M and N be von Neumann algebras. Suppose $1 \leqslant p < \infty$. An L^p-N-M-bimodule is a right L^p-M-module X equipped with a structure of left-N-module such that the associated $L^{\frac{p}{2}}(M)$-valued inner product $\langle \cdot, \cdot \rangle_X$ satisfies

$$\langle a^* x, y \rangle_X = \langle x, ay \rangle_X, \quad x, y \in X, a \in N. \tag{4.46}$$

An L^p-M-bimodule is an L^p-M-M-bimodule.

Suppose $-1 \leqslant q \leqslant 1$ and $2 \leqslant p < \infty$ (the case $p < 2$ is entirely left to the reader). If $\mathbb{E} \colon L^p(\Gamma_q(H) \rtimes_\alpha G) \to L^p(\mathrm{VN}(G))$ is the canonical conditional expectation, it is obvious that the formula

$$\langle \omega, \eta \rangle \overset{\mathrm{def}}{=} \mathbb{E}(\omega^* \eta) \tag{4.47}$$

defines an $L^{\frac{p}{2}}(\mathrm{VN}(G))$-valued inner product on $L^p(\Gamma_q(H) \rtimes_\alpha G)$. We can consider the associated right L^p-$\mathrm{VN}(G)$-module $L^p_c(\mathbb{E})$. It is easy to see that $L^p_c(\mathbb{E})$ is an L^p-$\mathrm{VN}(G)$-bimodule. We consider the closed subspace

$$\Omega'_{\psi,q,p,c} \overset{\mathrm{def}}{=} \overline{\mathrm{span}}\{\partial_{\psi,q,p}(x)a : x \in \mathrm{dom}\,\partial_{\psi,q,p}, a \in \mathrm{VN}(G)\} \tag{4.48}$$

of $L^p(\Gamma_q(H) \rtimes_\alpha G)$. For any $a, b \in \mathrm{VN}(G)$ and any $x \in \mathrm{dom}\,\partial_{\psi,q,p}$, note that

$$(\partial_{\psi,q,p}(x)a)b = \partial_{\psi,q,p}(x)ab.$$

Thus by linearity and density, $\Omega'_{\psi,q,p,c}$ is a right $\mathrm{VN}(G)$-module. Moreover, for any $a \in \mathrm{VN}(G)$ and any $x, b \in \mathcal{P}_G$, we have

$$b\partial_{\psi,q,p}(x)a \overset{(2.86)}{=} \big[\partial_{\psi,q,p}(bx) - \partial_{\psi,q,p}(b)x\big]a = \partial_{\psi,q,p}(bx)a - \partial_{\psi,q,p}(b)xa.$$

Thus $b\partial_{\psi,q,p}(x)a$ belongs to $\Omega'_{\psi,q,p,c}$. Since $\mathcal{P}_G$ is a core for $\partial_{\psi,q,p}$ according to Proposition 3.4, the same holds for $x \in \mathrm{dom}\,\partial_{\psi,q,p}$. If $b \in \mathrm{VN}(G)$ is a general element, we approximate it in the strong operator topology by a bounded net in $\mathcal{P}_G$ and obtain again that the same holds for $b \in \mathrm{VN}(G)$. By linearity and density, we deduce that $\Omega'_{\psi,q,p,c}$ is a left $\mathrm{VN}(G)$-module, so finally a $\mathrm{VN}(G)$-bimodule. It is obvious that the restriction of the bracket (4.47) defines an $L^{\frac{p}{2}}(\mathrm{VN}(G))$-valued inner product on this subspace. We can consider the associated right L^p-$\mathrm{VN}(G)$-module $\Omega_{\psi,q,p,c}$ which is also an L^p-$\mathrm{VN}(G)$-bimodule and which identifies canonically to a closed subspace of $L^p_c(\mathbb{E}) \overset{(2.76)}{=} L^p(\mathrm{VN}(G), L^2(\Gamma_q(H))_{c,p})$. Finally, we recall that H_ψ is the real Hilbert space generated by the $b_\psi(s)$'s where $s \in G$.

Lemma 4.11 *Let G be a discrete group.*

1. If $\xi \in H_\psi$ and if $s \in G$ then $s_q(\xi) \rtimes \lambda_s$ belongs to $\Omega_{\psi,q,p,c}$.
2. Moreover, we have

$$\Omega_{\psi,q,p,c} = \overline{\mathrm{span}}^{\Omega_{\psi,q,p,c}}\{s_q(\xi) \rtimes \lambda_s : \xi \in H_\psi, s \in G\}. \tag{4.49}$$

Proof

1. For any $s \in G$, we have

$$s_q(b_\psi(s)) \rtimes 1 \overset{(2.56)}{=} \big(s_q(b_\psi(s)) \rtimes \lambda_s\big)(1 \rtimes \lambda_{s^{-1}}) \overset{(2.84)}{=} \partial_{\psi,q,p}(\lambda_s)(1 \rtimes \lambda_{s^{-1}}).$$

Hence $s_q(b_\psi(s)) \rtimes 1$ belongs to (4.48). If ξ belongs to the span of the $b_\psi(s)$'s where $s \in G$, we deduce by linearity that $s_q(\xi) \rtimes 1$ belongs to $\Omega_{\psi,q,p,c}$. Now, for $\xi \in H_\psi$, there exists a sequence (ξ_n) of elements of the previous span such that $\xi_n \to \xi$ in H. By (2.41), we infer that $s_q(\xi_n) \to s_q(\xi)$ in $L^2(\Gamma_q(H))$. Hence $s_q(\xi_n) \rtimes 1 \to s_q(\xi) \rtimes 1$ in $L^p(\mathrm{VN}(G), L^2(\Gamma_q(H))_{c,p})$. We conclude that $s_q(\xi) \rtimes 1$ belongs to $\Omega_{\psi,q,p,c}$. Since $\Omega_{\psi,q,p,c}$ is a right $\mathrm{VN}(G)$-module, we conclude that $s_q(\xi) \rtimes \lambda_s = (s_q(\xi) \rtimes 1)\lambda_s$ belongs to $\Omega_{\psi,q,p,c}$.

2. For any $s, t \in G$, we have

$$\partial_{\psi,q,p}(\lambda_s)(1 \rtimes \lambda_t) \overset{(2.84)}{=} (s_q(b_\psi(s)) \rtimes \lambda_s)(1 \rtimes \lambda_t) \overset{(2.56)}{=} s_q(b_\psi(s)) \rtimes \lambda_{st}$$

which belongs to

$$\overline{\mathrm{span}}^{\Omega_{\psi,q,p,c}}\{s_q(\xi_s) \rtimes \lambda_s : \xi_s \in H_\psi, s \in G\}.$$

Since the closed span of the $\partial_{\psi,q,p}(\lambda_s)(1 \rtimes \lambda_t)$'s is $\Omega_{\psi,q,p,c}$,[6] the proof is complete.

$\square$

4.5 Hodge-Dirac Operators Associated to Semigroups of Markov Schur Multipliers

In this section, we consider some markovian semigroup $(T_t)_{t \geqslant 0}$ of Schur multipliers acting on $\mathrm{B}(\ell_I^2)$ that we defined in Proposition 2.4. If $1 \leqslant p < \infty$, we denote by A_p the (negative) infinitesimal generator on S_I^p which is defined as the closure of the unbounded operator $A \colon \mathrm{M}_{I,\mathrm{fin}} \to S_I^p$, $e_{ij} \mapsto \|\alpha_i - \alpha_j\|_H^2 e_{ij}$. So $\mathrm{M}_{I,\mathrm{fin}}$ is a core of A_p. By Junge et al. [128, (5.2)], we have $(A_p)^* = A_{p^*}$ if $1 < p < \infty$. If $-1 \leqslant q \leqslant 1$, recall that by Proposition 3.11, we have a closed operator

$$\partial_{\alpha,q,p} \colon \mathrm{dom}\, \partial_{\alpha,q,p} \subset S_I^p \to \mathrm{L}^p(\Gamma_q(H) \overline{\otimes} \mathrm{B}(\ell_I^2)), \quad e_{ij} \mapsto s_q(\alpha_i - \alpha_j) \otimes e_{ij}.$$

---

[6] It is clear that elements of the form $\partial_{\psi,q,p}(\lambda_s)(1 \rtimes \lambda_t)$ belong to $\Omega'_{\psi,q,p,c} \subset \Omega_{\psi,q,p,c}$. On the other hand, for the density, since $\Omega'_{\psi,q,p,c}$ is by definition dense in $\Omega_{\psi,q,p,c}$, it suffices to approximate $\partial_{\psi,q,p}(x)(1 \rtimes a)$ by elements of the span of the $\partial_{\psi,q,p}(\lambda_s)(1 \rtimes \lambda_t)$ for $x \in \mathrm{dom}\, \partial_{\psi,q,p}$ and $a \in \mathrm{VN}(G)$. The approximation needs to be in $L_c^p(\mathbb{E})$ norm, but according to Lemma 2.8, it suffices to approximate in $L^p(\Gamma_q(H) \rtimes_\alpha G)$ norm. By definition of $\mathrm{VN}(G)$ and Kaplansky's density theorem, there exists a bounded net (a_α) in $\mathcal{P}_G$ converging in the strong operator topology to a. Then it is not difficult to see that the net $(1 \rtimes a_\alpha)$ converges in the strong operator topology to $1 \rtimes a$. Thus we can approximate $\partial_{\psi,q,p}(x)(1 \rtimes a)$ in L^p norm by elements in the span of the $\partial_{\psi,q,p}(x)(1 \rtimes \lambda_t)$. Since $\mathcal{P}_G$ is a core of $\partial_{\psi,q,p}$, we can then approximate in turn by elements in the span of the $\partial_{\psi,q,p}(\lambda_s)(1 \rtimes \lambda_t)$.

Note that the adjoint operator $(\partial_{\alpha,q,p^*})^*$: $\mathrm{dom}(\partial_{\alpha,q,p^*})^* \subset L^p(\Gamma_q(H)\overline{\otimes}B(\ell_I^2)) \to S_I^p$ is closed by Kato [136, p. 168]. We will now define a Hodge-Dirac operator $D_{\alpha,q,p}$ in (4.59), relying on $\partial_{\alpha,q,p}$ and its adjoint. Then the main topic of this section will be to show that $D_{\alpha,q,p}$ is R-bisectorial (Theorem 4.7) and has a bounded H$^\infty$ functional calculus on $S_I^p \oplus \overline{\mathrm{Ran}\, \partial_{\alpha,q,p}}$ (Theorem 4.8). By Remark 4.4, this extends the Kato square root equivalence from Proposition 3.10.

Many arguments in this section are parallel to Sect. 4.1 in the Fourier multiplier case, though easier since there is no need any more for approximation properties as weak amenability, and $M_{I,\mathrm{fin}}$ is always an appropriate dense subspace of S_I^p together with completely contractive projections $\mathcal{T}_J$ from Definition 2.3.

Proposition 4.11 *Suppose* $1 < p < \infty$ *and* $-1 \leqslant q \leqslant 1$. *As unbounded operators, we have*

$$A_p = (\partial_{\alpha,q,p^*})^* \partial_{\alpha,q,p}. \qquad (4.50)$$

Proof For any $i, j \in I$, we have

$$(\partial_{\alpha,q,p^*})^* \partial_{\alpha,q,p}(e_{ij}) \overset{(2.95)}{=} (\partial_{\alpha,q,p^*})^*\big(s_q(\alpha_i - \alpha_j) \otimes e_{ij}\big) \qquad (4.51)$$

$$\overset{(3.86)}{=} \tau\big(s_q(\alpha_i - \alpha_j)s_q(\alpha_i - \alpha_j)\big)e_{ij}$$

$$\overset{(2.41)}{=} \|\alpha_i - \alpha_j\|_H^2\, e_{ij} = A_p(e_{ij}).$$

Argue as in Proposition 4.1, replacing Lemma 3.6 and Proposition 3.4 by Lemma 3.23 and Proposition 3.11. □

Now, we show that the noncommutative gradient $\partial_{\alpha,q,p}$ commutes with the semigroup and the resolvents of its generator.

Lemma 4.12 *Suppose* $1 < p < \infty$ *and* $-1 \leqslant q \leqslant 1$. *If* $x \in \mathrm{dom}\, \partial_{\alpha,q,p}$ *and* $t \geqslant 0$, *then* $T_{t,p}(x)$ *belongs to* $\mathrm{dom}\, \partial_{\alpha,q,p}$ *and we have*

$$\big(\mathrm{Id}_{L^p(\Gamma_q(H))} \otimes T_{t,p}\big)\partial_{\alpha,q,p}(x) = \partial_{\alpha,q,p}T_{t,p}(x). \qquad (4.52)$$

Proof For any $i, j \in I$, we have

$$\big(\mathrm{Id}_{L^p(\Gamma_q(H))} \otimes T_{t,p}\big)\partial_{\alpha,q,p}(e_{ij}) \overset{(2.95)}{=} \big(\mathrm{Id}_{L^p(\Gamma_q(H))} \otimes T_{t,p}\big)\big(s_q(\alpha_i - \alpha_j) \otimes e_{ij}\big)$$

$$= \mathrm{e}^{-t\|\alpha_i - \alpha_j\|^2} s_q(\alpha_i - \alpha_j) \otimes e_{ij}$$

$$\overset{(2.95)}{=} \mathrm{e}^{-t\|\alpha_i - \alpha_j\|^2} \partial_{\alpha,q,p}(e_{ij})$$

$$= \partial_{\alpha,q,p}\big(\mathrm{e}^{-t\|\alpha_i - \alpha_j\|^2} e_{ij}\big) = \partial_{\alpha,q,p}T_{t,p}(e_{ij}).$$

Argue as in Lemma 4.1, replacing Proposition 2.8 by Junge [117, p. 984]. □

Proposition 4.12 *Let* $1 < p < \infty$ *and* $-1 \leqslant q \leqslant 1$. *For any* $s \geqslant 0$ *and any* $x \in \text{dom } \partial_{\alpha,q,p}$, *we have* $\left(\text{Id}_{S_I^p} + sA_p\right)^{-1} x \in \text{dom } \partial_{\alpha,q,p}$ *and*

$$\left(\text{Id} \otimes (\text{Id} + sA_p)^{-1}\right)\partial_{\alpha,q,p}(x) = \partial_{\alpha,q,p}\left(\text{Id} + sA_p\right)^{-1}(x). \tag{4.53}$$

Proof Argue as in Proposition 4.2, replacing (4.3) by (4.52).   □

We have the following analogue of Proposition 4.3 which can be proved in a similar manner.

Proposition 4.13 *Suppose* $1 < p < \infty$ *and* $-1 \leqslant q \leqslant 1$. *The family*

$$\left\{t\partial_{\alpha,q,p}(\text{Id} + t^2 A_p)^{-1} : t > 0\right\} \tag{4.54}$$

of operators of $\text{B}(S_I^p, \text{L}^p(\Gamma_q(H)\overline{\otimes}\text{B}(\ell_I^2)))$ *is R-bounded.*

Proof Argue as in Proposition 4.3, replacing (3.31) by (3.88). Moreover, the argument yielding bounded H^∞ calculus of A_p is now [15] in place of [13].   □

Our Hodge-Dirac operator in (4.59) below will be constructed with $\partial_{\alpha,q,p}$ and the unbounded operator $(\partial_{\alpha,q,p*})^*|\overline{\text{Ran } \partial_{\alpha,q,p}}$. Note that the latter is by definition an unbounded operator on the Banach space $\overline{\text{Ran } \partial_{\alpha,q,p}}$ having domain $\text{dom}(\partial_{\alpha,q,p*})^* \cap \overline{\text{Ran } \partial_{\alpha,q,p}}$.

Lemma 4.13 *Let* $1 < p < \infty$ *and* $-1 \leqslant q \leqslant 1$. *The operator* $(\partial_{\alpha,q,p*})^*|\overline{\text{Ran } \partial_{\alpha,q,p}}$ *is densely defined and is closed. More precisely, the subspace* $\partial_{\alpha,q,p}(M_{I,\text{fin}})$ *of* $\text{dom}(\partial_{\alpha,q,p*})^*$ *is dense in* $\overline{\text{Ran } \partial_{\alpha,q,p}}$.

Proof Argue as in Lemma 4.2, replacing Proposition 3.4 resp. Proposition 4.1 by Proposition 3.11 resp. Proposition 4.11.   □

According to Lemma 4.12, $\text{Id} \otimes T_t$ leaves $\overline{\text{Ran } \partial_{\alpha,q,p}}$ invariant for any $t \geqslant 0$, so by continuity of $\text{Id} \otimes T_t$ also leaves $\overline{\text{Ran } \partial_{\alpha,q,p}}$ invariant. By Engel and Nagel [85, pp. 60-61], we can consider the generator

$$B_p \overset{\text{def}}{=} (\text{Id}_{\text{L}^p(\Gamma_q(H))} \otimes A_p)|\overline{\text{Ran } \partial_{\alpha,q,p}}. \tag{4.55}$$

of the restriction of $(\text{Id} \otimes T_t)_{t \geqslant 0}$ on $\overline{\text{Ran } \partial_{\alpha,q,p}}$.

Lemma 4.14 *Let* $1 < p < \infty$ *and* $-1 \leqslant q \leqslant 1$. *The operator* B_p *is injective and sectorial on* $\overline{\text{Ran } \partial_{\alpha,q,p}}$.

Proof Argue as in Lemma 4.3.   □

Lemma 4.15 *If* z *belongs to* $\text{L}^p(\Gamma_q(H)) \otimes M_{I,\text{fin}} \cap \overline{\text{Ran } \partial_{\alpha,q,p}}$ *then* $z \in \partial_{\alpha,q,p}(M_{I,\text{fin}})$.

Proof Argue as in Lemma 4.4. But now in place of a compactly supported cb. Fourier multiplier, apply the truncations $\mathcal{T}_J$ onto $J \times J$ matrices to a given sequence $(z_n)_n$ in dom $\partial_{\alpha,q,p}$.   □

Proposition 4.14 *Let* $1 < p < \infty$ *and* $-1 \leqslant q \leqslant 1$.

1. $L^p(\Gamma_q(H)) \otimes M_{I,\mathrm{fin}}$ is a core of $(\partial_{\alpha,q,p^*})^*$. Furthermore, if $y \in \mathrm{dom}(\partial_{\alpha,q,p^*})^*$ and if J is a subset of I, we have

$$(\partial_{\alpha,q,p^*})^*(\mathrm{Id} \otimes \mathcal{T}_J)(y) = \mathcal{T}_J(\partial_{\alpha,q,p^*})^*(y). \tag{4.56}$$

2. $\partial_{\alpha,q,p}(M_{I,\mathrm{fin}})$ is a core of $(\partial_{\alpha,q,p^*})^*|\overline{\mathrm{Ran}\,\partial_{\alpha,q,p}}$.
3. $\partial_{\alpha,q,p}(M_{I,\mathrm{fin}})$ is equally a core of $\partial_{\alpha,q,p}(\partial_{\alpha,q,p^*})^*|\overline{\mathrm{Ran}\,\partial_{\alpha,q,p}}$.
4. $\partial_{\alpha,q,p}(M_{I,\mathrm{fin}})$ is equally a core of B_p.

Proof Argue as in Proposition 4.4, replacing (3.33), (4.9), Lemma 4.1 resp. Lemma 4.4 by (3.90), (4.56), Lemma 4.12 resp. Lemma 4.15.   □

Proposition 4.15 *Let* $1 < p < \infty$ *and* $-1 \leqslant q \leqslant 1$.

1. For any $s > 0$, the operator $(\mathrm{Id} + sA_p)^{-1}(\partial_{\alpha,q,p^*})^*$ induces a bounded operator on $\overline{\mathrm{Ran}\,\partial_{\alpha,q,p}}$.
2. For any $s \geqslant 0$ and any $y \in \overline{\mathrm{Ran}\,\partial_{\alpha,q,p}} \cap \mathrm{dom}(\partial_{\alpha,q,p^*})^*$, the element $(\mathrm{Id} \otimes (\mathrm{Id} + sA_p)^{-1})(y)$ belongs to $\mathrm{dom}(\partial_{\alpha,q,p^*})^*$ and

$$(\mathrm{Id} + sA_p)^{-1}(\partial_{\alpha,q,p^*})^*(y) = (\partial_{\alpha,q,p^*})^*(\mathrm{Id} \otimes (\mathrm{Id} + sA_p)^{-1})(y). \tag{4.57}$$

3. For any $t \geqslant 0$ and any $y \in \overline{\mathrm{Ran}\,\partial_{\alpha,q,p}} \cap \mathrm{dom}(\partial_{\alpha,q,p^*})^*$, the element $(\mathrm{Id} \otimes T_t)(y)$ belongs to $\mathrm{dom}(\partial_{\alpha,q,p^*})^*$ and

$$T_t(\partial_{\alpha,q,p^*})^*(y) = (\partial_{\alpha,q,p^*})^*(\mathrm{Id} \otimes T_t)(y).$$

Proof Argue as in Proposition 4.5, replacing Proposition 4.3, Lemma 4.2, Proposition 4.1, (4.1), (4.3) resp. (4.10) by Proposition 4.13, Lemma 4.13, Proposition 4.11, (4.50), (4.52) resp. (4.57).   □

Proposition 4.14 enables us to identify B_p in terms of $\partial_{\alpha,q,p}$ and its adjoint.

Proposition 4.16 *Let* $1 < p < \infty$ *and* $-1 \leqslant q \leqslant 1$. *As unbounded operators, we have*

$$B_p = \partial_{\alpha,q,p}(\partial_{\alpha,q,p^*})^*|\overline{\mathrm{Ran}\,\partial_{\alpha,q,p}}. \tag{4.58}$$

Proof For any $i, j \in I$, we have

$$\partial_{\alpha,q,p}(\partial_{\alpha,q,p^*})^*\partial_{\alpha,q,p}(e_{ij}) \overset{(4.50)}{=} \partial_{\alpha,q,p}A_p(e_{ij}) = \|\alpha_i - \alpha_j\|_H^2 \partial_{\alpha,q,p}(e_{ij})$$

$$= \|\alpha_i - \alpha_j\|_H^2 (s_q(\alpha_i - \alpha_j) \otimes e_{ij})$$

$$= (\mathrm{Id}_{L^p(\Gamma_q(H))} \otimes A_p)(s_q(\alpha_i - \alpha_j) \otimes e_{ij})$$

$$= (\mathrm{Id}_{L^p(\Gamma_q(H))} \otimes A_p)(\partial_{\alpha,q,p}(e_{ij})).$$

Argue as in Proposition 4.6, replacing Proposition 4.4 by Proposition 4.14. $\square$

Note that the results of [15] gives the following result.

Proposition 4.17 *Suppose $1 < p < \infty$. The operators A_p and B_p have a bounded $\mathrm{H}^\infty(\Sigma_\omega)$ functional calculus of angle ω for some $\omega < \frac{\pi}{2}$.*

Suppose $1 < p < \infty$ and $-1 \leqslant q \leqslant 1$. We introduce the unbounded operator

$$D_{\alpha,q,p} \overset{\mathrm{def}}{=} \begin{bmatrix} 0 & (\partial_{\alpha,q,p^*})^*|\overline{\mathrm{Ran}\,\partial_{\alpha,q,p}} \\ \partial_{\alpha,q,p} & 0 \end{bmatrix} \tag{4.59}$$

on the Banach space $S_I^p \oplus_p \overline{\mathrm{Ran}\,\partial_{\alpha,q,p}}$ defined by

$$D_{\alpha,q,p}(x, y) \overset{\mathrm{def}}{=} ((\partial_{\alpha,q,p^*})^*(y), \partial_{\alpha,q,p}(x)) \tag{4.60}$$

for $x \in \mathrm{dom}\,\partial_{\alpha,q,p}$, $y \in \mathrm{dom}(\partial_{\alpha,q,p^*})^* \cap \overline{\mathrm{Ran}\,\partial_{\alpha,q,p}}$. By Lemma 4.13 and Proposition 3.11, this operator is a closed operator and can be seen as a differential square root of the generator of the semigroup $(T_{t,p})_{t \geqslant 0}$. We call it the Hodge-Dirac operator of the semigroup since we have Proposition 4.18 below.

Theorem 4.7 *Suppose $1 < p < \infty$ and $-1 \leqslant q \leqslant 1$. The Hodge-Dirac operator $D_{\alpha,q,p}$ is R-bisectorial on $S_I^p \oplus_p \overline{\mathrm{Ran}\,\partial_{\alpha,q,p}}$.*

Proof Argue as in Theorem 4.2, replacing Proposition 4.3, Theorem 4.1, (4.16), (4.1), (4.4), (4.10), (4.7) resp. (4.13) by Propositions 4.13, 4.17, (4.59), (4.50), (4.53), (4.57), (4.55) resp. (4.58). $\square$

Proposition 4.18 *Suppose $1 < p < \infty$ and $-1 \leqslant q \leqslant 1$. We have*

$$D_{\alpha,q,p}^2 = \begin{bmatrix} A_p & 0 \\ 0 & (\mathrm{Id}_{L^p(\Gamma_q(H))} \otimes A_p)|\overline{\mathrm{Ran}\,\partial_{\alpha,q,p}} \end{bmatrix}. \tag{4.61}$$

Proof Argue as in Proposition 4.7, replacing Proposition 4.6, (4.1), (4.13) resp. (4.16) by Proposition 4.16, (4.50), (4.58) resp. (4.59). $\square$

Now, we can state the following main result of this section.

Theorem 4.8 *Suppose* $1 < p < \infty$. *The Hodge-Dirac operator* $D_{\alpha,q,p}$ *has a bounded* H$^\infty(\Sigma_\omega^\pm)$ *functional calculus on a bisector, on the Banach space* $S_I^p \oplus_p \overline{\mathrm{Ran}\,\partial_{\alpha,q,p}}$.

Proof By Proposition 4.17, the operator $D_{\alpha,q,p}^2 \overset{(4.61)}{=} \begin{bmatrix} A_p & 0 \\ 0 & B_p \end{bmatrix}$ has a bounded H$^\infty$ functional calculus of angle $2\omega < \frac{\pi}{2}$. Since $D_{\alpha,q,p}$ is R-bisectorial by Theorem 4.7, we deduce by Proposition 2.1 that the operator $D_{\alpha,q,p}$ has a bounded H$^\infty(\Sigma_\omega^\pm)$ functional calculus on a bisector. $\qquad\square$

Remark 4.4 Similarly to Remark 4.1, the boundedness of the H$^\infty$ functional calculus of the operator $D_{\alpha,q,p}$ implies the boundedness of the Riesz transforms and this result may be thought of as a strengthening of the equivalence (3.88).

Remark 4.5 In a similar way to Remark 4.4, we also obtain for $(x, y) \in$ dom $\partial_{\alpha,q,p} \oplus \mathrm{dom}(\partial_{\alpha,q,p^*})^* | \overline{\mathrm{Ran}\,\partial_{\alpha,q,p}}$ that

$$\left\| (D_{\alpha,q,p}^2)^{\frac{1}{2}}(x, y) \right\|_p \cong \left\| A_p^{\frac{1}{2}}(x) \right\|_p + \left\| (\mathrm{Id}_{L^p(\Gamma_q(H))} \otimes A_p^{\frac{1}{2}})(y) \right\|_p$$

$$\cong \left\| \partial_{\alpha,q,p}(x) \right\|_p + \left\| (\partial_{\alpha,q,p^*})^*(y) \right\|_p \cong \left\| D_{\alpha,q,p}(x, y) \right\|_p .$$

Moreover, we have

$$\left\| (\mathrm{Id}_{L^p(\Gamma_q(H))} \otimes A_p^{\frac{1}{2}})(y) \right\|_p \cong \left\| (\partial_{\alpha,q,p^*})^*(y) \right\|_p, \quad y \in \mathrm{dom}(\partial_{\alpha,q,p^*})^* \cap \overline{\mathrm{Ran}\,\partial_{\alpha,q,p}}.$$
$$(4.62)$$

Proposition 4.19 *We have* $\overline{\mathrm{Ran}\,A_p} = \overline{\mathrm{Ran}(\partial_{\alpha,q,p^*})^*}$, $\overline{\mathrm{Ran}\,B_p} = \overline{\mathrm{Ran}\,\partial_{\alpha,q,p}}$, $\mathrm{Ker}\,A_p = \mathrm{Ker}\,\partial_{\alpha,q,p}$, $\mathrm{Ker}\,B_p = \mathrm{Ker}(\partial_{\alpha,q,p^*})^* = \{0\}$ *and*

$$S_I^p = \overline{\mathrm{Ran}(\partial_{\alpha,q,p^*})^*} \oplus \mathrm{Ker}\,\partial_{\alpha,q,p}. \qquad (4.63)$$

Here, by $(\partial_{\alpha,q,p^*})^*$ *we understand its restriction to* $\overline{\mathrm{Ran}\,\partial_{\alpha,q,p}}$. *However, we shall see in Corollary 4.2 that* $\mathrm{Ran}(\partial_{\alpha,q,p^*})^* = \mathrm{Ran}(\partial_{\alpha,q,p^*})^* | \overline{\mathrm{Ran}\,\partial_{\alpha,q,p}}$.

Proof By (2.20), we have $\overline{\mathrm{Ran}\,D_{\alpha,q,p}^2} = \overline{\mathrm{Ran}\,D_{\alpha,q,p}}$ and $\mathrm{Ker}\,D_{\alpha,q,p}^2 = \mathrm{Ker}\,D_{\alpha,q,p}$. It is not difficult to prove the first four equalities using (4.61) and (4.60). The last one is a consequence of the definition of A_p and of [110, p. 361]. $\qquad\square$

4.6 Extension to Full Hodge-Dirac Operator and Hodge Decomposition

We keep the standing assumptions of the previous section and thus we have a markovian semigroup $(T_t)_{t \geqslant 0}$ of Schur multipliers with generator A_p, the noncommutative gradient $\partial_{\alpha,q,p}$ and its adjoint $(\partial_{\alpha,q,p*})^*$, together with the Hodge-Dirac operator $D_{\alpha,q,p}$. We shall now extend the operator $D_{\alpha,q,p}$ to a densely defined bisectorial operator $\mathcal{D}_{\alpha,q,p}$ on $S_I^p \oplus L^p(\Gamma_q(H)\overline{\otimes}B(\ell_I^2))$ which will also be bisectorial and will have an $H^\infty(\Sigma_\omega^\pm)$ functional calculus on a bisector. The key will be Corollary 4.2 below. We let

$$\mathcal{D}_{\alpha,q,p} \overset{\text{def}}{=} \begin{bmatrix} 0 & (\partial_{\alpha,q,p*})^* \\ \partial_{\alpha,q,p} & 0 \end{bmatrix} \tag{4.64}$$

along the decomposition $S_I^p \oplus L^p(\Gamma_q(H)\overline{\otimes}B(\ell_I^2))$, with natural domains for $\partial_{\alpha,q,p}$ and $(\partial_{\alpha,q,p*})^*$.

Consider the sectorial operator $A_p^{\frac{1}{2}}$ on S_I^p. According to (2.16), we have the *topological* direct sum decomposition $S_I^p = \overline{\operatorname{Ran} A_p^{\frac{1}{2}}} \oplus \operatorname{Ker} A_p^{\frac{1}{2}}$. We define the operator $R_p \overset{\text{def}}{=} \partial_{\alpha,q,p} A_p^{-\frac{1}{2}} \colon \operatorname{Ran} A_p^{\frac{1}{2}} \to L^p(\Gamma_q(H)\overline{\otimes}B(\ell_I^2))$. According to the point 3 of Proposition 3.11, R_p is bounded on $\operatorname{Ran} A_p^{\frac{1}{2}}$, so extends to a bounded operator on $\overline{\operatorname{Ran} A_p^{\frac{1}{2}}} \overset{(2.21)}{=} \overline{\operatorname{Ran} A_p}$. We extend it to a bounded operator $R_p \colon S_I^p \to L^p(\Gamma_q(H)\overline{\otimes}B(\ell_I^2))$, called Riesz transform, by putting $R_p | \operatorname{Ker} A_p^{\frac{1}{2}} = 0$ along the preceding decomposition of S_I^p. We equally let $R_{p*}^* \overset{\text{def}}{=} (R_{p*})^*$.

Lemma 4.16 *Let* $-1 \leqslant q \leqslant 1$ *and* $1 < p < \infty$. *Then we have the subspace sum*

$$L^p(\Gamma_q(H)\overline{\otimes}B(\ell_I^2)) = \overline{\operatorname{Ran} \partial_{\alpha,q,p}} + \operatorname{Ker}(\partial_{\alpha,q,p*})^*. \tag{4.65}$$

Proof Argue as in Lemma 4.6. □

In the proof of Proposition 4.20 below, we shall need some information on the Wiener-Ito chaos decomposition for q-Gaussians. This is collected in the following lemma.

Lemma 4.17 *Let* $-1 \leqslant q \leqslant 1$ *and* $1 < p < \infty$.

1. *There exists a completely bounded projection* $\mathcal{P} \colon L^p(\Gamma_q(H)) \to L^p(\Gamma_q(H))$ *onto the closed space spanned by* $\{s_q(h) : h \in H\}$. *Moreover, the projections are compatible for different values of p.*
2. *For any finite subset J of I and* $y \in L^p(\Gamma_q(H)\overline{\otimes}B(\ell_I^2))$, $(\mathcal{P} \otimes T_J)(y)$ *can be written as* $\sum_{i,j \in J} s_q(h_{ij}) \otimes e_{ij}$ *for some* $h_{ij} \in H$.

3. Let J be a finite subset of I. Denoting temporarily $\mathcal{P}_p$ and $\mathcal{T}_{J,p}$ the operators $\mathcal{P}$ and $\mathcal{T}_J$ on the p-level, the identity mapping on $\mathrm{Gauss}_{q,p}(\mathbb{C}) \otimes S_J^p$ extends to an isomorphism

$$J_{p,2,J}: \mathrm{Ran}(\mathcal{P}_p \otimes \mathcal{T}_{J,p}) \to \mathrm{Ran}(\mathcal{P}_2 \otimes \mathcal{T}_{J,2}) \tag{4.66}$$

where $\mathrm{Ran}(\mathcal{P}_p \otimes \mathcal{T}_{J,p}) \subset L^p(\Gamma_q(H)\overline{\otimes}B(\ell_I^2))$.

Proof

1. Follows from Lemma 4.7.
2. This is easy and left to the reader.
3. Argue similarly to Lemma 4.7, replacing formally $\rtimes \lambda_s$ for $s \in G$ by $\otimes e_{ij}$ for $i, j \in I$.

□

Proposition 4.20 Let $-1 \leqslant q \leqslant 1$ and $1 < p < \infty$. Then the subspaces from Lemma 4.16 have trivial intersection, $\overline{\mathrm{Ran}\, \partial_{\alpha,q,p}} \cap \mathrm{Ker}(\partial_{\alpha,q,p*})^* = \{0\}$.

Proof Argue as in Proposition 4.9, replacing Theorem 5.4, Lemma 4.7 resp. Proposition 3.4 by Proposition 5.14, Lemma 4.17 resp. Proposition 3.11. Also replace the approximating net of Fourmer multipliers $(M_{\varphi_j})_j$ by the approximating net of Schur multipliers $(\mathcal{T}_J)_J$. □

Combining Lemma 4.16 and Proposition 4.20, we can now deduce the following corollary.

Corollary 4.2 Let $-1 \leqslant q \leqslant 1$ and $1 < p < \infty$. Then we have a topological direct sum decomposition

$$L^p(\Gamma_q(H)\overline{\otimes}B(\ell_I^2)) = \overline{\mathrm{Ran}\, \partial_{\alpha,q,p}} \oplus \mathrm{Ker}(\partial_{\alpha,q,p*})^* \tag{4.67}$$

where the associated first bounded projection is $R_p R_{p*}^*$. In particular, we have

$$\mathrm{Ran}(\partial_{\alpha,q,p*})^* = \mathrm{Ran}(\partial_{\alpha,q,p*})^* | \overline{\mathrm{Ran}\, \partial_{\alpha,q,p}}.$$

Proof Argue as in Corollary 4.1, replacing Lemma 4.6 resp. Proposition 4.9 by Lemma 4.16 resp. Proposition 4.20. □

Theorem 4.9 Let $-1 \leqslant q \leqslant 1$ and $1 < p < \infty$. Consider the operator $\mathcal{D}_{\alpha,q,p}$ from (4.64). Then $\mathcal{D}_{\alpha,q,p}$ is bisectorial and has a bounded H$^\infty(\Sigma_\omega^\pm)$ calculus.

Proof Argue as in Theorem 4.4, replacing Corollary 4.1, (4.33), (4.24) resp. (4.16) by Corollary 4.2, (4.67), (4.64) resp. (4.59). □

Theorem 4.10 (Hodge Decomposition) Suppose $1 < p < \infty$ and $-1 \leqslant q \leqslant 1$. If we identify $\overline{\mathrm{Ran}\, \partial_{\alpha,q,p}}$ and $\mathrm{Ran}(\partial_{\alpha,q,p*})^*$ as the closed subspaces $\{0\} \oplus \overline{\mathrm{Ran}\, \partial_{\alpha,q,p}}$

and $\overline{\operatorname{Ran}(\partial_{\alpha,q,p^*})^*} \oplus \{0\}$ of $S_I^p \oplus L^p(\Gamma_q(H)\overline{\otimes}B(\ell_I^2))$, we have

$$S_I^p \oplus L^p(\Gamma_q(H)\overline{\otimes}B(\ell_I^2)) = \overline{\operatorname{Ran}\partial_{\alpha,q,p}} \oplus \overline{\operatorname{Ran}(\partial_{\alpha,q,p^*})^*} \oplus \operatorname{Ker}\mathcal{D}_{\alpha,q,p}. \quad (4.68)$$

Proof Argue as in Theorem 4.5, replacing (4.24), (4.23) resp. (4.33) by (4.64), (4.63) resp. (4.67). □

4.7 Independence from H and α

In this short section, we again keep the standing assumptions from the two preceding sections and have a markovian semigroup of Schur multipliers. The main topic will be the proof of Theorem 4.11 showing that the bound of the bisectorial $H^\infty(\Sigma_\omega^\pm)$ functional calculus in Theorems 4.8 and 4.9 comes with a constant only depending on ω, q, p, but not on the markovian semigroup nor the associated Hilbert space H nor $\alpha \colon I \to H$. We show several intermediate lemmas before.

Lemma 4.18 Let $-1 \leqslant q \leqslant 1$. Let H be a Hilbert space and $i_1 \colon H_1 \subset H$ an embedding of a sub Hilbert space H_1. Moreover, let I be an index set and $I_1 \subset I$ a subset. Consider the mapping

$$J_1 \colon \begin{cases} \Gamma_q(H_1)\overline{\otimes}B(\ell_{I_1}^2) & \to \Gamma_q(H)\overline{\otimes}B(\ell_I^2) \\ a \otimes x & \mapsto \Gamma_q(i_1)(a) \otimes j_1(x) \end{cases},$$

where $j_1 \colon B(\ell_{I_1}^2) \to B(\ell_I^2)$, $x \mapsto P_1^* x P_1$ and $P_1 \colon \ell_I^2 \to \ell_{I_1}^2$, $(\xi_i)_{i\in I} \mapsto (\xi_i\delta_{i\in I_1})_{i\in I}$ is the canonical orthogonal projection. Then J_1 is a normal faithful trace preserving $*$-homomorphism and thus extends to a complete isometry on the L^p level, $1 \leqslant p \leqslant \infty$.

Proof According to [44, Theorem 2.11], since i_1 is an isometric embedding, $\Gamma_q(i_1)$ is a faithful $*$-homomorphism which preserves the traces. According to [128, p. 97], $\Gamma_q(i_1)$ is normal. Moreover, it is easy to check that j_1 is also a normal faithful $*$-homomorphism. Thus also J_1 is a normal faithful $*$-homomorphism, see also [189, p. 32]. Since $\Gamma_q(i_1)$ and j_1 preserve the traces, also J_1 preserves the trace. Now J_1 is a (complete) L^p isometry according to [128, p. 92]. □

In the following, we let $(H_n)_{n\in\mathbb{N}}$ be a sequence of mutually orthogonal sub Hilbert spaces of some big Hilbert space $H = \bigoplus_{n\in\mathbb{N}} H_n$ and let $I = \bigsqcup_{n\in\mathbb{N}} I_n$ be a partition of a big index set I into smaller pieces I_n.

Lemma 4.19 Let $1 < p < \infty$ and $-1 \leqslant q \leqslant 1$. Consider the mappings $\Psi_n \colon S_{I_n}^p \to S_I^p$, $x \mapsto P_n^* x P_n$, where $P_n \colon \ell_I^2 \to \ell_{I_n}^2$ is the canonical orthogonal *projection as previously,*

$$\Psi: \bigoplus_{n\in\mathbb{N}}^{p} S_{I_n}^{p} \to S_{I}^{p}, \ (x_n) \mapsto \sum_{n} \Psi_n(x_n)$$

and

$$J: \bigoplus_{n\in\mathbb{N}}^{p} L^p(\Gamma_q(H_n)\overline{\otimes}\mathrm{B}(\ell_{I_n}^2)) \to L^p(\Gamma_q(H)\overline{\otimes}\mathrm{B}(\ell_I^2)), \ (y_n) \mapsto \sum_{n} J_n(y_n).$$

1. *On the* L^∞ *level,* $\mathrm{Ran}(J_n) \cdot \mathrm{Ran}(J_m) = \{0\}$ *for* $n \neq m$.
2. *If the previous domains of* Ψ *and* J *that are exterior Banach space sums, are as indicated equipped with the* ℓ^p*-norms, then* Ψ *and* J *are isometries.*

Proof

1. By a density and normality argument, it suffices to pick $a \otimes x \in \Gamma_q(H_n)\otimes\mathrm{B}(\ell_{I_n}^2)$ and $b \otimes y \in \Gamma_q(H_m) \otimes \mathrm{B}(\ell_{I_m}^2)$ and calculate the product $J_n(a \otimes x)J_m(b \otimes y)$. We have

$$J_n(a \otimes x)J_m(b \otimes y) = \Gamma_q(i_n)(a)\Gamma_q(i_m)(b) \otimes P_n^* x P_n P_m^* y P_m = 0,$$

since $P_n P_m^*: \ell_{I_m}^2 \to \ell_{I_n}^2$ equals 0.

2. According to the first point, we have for any $x_n \in \Gamma_q(H_n)\overline{\otimes}\mathrm{B}(\ell_{I_n}^2)$,

$$\left(\sum_{n=1}^{N} J_n(x_n)\right)^* \left(\sum_{m=1}^{N} J_m(x_m)\right) = \left(\sum_{n=1}^{N} J_n(x_n^*)\right)\left(\sum_{m=1}^{N} J_m(x_m)\right) = \sum_{n=1}^{N} J_n(x_n^* x_n).$$

In the same way, by functional calculus of $\left|\sum_{n=1}^{N} J_n(x_n)\right|^2$ and $J_n(|x_n|^2)$, we obtain

$$\left|\sum_{n=1}^{N} J_n(x_n)\right|^p = \sum_{n=1}^{N} |J_n(x_n)|^p,$$

and thus, taking traces, we obtain

$$\left\|\sum_{n=1}^{N} J_n(x_n)\right\|_p = \left(\sum_{n=1}^{N} \|J_n(x_n)\|_p^p\right)^{\frac{1}{p}} = \left(\sum_{n=1}^{N} \|x_n\|_p^p\right)^{\frac{1}{p}}.$$

By a density argument, we infer that J is an isometry. The proof for Ψ is easier and left to the reader.

$\square$

In the following, we consider in addition the mappings $\alpha_n \colon I_n \to H_n$ and associate with it $\alpha \colon I \to H$ given by $\alpha(i) \overset{\text{def}}{=} \alpha_n(i)$ if $i \in I_n \subset \bigsqcup_{k \in \mathbb{N}} I_k$. We thus have noncommutative gradients $\partial_{\alpha_n,q,p}$ and $\partial_{\alpha,q,p}$.

Lemma 4.20 *Let $1 < p < \infty$ and $-1 \leqslant q \leqslant 1$. Recall the mappings J_n and Ψ_n from Lemma 4.19.*

1. For any $n \in \mathbb{N}$ and $x_n \in M_{I_n,\text{fin}}$, we have $\Psi_n(x_n) \in M_{I,\text{fin}}$ and

$$\partial_{\alpha,q,p} \Psi_n(x_n) = J_n \partial_{\alpha_n,q,p}(x_n).$$

2. For any $n \in \mathbb{N}$, $y_n \in \Gamma_q(H_n) \otimes M_{I_n,\text{fin}}$, we have $J_n(y_n) \in \text{dom}(\partial_{\alpha,q,p^*})^*$ and

$$(\partial_{\alpha,q,p^*})^* J_n(y_n) = \Psi_n(\partial_{\alpha_n,q,p^*})^*(y_n).$$

Proof

1. We calculate with $x_n = \sum_{i,j \in I_n} x_{ij} e_{ij}$ and explicit embedding $k_n \colon I_n \hookrightarrow I$,

$$\partial_{\alpha,q,p} \Psi_n(x_n) = \partial_{\alpha,q,p}(P_n^* x_n P_n) = \partial_{\alpha,q,p}\left(P_n^* \sum_{i,j} x_{ij} e_{ij} P_n \right)$$

$$= \sum_{i,j} x_{ij} \partial_{\alpha,q,p}(e_{k_n(i)k_n(j)})$$

$$= \sum_{i,j} x_{ij} s_q(\alpha(k_n(i)) - \alpha(k_n(j))) \otimes e_{k_n(i)k_n(j)}.$$

On the other hand, we have

$$J_n \partial_{\alpha_n,q,p}(x_n) = \sum_{i,j} x_{ij} J_n(s_q(\alpha_n(i) - \alpha_n(j)) \otimes e_{ij})$$

$$= \sum_{i,j} x_{ij} \Gamma_q(i_n)(s_q(\alpha_n(i) - \alpha_n(j))) \otimes P_n^* e_{ij} P_n$$

$$= \sum_{i,j} x_{ij} s_q(\alpha(k_n(i)) - \alpha(k_n(j))) \otimes e_{k_n(i)k_n(j)}.$$

2. First we note that $J_n(\Gamma_q(H_n) \otimes M_{I_n,\text{fin}}) \subset \text{dom}(\partial_{\alpha,q,p^*})^*$ and that $(\partial_{\alpha,q,p^*})^*$ maps $J_n(\Gamma_q(H_n) \otimes M_{I_n,\text{fin}})$ into $\Psi_n(S_{I_n}^p)$. Indeed for $s_q(h_n) \otimes e_{ij} \in \Gamma_q(H_n) \otimes M_{I_n,\text{fin}}$

and $x = \sum_{k,l} x_{kl} e_{kl} \in S_I^{p^*}$, we have

$$x \mapsto \tau_{\Gamma_q(H) \overline{\otimes} B(\ell_I^2)} \left((s_q(i_n(h_n)) \otimes e_{k_n(i)k_n(j)})^* \cdot \partial_{\alpha,q,p^*}(x) \right)$$

$$= \sum_{k,l} x_{kl} \tau_{\Gamma_q(H) \overline{\otimes} B(\ell_I^2)} \left(s_q(i_n(h_n)) \otimes e_{k_n(j)k_n(i)} \cdot s_q(\alpha(k) - \alpha(l)) \otimes e_{kl} \right)$$

$$= \tau_{\Gamma_q(H)} \left(s_q(i_n(h_n)) s_q(\alpha(k_n(j)) - \alpha(k_n(i))) \right) x_{k_n(i)k_n(j)}.$$

This defines clearly a linear form on $S_I^{p^*}$, so indeed $J_n(\Gamma_q(H_n) \otimes M_{I_n,\mathrm{fin}}) \subset \mathrm{dom}(\partial_{\alpha,q,p^*})^*$. Moreover, we see that if $x = e_{kl}$ and k, l not both in $k_n(I_n)$, then $\langle (\partial_{\alpha,q,p^*})^*(J_n(y_n)), e_{kl} \rangle = 0$. Therefore, $(\partial_{\alpha,q,p^*})^*$ maps $J_n(\Gamma_q(H_n) \otimes M_{I_n,\mathrm{fin}})$ into $\Psi_n(S_{I_n}^p)$. Now we check the claimed equality in the statement of the lemma by applying a dual element x_n to both sides. By density of $M_{I,\mathrm{fin}}$ in $S_I^{p^*}$, we can assume $x_n \in M_{I,\mathrm{fin}}$. Moreover, we can assume $x_n \in \Psi_n(M_{I_n,\mathrm{fin}})$, so that $x_n = \Psi_n \Psi_n^*(x_n)$. Then we have according to the first point of the lemma

$$\mathrm{Tr}_I(\Psi_n(\partial_{\alpha_n,q,p^*})^*(y_n)x_n^*)$$

$$= \mathrm{Tr}_{I_n}((\partial_{\alpha_n,q,p^*})^*(y_n)(\Psi_n^*(x_n))^*) = \tau_{\Gamma_q(H_n)\overline{\otimes}B(\ell_{I_n}^2)}(y_n(\partial_{\alpha_n,q,p^*}\Psi_n^*(x_n))^*)$$

$$= \tau_{\Gamma_q(H)\overline{\otimes}B(\ell_I^2)}(J_n(y_n(\partial_{\alpha_n,q,p^*}\Psi_n^*(x_n))^*))$$

$$= \tau_{\Gamma_q(H)\overline{\otimes}B(\ell_I^2)}(J_n(y_n)(J_n\partial_{\alpha_n,q,p^*}\Psi_n^*(x_n))^*)$$

$$= \tau_{\Gamma_q(H)\overline{\otimes}B(\ell_I^2)}(J_n(y_n)(\partial_{\alpha,q,p^*}\Psi_n\Psi_n^*(x_n))^*)$$

$$= \mathrm{Tr}_I((\partial_{\alpha,q,p^*})^*J_n(y_n)(\Psi_n\Psi_n^*(x_n))^*) = \mathrm{Tr}_I((\partial_{\alpha,q,p^*})^*J_n(y_n)x_n^*).$$

Here we have also used that J_n is trace preserving and multiplicative. The lemma is proved.

$\square$

Lemma 4.21 Let $1 < p < \infty$ and $-1 \leqslant q \leqslant 1$. Recall the mappings Ψ and J from Lemma 4.19.

1. Then for any $x = (x_n)$ with $x_n \in M_{I_n,\mathrm{fin}} \subset S_{I_n}^p$ and any $y = (y_n)$ with $y_n \in \overline{\Gamma_q(H_n) \otimes M_{I_n,\mathrm{fin}} \cap \mathrm{Ran}(\partial_{\alpha_n,q,p})}$ such that only finitely many x_n and y_n are nonzero, we have

$$D_{\alpha,q,p} \circ \begin{bmatrix} \Psi & 0 \\ 0 & J \end{bmatrix}(x,y) = \begin{bmatrix} \Psi & 0 \\ 0 & J \end{bmatrix} \circ \begin{bmatrix} D_{\alpha_1,q,p} & 0 & \cdots \\ 0 & D_{\alpha_2,q,p} & 0 & \cdots \\ \vdots & & 0 & \ddots \end{bmatrix}(x,y) \quad (4.69)$$

2. Let $\omega \in (0, \frac{\pi}{2})$ such that $D_{\alpha,q,p}$ and all $D_{\alpha_n,q,p}$ have an $H^\infty(\Sigma_\omega^\pm)$ calculus, e.g. according to Theorem 4.8, $\omega = \frac{\pi}{4}$. Then for any $m \in H^\infty(\Sigma_\omega^\pm)$, we have

$$
m(D_{\alpha,q,p}) \circ \begin{bmatrix} \Psi & 0 \\ 0 & J \end{bmatrix} = \begin{bmatrix} \Psi & 0 \\ 0 & J \end{bmatrix} \circ \begin{bmatrix} m(D_{\alpha_1,q,p}) & 0 & \cdots \\ 0 & m(D_{\alpha_2,q,p}) & 0 \\ \vdots & & 0 & \ddots \end{bmatrix} \tag{4.70}
$$

as bounded operators $\bigoplus_{n\in\mathbb{N}}^p \left(S_{I_n}^p \oplus_p \overline{\mathrm{Ran}(\partial_{\alpha_n,q,p})} \right) \to S_I^p \oplus_p \overline{\mathrm{Ran}(\partial_{\alpha,q,p})}$.

Proof

1. According to Lemma 4.20, we have

$$
D_{\alpha,q,p}(\Psi(x), J(y)) = D_{\alpha,q,p}\left(\sum_n \Psi_n(x_n), \sum_n J_n(y_n) \right)
$$

$$
= \left((\partial_{\alpha,q,p^*})^* \sum_n J_n(y_n), \partial_{\alpha,q,p} \sum_n \Psi_n(x_n) \right)
$$

$$
= \left(\sum_n \Psi_n(\partial_{\alpha_n,q,p^*})^* y_n, \sum_n J_n \partial_{\alpha_n,q,p} x_n \right)
$$

$$
= \left(\Psi \mathrm{diag}((\partial_{\alpha_n,q,p^*})^* : n \in \mathbb{N})y, J \mathrm{diag}(\partial_{\alpha_n,q,p} : n \in \mathbb{N})x \right).
$$

This shows (4.69).

2. According to (4.69), for any $\lambda \in \rho(D_{\alpha,q,p}) \cap \bigcap_{n\in\mathbb{N}} \rho(D_{\alpha_n,q,p})$, we have for (x, y) as in the first part of the lemma,

$$
(\lambda - D_{\alpha,q,p})^{-1} \begin{bmatrix} \Psi & 0 \\ 0 & J \end{bmatrix} \begin{bmatrix} (\lambda - D_{\alpha_1,q,p}) & 0 \\ 0 & \ddots \end{bmatrix} (x, y)
$$

$$
= (\lambda - D_{\alpha,q,p})^{-1}(\lambda - D_{\alpha,q,p}) \begin{bmatrix} \Psi & 0 \\ 0 & J \end{bmatrix} (x, y) = \begin{bmatrix} \Psi & 0 \\ 0 & J \end{bmatrix} (x, y)
$$

$$
= \begin{bmatrix} \Psi & 0 \\ 0 & J \end{bmatrix} \begin{bmatrix} (\lambda - D_{\alpha_1,q,p})^{-1} & 0 \\ 0 & \ddots \end{bmatrix} \begin{bmatrix} (\lambda - D_{\alpha_1,q,p}) & 0 \\ 0 & \ddots \end{bmatrix} (x, y).
$$

Thus, we obtain

$$
(\lambda - D_{\alpha,q,p})^{-1} \begin{bmatrix} \Psi & 0 \\ 0 & J \end{bmatrix} = \begin{bmatrix} \Psi & 0 \\ 0 & J \end{bmatrix} \begin{bmatrix} (\lambda - D_{\alpha_1,q,p})^{-1} & 0 \\ 0 & \ddots \end{bmatrix} \tag{4.71}
$$

on a dense subspace of $\bigoplus_{n\in\mathbb{N}}^{p}\left(S_{I_n}^{p}\oplus_p \overline{\mathrm{Ran}(\partial_{\alpha_n,q,p})}\right)$. By boundedness of both sides of (4.71), we obtain equality in (4.71) as bounded operators

$$\bigoplus_{n\in\mathbb{N}}^{p}\left(S_{I_n}^{p}\oplus_p \overline{\mathrm{Ran}(\partial_{\alpha_n,q,p})}\right) \to S_{I}^{p}\oplus_p \overline{\mathrm{Ran}(\partial_{\alpha,q,p})}.$$

Now by the Cauchy integral formula, we obtain (4.70) for $m \in H_0^\infty(\Sigma_\omega^\pm)$, and by the H$^\infty$ convergence lemma (see [68, Lemma 2.1] for the sectorial case) applied first for fixed (x,y) as in the first part of the lemma, also for $m \in H^\infty(\Sigma_\omega^\pm)$.

$\square$

Now, we can give an answer to a variant of [131, Problem C.5].

Theorem 4.11 *Let* $1 < p < \infty$, $-1 \leqslant q \leqslant 1$, $\omega = \frac{\pi}{4}$ *and* $D_{\alpha,q,p}$ *as in Theorem 4.8. Then the* H$^\infty(\Sigma_\omega^\pm)$ *calculus norm of* $D_{\alpha,q,p}$ *is controlled independently of* H *and* α, *that is,* $\|m(D_{\alpha,q,p})\| \leqslant C_{q,p}\|m\|_{H^\infty(\Sigma_\omega^\pm)}$. *In particular,* $\mathrm{sgn}(D_{\alpha,q,p}) = D_{\alpha,q,p}(D_{\alpha,q,p}^2)^{-\frac{1}{2}}$ *is bounded by a constant not depending on* H *or its dimension nor the mapping* $\alpha\colon I \to H$.

Proof Suppose that the statement of the theorem is false. For any $n \in \mathbb{N}$, then there exists a sequence $(D_{\alpha_n,q,p})$ of Hodge-Dirac operators such that the H$^\infty(\Sigma_\omega^\pm)$ functional calculus norm is bigger than n. We let $I \overset{\mathrm{def}}{=} \bigsqcup_{n\in\mathbb{N}} I_n$, $H \overset{\mathrm{def}}{=} \bigoplus_{n\in\mathbb{N}} H_n$ and consider $\alpha\colon I \to H$, $i \in I_n \mapsto \alpha_n(i)$. Let f_n be a function of H$^\infty(\Sigma_\omega^\pm)$ such that $\|f_n\|_{H^\infty(\Sigma_\omega^\pm)} = 1$ and $\|f_n(D_{\alpha_n,q,p})\| \geqslant n$. According to Theorem 4.8, the operator $D_{\alpha,q,p}$ has a bounded H$^\infty(\Sigma_\omega^\pm)$ functional calculus. Thus, for some constant $C < \infty$ and any f_n as above, according to Lemma 4.21,

$$C \geqslant \|f_n(D_{\alpha,q,p})\| \geqslant \left\|f_n(D_{\alpha,q,p})\circ \begin{bmatrix} \Psi & 0 \\ 0 & J \end{bmatrix}\right\|$$

$$= \left\|\begin{bmatrix} \Psi & 0 \\ 0 & J \end{bmatrix}\circ \begin{bmatrix} f_n(D_{\alpha_1,q,p}) & 0 \\ 0 & \ddots \end{bmatrix}\right\|$$

$$= \left\|\begin{bmatrix} f_n(D_{\alpha_1,q,p}) & 0 \\ 0 & \ddots \end{bmatrix}\right\| \geqslant \|f_n(D_{\alpha_n,q,p})\| \geqslant n.$$

Taking $n \to \infty$ yields a contradiction. The last sentence of Theorem 4.11 follows from taking $f(z) = 1_{\Sigma_\omega^+}(z) - 1_{\Sigma_\omega^-}(z)$.

$\square$

Remark 4.6 We have the following alternative proof of Theorem 4.11 using dimension-free estimates of the Riesz transform associated with A_p. It will show that the angle of the H$^\infty(\Sigma_\omega^\pm)$ calculus of $D_{\alpha,q,p}$ and $\mathcal{D}_{\alpha,q,p}$ can be chosen $\omega > \frac{\pi}{2}|\frac{1}{p} - \frac{1}{2}|$ and that the norm of the calculus is bounded by a constant K_ω

not depending on I nor the representation (α, H), in particular it is independent of the dimension of H.

Proof First note that since A_p has a (sectorial) completely bounded $\mathrm{H}^\infty(\Sigma_{2\omega})$ calculus with angle $2\omega > \pi|\frac{1}{p} - \frac{1}{2}|$, by the representation of Proposition 4.18, $\mathcal{D}^2_{\alpha,q,p}$ also has a sectorial $\mathrm{H}^\infty(\Sigma_{2\omega})$ calculus. For the rest of the proof, argue as in Remark 4.3, replacing [13, Theorem 4.1] by Arhancet [15]. This gives

$$\mathrm{Id}_{S^p} \otimes T_{t,p} = (\mathrm{Id}_{S^p} \otimes \mathbb{E}_p)(\mathrm{Id}_{S^p} \otimes U_{t,p})(\mathrm{Id}_{S^p} \otimes J_p).$$

Here, the linear maps $\mathbb{E}_p \colon \mathrm{L}^p(\Omega, S^p_I) \to S^p_I$ and J_p are complete contractions and $U_{t,p} \colon \mathrm{L}^p(\Omega, S^p_I) \to \mathrm{L}^p(\Omega, S^p_I)$ is a group of isometries. Then we need to control the UMD constant of $S^p(\mathrm{L}^p(M))$ with $M = \mathrm{L}^\infty(\Omega)\overline{\otimes}\mathrm{B}(\ell^2_I)$. According to [193, Corollary 4.5, Theorem 4.3], this constant is controlled by some universal bound times $p^2/(p-1)$. The rest of the proof is similar to that of Remark 4.3 where for the universal boundedness of the Riesz transform, we refer to Remark 3.4 in place of Proposition 3.3. $\qquad\square$

Chapter 5
Locally Compact Quantum Metric Spaces and Spectral Triples

Abstract In this chapter, we start by giving an overview of quantum (locally) compact metric spaces. Then, we show that we can associate quantum compact metric spaces to some Markov semigroups of Fourier multipliers satisfying additional conditions: an injectivity and a gap condition on the cocycle which represents the semigroup, and the finite dimensionality (with explicit control on p) of the cocycle Hilbert space. We show a similar result for semigroups of Schur multipliers and obtain a quantum locally compact metric space. We further explore the connections of our gap condition between Fourier multipliers and Schur multipliers with some examples. In the sequel, we introduce spectral triples (=noncommutative manifolds) associated to Markov semigroups of Fourier multipliers or Schur multipliers satisfying again some technical conditions, and in all we investigate four different settings. Along the way, we introduce a Banach space variant of the notion of spectral triple suitable for our context. Finally, we investigate the bisectoriality and the functional calculus of some Hodge-Dirac operators which are crucial in the noncommutative geometries which we introduce here.

5.1 Background on Quantum Locally Compact Metric Spaces

We recall definitions and characterizations of the notions that we need in this chapter. The main notion is that of quantum locally compact metric space. This concept has its origins in Connes' paper [59] of 1989, in which he shows that we can recover the geodesic distance dist of a compact oriented Riemannian spin manifold M using the Dirac operator D by the formula

$$\mathrm{dist}(x, y) = \sup_{f \in C(M),\, \|[D,f]\| \leqslant 1} |f(x) - f(y)|, \quad x, y \in M$$

where the commutator $[D, f] = Df - fD$ extends to a bounded operator.[1] See [60, Chapter 6] for more information and we refer to [222, Chapter 3] for a complete proof. Furthermore, it is known that the commutator $[D, f]$ induces a bounded operator if and only if f is a Lipschitz function and in this case the Lipschitz constant of f is equal to $\|[D, f]\|$. Moreover, this space of functions is norm dense in the space $C(M)$ of continuous functions. If we identify the points x, y as pure states ω_x and ω_y on the algebra $C(M)$, this formula can be seen as

$$\text{dist}(\omega_x, \omega_y) = \sup_{f \in C(M), \|[D, f]\| \leqslant 1} |\omega_x(f) - \omega_y(f)|, \quad x, y \in M.$$

Afterwards, Rieffel [197] axiomatized this formula replacing $C(M)$ by a unital C^*-algebra A (or even by an order-unit space $\mathcal{A}$), $f \mapsto \|[D, f]\|$ by a seminorm $\|\cdot\|$ defined on a subspace of A and ω_x, ω_y by arbitrary states obtaining essentially the formula (5.3) below and giving rise to a theory of *quantum* compact metric spaces. With this notion, Rieffel was able to define a quantum analogue of the Gromov-Hausdorff distance and to give a meaning to many approximations found in the physics literature, as the case of matrix algebras converging to the sphere. Moreover, this notion lead to many interesting new approximations as the case of fuzzy tori converging to the quantum tori. We refer to the surveys [147] and [197] and references therein for more information.

Recall that an order-unit space [2, p. 69], [3, Definition 1.8] is an ordered normed $\mathbb{R}$-vector space $\mathcal{A}$ with a closed positive cone and an element $1_{\mathcal{A}}$, satisfying $\|a\|_{\mathcal{A}} = \inf\{\lambda > 0 : -\lambda 1_{\mathcal{A}} \leqslant a \leqslant \lambda 1_{\mathcal{A}}\}$. The element $1_{\mathcal{A}}$ is called the distinguished order unit. The definition of an order-unit space is due to Kadison [133]. Important examples of order-unit spaces are given by real linear subspaces of selfadjoint elements containing the unit element in a unital C^*-algebra.

The following is a slight extension of [147, Definition 2.3] replacing selfadjoint parts of unital C^*-algebras by order-unit spaces.

Definition 5.1 A unital Lipschitz pair $(\mathcal{A}, \|\cdot\|)$ is a pair where $\mathcal{A}$ is an order-unit space and where $\|\cdot\|$ is a seminorm defined on a dense subspace dom $\|\cdot\|$ of $\mathcal{A}$ such that

$$\{a \in \text{dom} \|\cdot\| : \|a\| = 0\} = \mathbb{R}1_{\mathcal{A}}. \tag{5.1}$$

Remark 5.1 Note that if a seminorm $\|\cdot\|$ is defined on some subspace dom $\|\cdot\|$ of a unital C^*-algebra A such that $A_{\text{sa}} \cap \text{dom} \|\cdot\|$ is dense in A_{sa} and such that $\{a \in \text{dom} \|\cdot\| : \|a\| = 0\} = \mathbb{C}1_A$, then its restriction on $A_{\text{sa}} \cap \text{dom} \|\cdot\|$ defines a unital Lipschitz pair. In this case, we also say that $(A, \|\cdot\|)$ is a unital Lipschitz pair (or a compact quantum metric space if Definition 5.2 is satisfied).

---

[1] Recall that D is an unbounded operator acting on the Hilbert space of L^2-spinors and that the functions of $C(M)$ act on the same Hilbert space by multiplication operators.

If (X, dist) is a compact metric space, a fundamental example [147, Example 2.6], [148, Example 2.9] is given by $(C(X)_{\text{sa}}, \text{Lip})$ where $C(X)$ is the commutative unital C^*-algebra of continuous functions and where Lip is the Lipschitz seminorm, defined for any Lipschitz function $f : X \to \mathbb{C}$ by

$$\text{Lip}(f) \overset{\text{def}}{=} \sup \left\{ \frac{|f(x) - f(y)|}{\text{dist}(x, y)} : x, y \in X, x \neq y \right\}. \tag{5.2}$$

It is immediate that a function f has zero Lipschitz constant if and only if it is constant on X. Moreover, the algebra $\text{Lip}(X)$ of real Lipschitz functions is norm-dense in $C(X)_{\text{sa}}$ by the Stone-Weierstrass theorem.[2]

Now, following essentially [198, Definition 2.2] (see also [145, Definition 2.6], [146, Definition 1.2] or [147, Definition 2.42] for the case where $\mathcal{A}$ is the selfadjoint part of a unital C^*-algebra), we introduce a notion of quantum compact metric space. Recall that a linear functional φ on an order-unit space $\mathcal{A}$ is a state [2, p. 72] if $\|\varphi\| = \varphi(1_{\mathcal{A}}) = 1$.

Definition 5.2 A quantum compact metric space $(\mathcal{A}, \|\cdot\|)$ is a unital Lipschitz pair whose associated Monge-Kantorovich metric

$$\text{dist}_{\text{mk}}(\varphi, \psi) \overset{\text{def}}{=} \sup \left\{ |\varphi(a) - \psi(a)| : a \in \mathcal{A}, \|a\| \leqslant 1 \right\}, \quad \varphi, \psi \in S(\mathcal{A}) \tag{5.3}$$

metrizes the weak* topology restricted to the state space $S(\mathcal{A})$ of $\mathcal{A}$. When a Lipschitz pair $(\mathcal{A}, \|\cdot\|)$ is a quantum compact metric space, the seminorm $\|\cdot\|$ is referred to as a Lip-norm.

In the case of $(C(X)_{\text{sa}}, \text{Lip})$, the formula (5.3) gives by Villani [223, Remark 6.5] or Dudley [78, Theorem 11.8.2] the dual formulation of the classical Kantorovich-Rubinstein metric[3] between Borel probability measures μ and ν on X

$$\text{dist}_{\text{mk}}(\mu, \nu) \overset{\text{def}}{=} \sup \left\{ \left| \int_X f \, d\mu - \int_X f \, d\nu \right| : f \in C(X)_{\text{sa}}, \text{Lip}(f) \leqslant 1 \right\}, \tag{5.4}$$

which is a basic concept in optimal transport theory [223]. Considering the Dirac measures δ_x and δ_y at points $x, y \in X$ instead of μ and ν, we recover the distance $\text{dist}(x, y)$ with the formula (5.4).

---

[2] Indeed, $\text{Lip}(X)$ contains the constant functions. Moreover, $\text{Lip}(X)$ separates points in X. If $x_0, y_0 \in X$ with $x_0 \neq y_0$, we can use the lipschitz function $f : X \to \mathbb{R}, x \mapsto \text{dist}(x, y_0)$ since we have $f(x_0) > 0 = f(y_0)$.

[3] The original formulation [78, p. 329], [223, Definition 6.1] of the distance $\text{dist}_{\text{mk}}(\mu, \nu)$ between two Borel probability measures μ and ν on X is given by the infimum of $\int_{X \times X} \text{dist}(x, y) \, d\pi(x, y)$ over all Borel probability measures π on $X \times X$ whose marginals are given by μ and ν.

The compatibility of (5.4) with the weak* topology is well-known, see e.g. [39, Theorem 8.3.2]. This example is at the root of Definition 5.2. The theory of quantum compact metric spaces is the study of noncommutative generalizations of algebras of Lipschitz functions over compact metric spaces.

The compatibility of the Monge-Kantorovich metric with the weak* topology is hard to check directly in general. Fortunately, there exists a condition which is more practical. This condition is inspired by the fact that Arzéla-Ascoli's theorem shows that for any $x \in X$ the set

$$\{f \in C(X)_{\mathrm{sa}} : \mathrm{Lip}(f) \leqslant 1, f(x) = 0\}$$

is norm relatively compact and it is known that this property implies that (5.4) metrizes the weak* topology on the space of Borel probability measures on X. The noncommutative generalization relies on the following result which is a slight generalization of [182, Proposition 1.3] with a similar proof left to the reader.

Proposition 5.1 *Let $(\mathcal{A}, \|\cdot\|)$ be a unital Lipschitz pair. Let $\mu \in S(\mathcal{A})$ be a state of $\mathcal{A}$. If the set $\{a \in \mathrm{dom}\,\|\cdot\| : \|a\| \leqslant 1, \mu(a) = 0\}$ is norm relatively compact in $\mathcal{A}$ then $(\mathcal{A}, \|\cdot\|)$ is a quantum compact metric space.*

Note that in the case of selfadjoint parts of C*-algebras, this condition is also necessary, see [147, Theorem 2.43].

The Lipschitz seminorm Lip associated to a compact metric space (X, dist) enjoys a natural property with respect to the multiplication of functions in $C(X)$, called the Leibniz property [78, page 306], [226, Proposition 1.30]. Indeed, for any Lipschitz functions $f, g \colon X \to \mathbb{C}$ we have

$$\mathrm{Lip}(fg) \leqslant \|f\|_{C(X)} \mathrm{Lip}(g) + \mathrm{Lip}(f) \|g\|_{C(X)}. \tag{5.5}$$

Moreover, the Lipschitz seminorm is lower-semicontinuous with respect to the C*-norm of $C(X)$, i.e. the uniform convergence norm on X. These two additional properties were not assumed in the previous Definition 5.2, yet they are quite natural. However, as research in noncommutative metric geometry progressed, the need for a noncommutative analogue of these properties for some developments became evident. So, some additional conditions are often added to Definition 5.2 which brings us to the following definition which is a slight generalization of [147, Definition 2.21] for order-unit spaces embedded in unital C*-algebras, see also [149, Definition 1.3], [144, Definition 2.2.2], [145, Definition 2.19], [147, Definition 2.45] and [148, Definition 2.19]. Here $a \circ b \overset{\text{def}}{=} \frac{1}{2}(ab + ba)$ and $\{a, b\} \overset{\text{def}}{=} \frac{1}{2i}(ab - ba)$.

Definition 5.3

1. A unital Leibniz pair $(\mathcal{A}, \|\cdot\|)$ is a unital Lipschitz pair where $\mathcal{A}$ is a real linear subspace of selfadjoint elements containing the unit element in a unital C*-algebra A such that:

   a. the domain of $\|\cdot\|$ is a Jordan-Lie subalgebra of A_{sa},

b. for any $a, b \in \text{dom } \|\cdot\|$, we have:

$$\|a \circ b\| \leqslant \|a\|_A \|b\| + \|a\| \|b\|_A \quad \text{and} \quad \|\{a, b\}\| \leqslant \|a\|_A \|b\| + \|a\| \|b\|_A .$$

2. A unital Leibniz pair $(\mathcal{A}, \|\cdot\|)$ is a Leibniz quantum compact metric space when $\|\cdot\|$ is lower semicontinuous.

Note that neither $\mathcal{A}$ nor $\mathcal{A} \cap \text{dom } \|\cdot\|$ are in general a Jordan or Lie or a usual algebra. We continue with a useful observation [147, p. 18], [148, Proposition 2.17] in the spirit of Remark 5.1.

Proposition 5.2 *Let A be a unital C^*-algebra and $\|\cdot\|$ be a seminorm defined on a dense $\mathbb{C}$-subspace $\text{dom } \|\cdot\|$ of A, such that $\text{dom } \|\cdot\|$ is closed under the adjoint operation, satisfying $\{a \in \text{dom } \|\cdot\| : \|a\| = 0\} = \mathbb{C}1_{\mathcal{A}}$ and such that*

$$\|ab\| \leqslant \|a\|_A \|b\| + \|a\| \|b\|_A , \quad a, b \in \text{dom } \|\cdot\| .$$

If $\|\cdot\|_{\text{sa}}$ is the restriction of $\|\cdot\|$ to $A_{\text{sa}} \cap \text{dom } \|\cdot\|$, then $(A_{\text{sa}}, \|\cdot\|_{\text{sa}})$ is a unital Leibniz pair.

We also need a notion of quantum *locally* compact metric spaces. The paper [143] gives such a definition in the case of a Lipschitz pair $(\mathcal{A}, \|\cdot\|)$ where $\mathcal{A}$ is a C^*-algebra. However, for semigroups of Schur multipliers, we need a version for order-unit spaces unfortunately not covered by Latrémolière [143]. Consequently, in the sequel, we try to generalize some notions of [143]. The following is a variant of [143, Definition 2.3] and [143, Definition 2.27]. Here uA is the unitization of the algebra A.

Definition 5.4 A Lipschitz pair $(\mathcal{A}, \|\cdot\|)$ is a closed subspace $\mathcal{A}$ of selfadjoint elements of a non-unital C^*-algebra A and a seminorm $\|\cdot\|$ defined on a dense subspace of $\mathcal{A} \oplus \mathbb{R}1_{uA}$ such that

$$\{x \in \mathcal{A} \oplus \mathbb{R}1_{uA} : \|x\| = 0\} = \mathbb{R}1_{uA}.$$

A Lipschitz triple $(\mathcal{A}, \|\cdot\|, \mathfrak{M})$ consists of a Lipschitz pair $(\mathcal{A}, \|\cdot\|)$ and an abelian C^*-algebra $\mathfrak{M}$ of A such that $\mathfrak{M}$ contains an approximate unit of A.

The following is [143, Definition 2.23].

Definition 5.5 Let A be a non-unital C^*-algebra and $\mathfrak{M}$ be an abelian C^*-subalgebra of A containing an approximate unit of A. Let $\mu: A \to \mathbb{C}$ be a state of A. We call μ a local state (of $(A, \mathfrak{M})$) provided that there exists a projection e in $\mathfrak{M}$ of compact support[4] in the Gelfand spectrum of $\mathfrak{M}$ such that $\mu(e) = 1$.

---

[4] It is not clear if the support must be in addition open in [143] since the indicator 1_A of a subset A is continuous if and only if A is both open and closed.

Inspired by Latrémolière [143, Theorem 3.10], Latrémolière [147, Theorem 2.73], we introduce the following definition. Of course, we recognize that this definition is a bit artificial. A better choice would be to generalize the results and definitions of [143] to a larger context. Here μ is extended on the unitization as in [143, Notation 2.2].

Definition 5.6 We say that a Lipschitz triple $(\mathcal{A}, \|\cdot\|, \mathfrak{M})$ is a quantum locally compact metric space if for a local state μ of $(A, \mathfrak{M})$ and any compactly supported $a, b \in \mathfrak{M}$, the set

$$a\{x \in \mathcal{A} \oplus \mathbb{R}1_{uA} : \|x\| \leqslant 1, \mu(x) = 0\}b$$

is totally bounded (i.e. relatively compact) in the norm topology of uA.

Now, we give a variant of Definition 5.3.

Definition 5.7 Let $(\mathcal{A}, \|\cdot\|, \mathfrak{M})$ be a quantum locally compact metric space. We call it a Leibniz quantum locally compact metric space if $(\mathcal{A} \oplus \mathbb{R}1_{uA}, \|\cdot\|)$ is a unital Leibniz pair in the sense of Definition 5.3 and if the seminorm $\|\cdot\|$ is lower semicontinuous with respect to the C^*-norm $\|\cdot\|_{uA}$.

Finally, following [143, Condition 4.3] (see also [32, page 4] for a related discussion), we introduce the following definition.

Definition 5.8 We say that a quantum locally compact metric space $(\mathcal{A}, \|\cdot\|, \mathfrak{M})$ in the sense of Definition 5.6 satisfies the boundedness condition if the Lipschitz ball $\{a \in \mathcal{A} : \|a\| \leqslant 1\}$ is norm bounded.

5.2 Quantum Compact Metric Spaces Associated to Semigroups of Fourier Multipliers

In this section, we consider a markovian semigroup $(T_t)_{t \geqslant 0}$ of Fourier multipliers on a group von Neumann algebra $\mathrm{VN}(G)$ where G is a discrete group, as in Proposition 2.3. We introduce new compact quantum metric spaces in the spirit of the ones of [119]. We also add to the picture the lower semicontinuity and a careful examination of the domains. By Lemma 3.8, the following definition is correct. It is far from clear if it is possible to do the same analysis at the level $p = \infty$ considered in [119, Section 1.2].

Definition 5.9 Suppose $2 \leqslant p < \infty$. Let G be a discrete group. Let A_p denote the L^p realization of the (negative) generator of $(T_t)_{t \geqslant 0}$. For any $x \in \mathrm{dom}\, A_p^{\frac{1}{2}}$ we let

$$\|x\|_{\Gamma, p} \overset{\mathrm{def}}{=} \max\left\{ \left\|\Gamma(x, x)^{\frac{1}{2}}\right\|_{L^p(\mathrm{VN}(G))}, \left\|\Gamma(x^*, x^*)^{\frac{1}{2}}\right\|_{L^p(\mathrm{VN}(G))} \right\}. \tag{5.6}$$

We start with an elementary fact.

Proposition 5.3 *Suppose* $2 \leqslant p < \infty$. *Let G be a discrete group. Then $\|\cdot\|_{\Gamma,p}$ is a seminorm on the subspace* dom $A_p^{\frac{1}{2}}$ *of* $L^p(\mathrm{VN}(G))$.

Proof 1. If x belongs to dom $A_p^{\frac{1}{2}}$ and if $k \in \mathbb{C}$, we have

$$\|kx\|_{\Gamma,p} \overset{(5.6)}{=} \max\left\{ \left\|\Gamma(kx, kx)^{\frac{1}{2}}\right\|_{L^p(\mathrm{VN}(G))}, \left\|\Gamma((kx)^*, (kx)^*)^{\frac{1}{2}}\right\|_{L^p(\mathrm{VN}(G))} \right\}$$

$$= \max\left\{ \left\|\Gamma(kx, kx)^{\frac{1}{2}}\right\|_{L^p(\mathrm{VN}(G))}, \left\|\Gamma(\bar{k}x^*, \bar{k}x^*)^{\frac{1}{2}}\right\|_{L^p(\mathrm{VN}(G))} \right\}$$

$$= \max\left\{ |k|\left\|\Gamma(x, x)^{\frac{1}{2}}\right\|_{L^p(\mathrm{VN}(G))}, |k|\left\|\Gamma(x^*, x^*)^{\frac{1}{2}}\right\|_{L^p(\mathrm{VN}(G))} \right\}$$

$$= |k| \max\left\{ \left\|\Gamma(x, x)^{\frac{1}{2}}\right\|_{L^p(\mathrm{VN}(G))}, \left\|\Gamma(x^*, x^*)^{\frac{1}{2}}\right\|_{L^p(\mathrm{VN}(G))} \right\} \overset{(5.6)}{=} |k| \, \|x\|_{\Gamma,p}.$$

Let us turn to the triangular inequality. Assume that $x, y \in \mathcal{P}_G$. Note that $L^{\frac{p}{2}}(\mathrm{VN}(G))$ is a normed space since $p \geqslant 2$. According to the part 4 of Lemma 2.15 and (2.50) applied with $Z = L^{\frac{p}{2}}(\mathrm{VN}(G))$, we have

$$\|\Gamma(x + y, x + y)\|_{L^{\frac{p}{2}}(\mathrm{VN}(G))}^{\frac{1}{2}} \overset{(2.50)}{\leqslant} \|\Gamma(x, x)\|_{L^{\frac{p}{2}}(\mathrm{VN}(G))}^{\frac{1}{2}} + \|\Gamma(y, y)\|_{L^{\frac{p}{2}}(\mathrm{VN}(G))}^{\frac{1}{2}}.$$

Using (2.25), we can rewrite this inequality under the form

$$\left\|\Gamma(x + y, x + y)^{\frac{1}{2}}\right\|_{L^p(\mathrm{VN}(G))} \leqslant \left\|\Gamma(x, x)^{\frac{1}{2}}\right\|_{L^p(\mathrm{VN}(G))} + \left\|\Gamma(y, y)^{\frac{1}{2}}\right\|_{L^p(\mathrm{VN}(G))}. \tag{5.7}$$

Then, for the general case where $x, y \in$ dom $A_p^{\frac{1}{2}}$, consider some sequences (x_n) and (y_n) of $\mathcal{P}_G$ such that $x_n \overset{a}{\to} x$ and $y_n \overset{a}{\to} y$ with Lemma 3.8 and (2.52). By Remark 2.1, we have $x_n + y_n \overset{a}{\to} x + y$ and thus

$$\left\|\Gamma(x + y, x + y)^{\frac{1}{2}}\right\|_{L^p(\mathrm{VN}(G))} \overset{(2.25)}{=} \left\|\Gamma(x + y, x + y)\right\|_{L^{\frac{p}{2}}(\mathrm{VN}(G))}^{\frac{1}{2}} \tag{5.8}$$

$$\overset{(2.53)}{=} \lim_n \left\|\Gamma(x_n + y_n, x_n + y_n)\right\|_{L^{\frac{p}{2}}(\mathrm{VN}(G))}^{\frac{1}{2}} \tag{5.9}$$

$$\overset{(2.25)}{=} \lim_n \left\|\Gamma(x_n + y_n, x_n + y_n)^{\frac{1}{2}}\right\|_{L^p(\mathrm{VN}(G))}$$

$$\overset{(5.7)}{\leqslant} \lim_n \left\|\Gamma(x_n, x_n)^{\frac{1}{2}}\right\|_{L^p(\mathrm{VN}(G))} + \lim_n \left\|\Gamma(y_n, y_n)^{\frac{1}{2}}\right\|_{L^p(\mathrm{VN}(G))}$$

$$\overset{(2.53)}{=} \left\|\Gamma(x, x)^{\frac{1}{2}}\right\|_{L^p(\mathrm{VN}(G))} + \left\|\Gamma(y, y)^{\frac{1}{2}}\right\|_{L^p(\mathrm{VN}(G))}.$$

A similar reasoning for x, y replaced by x^*, y^* gives

$$\left\|\Gamma(x^* + y^*, x^* + y^*)^{\frac{1}{2}}\right\|_{L^p(VN(G))} \leqslant \left\|\Gamma(x^*, x^*)^{\frac{1}{2}}\right\|_{L^p(VN(G))}$$
$$+ \left\|\Gamma(y^*, y^*)^{\frac{1}{2}}\right\|_{L^p(VN(G))}. \qquad (5.10)$$

Finally, we obtain

$$\|x + y\|_{\Gamma,p} \overset{(5.6)}{=} \max\left\{\left\|\Gamma(x + y, x + y)^{\frac{1}{2}}\right\|_{L^p}, \left\|\Gamma(x^* + y^*, x^* + y^*)^{\frac{1}{2}}\right\|_{L^p}\right\}$$

$$\overset{(5.8)(5.10)}{\leqslant} \max\left\{\left\|\Gamma(x, x)^{\frac{1}{2}}\right\|_{L^p(VN(G))} + \left\|\Gamma(y, y)^{\frac{1}{2}}\right\|_{L^p(VN(G))},\right.$$

$$\left.\left\|\Gamma(x^*, x^*)^{\frac{1}{2}}\right\|_{L^p(VN(G))} + \left\|\Gamma(y^*, y^*)^{\frac{1}{2}}\right\|_{L^p(VN(G))}\right\}$$

$$\leqslant \max\left\{\left\|\Gamma(x, x)^{\frac{1}{2}}\right\|_{L^p(VN(G))}, \left\|\Gamma(x^*, x^*)^{\frac{1}{2}}\right\|_{L^p(VN(G))}\right\}$$

$$+ \max\left\{\left\|\Gamma(y, y)^{\frac{1}{2}}\right\|_{L^p(VN(G))}, \left\|\Gamma(y^*, y^*)^{\frac{1}{2}}\right\|_{L^p(VN(G))}\right\}$$

$$\overset{(5.6)}{=} \|x\|_{\Gamma,p} + \|y\|_{\Gamma,p}.$$

$$\square$$

Now, we prove the Leibniz property of these seminorms. We recall that $C_r^*(G)$ is the reduced group C*-algebra of the discrete group G containing $\mathcal{P}_G$ as a dense subspace.

Proposition 5.4 *Suppose* $2 \leqslant p < \infty$. *Let* G *be a discrete group. For any* $x, y \in \mathcal{P}_G$, *we have*

$$\|xy\|_{\Gamma,p} \leqslant \|x\|_{C_r^*(G)} \|y\|_{\Gamma,p} + \|x\|_{\Gamma,p} \|y\|_{C_r^*(G)}. \qquad (5.11)$$

Proof Let $x, y \in \mathcal{P}_G$. Using the structure of bimodule of $L^p(VN(G), L^2(\Omega)_{c,p})$ in the second inequality, we see that

$$\left\|\Gamma(xy, xy)^{\frac{1}{2}}\right\|_{L^p(VN(G))}$$

$$\overset{(3.37)}{=} \left\|\partial_\psi(xy)\right\|_{L^p(VN(G), L^2(\Omega)_{c,p})} \overset{(2.86)}{=} \left\|x\partial_\psi(y) + \partial_\psi(x)y\right\|_{L^p(VN(G), L^2(\Omega)_{c,p})}$$

$$\leqslant \left\|x\partial_\psi(y)\right\|_{L^p(VN(G), L^2(\Omega)_{c,p})} + \left\|\partial_\psi(x)y\right\|_{L^p(VN(G), L^2(\Omega)_{c,p})}$$

$$\leqslant \|x\|_{C_r^*(G)} \left\|\partial_\psi(y)\right\|_{L^p(VN(G), L^2(\Omega)_{c,p})} + \left\|\partial_\psi(x)\right\|_{L^p(VN(G), L^2(\Omega)_{c,p})} \|y\|_{C_r^*(G)}$$

$$\overset{(3.37)}{=} \|x\|_{C_r^*(G)} \left\|\Gamma(y, y)^{\frac{1}{2}}\right\|_{L^p(VN(G))} + \left\|\Gamma(x, x)^{\frac{1}{2}}\right\|_{L^p(VN(G))} \|y\|_{C_r^*(G)}.$$

Similarly, we have

$$\left\| \Gamma(y^* x^*, y^* x^*)^{\frac{1}{2}} \right\|_{L^p(\mathrm{VN}(G))}$$

$$\leqslant \|x^*\|_{C_r^*(G)} \left\| \Gamma(y^*, y^*)^{\frac{1}{2}} \right\|_{L^p(\mathrm{VN}(G))} + \left\| \Gamma(x^*, x^*)^{\frac{1}{2}} \right\|_{L^p(\mathrm{VN}(G))} \|y^*\|_{C_r^*(G)}$$

$$= \|x\|_{C_r^*(G)} \left\| \Gamma(y^*, y^*)^{\frac{1}{2}} \right\|_{L^p(\mathrm{VN}(G))} + \left\| \Gamma(x^*, x^*)^{\frac{1}{2}} \right\|_{L^p(\mathrm{VN}(G))} \|y\|_{C_r^*(G)}.$$

Therefore, $\|\cdot\|_{\Gamma, p}$ satisfies the Leibniz property.   □

For the proof of Theorem 5.1, we shall need the following lemma. Note that $\mathrm{dom}\, \partial_{\psi, q, p}$ was defined in Proposition 3.4. See Theorem 3.1 for the notation $L_{cr}^p(\mathbb{E})$.

Lemma 5.1 *Let* $2 \leqslant p < \infty$ *and* $-1 \leqslant q \leqslant 1$. *Let* G *be a discrete group. Suppose that* $L^p(\mathrm{VN}(G))$ *has CCAP and that* $\mathrm{VN}(G)$ *has QWEP. The map* $\partial_{\psi, q, p} \colon \mathrm{dom}\, \partial_{\psi, q, p} \subset L^p(\mathrm{VN}(G)) \to L_{cr}^p(\mathbb{E})$ *is closed.*

Proof Let (x_n) be a sequence in $\mathrm{dom}\, \partial_{\psi, q, p}$ such that $x_n \to x$ where x is some element of $L^p(\mathrm{VN}(G))$ and $\partial_{\psi, q, p}(x_n) \to z$ where $z \in L_{cr}^p(\mathbb{E})$. We have

$$\left\| \partial_{\psi, q, p}(y) \right\|_{L_{cr}^p(\mathbb{E})} \overset{(3.36)(3.32)(5.6)}{=} \|y\|_{\Gamma, p} \overset{(3.38)}{\approx_p} \left\| A_p^{\frac{1}{2}}(y) \right\|_{L^p(\mathrm{VN}(G))}.$$ Thus the sequence

$(A_p^{\frac{1}{2}}(x_n))$ converges in $L^p(\mathrm{VN}(G))$. Since $A_p^{\frac{1}{2}}$ is closed, x belongs to $\mathrm{dom}\, A_p^{\frac{1}{2}} = \mathrm{dom}\, \partial_{\psi, q, p}$ and $A_p^{\frac{1}{2}}(x) = \lim_n A_p^{\frac{1}{2}}(x_n)$. By the previous equivalence, we deduce that $\partial_{\psi, q, p}(x_n) \to \partial_{\psi, q, p}(x)$ in $L_{cr}^p(\mathbb{E})$. We infer that $z = \partial_{\psi, q, p}(x)$. We conclude with (2.1).   □

In the next theorem, recall that if A is the generator of a markovian semigroup of Fourier multipliers, then there exists a real Hilbert space H together with a mapping $b_\psi \colon G \to H$ such that the symbol $\psi \colon G \to \mathbb{C}$ of A satisfies $\psi(s) = \|b_\psi(s)\|_H^2$. Following [129, p. 1962], we define

$$\mathrm{Gap}_\psi \overset{\mathrm{def}}{=} \inf_{b_\psi(s) \neq b_\psi(t)} \left\| b_\psi(s) - b_\psi(t) \right\|_H^2. \tag{5.12}$$

By [31, Proposition 2.10.2], note that Gap_ψ is independent of b_ψ, that is, if $b_\psi \colon I \to H_b$ and $c_\psi \colon I \to H_c$ are two cocycles such that $\psi(s) = \|b_\psi(s)\|_{H_b}^2 = \|c_\psi(s)\|_{H_c}^2$ for all $s \in G$, then Gap_ψ takes the same value in (5.12), whether it is defined via b_ψ or c_ψ.

Recall that $C_r^*(G)$ is the reduced group C^*-algebra associated with the discrete group G. We also write $C_r^*(G)_0$ for the range of the bounded projection $C_r^*(G) \to C_r^*(G)$, $\lambda_s \mapsto (1 - \delta_{s=e})\lambda_s$, that is the space of elements of $C_r^*(G)$ with vanishing trace. Note that $L_0^p(\mathrm{VN}(G))$ was defined before Proposition 2.2. In the sequel, we denote by $\|\cdot\|_{\Gamma, p}$ the restriction of (5.6) on $\mathrm{dom}\, A_p^{\frac{1}{2}} \cap C_r^*(G)$. In the following theorem, note that G satisfies the assumptions of the second point if G is a free group [46, Corollary 12.3.5] or an amenable group.

Theorem 5.1 *Let* $2 \leqslant p < \infty$. *Let* G *be a discrete group. Let* $b_\psi : G \to H$ *be an injective cocycle with values in a finite-dimensional Hilbert space of dimension* $n < p$. *Assume that* $\mathrm{Gap}_\psi > 0$.

1. *If* $\mathrm{L}^p(\mathrm{VN}(G))$ *has* CCAP *and* $\mathrm{VN}(G)$ *has* QWEP *then* $(\mathrm{C}_r^*(G), \|\cdot\|_{\Gamma,p})$ *is a quantum compact metric space.*
2. *If* G *is in addition weakly amenable with the approximating net* $\varphi_j : G \to \mathbb{C}$ *satisfying* $\sup_j \|M_{\varphi_j}\|_{\mathrm{C}_r^*(G) \to \mathrm{C}_r^*(G)} \leqslant 1$ *then* $(\mathrm{C}_r^*(G), \|\cdot\|_{\Gamma,p})$ *is a Leibniz quantum compact metric space.*

Proof

1. Since $\mathrm{dom}\, \|\cdot\|_{\Gamma,p} = \mathrm{dom}\, A_p^{\frac{1}{2}} \cap \mathrm{C}_r^*(G)$ contains the dense subspace $\mathcal{P}_G$ of $\mathrm{C}_r^*(G)$, the $\mathbb{C}$-subspace $\mathrm{dom}\, \|\cdot\|_{\Gamma,p}$ is dense in $\mathrm{C}_r^*(G)$. By the first part of Lemma 3.11 and Proposition 3.4, $\mathrm{dom}\, \|\cdot\|_{\Gamma,p}$ is closed under the adjoint operation. According to Proposition 5.3, $\|\cdot\|_{\Gamma,p}$ is a seminorm. We will show that

$$\left\{ x \in \mathrm{dom}\, \|\cdot\|_{\Gamma,p} : \|x\|_{\Gamma,p} = 0 \right\} = \mathbb{C}1. \tag{5.13}$$

Indeed, we have $A(1) = \psi(e)1 = 0$, so that $\Gamma(1,1) \overset{(2.78)}{=} \frac{1}{2}\left[A(1^*)1 + 1^*A(1) - A(1^*1)\right] = 0$ and consequently

$$\|1\|_{\Gamma,p} \overset{(5.6)}{=} \max\left\{ \left\|\Gamma(1,1)^{\frac{1}{2}}\right\|_{\mathrm{L}^p(\mathrm{VN}(G))}, \left\|\Gamma(1^*,1^*)^{\frac{1}{2}}\right\|_{\mathrm{L}^p(\mathrm{VN}(G))} \right\} = 0.$$

In the other direction, if $\|x\|_{\Gamma,p} = 0$, then according to (3.38), we have $A_p^{\frac{1}{2}}(x) = 0$. By (2.21), we deduce that x belongs to $\mathrm{Ker}\, A_p$. For any $s \in G$, we infer that

$$0 = \tau_G\big(A_p(x)\lambda_s^*\big) = \tau_G\big(x A_{p^*}(\lambda_{s^{-1}})\big) = \psi(s^{-1})\tau_G(x\lambda_{s^{-1}}).$$

Note that by (2.34), the injectivity of b_ψ and $\psi(e) = 0$ we have $\psi(s^{-1}) \neq 0$ for $s \neq e$. We deduce that $\tau_G(x\lambda_s^*) = 0$ for these s. Finally, we obtain $x \in \mathbb{C}1$ by approximation using CCAP. We conclude that (5.13) is true.

Since $\tau_G : \mathrm{C}_r^*(G) \to \mathbb{C}$ is a state, with Proposition 5.1 it suffices to show that

$$\left\{ x \in \mathrm{dom}\, \|\cdot\|_{\Gamma,p} : \|x\|_{\Gamma,p} \leqslant 1, \ \tau_G(x) = 0 \right\} \text{ is relatively compact in } \mathrm{C}_r^*(G). \tag{5.14}$$

Note that $A_2^{-1} : \mathrm{L}_0^2(\mathrm{VN}(G)) \to \mathrm{L}_0^2(\mathrm{VN}(G))$ is compact. Indeed, $(\lambda_s)_{s \in G \setminus \{e\}}$ is an orthonormal basis of $\mathrm{L}_0^2(\mathrm{VN}(G))$ consisting of eigenvectors of A_2^{-1}. Moreover, the corresponding eigenvalues are $\|b_\psi(s)\|_H^{-2}$ and this family vanish at infinity.[5] For the latter, note that the condition $\mathrm{Gap}_\psi = \inf_{b_\psi(s) \neq b_\psi(t)} \|b_\psi(s) - b_\psi(t)\|_H^2 > 0$ together with the injectivity of b_ψ imply that any compact subset of H meets the $b_\psi(s)$ only for a finite number of $s \in G$. As H is finite-dimensional by assumption, the closed ball $B(0, \frac{1}{\sqrt{\varepsilon}})$ for $\varepsilon > 0$ of H is compact, so contains a finite number of $b_\psi(s)$. We deduce that $\|b_\psi(s)\|_H^{-2} \to 0$. We have shown that A_2^{-1} is compact. Now, according to [129, Lemma 5.8], we have

$$\|T_t\|_{L_0^1(\mathrm{VN}(G)) \to L^\infty(\mathrm{VN}(G))} \lesssim \frac{1}{t^{\frac{n}{2}}}, \quad t > 0.$$

By Proposition 2.2 applied with $z = \frac{1}{2}$ and $q = \infty$, we have a well-defined compact operator $A^{-\frac{1}{2}} : L_0^p(\mathrm{VN}(G)) \to L_0^\infty(\mathrm{VN}(G))$. So the image $\mathcal{I}$ by $A^{-\frac{1}{2}}$ of the closed unit ball of $L_0^p(\mathrm{VN}(G)) = \overline{\mathrm{Ran}\, A_p}$ is compact. Note that $\overline{\mathrm{Ran}\, A_p^{\frac{1}{2}}} \subset \overline{\mathrm{Ran}\, A_p}$ by (2.21). Hence the subset[6]

$$\left\{ x \in C_r^*(G)_0 \cap \mathrm{dom}\, A_p^{\frac{1}{2}} : \left\| A_p^{\frac{1}{2}}(x) \right\|_{L^p(\mathrm{VN}(G))} \leqslant 1 \right\}$$

of $\mathcal{I}$ is relatively compact in $C_r^*(G)$.

Note that for any $x \in \mathrm{dom}\, A_p^{\frac{1}{2}}$ we have

$$\left\| A_p^{\frac{1}{2}}(x) \right\|_{L^p(\mathrm{VN}(G))} \overset{(3.38)}{\lesssim_p} \max\left\{ \left\| \Gamma(x,x)^{\frac{1}{2}} \right\|_{L^p(\mathrm{VN}(G))}, \left\| \Gamma(x^*,x^*)^{\frac{1}{2}} \right\|_{L^p(\mathrm{VN}(G))} \right\}$$

$$\overset{(5.6)}{=} \|x\|_{\Gamma,p}.$$

We deduce from this inequality that

$$\left\{ x \in C_r^*(G)_0 \cap \mathrm{dom}\, A_p^{\frac{1}{2}} : \|x\|_{\Gamma,p} \leqslant 1 \right\}$$

is relatively compact in $C_r^*(G)$. Now, we have

$$\left\{ x \in C_r^*(G)_0 \cap \mathrm{dom}\, A_p^{\frac{1}{2}} : \|x\|_{\Gamma,p} \leqslant 1 \right\}$$

$$= \left\{ x \in C_r^*(G) \cap \mathrm{dom}\, A_p^{\frac{1}{2}} : \|x\|_{\Gamma,p} \leqslant 1, \ \tau_G(x) = 0 \right\}.$$

We deduce (5.14).

---

[5] Recall that a family $(x_s)_{s \in I}$ vanishes at infinity means that for any $\varepsilon > 0$, there exists a finite subset J of I such that for any $s \in I - J$ we have $|x_s| < \varepsilon$.

[6] Write $x = A^{-\frac{1}{2}} A_p^{\frac{1}{2}} x$.

2. Let $x \in C_r^*(G)$ and (x_j) be a net of elements of dom $A_p^{\frac{1}{2}} \cap C_r^*(G)$ such that $\|x_j - x\|_{C_r^*(G)} \to 0$ and $\|x_j\|_{\Gamma,p} \leqslant 1$ for any n. By (3.36), (3.32) and (5.6), this means that $\left\|\partial_{\psi,1,p}(x_j)\right\|_{L_{cr}^p(\mathbb{E})} \leqslant 1$. We have to show that x belongs to dom $A_p^{\frac{1}{2}}$ and that $\|x\|_{\Gamma,p} \leqslant 1$, i.e. $\left\|\partial_{\psi,1,p}(x)\right\|_{L_{cr}^p(\mathbb{E})} \leqslant 1$. Since $\|\cdot\|_{L^p(VN(G))} \leqslant \|\cdot\|_{C_r^*(G)}$, note that

$$\|x_j - x\|_{L^p(VN(G))} \to 0.$$

Then $\left(x_j, \partial_{\psi,1,p}(x_j)\right)$ is a net of elements of the graph of $\partial_{\psi,1,p}$, which is bounded for the graph norm. Note that this graph is closed by Lemma 5.1 and convex, hence weakly closed by [167, Theorem 2.5.16]. Since bounded sets are weakly relatively compact by [167, Theorem 2.8.2], there exists a subnet $\left(x_{j_i}, \partial_{\psi,1,p}(x_{j_i})\right)$ which converges weakly to an element $\left(z, \partial_{\psi,1,p}(z)\right)$ of the graph of $\partial_{\psi,1,p}$. In particular, the net (x_{j_i}) converges weakly to z and the net $\left(\partial_{\psi,1,p}(x_{j_i})\right)$ converges weakly to $\partial_{\psi,1,p}(z)$. Then necessarily $x = z$. By (2.1), we conclude that x belongs to dom $\partial_{\psi,1,p} = \text{dom } A_p^{\frac{1}{2}}$. Moreover, with [167, Theorem 2.5.21], we have by weak convergence

$$\left\|\partial_{\psi,1,p}(x)\right\|_{L^p(VN(G))} \leqslant \liminf_i \left\|\partial_{\psi,1,p}(x_{j_i})\right\|_{L^p(VN(G))} \leqslant 1.$$

Now, suppose that G is weakly amenable and consider an approximating net (M_{φ_j}) satisfying the assumptions of the second part and the properties following Lemma 2.5. We will extend the Leibniz property (5.11) to elements x, y of dom $A_p^{\frac{1}{2}} \cap C_r^*(G)$. In particular, the net (M_{φ_j}) converges to $\text{Id}_{L^p(VN(G))}$ and to $\text{Id}_{C_r^*(G)}$ in the point-norm topologies with $\sup_j \left\|M_{\varphi_j}\right\|_{cb, L^p(VN(G)) \to L^p(VN(G))} \leqslant 1$. We have $x = \lim_j M_{\varphi_j}(x)$ and $y = \lim_j M_{\varphi_j}(y)$ where the convergence holds in $C_r^*(G)$ and in $L^p(VN(G))$.

Thus also $xy = \lim_j M_{\varphi_j}(x)M_{\varphi_j}(y)$ in $C_r^*(G)$ since $C_r^*(G)$ is a Banach algebra. Using (3.33) and (3.32) in last inequality, we obtain

$$\left\|M_{\varphi_j}(x)M_{\varphi_j}(y)\right\|_{\Gamma,p}$$

$$\overset{(5.11)}{\leqslant} \left\|M_{\varphi_j}(x)\right\|_{C_r^*(G)} \left\|M_{\varphi_j}(y)\right\|_{\Gamma,p} + \left\|M_{\varphi_j}(y)\right\|_{C_r^*(G)} \left\|M_{\varphi_j}(x)\right\|_{\Gamma,p}$$

$$\overset{(5.6)}{\leqslant} \|x\|_{C_r^*(G)} \max\left\{\left\|\Gamma(M_{\varphi_j}(y), M_{\varphi_j}(y))^{\frac{1}{2}}\right\|_{L^p}, \left\|\Gamma(M_{\varphi_j}(y)^*, M_{\varphi_j}(y)^*)^{\frac{1}{2}}\right\|_{L^p}\right\}$$

$$+ \|y\|_{C_r^*(G)} \max\left\{\left\|\Gamma(M_{\varphi_j}(x), M_{\varphi_j}(x))^{\frac{1}{2}}\right\|_{L^p}, \left\|\Gamma(M_{\varphi_j}(x)^*, M_{\varphi_j}(x)^*)^{\frac{1}{2}}\right\|_{L^p}\right\}$$

$$\overset{(3.36)(2.72)(2.77)}{\leqslant} \|x\|_{C_r^*(G)} \|y\|_{\Gamma,p} + \|y\|_{C_r^*(G)} \|x\|_{\Gamma,p}.$$

Using the lower semicontinuity of $\|\cdot\|_{\Gamma,p}$, we conclude that $xy \in \mathrm{dom}\, A_p^{\frac{1}{2}}$ and that

$$\|xy\|_{\Gamma,p} \stackrel{(3.35)}{\leqslant} \|x\|_{C_r^*(G)} \|y\|_{\Gamma,p} + \|y\|_{C_r^*(G)} \|x\|_{\Gamma,p}.$$

The theorem is proved.

$\square$

Remark 5.2 In the part 2, the argument proves that $\mathrm{dom}\, A_p^{\frac{1}{2}} \cap C_r^*(G)$ is a $*$-subalgebra of $C_r^*(G)$.

Remark 5.3 Note that in Theorem 5.1, in contrast to its counterpart for Schur multipliers, Theorem 5.3 below, we have a restriction of the exponent $p > n$, where n denotes the dimension of the representing Hilbert space. This restriction is in fact necessary in general as the following example shows.

Consider the abelian discrete group $G = \mathbb{Z}^2$, the canonical identification $b_\psi : \mathbb{Z}^2 \to \mathbb{R}^2$ of $\mathbb{Z}^2$ into $\mathbb{R}^2$ and the trivial homomorphism $\alpha : \mathbb{Z}^2 \to O(\mathbb{R}^2)$, $(n, m) \mapsto \mathrm{Id}_{\mathbb{R}^2}$. The associated length function is $\psi : \mathbb{Z}^2 \to \mathbb{R}_+$, $(n, m) \mapsto \psi(n, m) = \|b_\psi(n, m)\|_{\mathbb{R}^2}^2 = n^2 + m^2$. Now pick the critical exponent $p = 2 = \dim \mathbb{R}^2$. Note that the operator space $L^2(\mathrm{VN}(\mathbb{Z}^2)) = L^2(\mathbb{T}^2)$ is an operator Hilbert space by [188, Proposition 2.1 (iii)] and consequently has CCAP.[7] We also calculate

$$\mathrm{Gap}_\psi \stackrel{(5.12)}{=} \inf_{b_\psi(n,m)\neq b_\psi(n',m')} \|b_\psi(n, m) - b_\psi(n', m')\|_{\mathbb{R}^2}^2$$

$$= \inf_{(n,m)\neq(n',m')} \|(n, m) - (n', m')\|_{\mathbb{R}^2}^2 = 1 > 0.$$

However, $(C_r^*(\mathbb{Z}^2), \|\cdot\|_{\Gamma,2})$ is not a quantum compact metric space

Proof We will use [147, Theorem 2.4.3]. Indeed, the set

$$\left\{ x \in C_r^*(\mathbb{Z}^2)_0 \cap \mathrm{dom}\, A_2^{\frac{1}{2}} : \|x\|_{\Gamma,2} \leqslant 1 \right\}$$

is not bounded, so in particular, not relatively compact. Indeed, consider the double sequence $(\alpha_{n,m})_{n,m\in\mathbb{Z}}$ defined by

$$\alpha_{n,m} \stackrel{\mathrm{def}}{=} \frac{1}{(1 + n^2 + m^2) \log(2 + n^2 + m^2)} \delta_{(n,m)\neq(0,0)}. \tag{5.15}$$

---

[7] More generally, if $1 < p < \infty$ the operator space $L^p(\mathbb{T}^2)$ has CCAP by Junge and Ruan [121, Proposition 3.5] since an abelian group is weakly amenable with Cowling-Haagerup constant equal to 1.

Fix a number $N \in \mathbb{N}$ and consider the selfadjoint element

$$x_N \overset{\text{def}}{=} \sum_{n^2 + m^2 \leqslant N} \alpha_{n,m} e^{2\pi i \langle (n,m), \cdot \rangle}$$

of $C_r^*(\mathbb{Z}^2) = C(\mathbb{T}^2)$. We claim that (x_N) is an unbounded sequence in $C_r^*(\mathbb{Z}^2)_0 = C(\mathbb{T}^2)_0$ but bounded in $\|\cdot\|_{\Gamma,2}$-seminorm. Indeed, observe that x_N is a trigonometric polynomial without constant term, so belonging indeed to $C_r^*(\mathbb{Z}^2)_0 = C(\mathbb{T}^2)_0$. Moreover, $\|x_N\|_{C(\mathbb{T}^2)} \geqslant |x_N(0)| = \sum_{n^2+m^2 \leqslant N} \alpha_{n,m}$. But the series $\sum_{(n,m) \in \mathbb{Z}^2} \alpha_{n,m}$ diverges since using an integral test [89, Proposition 7.57] and a change of variables to polar coordinates we have

$$\sum_{(n,m) \in \mathbb{Z}^2} \alpha_{n,m} \cong \int_{\mathbb{R}^2} \frac{1}{(1 + x^2 + y^2) \log(2 + x^2 + y^2)} \, dx \, dy$$

$$= C \int_0^\infty \frac{1}{(1 + r^2) \log(2 + r^2)} r \, dr = \infty.$$

On the other hand, using Plancherel theorem in the second equality and again an integral test and a change of variables to polar coordinates, we obtain

$$\|x_N\|_{\Gamma,2}^2 \overset{(3.38)}{\cong} \left\| A^{\frac{1}{2}}(x_N) \right\|_{L^2(\mathbb{T}^2)}^2 = \left\| \left(\delta_{n^2+m^2 \leqslant N} \, \psi(n,m)^{\frac{1}{2}} \alpha_{n,m} \right)_{(n,m) \in \mathbb{Z}^2} \right\|_{\ell_{\mathbb{Z}^2}^2}^2$$

$$= \sum_{n^2+m^2 \leqslant N} (n^2 + m^2) \alpha_{n,m}^2 \overset{(5.15)}{\leqslant} \sum_{(n,m) \in \mathbb{Z}^2} \frac{1}{(1 + n^2 + m^2) \left[\log(2 + n^2 + m^2) \right]^2}$$

$$\cong \int_{\mathbb{R}^2} \frac{1}{(1 + x^2 + y^2) \left[\log(2 + x^2 + y^2) \right]^2} \, dx \, dy$$

$$= C \int_0^\infty \frac{1}{(1 + r^2) \left[\log(2 + r^2) \right]^2} r \, dr < \infty.$$

□

We refer to the survey [54] and to [119, pp. 628-629] for more information. Recall that a finitely generated discrete group G has rapid decay of order $r \geqslant 0$ if we have an estimate

$$\|x\|_{\mathrm{VN}(G)} \lesssim k^r \|x\|_{L^2(\mathrm{VN}(G))}$$

for any $x = \sum_{|s| \leqslant k} a_s \lambda_s$. Similarly using [119, Lemma 1.3.1], we can obtain the following result.

Theorem 5.2 *Let G be a finitely generated discrete group with rapid r-decay. Suppose that there exists $\beta > 0$ such that $2\frac{2r+1}{\beta} \leqslant p < \infty$ and that $\inf_{|s|=k} \psi(s) \gtrsim k^\beta$.*

1. *If $L^p(\mathrm{VN}(G))$ has CCAP and $\mathrm{VN}(G)$ has QWEP then $(C_r^*(G), \|\cdot\|_{\Gamma,p})$ is a quantum compact metric space.*
2. *If G is in addition weakly amenable with the approximating net $\varphi_j : G \to \mathbb{C}$ satisfying $\sup_j \|M_{\varphi_j}\|_{C_r^*(G)\to C_r^*(G)} < \infty$ and $\sup_j \|M_{\varphi_j}\|_{\mathrm{cb},L^p(\mathrm{VN}(G))\to L^p(\mathrm{VN}(G))} = 1$, then $(C_r^*(G), \|\cdot\|_{\Gamma,p})$ is a Leibniz quantum compact metric space.*

Proof Using the notation $n \overset{\mathrm{def}}{=} 2\frac{2r+1}{\beta}$, we have by [119, Lemma 1.3.1] the estimate

$$\|T_t\|_{L_0^2(\mathrm{VN}(G))\to\mathrm{VN}(G)} \lesssim \frac{1}{t^{\frac{2r+1}{2\beta}}} = \frac{1}{t^{\frac{n}{4}}}, \quad t > 0.$$

By Junge and Mei [119, Lemma 1.1.2], we deduce that $\|T_t\|_{L_0^1(\mathrm{VN}(G))\to\mathrm{VN}(G)} \lesssim \frac{1}{t^{\frac{n}{2}}}$. Moreover, by the argument of [119, p. 628] (using [74, (15.12.8.1)]), the operator A^{-1} is compact on $L_0^2(\mathrm{VN}(G))$. Since $2\frac{2r+1}{\beta} \leqslant p$, we can use Proposition 2.2 with $z = \frac{1}{2}$ and $q = \infty$. We deduce that $A^{-\frac{1}{2}} : L_0^p(\mathrm{VN}(G)) \to \mathrm{VN}(G)_0$ is compact. The end of the proof is similar to the proof of Theorem 5.1. $\qquad\square$

5.3 Gaps and Estimates of Norms of Schur Multipliers

Consider a semigroup $(T_t)_{t \geqslant 0}$ of selfadjoint unital completely positive Schur multipliers on $B(\ell_I^2)$ as in (2.37). In order to find quantum compact metric spaces associated with such a semigroup, we need some supplementary information on the semigroup. One important of them is the notion of the gap, which we study in this section. In all the section, we suppose $\dim H < \infty$. We define the gap of α by

$$\mathrm{Gap}_\alpha \overset{\mathrm{def}}{=} \inf_{\alpha_i-\alpha_j \neq \alpha_k-\alpha_l} \left\|(\alpha_i - \alpha_j) - (\alpha_k - \alpha_l)\right\|_H^2. \tag{5.16}$$

Note that by the proof of [10, Proposition 5.4] and [31, Theorem C.2.3], Gap_α is then independent of α, that is, if $\alpha : I \to H_\alpha$, $\beta : I \to H_\beta$ are two families such that (2.36) and (2.37) hold for both, then $\mathrm{Gap}_\alpha = \mathrm{Gap}_\beta$.

Lemma 5.2 *If $\dim H = n$ and $\mathrm{Gap}_\alpha > 0$, for any integer $k \geqslant 1$, we have*

$$\mathrm{card}\left\{\alpha_i - \alpha_j : k^2\mathrm{Gap}_\alpha \leqslant \|\alpha_i - \alpha_j\|_H^2 \leqslant (k+1)^2\mathrm{Gap}_\alpha\right\} \leqslant (5^n - 1)k^{n-1}.$$

Proof If B_n denotes the open Euclidean unit ball in H and if ξ_1, ξ_2 are distinct and can be written $\xi_1 = \alpha_i - \alpha_j$ and $\xi_2 = \alpha_k - \alpha_l$ for some $i, j, k, l \in I$, we have[8]

$$\left(\xi_1 + \frac{\sqrt{\text{Gap}_\alpha}}{2} B_n\right) \cap \left(\xi_2 + \frac{\sqrt{\text{Gap}_\alpha}}{2} B_n\right) = \emptyset.$$

Counting the maximum number of disjoint balls of radius $\frac{\sqrt{\text{Gap}_\alpha}}{2}$ in the annulus $\left((k+1)\sqrt{\text{Gap}_\alpha} + \frac{\sqrt{\text{Gap}_\alpha}}{2}\right) B_n - \left(k\sqrt{\text{Gap}_\alpha} - \frac{\sqrt{\text{Gap}_\alpha}}{2}\right) B_n$ combined with the binomial theorem, we obtain

$$\text{card}\left\{\alpha_i - \alpha_j : k^2\text{Gap}_\alpha \leqslant \|\alpha_i - \alpha_j\|_H^2 \leqslant (k+1)^2\text{Gap}_\alpha\right\}$$

$$\leqslant \frac{\text{vol}\left(\left((k+1)\sqrt{\text{Gap}_\alpha} + \frac{\sqrt{\text{Gap}_\alpha}}{2}\right) B_n\right) - \text{vol}\left(\left(k\sqrt{\text{Gap}_\alpha} - \frac{\sqrt{\text{Gap}_\alpha}}{2}\right) B_n\right)}{\text{vol}\left(\frac{\sqrt{\text{Gap}_\alpha}}{2} B_n\right)}$$

$$= (2k+3)^n - (2k-1)^n = \sum_{j=0}^{n}\binom{n}{j}(2k)^{n-j}\left(3^j - (-1)^j\right)$$

$$\leqslant k^{n-1}\sum_{j=1}^{n}\binom{n}{j}2^{n-j}\left(3^j - (-1)^j\right)$$

$$\leqslant k^{n-1}\sum_{j=0}^{n}\binom{n}{j}2^{n-j}3^j - k^{n-1}\sum_{j=0}^{n}2^{n-j}(-1)^j = (5^n - 1)k^{n-1}.$$

$\square$

We need the following lemma which says that Schur multipliers with $0-1$ entries of diagonal block rectangular shape are completely contractive.

Lemma 5.3 *Let I be a non-empty index set. Let $\{I_1, \ldots, I_N\}$ and $\{J_1, \ldots, J_N\}$ be two subpartitions of I, i.e. $I_k \subset I$ and $I_k \cap I_l = \emptyset$ for $k \neq l$ (and similarly for $J_1, \ldots, J_N$). The matrix $B = [b_{ij}]$, where*

$$b_{ij} = \begin{cases} 1 & \text{if } i \in I_k, \ j \in J_k \text{ for the same } k \\ 0 & \text{otherwise} \end{cases},$$

induces a completely contractive Schur multiplier $M_B : S_I^\infty \to S_I^\infty$ and is also completely contractive on S_I^p for any $1 \leqslant p \leqslant \infty$.

---

[8] If $\eta \in \left(\xi_1 + \frac{\sqrt{\text{Gap}_\alpha}}{2} B_n\right) \cap \left(\xi_2 + \frac{\sqrt{\text{Gap}_\alpha}}{2} B_n\right)$ then $|\xi_1 - \xi_2| \leqslant |\xi_1 - \eta| + |\eta - \xi_2| < \frac{\sqrt{\text{Gap}_\alpha}}{2} + \frac{\sqrt{\text{Gap}_\alpha}}{2} = \sqrt{\text{Gap}_\alpha}$ which is impossible.

Proof For any $i \in \{1, \ldots, N\}$ consider the orthogonal projections $P_i \colon \ell_I^2 \to \ell_I^2$, $(\xi_k)_{k \in I} \to (\xi_k 1_{k \subset I_i})_{k \in I}$ and $Q_i \colon \ell_I^2 \to \ell_I^2$, $(\xi_k)_{k \in I} \to (\xi_k 1_{k \in J_i})_{k \in I}$. If $i \neq j$, the ranges $P_i(\ell_I^2)$ and $P_j(\ell_I^2)$ (resp. $Q_i(\ell_I^2)$ and $Q_j(\ell_I^2)$) are orthogonal. Moreover, if $C \in S_I^\infty$ it is obvious that $M_B(C) = \sum_{i=1}^N P_i C Q_i$. Then for any $\xi, \eta \in \ell_I^2$, using the Cauchy-Schwarz Inequality in the second inequality, we obtain

$$\left| \langle M_B(C)\xi, \eta \rangle_{\ell_I^2} \right| = \left| \left\langle \sum_{i=1}^N P_i C Q_i \xi, \eta \right\rangle_{\ell_I^2} \right|$$

$$= \left| \sum_{i=1}^N \langle C Q_i \xi, P_i \eta \rangle_{\ell_I^2} \right| \leqslant \sum_{i=1}^N \left| \langle C Q_i \xi, P_i \eta \rangle_{\ell_I^2} \right|$$

$$\leqslant \left(\sum_{i=1}^N \| C Q_i \xi \|_{\ell_I^2}^2 \right)^{\frac{1}{2}} \left(\sum_{i=1}^N \| P_i \eta \|_{\ell_I^2}^2 \right)^{\frac{1}{2}}$$

$$\leqslant \| C \|_{S_I^\infty} \left(\sum_{i=1}^N \| Q_i \xi \|_{\ell_I^2}^2 \right)^{\frac{1}{2}} \left(\sum_{i=1}^N \| P_i \eta \|_{\ell_I^2}^2 \right)^{\frac{1}{2}}$$

$$= \| C \|_{S_I^\infty} \left\| \sum_{i=1}^N Q_i \xi \right\|_{\ell_I^2} \left\| \sum_{i=1}^N P_i \eta \right\|_{\ell_I^2} \leqslant \| C \|_{S_I^\infty} \| \xi \|_{\ell_I^2} \| \eta \|_{\ell_I^2}.$$

We infer that $\| M_B(C) \|_{S_I^\infty} \leqslant \| C \|_{S_I^\infty}$. Thus $M_B \colon S_I^\infty \to S_I^\infty$ is a contraction, and since it is a Schur multiplier, it is even a complete contraction [185, Corollary 8.8]. Moreover, since B has only real entries, M_B is symmetric, so also (completely) contractive on S_I^1, and by interpolation also on S_I^p for $1 \leqslant p \leqslant \infty$. $\qquad\square$

Recall that $\overline{\mathrm{Ran}\, A_\infty}$ satisfies (3.98).

Lemma 5.4 *If* $\dim H = n$, $\mathrm{Gap}_\alpha > 0$ *and if the function* $\widetilde{m} \colon H \to \mathbb{C}$ *satisfies*

$$\left| \widetilde{m}(\xi) \right| \leqslant c_n \| \xi \|_H^{-(n+\varepsilon)} \quad \text{for some } \varepsilon > 0,$$

then the Schur multiplier $M_{[\widetilde{m}(\alpha_i - \alpha_j)]} \colon S_I^\infty \to S_I^\infty$ *is completely bounded. Moreover, for any* $t > 0$, *we have*

$$\| T_t \|_{\mathrm{cb}, \overline{\mathrm{Ran}\, A_\infty} \to \overline{\mathrm{Ran}\, A_\infty}} \leqslant \frac{c(n)}{(\mathrm{Gap}_\alpha\, t)^{\frac{n}{2}}}.$$

Proof Let $x = \sum_{i,j \in I'} x_{ij} \otimes e_{ij} \in S_\ell^\infty(S_{I'}^\infty) \subset S_\ell^\infty(S_I^\infty)$, where $I' \subset I$ is finite and ℓ is an integer. Let $\{I_1, \ldots, I_N\}$ be the partition of I' corresponding to the equivalence relation $i \cong j \overset{\text{def}}{\Longleftrightarrow} \alpha_i = \alpha_j$, and let $J_1 = I_1, \ldots, J_N = I_N$. Let B_0 be the Schur multiplier symbol from Lemma 5.3 associated with these partitions.

Moreover for $\xi \in \alpha(I') - \alpha(I')$, we let $\{I_1^\xi, \ldots, I_M^\xi\}$ and $\{J_1^\xi, \ldots, J_M^\xi\}$ be the subsets of $\{I_1, \ldots, I_N\}$ such that $\alpha_i - \alpha_j = \xi \Leftrightarrow i \in I_k^\xi$ and $j \in J_k^\xi$ for the same k. We let B_ξ be the Schur multiplier symbol from Lemma 5.3 associated with these subpartitions. Then, using Lemma 5.2 and our growth assumption on $\tilde{m}$ in the third inequality, we obtain

$$\left\| \sum_{i,j \in I'} \tilde{m}(\alpha_i - \alpha_j) x_{ij} \otimes e_{ij} \right\|_{S_\ell^\infty(S_I^\infty)}$$

$$= \left\| \sum_{\substack{i,j \in I' \\ \alpha_i = \alpha_j}} \tilde{m}(\alpha_i - \alpha_j) x_{ij} \otimes e_{ij} + \sum_{k \geqslant 1} \sum_{\substack{\xi \in \alpha(I') - \alpha(I') \\ k^2 \mathrm{Gap}_\alpha \leqslant \|\xi\|^2 < (k+1)^2 \mathrm{Gap}_\alpha}} \tilde{m}(\alpha_i - \alpha_j) x_{ij} \otimes e_{ij} \right\|$$

$$\leqslant |\tilde{m}(0)| \left\| \sum_{\substack{i,j \in I' \\ \alpha_i = \alpha_j}} x_{ij} \otimes e_{ij} \right\|_{S_\ell^\infty(S_I^\infty)}$$

$$+ \sum_{k \geqslant 1} \sum_{\substack{\xi \in \alpha(I') - \alpha(I') \\ k^2 \mathrm{Gap}_\alpha \leqslant \|\xi\|^2 < (k+1)^2 \mathrm{Gap}_\alpha}} |\tilde{m}(\xi)| \left\| \sum_{\substack{i,j \in I' \\ \alpha_i - \alpha_j = \xi}} x_{ij} \otimes e_{ij} \right\|_{S_\ell^\infty(S_I^\infty)}$$

$$= |\tilde{m}(0)| \left\| (\mathrm{Id}_{S_\ell^\infty} \otimes M_{B_0})(x) \right\|_{S_\ell^\infty(S_I^\infty)}$$

$$+ \sum_{k \geqslant 1} \sum_{\substack{\xi \in \alpha(I') - \alpha(I') \\ k^2 \mathrm{Gap}_\alpha \leqslant \|\xi\|^2 < (k+1)^2 \mathrm{Gap}_\alpha}} |\tilde{m}(\xi)| \left\| (\mathrm{Id}_{S_\ell^\infty} \otimes M_{B_\xi})(x) \right\|_{S_\ell^\infty(S_I^\infty)} \qquad (5.17)$$

$$\leqslant \left(|\tilde{m}(0)| + \sum_{k \geqslant 1} \sum_{\substack{\xi \in \alpha(I') - \alpha(I') \\ k^2 \mathrm{Gap}_\alpha \leqslant \|\xi\|^2 < (k+1)^2 \mathrm{Gap}_\alpha}} |\tilde{m}(\xi)| \right) \|x\|_{S_\ell^\infty(S_I^\infty)}$$

$$\leqslant 5^n \left(|\tilde{m}(0)| + \sum_{k \geqslant 1} k^{n-1} \left(k \sqrt{\mathrm{Gap}_\alpha} \right)^{-(n+\varepsilon)} \right) \|x\|_{S_\ell^\infty(S_I^\infty)}$$

$$= 5^n \left(|\tilde{m}(0)| + \sum_{k \geqslant 1} k^{-1-\varepsilon} \left(\mathrm{Gap}_\alpha \right)^{-\frac{n-\varepsilon}{2}} \right) \|x\|_{S_\ell^\infty(S_I^\infty)} = c_{n,\varepsilon}(\mathrm{Gap}_\alpha) \|x\|_{S_\ell^\infty(S_I^\infty)}.$$

Note that the Riemann series $\sum_{k \geqslant 1} k^{-1-\varepsilon}$ converges.

For the second assertion, we use the function $\widetilde{m} = e^{-t\|\cdot\|_H}$. Since $x \in \overline{\operatorname{Ran} A_\infty}$ we may ignore the previous term $|\widetilde{m}(0)|$ and from (5.17), it suffices to use the inequality

$$\sum_{k=1}^{\infty} k^{n-1} e^{-t \operatorname{Gap}_\alpha k^2} \leqslant c(n)(\operatorname{Gap}_\alpha t)^{-\frac{n}{2}}.$$

$\square$

5.4 Seminorms Associated to Semigroups of Schur Multipliers

In this section, we consider again a markovian semigroup $(T_t)_{t \geqslant 0}$ of Schur multipliers on $B(\ell_I^2)$ satisfying Proposition 2.4. Recall from Sect. 2.7 that we have a generator A_p and also a carré du champ Γ. We shall introduce a family of seminorms that will give rise in Sect. 5.5 below to some quantum locally compact metric spaces. First suppose $2 \leqslant p < \infty$. For any $x \in \operatorname{dom} A_p^{\frac{1}{2}}$ we let

$$\|x\|_{\Gamma,p} \overset{\text{def}}{=} \max\left\{\left\|\Gamma(x,x)^{\frac{1}{2}}\right\|_{S_I^p}, \left\|\Gamma(x^*,x^*)^{\frac{1}{2}}\right\|_{S_I^p}\right\}. \tag{5.18}$$

We recall that we use here the definition of Γ from (2.91) extended to $\operatorname{dom} A_p^{\frac{1}{2}}$ in Lemma 3.25. We use $L_{cr}^p(\mathbb{E}) \overset{\text{def}}{=} L_c^p(\mathbb{E}) \cap L_r^p(\mathbb{E})$. If $x \in \operatorname{dom} A_p^{\frac{1}{2}}$, note that

$$\|x\|_{\Gamma,p} \tag{5.19}$$

$$\overset{(5.18)(3.93)}{=} \max\left\{\left\|(\mathbb{E}[\partial_{\alpha,q,p}(x)^* \partial_{\alpha,q,p}(x)])^{\frac{1}{2}}\right\|_{S_I^p}, \left\|(\mathbb{E}[\partial_{\alpha,q,p}(x^*)^* \partial_{\alpha,q,p}(x^*)])^{\frac{1}{2}}\right\|_{S_I^p}\right\}$$

$$\overset{(3.89)}{=} \max\left\{\left\|(\mathbb{E}[\partial_{\alpha,q,p}(x)^* \partial_{\alpha,q,p}(x)])^{\frac{1}{2}}\right\|_{S_I^p}, \left\|(\mathbb{E}[\partial_{\alpha,q,p}(x)\partial_{\alpha,q,p}(x)^*])^{\frac{1}{2}}\right\|_{S_I^p}\right\}$$

$$\overset{(2.72)}{=} \left\|\partial_{\alpha,q,p}(x)\right\|_{L_{cr}^p(\mathbb{E})}.$$

For the proof of Proposition 5.5, we shall need the following lemma which can be proved as Lemma 5.1. Note that $\operatorname{dom} \partial_{\alpha,q,p}$ was defined in Proposition 3.11.

Lemma 5.5 Let $2 \leqslant p < \infty$ and $-1 \leqslant q \leqslant 1$. The mapping $\partial_{\alpha,q,p} \colon \operatorname{dom} \partial_{\alpha,q,p} \subset S_I^p \to L_{cr}^p(\mathbb{E})$ is closed.

Now, we can state the following result.

Proposition 5.5 *Suppose* $2 \leqslant p < \infty$.

1. $\|\cdot\|_{\Gamma,p}$ *is a seminorm on* $\operatorname{dom} A_p^{\frac{1}{2}}$.

2. $\|\cdot\|_{\Gamma,p}$ *is lower semicontinuous on* $\operatorname{dom} A_p^{\frac{1}{2}}$ *equipped with the topology induced by the weak topology of* S_I^p.

3. *For any* $x, y \in \operatorname{dom} A_p^{\frac{1}{2}}$, *we have*

$$\|xy\|_{\Gamma,p} \leqslant \|x\|_{S_I^\infty} \|y\|_{\Gamma,p} + \|y\|_{S_I^\infty} \|x\|_{\Gamma,p}. \tag{5.20}$$

Proof 1. The proof is similar to the one of Proposition 5.3.

2. Let $x \in S_I^p$ and (x_β) be a net of elements of $\operatorname{dom} A_p^{\frac{1}{2}}$ such that (x_β) converges to x for the weak topology of S_I^p with

$$\left\|\partial_{\alpha,q,p}(x_\beta)\right\|_{L_{cr}^p(\mathbb{E})} \overset{(5.19)}{=} \|x_\beta\|_{\Gamma,p} \leqslant 1.$$

We have to show that x belongs to $\operatorname{dom} A_p^{\frac{1}{2}}$ and that $\left\|\partial_{\alpha,q,p}(x)\right\|_{L_{cr}^p(\mathbb{E})} \leqslant 1$. Note that the net $(\partial_{\alpha,q,p}(x_\beta))$ is bounded in $L_{cr}^p(\mathbb{E})$ so admits a weakly convergent subnet $(\partial_{\alpha,q,p}(x_{\beta_i}))$ by [167, Theorem 2.5.16]. Then $(x_{\beta_i}, \partial_{\alpha,q,p}(x_{\beta_i}))$ is a weakly convergent net in the graph of $\partial_{\alpha,q,p}$. Note that this graph is closed according to Lemma 5.5 and convex, hence weakly closed by [167, Theorem 2.5.16]. Thus the limit of $(x_{\beta_i}, \partial_{\alpha,q,p}(x_{\beta_i}))$ belongs again to the graph and is of the form $(z, \partial_{\alpha,q,p}(z))$ for some $z \in \operatorname{dom} \partial_{\alpha,q,p}$. In particular, (x_{β_i}) converges weakly to z and $\partial_{\alpha,q,p}(x_{\beta_i})$ converges weakly to $\partial_{\alpha,q,p}(z)$. We infer that $x = z$. We deduce by (2.1), that x belongs to $\operatorname{dom} \partial_{\alpha,q,p} = \operatorname{dom} A_p^{\frac{1}{2}}$. Moreover, with [167, Theorem 2.5.21], we have

$$\left\|\partial_{\alpha,q,p}(x)\right\|_{L_{cr}^p(\mathbb{E})} \leqslant \liminf_i \left\|\partial_{\alpha,q,p}(x_{\beta_i})\right\|_{L_{cr}^p(\mathbb{E})} \leqslant 1.$$

3. For elements $x, y \in M_{I,\mathrm{fin}}$, the proof is identical to the one of Proposition 5.4. We will extend the Leibniz property (5.20) to elements x, y of $\operatorname{dom} A_p^{\frac{1}{2}}$. We have $x = \lim_J \mathcal{T}_J(x)$ and $y = \lim_J \mathcal{T}_J(y)$ where the convergence holds in S_I^∞ and in S_I^p. Thus also $xy = \lim_J \mathcal{T}_J(x)\mathcal{T}_J(y)$ in S_I^p since S_I^p is a Banach algebra for the usual product of operators by [38, p. 225]. We obtain

$$\|\mathcal{T}_J(x)\mathcal{T}_J(y)\|_{\Gamma,p} \overset{(5.20)}{\leqslant} \|\mathcal{T}_J(x)\|_{S_I^\infty} \|\mathcal{T}_J(y)\|_{\Gamma,p} + \|\mathcal{T}_J(y)\|_{S_I^\infty} \|\mathcal{T}_J(x)\|_{\Gamma,p}$$

$$\overset{(5.19)}{\leqslant} \|x\|_{S_I^\infty} \left\|\partial_{\alpha,q,p}(\mathcal{T}_J(y))\right\|_{L_{cr}^p(\mathbb{E})} + \|y\|_{S_I^\infty} \left\|\partial_{\alpha,q,p}(\mathcal{T}_J(x))\right\|_{L_{cr}^p(\mathbb{E})}$$

$$\overset{(3.90)(2.74)(2.70)}{\leqslant} \|x\|_{S_I^\infty} \|y\|_{\Gamma,p} + \|y\|_{S_I^\infty} \|x\|_{\Gamma,p}. \tag{5.21}$$

Using the lower semicontinuity of $\|\cdot\|_{\Gamma,p}$, we conclude that $xy \in \operatorname{dom} A_p^{\frac{1}{2}}$ and that

$$\|xy\|_{\Gamma,p} \leqslant \|x\|_{S_I^\infty} \|y\|_{\Gamma,p} + \|y\|_{S_I^\infty} \|x\|_{\Gamma,p}.$$

$\square$

Remark 5.4   The argument proves that $\operatorname{dom} A_p^{\frac{1}{2}}$ is a $*$-subalgebra of S_I^∞.

Suppose $1 < p \leqslant 2$. For any element x of $\mathrm{M}_{I,\mathrm{fin}}$, we let

$$\|x\|_{\Gamma,p} \stackrel{\mathrm{def}}{=} \inf_{x=y+z} \left\{ \left\|\Gamma(y,y)^{\frac{1}{2}}\right\|_{S_I^p}^{\frac{p}{2}} + \left\|\Gamma(z^*,z^*)^{\frac{1}{2}}\right\|_{S_I^p}^{\frac{p}{2}} \right\}^{\frac{2}{p}} \qquad (5.22)$$

where the infimum is taken over all $y, z \in \mathrm{M}_{I,\mathrm{fin}}$ such that $x = y + z$.

Proposition 5.6   *For $1 < p \leqslant 2$, $\|\cdot\|_{\Gamma,p}$ is a $\frac{p}{2}$-seminorm on $\mathrm{M}_{I,\mathrm{fin}}$, that is, the triangle inequality holds under the form $\|x + y\|_{\Gamma,p}^{\frac{p}{2}} \leqslant \|x\|_{\Gamma,p}^{\frac{p}{2}} + \|y\|_{\Gamma,p}^{\frac{p}{2}}$.*

Proof   The homogeneity of $\|\cdot\|_{\Gamma,p}$ can be shown as in the case $p \geqslant 2$ and is left to the reader. Let us turn to the triangle inequality. For any $y, y' \in \mathrm{M}_{I,\mathrm{fin}}$, the part 4 of Lemma 2.17 says that

$$\left\|\Gamma(y,y')\right\|_{S_I^{\frac{p}{2}}}^{\frac{p}{2}} \leqslant \left\|\Gamma(y,y)^{\frac{1}{2}}\right\|_{S_I^p}^{\frac{p}{2}} \left\|\Gamma(y',y')^{\frac{1}{2}}\right\|_{S_I^p}^{\frac{p}{2}}. \qquad (5.23)$$

Recall that $S_I^{\frac{p}{2}}$ is a $\frac{p}{2}$-normed space for $p < 2$. With this inequality, we can now estimate for arbitrary $y, y' \in \mathrm{M}_{I,\mathrm{fin}}$

$$\left\|\Gamma(y+y',y+y')^{\frac{1}{2}}\right\|_{S_I^p}^p \stackrel{(2.25)}{=} \left\|\Gamma(y+y',y+y')\right\|_{S_I^{\frac{p}{2}}}^{\frac{p}{2}}$$

$$= \left\|\Gamma(y,y) + \Gamma(y',y') + \Gamma(y,y') + \Gamma(y',y)\right\|_{S_I^{\frac{p}{2}}}^{\frac{p}{2}}$$

$$\leqslant \left\|\Gamma(y,y)\right\|_{S_I^{\frac{p}{2}}}^{\frac{p}{2}} + \left\|\Gamma(y',y')\right\|_{S_I^{\frac{p}{2}}}^{\frac{p}{2}} + \left\|\Gamma(y,y')\right\|_{S_I^{\frac{p}{2}}}^{\frac{p}{2}} + \left\|\Gamma(y',y)\right\|_{S_I^{\frac{p}{2}}}^{\frac{p}{2}}$$

$$\stackrel{(5.23)}{\leqslant} \left\|\Gamma(y,y)^{\frac{1}{2}}\right\|_{S_I^p}^p + \left\|\Gamma(y',y')^{\frac{1}{2}}\right\|_{S_I^p}^p + 2\left\|\Gamma(y,y)^{\frac{1}{2}}\right\|_{S_I^p}^{\frac{p}{2}} \left\|\Gamma(y',y')^{\frac{1}{2}}\right\|_{S_I^p}^{\frac{p}{2}}$$

$$= \left(\left\|\Gamma(y,y)^{\frac{1}{2}}\right\|_{S_I^p}^{\frac{p}{2}} + \left\|\Gamma(y',y')^{\frac{1}{2}}\right\|_{S_I^p}^{\frac{p}{2}} \right)^2.$$

Taking the square roots, we obtain

$$\left\| \Gamma(y+y', y+y')^{\frac{1}{2}} \right\|_{S_I^p}^{\frac{p}{2}} \leqslant \left\| \Gamma(y,y)^{\frac{1}{2}} \right\|_{S_I^p}^{\frac{p}{2}} + \left\| \Gamma(y',y')^{\frac{1}{2}} \right\|_{S_I^p}^{\frac{p}{2}}. \tag{5.24}$$

Now consider some elements x and x' of $M_{I,\mathrm{fin}}$ and some decompositions $x = y+z$ and $x' = y' + z'$ where $y, z, y', z' \in M_{I,\mathrm{fin}}$. We have $x + x' = y + y' + z + z'$. We obtain

$$\left\| x + x' \right\|_{\Gamma,p}^{\frac{p}{2}} \overset{(5.22)}{\leqslant} \left\| \Gamma(y+y', y+y')^{\frac{1}{2}} \right\|_{S_I^p}^{\frac{p}{2}} + \left\| \Gamma((z+z')^*, (z+z')^*)^{\frac{1}{2}} \right\|_{S_I^p}^{\frac{p}{2}}$$

$$\overset{(5.24)}{\leqslant} \left\| \Gamma(y,y)^{\frac{1}{2}} \right\|_{S_I^p}^{\frac{p}{2}} + \left\| \Gamma(y',y')^{\frac{1}{2}} \right\|_{S_I^p}^{\frac{p}{2}} + \left\| \Gamma(z^*,z^*)^{\frac{1}{2}} \right\|_{S_I^p}^{\frac{p}{2}} + \left\| \Gamma(z'^*, z'^*)^{\frac{1}{2}} \right\|_{S_I^p}^{\frac{p}{2}}.$$

Passing to the infimum, we conclude that

$$\left\| x + x' \right\|_{\Gamma,p}^{\frac{p}{2}} \leqslant \left\| x \right\|_{\Gamma,p}^{\frac{p}{2}} + \left\| x' \right\|_{\Gamma,p}^{\frac{p}{2}}.$$

$\square$

Remark 5.5 It is not clear if the results of [123] could be used to replace "$\frac{p}{2}$-seminorm" by "seminorm". No attempts were made.

Suppose $1 < p \leqslant \infty$. Let I be an index set. Recall that $\overline{\mathrm{Ran}\, A_p}$ satisfies (3.98). Suppose that the map $\alpha \colon I \to H$ of (2.36) is injective. In this case, $\overline{\mathrm{Ran}\, A_p}$ is the subspace of elements of S_I^p with null diagonal. Furthermore, $\mathrm{Ker}\, A_p$ is the diagonal of S_I^p. We will also use the space $\left(\mathrm{dom}\, A_p^{\frac{1}{2}} \right)_0 = \mathrm{dom}\, A_p^{\frac{1}{2}} \cap \overline{\mathrm{Ran}\, A_p}$ of those matrices in $\mathrm{dom}\, A_p^{\frac{1}{2}}$ that have null diagonal.

We define the seminorm $\| \cdot \|_{\Gamma,\alpha,p} \overset{\mathrm{def}}{=} \| \cdot \|_{\Gamma,p} \oplus 0$ on the space $\left(\mathrm{dom}\, A_p^{\frac{1}{2}} \right)_0 \oplus \mathbb{C}\mathrm{Id}_{\ell_I^2}$ (on $M_{I,\mathrm{fin}} \oplus \mathbb{C}\mathrm{Id}_{\ell_I^2}$ if $1 < p < 2$). That is, for any element $x = x_0 + \lambda \mathrm{Id}_{\ell_I^2}$ of the space $\left(\mathrm{dom}\, A_p^{\frac{1}{2}} \right)_0 \oplus \mathbb{C}\mathrm{Id}_{\ell_I^2}$ (with in addition $x_0 \in M_{I,\mathrm{fin}}$ in case $1 < p < 2$), we let

$$\| x \|_{\Gamma,\alpha,p} \overset{\mathrm{def}}{=} \begin{cases} \inf_{x_0=y+z} \left\{ \left\| \Gamma(y,y)^{\frac{1}{2}} \right\|_{S_I^p}^{\frac{p}{2}} + \left\| \Gamma(z^*,z^*)^{\frac{1}{2}} \right\|_{S_I^p}^{\frac{p}{2}} \right\}^{\frac{2}{p}} & \text{if } 1 < p \leqslant 2 \\[2ex] \max \left\{ \left\| \Gamma(x_0,x_0)^{\frac{1}{2}} \right\|_{S_I^p}, \left\| \Gamma(x_0^*, x_0^*)^{\frac{1}{2}} \right\|_{S_I^p} \right\} & \text{if } p \geqslant 2 \end{cases}, \tag{5.25}$$

where the infimum is taken over all $y, z \in M_{I,\mathrm{fin}}$ such that $x_0 = y + z$. Note that when $p = \infty$, we suppose in addition that I is finite, in order to have a well-defined domain of the previous seminorm.

Proposition 5.7 *Suppose* $2 \leqslant p \leqslant \infty$. *If* $p = \infty$, *we suppose that* I *is finite. We have*

$$\left\{ x \in \left(\operatorname{dom} A_p^{\frac{1}{2}} \right)_0 \oplus \mathbb{C}\operatorname{Id}_{\ell_I^2} : \|x\|_{\Gamma,\alpha,p} = 0 \right\} = \mathbb{C}\operatorname{Id}_{\ell_I^2}. \tag{5.26}$$

Proof *Case* $2 \leqslant p < \infty$. Let $x = x_0 + \lambda \operatorname{Id}_{\ell_I^2}$ be an element of $\left(\operatorname{dom} A_p^{\frac{1}{2}} \right)_0 \oplus \mathbb{C}\operatorname{Id}_{\ell_I^2}$. According to Theorem 3.5, we have

$$\left\| A_p^{\frac{1}{2}}(x_0) \right\|_{S_I^p} \overset{(3.92)}{\lesssim} \max \left\{ \left\| \Gamma(x_0, x_0)^{\frac{1}{2}} \right\|_{S_I^p}, \left\| \Gamma(x_0^*, x_0^*)^{\frac{1}{2}} \right\|_{S_I^p} \right\} \overset{(5.18)}{=} \|x_0\|_{\Gamma,p}.$$

This implies that when $\|x\|_{\Gamma,\alpha,p} = 0$, that is $\|x_0\|_{\Gamma,p} = 0$, we have $\left\| A_p^{\frac{1}{2}}(x_0) \right\|_{S_I^p} = 0$ and finally $A_p^{\frac{1}{2}}(x_0) = 0$. By (2.21), we deduce that x_0 belongs to $\operatorname{Ker} A_p$, hence that x_0 is diagonal. Since x_0 has a null diagonal, we conclude that $x_0 = 0$. The reverse inclusion is true by (5.25).

Case $p = \infty$ *and* I *finite.* Fix some $2 < p_0 < \infty$. Let $x = x_0 + \lambda \operatorname{Id}_{\ell_I^2}$ be an element of $\left(\operatorname{dom} A_p^{\frac{1}{2}} \right)_0 \oplus \mathbb{C}\operatorname{Id}_{\ell_I^2}$. Using Theorem 3.5 in the first inequality, we can write

$$\left\| A^{\frac{1}{2}}(x_0) \right\|_{S_I^{p_0}} \lesssim_{p_0} \max \left\{ \left\| \Gamma(x_0, x_0)^{\frac{1}{2}} \right\|_{S_I^{p_0}}, \left\| \Gamma(x_0^*, x_0^*)^{\frac{1}{2}} \right\|_{S_I^{p_0}} \right\} \tag{5.27}$$

$$\lesssim_I \max \left\{ \left\| \Gamma(x_0, x_0)^{\frac{1}{2}} \right\|_{S_I^\infty}, \left\| \Gamma(x_0^*, x_0^*)^{\frac{1}{2}} \right\|_{S_I^\infty} \right\} \overset{(5.25)}{=} \|x_0\|_{\Gamma,\alpha,\infty}.$$

The end of the proof is similar to the case $2 \leqslant p < \infty$. $\square$

5.5   Quantum Metric Spaces Associated to Semigroups of Schur Multipliers

As observed in [196, p. 3], the linear space of all selfadjoint elements of a unital C^*-algebra is an order-unit space. So if I is an index set (finite or infinite) then $(S_I^\infty)_{\mathrm{sa}} \oplus \mathbb{R}\operatorname{Id}_{\ell_I^2}$ is an order-unit space. Moreover, by Rieffel [197, Proposition 2.3], a subspace of an order-unit space containing the order-unit is also an order-unit space. We conclude that $(\overline{\operatorname{Ran} A_\infty})_{\mathrm{sa}} \oplus \mathbb{R}\operatorname{Id}_{\ell_I^2}$ is an order-unit space. Now, we will prove in this section that $(\overline{\operatorname{Ran} A_\infty})_{\mathrm{sa}} \oplus \mathbb{R}\operatorname{Id}_{\ell_I^2}$ equipped with the restriction of $\|\cdot\|_{\Gamma,\alpha,p}$ defined by (5.25) on this subspace (also denoted by $\|\cdot\|_{\Gamma,\alpha,p}$) is a quantum locally compact metric space.

Moreover, since Lipschitz pairs in the sense of Definition 5.1 are only defined in the literature for seminorms and not for quasi-seminorms, we restrict our investigation on $\|\cdot\|_{\Gamma,\alpha,p}$ in the next Theorem 5.3 to the case $2 \leqslant p \leqslant \infty$.

Recall that if $(T_t)_{t \geqslant 0}$ is a strongly continuous bounded semigroup on a Banach space X with negative generator A, then a particular case of [98, Proposition 3.3.5] gives the following representation for the operator $(A + \varepsilon)^{-\beta}$

$$(A + \varepsilon)^{-\beta}(x) = \frac{1}{\Gamma(\beta)} \int_0^\infty t^{\beta-1} e^{-\varepsilon t} T_t(x) \, dt, \quad x \in \text{Ran } X, \ \varepsilon > 0 \qquad (5.28)$$

where $\text{Re } \beta > 0$.

Lemma 5.6 *Let $(T_t)_{t \geqslant 0}$ be a strongly continuous bounded semigroup on a Banach space X with negative generator A. Assume that $\|T_t\|_{\text{Ran}(A) \to X} \lesssim \frac{1}{t^d}$ for some $d > \frac{1}{2}$, $t \geqslant 1$, where $\text{Ran}(A)$ is normed as a subspace of X. Then $A^{-\frac{1}{2}}$, initially defined on $\text{Ran } A$, extends to a bounded operator on $\overline{\text{Ran } A}$.*

Proof Note that according to Sect. 2.1, the projection onto the null space of A is given by $P(x) = \lim_{t \to \infty} \frac{1}{t} \int_0^t T_s(x) \, ds$. If $x \in X$, we obtain

$$\|P(x)\| \leqslant \limsup_{t \to \infty} \frac{1}{t} \int_0^1 \|T_s(x)\| \, ds + \frac{1}{t} \int_1^t \|T_s(x)\| \, ds$$

$$\leqslant \limsup_{t \to \infty} \frac{1}{t} \int_1^t s^{-d} \, ds \, \|x\| \lesssim \limsup_{t \to \infty} \frac{1}{t} t^{-d+1}$$

$$= 0.$$

Thus, A is injective and $A^{-1} : \text{Ran } A \to X$ is well-defined. If $x \in \text{Ran } A$, we have

$$\left\| (A + \varepsilon)^{-\frac{1}{2}}(x) \right\|_X \overset{(5.28)}{=} \left\| \frac{1}{\Gamma(\frac{1}{2})} \int_0^\infty t^{-\frac{1}{2}} e^{-\varepsilon t} T_t(x) \, dt \right\|_X$$

$$\leqslant \frac{1}{\Gamma(\frac{1}{2})} \int_0^\infty t^{-\frac{1}{2}} e^{-\varepsilon t} \|T_t(x)\|_X \, dt$$

$$\lesssim \int_0^1 t^{-\frac{1}{2}} e^{-\varepsilon t} \|T_t(x)\|_X \, dt + \int_1^\infty t^{-\frac{1}{2}} e^{-\varepsilon t} \|T_t(x)\|_X \, dt$$

$$\lesssim \int_0^1 t^{-\frac{1}{2}} e^{-\varepsilon t} \|x\|_X \, dt + \int_1^\infty t^{-\frac{1}{2}-d} e^{-\varepsilon t} \|x\|_X \, dt$$

$$\leqslant \left(\int_0^1 t^{-\frac{1}{2}} e^{-\varepsilon t} \, dt + \int_1^\infty t^{-(d+\frac{1}{2})} e^{-\varepsilon t} \, dt \right) \|x\|_X .$$

The quantity is uniformly bounded in ε. Hence we obtain $\left\| (A + \varepsilon)^{-1} \right\|_{\mathrm{Ran}\, A \to X} \lesssim 1$. Moreover, if $x \in \mathrm{Ran}\, A$ we have

$$(A + \varepsilon)^{-1} x - A^{-1} x = (A + \varepsilon)^{-1} \big(A - (A + \varepsilon) \big) A^{-1} x$$

$$= -\varepsilon (A + \varepsilon)^{-1} A^{-1} x \xrightarrow[\varepsilon \to 0]{} 0.$$

By [167, Corollary 2.3.34], the operator A^{-1}, hence $A^{-\frac{1}{2}}$, extends to a bounded operator on $\overline{\mathrm{Ran}\, A}$. $\square$

In the proof of the following result, we consider $\overline{\mathrm{Ran}\, A_\infty}$ living in the non-unital C^*-algebra S_I^∞.

Lemma 5.7 *Let* $1 < p < \infty$*. Assume that* I *is infinite, that the Hilbert space* H *is of finite dimension* $n \in \mathbb{N}$ *and that* $\mathrm{Gap}_\alpha > 0$*.*

1. *The operator* $A_p^{-\frac{1}{2}} : \overline{\mathrm{Ran}\, A_p} \to \overline{\mathrm{Ran}\, A_p} \subset \overline{\mathrm{Ran}\, A_\infty}$ *is bounded.*
2. *Suppose* $-1 \leqslant q \leqslant 1$*. Let* $B_p = (\mathrm{Id}_{L^p(\Gamma_q(H))} \otimes A_p)|_{\overline{\mathrm{Ran}\, \partial_{\alpha,q,p}}}$*:* $\mathrm{dom}\, B_p \subset$ $\overline{\mathrm{Ran}\, \partial_{\alpha,q,p}} \to \overline{\mathrm{Ran}\, \partial_{\alpha,q,p}}$*. Then the operator* $B_p^{-\frac{1}{2}}$ *is bounded.*

Proof

1. We begin by showing that $A_2^{-\frac{1}{2}} : \overline{\mathrm{Ran}\, A_2} \to \overline{\mathrm{Ran}\, A_2}$ is bounded. For any $i, j \in I$ such that $\alpha_i - \alpha_j \neq 0$, we have by the gap condition $\| \alpha_i - \alpha_j \|_H = \| \alpha_i - \alpha_j - 0 \| \geqslant \mathrm{Gap}_\alpha$. Since $a_{ij} = \| \alpha_i - \alpha_j \|_H^2$, we deduce that the diagonal operator $A_2^{-\frac{1}{2}}$ is bounded on the Hilbert space $\overline{\mathrm{Ran}\, A_2}$.

   Next we show that $A^{-\frac{1}{2}} : \overline{\mathrm{Ran}\, A_\infty} \to \overline{\mathrm{Ran}\, A_\infty}$ is bounded. Indeed, note that $(T_t)_{t \geqslant 0}$ extends to a bounded strongly continuous semigroup on $\overline{\mathrm{Ran}\, A_\infty}$. Moreover, Lemma 5.4 (in the case where H is one-dimensional, we inject H beforehand into a two-dimensional Hilbert space so that this lemma yields an estimate $\| T_t \|_{\infty \to \infty} \leqslant C t^{-\frac{n}{2}}$ with $\frac{n}{2} > \frac{1}{2}$) together with Lemma 5.6 yield that $A^{-\frac{1}{2}}$ is bounded on $\overline{\mathrm{Ran}\, A_\infty}$.

   Now, if $2 < p < \infty$, it suffices to interpolate the operator $A^{-\frac{1}{2}}$ between levels 2 and ∞ and to note that we have $\overline{\mathrm{Ran}\, A_p} = (\overline{\mathrm{Ran}\, A_2}, \overline{\mathrm{Ran}\, A_\infty})_\theta$ for the right $\theta \in [0, 1)$. Indeed, the interpolation identity follows from the fact that the $\overline{\mathrm{Ran}\, A_p}$ are complemented subspaces of S_I^p by the complementary projections of the projections of the one of (2.17) (note that $(T_t)_{t \geqslant 0}$ is a bounded semigroup on S_I^p for $2 \leqslant p \leqslant \infty$), and that these spectral projections are compatible for different values of $2 \leqslant p \leqslant \infty$. Then if $1 < p < 2$, we use that $\overline{\mathrm{Ran}\, A_p}$ is the dual space of $\overline{\mathrm{Ran}\, A_{p^*}}$ via the usual duality bracket using (2.8) and [167, page 94], and that $A_p^{-\frac{1}{2}}$ on the first space is the adjoint of $A_{p^*}^{-\frac{1}{2}}$ on the second space.
2. Note that by (4.55), B_p is the generator of the (strongly continuous) semigroup $((\mathrm{Id} \otimes T_t)_{t \geqslant 0} | \overline{\mathrm{Ran}\, \partial_{\alpha,q,p}})_{t \geqslant 0}$. Now, we use Lemma 5.4 to obtain the bound

$\|\mathrm{Id} \otimes T_t\|_{\infty \to \infty} \lesssim \frac{1}{t^d}$ for some $d > \frac{1}{2}$. With Proposition 4.19, we have $\overline{\mathrm{Ran}\, B_p} = \overline{\mathrm{Ran}\, \partial_{\alpha,q,p}}$. By Lemma 5.6 and an argument as in the first part of the proof, we conclude that $B_p^{-\frac{1}{2}} : \overline{\mathrm{Ran}\, \partial_{\alpha,q,p}} \to \overline{\mathrm{Ran}\, \partial_{\alpha,q,p}}$ is bounded.

$\square$

In the following theorem, we consider the abelian C*-subalgebra $\mathfrak{M}$ of S_I^∞ consisting of its diagonal operators. We recall that we restrict to $p \geqslant 2$ to have a seminorm $\|\cdot\|_{\Gamma,\alpha,p}$ in the usual sense.

Theorem 5.3 *Assume that the Hilbert space H is of finite dimension and that* $\mathrm{Gap}_\alpha > 0$.

1. *Suppose $2 \leqslant p \leqslant \infty$ and that I is finite. Then $\big((\mathrm{Ran}\, A_\infty)_{\mathrm{sa}} \oplus \mathbb{R}\mathrm{Id}_{\ell_I^2}, \|\cdot\|_{\Gamma,\alpha,p}\big)$ is a Leibniz quantum compact metric space.*

2. *Suppose $2 \leqslant p < \infty$ and that I is infinite. We consider the abelian C*-subalgebra $\mathfrak{M}$ of diagonal operators of S_I^∞. Then $\big((\overline{\mathrm{Ran}\, A_\infty})_{\mathrm{sa}} \oplus \mathbb{R}\mathrm{Id}_{\ell_I^2}, \|\cdot\|_{\Gamma,\alpha,p}, \mathfrak{M}\big)$ is a Leibniz quantum locally compact metric space satisfying the boundedness condition.*

Proof

1. Note that $\mathrm{Ran}\, A_\infty$ is a subspace of the space of matrices of S_I^∞ with null diagonal and that we have a contractive projection from S_I^∞ onto the diagonal of S_I^∞. So it is easy to check that $\mathrm{Tr} \oplus \mathrm{Id}_\mathbb{R} : \mathrm{Ran}\, A_\infty \oplus \mathbb{R}\mathrm{Id}_{\ell_I^2} \to \mathbb{R}$, $x + \lambda \mapsto \lambda$ is a state of the order-unit space $(\mathrm{Ran}\, A_\infty)_{\mathrm{sa}} \oplus \mathbb{R}\mathrm{Id}_{\ell_I^2}$. By Proposition 5.1, it suffices to show that

$$\big\{x \in \mathrm{dom}\, \|\cdot\|_{\Gamma,\alpha,p} : \|x\|_{\Gamma,\alpha,p} \leqslant 1, \ (\mathrm{Tr} \oplus \mathrm{Id}_\mathbb{C})(x) = 0\big\} \text{ is rel. compact in}$$

$$S_I^\infty \oplus \mathbb{C}\mathrm{Id}_{\ell_I^2}. \tag{5.29}$$

Case $2 \leqslant p < \infty$. The operator $A^{-\frac{1}{2}} : \mathrm{Ran}\, A_p \to \mathrm{Ran}\, A_\infty$ is bounded by finite-dimensionality. Applying this operator to the closed unit ball of $\mathrm{Ran}\, A_p = \mathrm{Ran}\, A_p^{\frac{1}{2}}$ equipped with the norm $\|\cdot\|_{S_I^p}$, we obtain that the set[9]

$$\big\{x \in \mathrm{Ran}\, A_\infty : \big\|A_p^{\frac{1}{2}}(x)\big\|_{S_I^p} \leqslant 1\big\}$$

---

[9] Since I is finite, we have $\mathrm{dom}\, A_p^{\frac{1}{2}} = S_I^p$. Write $x = A^{-\frac{1}{2}} A_p^{\frac{1}{2}} x$.

is bounded in S_I^∞. Note that for any $x \in S_I^p$ we have

$$\left\|A^{\frac{1}{2}}(x)\right\|_{S_I^p} \overset{(3.95)}{\lesssim_p} \max\left\{\left\|\Gamma(x,x)^{\frac{1}{2}}\right\|_{S_I^p}, \left\|\Gamma(x^*,x^*)^{\frac{1}{2}}\right\|_{S_I^p}\right\} \overset{(5.18)}{=} \|x\|_{\Gamma,\alpha,p}.$$

(5.30)

We deduce from this inequality that

$$\left\{x \in \operatorname{Ran} A_\infty : \|x\|_{\Gamma,\alpha,p} \leqslant 1\right\}$$

(5.31)

is bounded, hence by finite dimensionality, relatively compact in S_I^∞. We have

$$\left\{x \in \operatorname{Ran} A_\infty : \|x\|_{\Gamma,\alpha,p} \leqslant 1\right\}$$
$$= \left\{x \in \operatorname{dom} \|\cdot\|_{\Gamma,\alpha,p} : \|x\|_{\Gamma,\alpha,p} \leqslant 1,\ (\operatorname{Tr} \oplus \operatorname{Id}_{\mathbb{C}})(x) = 0\right\}.$$

We deduce (5.29).

Case $p = \infty$. Fix some $2 < p_0 < \infty$. By finite-dimensionality, the operator $A^{-\frac{1}{2}}: \operatorname{Ran} A_{p_0} \to \operatorname{Ran} A_\infty$ is bounded. This implies that the set[10]

$$\left\{x \in \operatorname{Ran} A_\infty : \left\|A_{p_0}^{\frac{1}{2}}(x)\right\|_{S_I^{p_0}} \leqslant 1\right\}$$

is bounded in S_I^∞. The inequality (5.27) says that the set

$$\left\{x \in \operatorname{Ran} A_\infty \oplus \mathbb{C}\operatorname{Id}_{\ell_I^2} : \|x\|_{\Gamma,\alpha,\infty} \leqslant 1 \text{ and } (\operatorname{Tr} \oplus \operatorname{Id}_{\mathbb{C}})(x) = 0\right\}$$
$$= \left\{x \in \operatorname{Ran} A_\infty : \|x\|_{\Gamma,\alpha,\infty} \leqslant 1\right\}$$

is bounded, i.e. relatively compact in S_I^∞. The desired result follows from Proposition 5.1.

For the Leibniz rule, see the end of the proof.

2. Suppose that J is a finite subset of the infinite set I. We denote by $\mu_J : S_I^\infty \to \mathbb{C}$ the *normalized* "partial trace" where $\mu_J(x)$ is $1/|J|$ times the sum of the diagonal entries of index belonging to J. This state is local since $\mu_J(\operatorname{diag}(0,\ldots,0,1,\ldots,1,0,\ldots,0)) = 1$ and since $\operatorname{diag}(0,\ldots,0,1,\ldots,1,0,\ldots,0)$ belongs to $\mathfrak{M}$. This state extends to the unitization by $\mu \overset{\text{def}}{=} \mu_J \oplus \operatorname{Id}_{\mathbb{C}}$. We claim that if $a,b \in \mathfrak{M}$ are compactly supported then the set

$$a\left\{x \in \overline{\operatorname{Ran} A_\infty} \oplus \mathbb{C}\operatorname{Id}_{\ell_I^2} : \|x\|_{\Gamma,\alpha,p} \leqslant 1,\ \mu_J(x) = 0\right\}b$$

(5.32)

---

[10] Since I is finite, we have $\operatorname{dom} A_{p_0}^{\frac{1}{2}} = S_I^{p_0}$.

is relatively compact. Recall that the Gelfand spectrum of $\mathfrak{M}$ is I since we can identify $\mathfrak{M}$ with $c_0(I)$. Thus, compactly supported elements are those lying in span$\{e_{ii} : i \in I\}$. So we can write $a = \sum_{i \in J_a} a_i e_{ii}$ and $b = \sum_{j \in J_b} b_j e_{jj}$ where J_a and J_b are finite subsets of I. Now, we see that the previous set is contained in span$\{e_{ij} : i \in J_a,\ j \in J_b\}$, which is a finite-dimensional space since J_a and J_b are finite. Thus it suffices to show boundedness of the previous set (5.32). Using the fact that matrices in $\overline{\mathrm{Ran}\,A_\infty}$ have null diagonal, Lemma 5.7 and Leibniz' rule (5.20) it is easy to check that (5.32) is equal to

$$a\{x \in \overline{\mathrm{Ran}\,A_\infty} \cap \mathrm{M}_{J_a, J_b} : \|x\|_{\Gamma, \alpha, p} \leqslant 1\}b \tag{5.33}$$

where M_{J_a, J_b} is the space of $J_a \times J_b$ matrices. Essentially by mimicking the proof of (5.31) with Lemma 5.7, the subset $\{x \in \overline{\mathrm{Ran}\,A_\infty} \cap \mathrm{M}_{J_a, J_b} : \|x\|_{\Gamma, \alpha, p} \leqslant 1\}$ is bounded. Hence the subset (5.33) is also bounded.

Note we show that this quantum locally compact metric space satisfies Definition 5.8. According to Lemma 5.7, the operator $A_p^{-\frac{1}{2}} : \overline{\mathrm{Ran}\,A_p} \to \overline{\mathrm{Ran}\,A_p} \subset \overline{\mathrm{Ran}\,A_\infty}$ is bounded. So the set

$$\left\{x \in \overline{\mathrm{Ran}\,A_\infty} \cap \mathrm{dom}\,A_p^{\frac{1}{2}} : \left\|A_p^{\frac{1}{2}}(x)\right\|_{S_I^p} \leqslant 1\right\}$$

is bounded in S_I^∞. By (5.30), we deduce that

$$\left\{x \in \overline{\mathrm{Ran}\,A_\infty} \cap \mathrm{dom}\,A_p^{\frac{1}{2}} : \|x\|_{\Gamma, \alpha, p} \leqslant 1\right\}$$

is bounded in S_I^∞. So Definition 5.8 is satisfied.

We check that the seminorm $\|\cdot\|_{\Gamma, \alpha, p}$ equally satisfies the Leibniz inequality

$$\|xy\|_{\Gamma, \alpha, p} \leqslant \|x\|_{\mathrm{B}(\ell_I^2)} \|y\|_{\Gamma, \alpha, p} + \|x\|_{\Gamma, \alpha, p} \|y\|_{\mathrm{B}(\ell_I^2)}, \quad x, y \in \left(\mathrm{dom}\,A_p^{\frac{1}{2}}\right)_0 \oplus \mathbb{C}\mathrm{Id}_{\ell_I^2}. \tag{5.34}$$

Indeed, if we write $x = x_0 + \lambda \mathrm{Id}_{\ell_I^2}$ and $y = y_0 + \mu \mathrm{Id}_{\ell_I^2}$ then we have

$$xy = x_0 y_0 + \lambda y_0 + \mu x_0 + \lambda \mu \mathrm{Id}_{\ell_I^2}. \tag{5.35}$$

Thus, we have with limits in S_I^p norm, observing that

$$\mathcal{T}_J(\mathrm{Id}_{\ell_I^2})\mathcal{T}_J(z) = \mathcal{T}_J(z)\mathcal{T}_J(\mathrm{Id}_{\ell_I^2}) = \mathcal{T}_J(z)$$

and that the product $S_I^p \times S_I^p \to S_I^p$ is continuous [38, p. 225],

$$xy - \lambda\mu\mathrm{Id}_{\ell_I^2} = x_0 y_0 + \lambda y_0 + \mu x_0 = \lim_J \left[\mathcal{T}_J(x_0)\mathcal{T}_J(y_0) + \lambda\mathcal{T}_J(y_0) + \mu\mathcal{T}_J(x_0) \right]$$

$$= \lim_J \left[\mathcal{T}_J(x_0 + \lambda\mathrm{Id}_{\ell_I^2})\mathcal{T}_J(y_0) + \mu\mathcal{T}_J(x_0) \right]$$

$$= \lim_J \left[\mathcal{T}_J(x_0 + \lambda\mathrm{Id}_{\ell_I^2})\mathcal{T}_J(y_0 + \mu\mathrm{Id}_{\ell_I^2}) - \lambda\mu\mathcal{T}_J(\mathrm{Id}_{\ell_I^2})\mathcal{T}_J(\mathrm{Id}_{\ell_I^2}) \right].$$

Therefore, we have $xy - \lambda\mu\mathrm{Id}_{\ell_I^2} = \lim_J \mathcal{T}_J(x)\mathcal{T}_J(y) - \mathcal{T}_J(\lambda\mu\mathrm{Id}_{\ell_I^2})$, limit in S_I^p. Moreover, similarly to (5.21) we have

$$\left\| \mathcal{T}_J(x)\mathcal{T}_J(y) - \mathcal{T}_J(\lambda\mu\mathrm{Id}_{\ell_I^2}) \right\|_{\Gamma,p} \leqslant \left\| \mathcal{T}_J(x)\mathcal{T}_J(y) \right\|_{\Gamma,p} + \left\| \mathcal{T}_J(\lambda\mu\mathrm{Id}_{\ell_I^2}) \right\|_{\Gamma,p}$$

$$\overset{(5.20)(2.92)}{\leqslant} \left\| \mathcal{T}_J(x) \right\|_{S_I^\infty} \left\| \mathcal{T}_J(y) \right\|_{\Gamma,p} + \left\| \mathcal{T}_J(y) \right\|_{S_I^\infty} \left\| \mathcal{T}_J(x) \right\|_{\Gamma,p}$$

$$\overset{(2.92)}{\leqslant} \left\| x \right\|_{\mathrm{B}(\ell_I^2)} \left\| \mathcal{T}_J(y_0) \right\|_{\Gamma,p} + \left\| y \right\|_{\mathrm{B}(\ell_I^2)} \left\| \mathcal{T}_J(x_0) \right\|_{\Gamma,p}$$

$$\leqslant \left\| x \right\|_{\mathrm{B}(\ell_I^2)} \left\| y_0 \right\|_{\Gamma,p} + \left\| y \right\|_{\mathrm{B}(\ell_I^2)} \left\| x_0 \right\|_{\Gamma,p}$$

$$\overset{(5.25)}{=} \left\| x \right\|_{\mathrm{B}(\ell_I^2)} \left\| y \right\|_{\Gamma,\alpha,p} + \left\| y \right\|_{\mathrm{B}(\ell_I^2)} \left\| x \right\|_{\Gamma,\alpha,p}.$$

Then by the lower semicontinuity of the seminorm $\|\cdot\|_{\Gamma,p}$ from Proposition 5.5, we deduce that $xy - \lambda\mu\mathrm{Id}_{\ell_I^2}$ belongs to $\mathrm{dom}\, A_p^{\frac{1}{2}}$ and that

$$\left\| xy \right\|_{\Gamma,\alpha,p} \overset{(5.25)(5.35)}{=} \left\| xy - \lambda\mu\mathrm{Id}_{\ell_I^2} \right\|_{\Gamma,p} \leqslant \left\| x \right\|_{\mathrm{B}(\ell_I^2)} \left\| y \right\|_{\Gamma,\alpha,p} + \left\| y \right\|_{\mathrm{B}(\ell_I^2)} \left\| x \right\|_{\Gamma,\alpha,p}.$$

Finally, we check both properties of the point 1.(b) of Definition 5.3 as in the proof of [148, Proposition 2.17]. Thus, $(\overline{\mathrm{Ran}\, A_\infty} \oplus \mathbb{R}\mathrm{Id}_{\ell_I^2}, \|\cdot\|)$ is a unital Leibniz pair.

We check the semicontinuity property. To this end, let (x_n) be a sequence in $\left(\mathrm{dom}\, A_p^{\frac{1}{2}} \right)_0 \oplus \mathbb{C}\mathrm{Id}_{\ell_I^2} \subset \mathrm{B}(\ell_I^2)$ such that $x_n \to x$ in $\mathrm{B}(\ell_I^2)$ with $\|x_n\|_{\Gamma,\alpha,p} \leqslant 1$ for all n. We can write $x_n = x_{n,0} + \lambda_n \mathrm{Id}_{\ell_I^2}$ with $x_{n,0} \in \left(\mathrm{dom}\, A_p^{\frac{1}{2}} \right)_0$ and $\lambda_n \in \mathbb{C}$. Since $\overline{\mathrm{Ran}\, A_\infty} \oplus \mathbb{C}\mathrm{Id}_{\ell_I^2}$ is a closed subspace of $\mathrm{B}(\ell_I^2)$, we have $x = x_0 + \lambda\mathrm{Id}_{\ell_I^2}$ for some $x_0 \in \overline{\mathrm{Ran}\, A_\infty}$ and $\lambda \in \mathbb{C}$. Since the map $\overline{\mathrm{Ran}\, A_\infty} \oplus \mathbb{C}\mathrm{Id}_{\ell_I^2} \to \mathbb{C}$, $x_0 + \lambda\mathrm{Id}_{\ell_I^2} \mapsto \lambda$ is continuous, we have $\lambda_n \to \lambda$, and therefore also $x_{n,0} \to x_0$ in S_I^∞. Appealing to Proposition 5.5, it suffices to show that $x_{n,0}$ converges weakly to x_0 in S_I^p. Since the convergence already holds in S_I^∞ norm, it suffices to show that $x_{n,0}$ is bounded in S_I^p. But this follows from $\left\| A_p^{\frac{1}{2}}(x_{n,0}) \right\|_{S_I^p} \lesssim \|x_{n,0}\|_{\Gamma,p} \overset{(5.25)}{=} \|x_n\|_{\Gamma,\alpha,p} \leqslant 1$ and the fact that $A_p^{-\frac{1}{2}}: \overline{\mathrm{Ran}\, A_p} \to \overline{\mathrm{Ran}\, A_p}$ is bounded according to Lemma 5.7.

1. (Leibniz rule) Note that if I is finite, then the previous proof of Leibniz quantum locally compact metric space shows that $\left(\left(\overline{\operatorname{Ran} A_\infty}\right)_{\mathrm{sa}} \oplus \mathbb{R}\mathrm{Id}_{\ell_I^2}, \|\cdot\|_{\Gamma,\alpha,p}\right)$ is a Leibniz quantum compact metric space, in case $2 \leqslant p < \infty$. We indicate how the same proof also works for $p = \infty$. Note first that when I is finite, we have $S_I^\infty = \mathrm{M}_{I,\mathrm{fin}}$, so that the domain of $\|\cdot\|_{\Gamma,\alpha,\infty}$ is the full space $\overline{\operatorname{Ran} A_\infty} \oplus \mathbb{C}\mathrm{Id}_{\ell_I^2}$: as in the proof of Proposition 5.5, one can show that for any $x, y \in \mathrm{M}_{I,\mathrm{fin}}$, one has

$$\|xy\|_{\Gamma,\infty} \leqslant \|x\|_{S_I^\infty} \|y\|_{\Gamma,\alpha,\infty} + \|y\|_{S_I^\infty} \|x\|_{\Gamma,\alpha,\infty}.$$

Thus, in particular the same holds if $x, y \in \mathrm{dom}\,\|\cdot\|_{\Gamma,\alpha,\infty} = \overline{\operatorname{Ran} A_\infty} \oplus \mathbb{C}\mathrm{Id}_{\ell_I^2}$. For the lower semicontinuity, if suffices to note that the seminorm satisfies $\|x\|_{\Gamma,\alpha,\infty} \lesssim \|x\|_{S_I^\infty}$, and by the reversed triangle inequality, we also have $\big|\, \|x\|_{\Gamma,\alpha,\infty} - \|y\|_{\Gamma,\alpha,\infty}\, \big| \leqslant \|x - y\|_{\Gamma,\alpha,\infty}$.

$\square$

5.6 Gaps of Some Markovian Semigroups of Schur and Fourier Multipliers

In this section, we study some typical examples of markovian semigroups of Schur and Fourier multipliers. We equally calculate their gaps (5.12) and (5.16) and examine the injectivity of their Hilbert space representation. This information is important for applications to compact quantum metric spaces in Sects. 5.2 and 5.5.

Heat Schur Semigroup and Poisson Schur Semigroup In the following, I is equal to $\{1, \ldots, n\}$, $\mathbb{N}$ or $\mathbb{Z}$. We consider the heat Schur semigroup $(T_t)_{t \geqslant 0}$ acting on $\mathrm{B}(\ell_I^2)$ defined by

$$T_t : [x_{ij}] \mapsto \left[\mathrm{e}^{-|i-j|^2 t} x_{ij}\right]. \tag{5.36}$$

Moreover, we also consider the Poisson Schur semigroup $(T_t)_{t \geqslant 0}$ acting on $\mathrm{B}(\ell_I^2)$ defined by

$$T_t : [x_{ij}] \mapsto \left[\mathrm{e}^{-|i-j|t} x_{ij}\right]. \tag{5.37}$$

These two semigroups are examples of noncommutative diffusion semigroups consisting of Schur multipliers.

Indeed, for the first one, we can take the real Hilbert space $H = \mathbb{R}$ and put $\alpha_i = i$ for $\alpha \in I$. So we have $\|\alpha_i - \alpha_j\|_{\mathbb{R}}^2 = |i - j|^2$. For the second one, we can consider the real Hilbert space $H = \ell_I^2$ and put $\alpha_i = \sum_{k=0}^i e_k$ for $i \in I$ with $i \geqslant 0$ and $\alpha_i = \sum_{k=0}^{|i|} e_{-k}$ for $i \in I$ with $i < 0$ (if I contains negative elements). Then for

$i > j \geqslant 0$, using the orthogonality of the e_l's in the third equality, we have

$$\|\alpha_i - \alpha_j\|_{\ell_I^2}^2 = \left\| \sum_{k=0}^{i} e_k - \sum_{k=0}^{j} e_k \right\|_{\ell_I^2}^2 = \left\| \sum_{k=j+1}^{i} e_k \right\|_{\ell_I^2}^2$$

$$= \sum_{k=j+1}^{i} \|e_k\|_{\ell_I^2}^2 = \sum_{k=j+1}^{i} 1 = i - j = |i - j|.$$

In a similar way, we obtain for i, j in general position that $\|\alpha_i - \alpha_j\|_{\ell_I^2}^2 = |i - j|$. Note that both mappings α are injective.

Lemma 5.8 *Consider the previous heat Schur semigroup from (5.36). We have* $\mathrm{Gap}_\alpha = 1$.

Proof We have

$$\mathrm{Gap}_\alpha \overset{(5.16)}{=} \inf_{\alpha_i - \alpha_j \neq \alpha_k - \alpha_l} \left\| (\alpha_i - \alpha_j) - (\alpha_k - \alpha_l) \right\|_{\mathbb{R}}^2 = \inf_{i-j \neq k-l} \left\| (i - j) - (k - l) \right\|_{\mathbb{R}}^2$$

which is clearly equal to 1.   □

We will now calculate Gap_α for the previous Poisson Schur semigroup. First we have

Lemma 5.9 *Consider the previous Poisson Schur semigroup from (5.37) with $I = \mathbb{Z}$. Then for any $i, j, k, l \in \mathbb{Z}$, $i \geqslant j \geqslant k \geqslant l$, we have*

$$\langle \alpha_i - \alpha_j, \alpha_k - \alpha_l \rangle_H = 0.$$

Proof Indeed, we have

$$\langle \alpha_i - \alpha_j, \alpha_k - \alpha_l \rangle_H = \left\langle \sum_{r=0}^{i} e_r - \sum_{r=0}^{j} e_r, \sum_{r=0}^{k} e_r - \sum_{r=0}^{l} e_r \right\rangle_H = \left\langle \sum_{r=j+1}^{i} e_r, \sum_{r=l+1}^{k} e_r \right\rangle_H = 0$$

□

Lemma 5.10 *Consider again the previous Poisson Schur semigroup from (5.37). We have* $\mathrm{Gap}_\alpha = 1$.

Proof It is clear that it suffices to examine the case $I = \mathbb{Z}$.

For the inequality $\mathrm{Gap}_\alpha \leqslant 1$, it suffices to take $i = 1$, $j = k = l = 0$, in which case we have

$$\left\| (\alpha_i - \alpha_j) - (\alpha_k - \alpha_l) \right\|_H^2 = \|\alpha_i - \alpha_j\|_H^2 = |i - j| = 1.$$

For the reverse inequality $\mathrm{Gap}_\alpha \geqslant 1$, consider any i, j, k, l such that $\alpha_i - \alpha_j \neq \alpha_k - \alpha_l$. We want to estimate $\left\| (\alpha_i - \alpha_j) - (\alpha_k - \alpha_l) \right\|^2$ from below. Using that $\|x\| = \|-x\|$ and exchanging names of indices, we can assume without loss of generality that $\max(i, j, k, l) = i$.

First case: We have $l = \min(i, j, k, l)$. Then exchanging the names of indices j and k if necessary, we have $i \geqslant j \geqslant k \geqslant l$. Thus, according to Lemma 5.9, we have $\langle \alpha_i - \alpha_j, \alpha_k - \alpha_l \rangle = 0$. Consequently,

$$\left\| (\alpha_i - \alpha_j) - (\alpha_k - \alpha_l) \right\|^2 = \|\alpha_i - \alpha_j\|^2 + \|\alpha_k - \alpha_l\|^2 - 2\langle \alpha_i - \alpha_j, \alpha_k - \alpha_l \rangle$$
$$= |i - j| + |k - l| - 0.$$

Clearly, if this expression is 0, then $i = j$ and $k = l$, which is excluded by $\alpha_i - \alpha_j \neq \alpha_k - \alpha_l$, or $\alpha_i - \alpha_k \neq \alpha_j - \alpha_l$ in case that we had exchanged names of indices. In any other case, this expression is $\geqslant 1$ since i, j, k, l take entire values.

Second case: We have $\min(i, j, k, l) \in \{j, k\}$ and l is the second smallest value among i, j, k, l. Then exchanging the names of indices j and k if necessary, we can suppose $\min(i, j, k, l) = k$. So we have $i \geqslant j \geqslant l \geqslant k$, and thus by Lemma 5.9 $\langle \alpha_i - \alpha_j, \alpha_l - \alpha_k \rangle = 0$. We calculate

$$\left\| (\alpha_i - \alpha_j) - (\alpha_k - \alpha_l) \right\|^2 = \|\alpha_i - \alpha_j\|^2 + \|\alpha_k - \alpha_l\|^2 + 2\langle \alpha_i - \alpha_j, \alpha_l - \alpha_k \rangle$$
$$= |i - j| + |k - l| + 0.$$

We argue as before to see that this quantity is $\geqslant 1$.

Third case: We have $\min(i, j, k, l) \in \{j, k\}$ and l is the second biggest value among i, j, k, l. Then exchanging the names of indices j and k if necessary, we can suppose $\min(i, j, k, l) = k$. So we have $i \geqslant l \geqslant j \geqslant k$, and thus by Lemma 5.9

$$\langle \alpha_i - \alpha_j, \alpha_k - \alpha_l \rangle = \langle \alpha_i - \alpha_l, \alpha_k - \alpha_l \rangle + \langle \alpha_l - \alpha_j, \alpha_k - \alpha_l \rangle$$
$$= -0 + \langle \alpha_l - \alpha_j, \alpha_k - \alpha_j \rangle + \langle \alpha_l - \alpha_j, \alpha_j - \alpha_l \rangle$$
$$= -0 - 0 - \|\alpha_l - \alpha_j\|^2 = -|l - j|.$$

Then we calculate

$$\left\| (\alpha_i - \alpha_j) - (\alpha_k - \alpha_l) \right\|^2 = \|\alpha_i - \alpha_j\|^2 + \|\alpha_k - \alpha_l\|^2 - 2\langle \alpha_i - \alpha_j, \alpha_k - \alpha_l \rangle$$
$$= |i - j| + |k - l| + 2|l - j| = i - j + l - k + 2l - 2j = i - k + 3(l - j).$$

Again we argue as before to see that this quantity is $\geqslant 1$.   □

Markovian Semigroups of Herz-Schur Multipliers vs. Markovian Semigroups of Fourier Multipliers Let G be a discrete group and $\psi : G \to \mathbb{R}$ be a function. Suppose that $\psi(e) = 0$. Recall that by [31, Corollary C.4.19], the function ψ is conditionally negative definite if and only if for any $t \geqslant 0$, the function $\mathrm{e}^{-t\psi}$ is of positive type. On the one hand, by [71, Proposition 4.2] that exactly means that ψ induces a completely positive Fourier multiplier $T_t \overset{\text{def}}{=} M_{\exp(-t\psi)} \colon \text{VN}(G) \to \text{VN}(G)$ for any $t \geq 0$. On the other hand, by [31, Definition C.4.1] and [46, Theorem D.3] that is equivalent to say that ψ induces a completely positive Herz-Schur multiplier $T_t^{\text{HS}} \colon B(\ell_G^2) \to B(\ell_G^2)$ for any $t \geq 0$ (whose symbol is $\left[\exp(-t\phi(s,r))\right]_{s,r \in G}$ where $\phi(s,r) \overset{\text{def}}{=} \psi(s^{-1}r)$ for any $s,r \in G$). In this case, by Proposition 2.3, we obtain a markovian semigroup $(T_t)_{t \geq 0}$ of Fourier multipliers and it is easy to check and well-known that we obtain a markovian semigroup $(T_t^{\text{HS}})_{t \geq 0}$ of Herz-Schur multipliers. Hence there is a bijective correspondence between markovian semigroups of Fourier multipliers on $\text{VN}(G)$ and markovian semigroups of Herz-Schur multipliers on $B(\ell_G^2)$. So any triple (b, π, H) associated to a markovian semigroup $(T_t)_{t \geq 0}$ of Fourier multipliers by Proposition 2.3 gives a couple (α, H) associated to $(T_t^{\text{HS}})_{t \geq 0}$ and conversely. More precisely, if $\psi(s) = \left\|b_\psi(s)\right\|_H^2$ then for any $s,r \in G$

$$\phi(s,r) = \psi(s^{-1}r) = \left\|b_\psi(s^{-1}r)\right\|_H^2 \overset{(2.33)}{=} \left\|b_\psi(s^{-1}) + \pi_{s^{-1}}(b_\psi(r))\right\|_H^2$$

$$\overset{(2.35)}{=} \left\|-\pi_{s^{-1}}(b_\psi(s)) + \pi_{s^{-1}}(b_\psi(r))\right\|_H^2 = \left\|b_\psi(r) - b_\psi(s)\right\|_H^2.$$

So we can consider the couple (b_ψ, H) for the semigroup $(T_t^{\text{HS}})_{t \geq 0}$.

Next we compare the gaps of Herz-Schur and Fourier markovian semigroups as we encountered them in Sects. 5.5 and 5.2. We will see in Proposition 5.10 that a strict inequality may occur in the following result.

Proposition 5.8 *Let $(T_t)_{t \geq 0}$ and $(T_t^{\text{HS}})_{t \geq 0}$ be markovian semigroups of Fourier multipliers and Herz-Schur multipliers as previously. Consider a triple (b, π, H) for the first semigroup and the couple (α, H) such that $\alpha \overset{\text{def}}{=} b$ for the second semigroup. We have*

$$\text{Gap}_\alpha \leq \text{Gap}_\psi \tag{5.38}$$

where Gap_α and Gap_ψ are defined in (5.16) and (5.12).

Proof For any $i, j, l, k \in G$, we have $\alpha_i - \alpha_j = b(i) - b(j) \overset{(2.33)}{=} \pi_j(b(j^{-1}i)) = \pi_j(\alpha_{j^{-1}i})$ and $\alpha_k - \alpha_l = \pi_l(\alpha_{l^{-1}k})$. Thus, we have

$$\text{Gap}_\alpha \overset{(5.16)}{=} \inf_{\alpha_i - \alpha_j \neq \alpha_k - \alpha_l} \left\|(\alpha_i - \alpha_j) - (\alpha_k - \alpha_l)\right\|_H^2$$

$$= \inf_{\alpha_i - \alpha_j \neq \alpha_k - \alpha_l} \left\|\pi_j(\alpha_{j^{-1}i}) - \pi_l(\alpha_{l^{-1}k})\right\|_H^2$$

$$= \inf_{i,j,k,l} \left\|\pi_{l^{-1}j}(\alpha_{j^{-1}i}) - \alpha_{l^{-1}k}\right\|_H^2 = \inf_{j,s,r} \left\|\pi_j(\alpha_s) - \alpha_r\right\|_H^2,$$

where the infimum is taken over those $j, s, r \in G$ such that the considered norm is $\neq 0$. On the other hand, we have, since $b(s) = \alpha_s$ and by considering $j = e$ and $\pi_e = \mathrm{Id}_H$ in the inequality

$$\mathrm{Gap}_\psi \overset{(5.12)}{=} \inf_{b(s) \neq b(t)} \|b(s) - b(t)\|_H^2 = \inf_{\alpha_s \neq \alpha_r} \|\alpha_s - \alpha_r\|_H^2$$

$$\geqslant \inf_{j,s,r} \|\pi_j(\alpha_s) - \alpha_r\|_H^2 = \mathrm{Gap}_\alpha,$$

(again infimum over non-zero quantities).   □

Finite-Dimensional Hilbert Spaces and Coboundary Cocycles over Infinite Groups: A Dichotomy Between Non-injective Cocycles and the Condition $\mathrm{Gap}_\psi = 0$ Now, we consider particular markovian semigroups of Fourier multipliers and their corresponding semigroups of Herz-Schur multipliers and ask whether the representations $b = \alpha$ are injective and calculate the gaps Gap_ψ and Gap_α defined in (5.16) and (5.12). We start with a general observation. Consider an infinite discrete group G and a finite-dimensional orthogonal representation $\pi : G \to O(H)$. Consider a 1-cocycle $b : G \to H$ with respect to π. So we have a markovian semigroup of Fourier multipliers on $\mathrm{VN}(G)$ and a semigroup of Herz-Schur multipliers on $\mathrm{B}(\ell_G^2)$.

For 1-coboundaries, the situation is not as nice as in Sect. 5.2. Indeed, we have the following proposition.

Proposition 5.9 *Let G be an infinite discrete group and $b : G \to H$ be a 1-coboundary with respect to some finite-dimensional orthogonal representation $\pi : G \to O(H)$. If b is injective, then $\mathrm{Gap}_\psi = \mathrm{Gap}_\alpha = 0$. So if $\mathrm{Gap}_\psi > 0$ or $\mathrm{Gap}_\alpha > 0$ then b is non-injective.*

Proof By definition [31, Definition 2.2.3], there exists $\xi \in H$ such that $b(s) = \pi_s(\xi) - \xi$ for any $s \in G$. Clearly, the formula $b(s) = \pi_s(\xi) - \xi$ implies that $\xi \neq 0$ in the case where b is injective. For simplicity, we assume that $\|\xi\|_H = 1$. Since b is injective, $O = \{\pi_s(\xi) : s \in G\}$ is an infinite subset of the sphere S of H. Since H is finite-dimensional, S is compact. So there exists some accumulation point η in the sphere of the orbit O. We have thus a convergent sequence $(\pi_{s_n}(\xi))$ consisting by injectivity of b of different points. We infer that

$$0 \leqslant \mathrm{Gap}_\psi \overset{(5.12)}{\leqslant} \lim_{n \to \infty} \|b(s_n) - b(s_{n+1})\|^2 = \lim_{n \to \infty} \|\pi_{s_{n+1}}(\xi) - \pi_{s_n}(\xi)\|^2 = 0.$$

Thus $0 \leqslant \mathrm{Gap}_\alpha \overset{(5.38)}{\leqslant} \mathrm{Gap}_\psi = 0.$ □

Remark 5.6 In the paper [129, p. 1967], the authors were able to find the original paper which contains the famous Bieberbach Theorem. Unfortunately, this theorem is badly written in [129, p. 1967] (since a crucial assumption is missing; to compare with a textbook, e.g. [194, Th 7.2.4 p. 306]). So the proof of [129, Theorem 6.4] is doubtful since this "more general version" of Bieberbach Theorem is used in the proof. Nevertheless, we think that an additional "proper" assumption could lead to a correct statement of this interesting idea.

Recall that a topological group has property (FH) if every affine isometric action of G on a real Hilbert space has a fixed point, see [31, Definition 2.1.4]. Note that by [31, Theorem 2.12.4], if a discrete group G has (T) then G has (FH) and the converse is true if G is countable. A finite group has (FH). The groups $\mathbb{Z}^k$ and free groups $\mathbb{F}_k$ do not have (FH) if $k \geq 1$. Using [31, Proposition 2.2.10], we deduce the following result. In [129, p. 1968], it is written that "infinite groups satisfying Kazhdan property (T) do not admit finite-dimensional standard cocycles" (i.e. injective, finite-dimensional with $\mathrm{Gap}_\psi > 0$). But from our point of view, the proof is missing. Proposition 5.9 allows us to give a proof and to obtain a slightly more general version.

Corollary 5.1 *Let G be an infinite discrete group with property (FH) and $b \colon G \to H$ be a 1-cocycle with respect to some finite-dimensional orthogonal representation $\pi \colon G \to \mathrm{O}(H)$. If b is injective, then $\mathrm{Gap}_\psi = \mathrm{Gap}_\alpha = 0$.*

Semigroups on Finite Groups Let G be a finite group. By [129, pp. 1970-1971], there always exists an orthogonal representation $\pi \colon G \to H$ on some finite-dimensional real Hilbert space H and an injective 1-cocycle $b \colon G \to H$ with respect to π. In this case, since G is finite, b only takes a finite number of values. This implies that $\mathrm{Gap}_\psi \overset{(5.12)}{=} \inf_{b(s) \neq b(t)} \|b(s) - b(t)\|_H^2 > 0$. For example, we can consider the left regular representation $\pi \colon G \to \mathrm{B}(\ell_G^2)$ defined by $\pi_s(e_t) = e_{st}$ for any $s, t \in G$ and the cocycle $b \colon G \to \ell_G^2$, $s \mapsto \pi_s(\xi) - \xi$ where ξ is some vector of ℓ_G^2 satisfying $\pi_s(\xi) \neq \xi$ for any element s of $G - \{e_G\}$. We refer to [129, a) and b) p. 1971] for other interesting examples of 1-cocycles. Note in addition that in the context of Schur multipliers, we also have $\mathrm{Gap}_\alpha > 0$ if $\alpha = b$.

Heat Semigroup on $\mathbb{T}^n$ Here $G = \mathbb{Z}^n$. We consider the Heat semigroup $(T_t)_{t \geq 0}$ on $\mathrm{L}^\infty(\mathbb{T}^n) = \mathrm{VN}(\mathbb{Z}^n)$ defined by $T_t \colon \mathrm{L}^\infty(\mathbb{T}^n) \to \mathrm{L}^\infty(\mathbb{T}^n)$, $e^{ik\cdot} \mapsto e^{-t|k|^2}e^{ik\cdot}$. The associated finite-dimensional injective cocycle is given by the canonical inclusion $b \colon \mathbb{Z}^n \to \mathbb{R}^n$ equipped with the trivial action $\pi_k = \mathrm{Id}_{\mathbb{R}^n}$ for all $k \in \mathbb{Z}^n$.

Lemma 5.11 *Consider the previous heat semigroup. We have $\mathrm{Gap}_\psi = 1$.*

Proof Indeed, we have

$$\mathrm{Gap}_\psi \overset{(5.12)}{=} \inf_{b_\psi(k) \neq b_\psi(l)} \|b_\psi(k) - b_\psi(l)\|_{\mathbb{R}^n}^2 = \inf_{k \neq l, k, l \in \mathbb{Z}^n} |k - l|^2 = 1.$$

$\square$

The Donut Type Markovian Semigroup of Fourier and Herz-Schur Multipliers
Consider now the example of donut type Fourier multipliers in the spirit of [129,

Section 5.3]. That is, we consider the group $G = \mathbb{Z}$ with cocycle

$$b(n) \overset{\text{def}}{=} \left(e^{2\pi i\alpha n}, e^{2\pi i\beta n}\right) - (1, 1) \in \mathbb{C}^2 = \mathbb{R}^4, \tag{5.39}$$

where $\alpha, \beta \in \mathbb{R}$ and we consider $\mathbb{C}^2$ as the real Hilbert space $\mathbb{R}^4$. The associated cocycle orthogonal representation is[11]

$$\pi_n(x, y) \overset{\text{def}}{=} \left(e^{2\pi i\alpha n}x, e^{2\pi i\beta n}y\right) \qquad n \in \mathbb{Z}, \ x, y \in \mathbb{C}. \tag{5.40}$$

Proposition 5.10 *Consider the case that both α and β take rational values. Then $b: \mathbb{Z} \to \mathbb{C}^2$ is not injective, but $\mathrm{Gap}_\psi, \mathrm{Gap}_\alpha > 0$. Moreover, the strict inequality $\mathrm{Gap}_\alpha < \mathrm{Gap}_\psi$ may happen.*

Proof Consider $p, q \in \mathbb{Z}$ and $N \in \mathbb{N}^*$ such that $\alpha = \frac{p}{N}$ and $\beta = \frac{q}{N}$. We have

$$b(n + N) = \left(e^{2\pi i(p+\alpha n)}, e^{2\pi i(q+\beta n)}\right) - (1, 1) = \left(e^{2\pi i\alpha n}, e^{2\pi i\beta n}\right) - (1, 1) = b(n),$$

so that b is N-periodic, and hence only takes a finite number of values. In particular, the function $b: \mathbb{Z} \to \mathbb{C}^2$ is not injective and the set $\{b(n) - b(m) : n, m \in \mathbb{Z}\}$ is finite, which readily implies that $\mathrm{Gap}_\psi = \inf_{b(n) \neq b(m)} \|b(n) - b(m)\|_{\mathbb{R}^4}^2 > 0$. In the same manner, we obtain $\mathrm{Gap}_\alpha = \inf_{b(i)-b(j) \neq b(k)-b(l)} \|b(i) - b(j) - (b(k) - b(l))\|_{\mathbb{R}^4}^2 > 0$.

We turn to the statement of strict inequality. To this end, we take $\alpha = \beta = \frac{1}{8}$. Then

$$\mathrm{Gap}_\psi = 2 \inf \left\| (e^{2\pi i\alpha n} - 1) - (e^{2\pi i\alpha m} - 1) \right\|_{\mathbb{C}}^2$$

$$= 2 \inf \left\| e^{2\pi i\alpha(n-m)} - 1 \right\|_{\mathbb{C}}^2$$

$$= 2 \left| e^{2\pi i\frac{1}{8}} - 1 \right|^2 = 2 \cdot \left| \frac{1+i}{\sqrt{2}} - 1 \right|^2$$

---

[11] For any $n, m \in \mathbb{Z}$, we have the cocycle law

$$b(n) + \pi_n(b_m) = \left(e^{2\pi i\alpha n}, e^{2\pi i\beta n}\right) - (1, 1) + \pi_n\left[\left(e^{2\pi i\alpha m}, e^{2\pi i\beta m}\right) - (1, 1)\right]$$

$$= \left(e^{2\pi i\alpha n}, e^{2\pi i\beta n}\right) - (1, 1) + \left(e^{2\pi i\alpha(n+m)}, e^{2\pi i\beta(n+m)}\right) - \left(e^{2\pi i\alpha n}, e^{2\pi i\beta n}\right)$$

$$= \left(e^{2\pi i\alpha(n+m)}, e^{2\pi i\beta(n+m)}\right) - (1, 1) = b(n + m).$$

$$= 2 \cdot \left(\left(\frac{1}{\sqrt{2}} - 1 \right)^2 + \frac{1}{(\sqrt{2})^2} \right)$$

$$= 4 \left(1 - \frac{1}{\sqrt{2}} \right).$$

On the other hand, according to the proof of Proposition 5.8, we have

$$\mathrm{Gap}_\alpha = 2 \inf \left\| e^{2\pi i \alpha n} (e^{2\pi i \alpha r} - 1) - (e^{2\pi i \alpha s} - 1) \right\|_\mathbb{C}^2$$

$$\leqslant 2 \left\| e^{2\pi i \alpha \cdot 1} (e^{2\pi i \alpha \cdot 2} - 1) - (e^{2\pi i \alpha \cdot 4} - 1) \right\|_\mathbb{C}^2$$

$$= 2 \left\| \frac{1 + i}{\sqrt{2}} (i - 1) - (-1 - 1) \right\|_\mathbb{C}^2 = 2 \left| \frac{1}{\sqrt{2}} (i - 1 - 1 - i) + 2 \right|^2$$

$$= 2 \left| 2 \left(1 - \frac{1}{\sqrt{2}} \right) \right|^2$$

$$= 8 \left(1 - \frac{1}{\sqrt{2}} \right)^2 < 4 \left(1 - \frac{1}{\sqrt{2}} \right) = \mathrm{Gap}_\psi,$$

since $1 - \frac{1}{\sqrt{2}} \in \left(0, \frac{1}{2} \right)$.   □

Lemma 5.12 *Suppose that at least one of the numbers α and β is irrational. Then the cocycle b is injective, and $\mathrm{Gap}_\alpha = \mathrm{Gap}_\psi = 0$.*

Proof If, say α, is irrational, then $e^{2\pi i \alpha n} = 1$ for some $n \in \mathbb{Z}$ implies that $\alpha n \in \mathbb{Z}$, so $n = 0$. Then $b(n) = b(m)$ for some $n, m \in \mathbb{Z}$ implies that $e^{2\pi i \alpha n} - 1 = e^{2\pi i \alpha m} - 1$, so $e^{2\pi i \alpha (n-m)} = 1$, and consequently, $n - m = 0$. We infer that b is injective. Note that $G = \mathbb{Z}$ is infinite and $H = \mathbb{R}^4$ is finite-dimensional. Furthermore, since $b(n) \overset{(5.39)}{=} \left(e^{2\pi i \alpha n}, e^{2\pi i \beta n} \right) - (1, 1) \overset{(5.40)}{=} \pi_n(1, 1) - (1, 1)$, the cocycle b is a coboundary. Thus the assumptions of Proposition 5.9 are fulfilled. This proposition implies that $\mathrm{Gap}_\alpha = \mathrm{Gap}_\psi = 0$.   □

The Free Group $\mathbb{F}_2$ Consider now the example of the action of the free group $\mathbb{F}_2$ with two generators a_1 and a_2 on $\mathbb{R}^3$ from [129, Section 5.5]. That is, we consider the representation $\pi: \mathbb{F}_2 \to O(\mathbb{R}^3)$ from the Banach-Tarski paradox. Take an angle $\theta \in \mathbb{R} \backslash 2\pi \mathbb{Q}$ and define π uniquely by putting

$$\pi_{a_1} = \begin{bmatrix} \cos \theta & -\sin \theta & 0 \\ \sin \theta & \cos \theta & 0 \\ 0 & 0 & 1 \end{bmatrix} \quad \text{and} \quad \pi_{a_2} = \begin{bmatrix} 1 & 0 & 0 \\ 0 & \cos \theta & -\sin \theta \\ 0 & \sin \theta & \cos \theta \end{bmatrix}.$$

Then it is known that $\pi \colon \mathbb{F}_2 \to O(\mathbb{R}^3)$ is injective. Further it is shown in [129, Section 5.5] that there exists some $\xi \in \mathbb{R}^3 \backslash \{0\}$ such that $b \colon \mathbb{F}_2 \to \mathbb{R}^3$, $b(s) = \pi_s(\xi) - \xi$ is an injective coboundary cocycle. We appeal again to Proposition 5.9 to deduce that in this case $\mathrm{Gap}_\alpha = \mathrm{Gap}_\psi = 0$.

5.7 Banach Spectral Triples

In the remainder of Sect. 5, we will establish that our Hodge-Dirac operators defined in (4.24) and in (4.64) associated with markovian semigroups of Fourier multipliers or Schur multipliers give rise to (locally) compact Banach spectral triples, see Proposition 5.14, Proposition 5.16 and Theorem 5.4. We will also introduce such Banach spectral triples for two other Hodge-Dirac operators, see Theorem 5.6 and Theorem 5.5. We refer to [49, 58, 60, 90], and [222] for more information on spectral triples. Let us recall this notion. A (possibly kernel-degenerate, compact) spectral triple (A, H, D) consists of a unital C*-algebra A, a Hilbert space H, a (densely defined, unbounded) selfadjoint operator D and a *-representation $\pi \colon A \to \mathrm{B}(H)$ which satisfy the following properties.

1. D^{-1} is compact on $\overline{\mathrm{Ran}\, D}$.
2. The set

$$\mathrm{Lip}_D(A) \overset{\mathrm{def}}{=} \big\{ a \in A : \pi(a) \cdot \mathrm{dom}\, D \subset \mathrm{dom}\, D \text{ and the unbounded operator}$$

$$[D, \pi(a)] \colon \mathrm{dom}\, D \subset H \to H \text{ extends to an element of } \mathrm{B}(H) \big\}$$

is dense in A.

Sometimes the condition that π is unital or faithful is added. There exist variations of this definition where the operator D^{-1} is replaced by another operator.

In the next sections, we give new examples of spectral triples. Our examples can be generalized to the context of L^p-spaces instead of Hilbert spaces. So, it is natural to state the following definition.

Definition 5.10 Consider a triple (A, X, D) constituted of the following data: a Banach space X, a closed unbounded operator D on X with dense domain $\mathrm{dom}\, D \subset X$, and an algebra A equipped with a homomorphism $\pi \colon A \to \mathrm{B}(X)$. In this case, we define the Lipschitz algebra

$$\mathrm{Lip}_D(A) \overset{\mathrm{def}}{=} \big\{ a \in A : \pi(a) \cdot \mathrm{dom}\, D \subset \mathrm{dom}\, D \text{ and the unbounded operator} \tag{5.41}$$

$$[D, \pi(a)] \colon \mathrm{dom}\, D \subset X \to X \text{ extends to an element of } \mathrm{B}(X) \big\}.$$

We say that (A, X, D) is a (compact) Banach spectral triple if in addition X is reflexive,[12] A is a Banach algebra, D is a bisectorial operator on X and if we have

1. D admits a bounded H^∞ functional calculus on a bisector $\Sigma_\omega^\pm$.
2. D^{-1} is a compact operator on $\overline{\operatorname{Ran} D}$.
3. The subset $\operatorname{Lip}_D(A)$ is dense in A.

In this situation, we have by [110, p. 448] a direct sum decomposition $X = \overline{\operatorname{Ran} D} \oplus \operatorname{Ker} D$. Since $|D| = \operatorname{sgn}(D)D$ and $D = \operatorname{sgn}(D)|D|$, we can replace D^{-1} by $|D|^{-1}$ in the second point since $\operatorname{sgn}(D)$ is a bounded operator, see Remark 4.1.

Finally, note that a spectral triple is a Banach spectral triple.

In the sequel, we will use sometimes the notation a for $\pi(a)$. If $a \in \operatorname{Lip}_D(A)$, we let

$$\|a\|_D \stackrel{\text{def}}{=} \big\|[D, a]\big\|_{X \to X}. \tag{5.42}$$

In the following, at several places we will use ideas of the proofs of [150, Proposition 1.6] and [196, Proposition 3.7]. We refer to [225] for related things. The first point says that the map $\partial_D: \operatorname{Lip}_D(A) \to \mathrm{B}(X)$, $a \mapsto [D, a]$ is a derivation.

Proposition 5.11 *Let X be a reflexive Banach space. Consider a closed linear operator D on X with dense domain $\operatorname{dom} D \subset X$, a Banach algebra A and a homomorphism $\pi: A \to \mathrm{B}(X)$.*

1. The space $\operatorname{Lip}_D(A)$ is a subalgebra of A. Moreover, if $a, b \in \operatorname{Lip}_D(A)$, we have

$$[D, ab] = a[D, b] + [D, a]b$$

and

$$\|ab\|_D \leqslant \|a\|_D \|\pi(b)\|_{\mathrm{B}(X)} + \|\pi(a)\|_{\mathrm{B}(X)} \|b\|_D.$$

2. $\|\cdot\|_D$ is a seminorm on $\operatorname{Lip}_D(A)$.
3. For any $a \in \operatorname{Lip}_D(A)$ we have $\pi(a)^* \cdot \operatorname{dom} D^* \subset \operatorname{dom} D^*$ and the linear operator

$$[D^*, \pi(a)^*]: \operatorname{dom} D^* \subset X^* \to X^*$$

extends to a bounded operator on X^* denoted with the same notation and we have

$$[D, \pi(a)]^* = -[D^*, \pi(a)^*]. \tag{5.43}$$

---

[12] It may perhaps be possible to replace the reflexivity by an assumption of weak compactness, see [110, p. 361].

4. *Suppose that $\pi : A \to B(X)$ is continuous when A is equipped with a topology $\mathcal{T}$ (which may be different from the norm topology) and when $B(X)$ is equipped with the weak operator topology. Then $\|\cdot\|_D$ is lower semicontinuous on $\mathrm{Lip}_D(A)$ when $\mathrm{Lip}_D(A)$ is equipped with the induced topology by the topology $\mathcal{T}$ of A. In particular, if a net $(a_j)_j$ of $\mathrm{Lip}_D(A)$ converges in the topology $\mathcal{T}$ to some $a \in A$, and the net is bounded in $\|\cdot\|_D$ seminorm, then a belongs to $\mathrm{Lip}_D(A)$.*

Proof

1. If $a, b \in \mathrm{Lip}_D(A)$ then $ab \cdot \mathrm{dom}\, D = a \cdot (b \cdot \mathrm{dom}\, D) \subset a \cdot \mathrm{dom}\, D \subset \mathrm{dom}\, D$. Moreover, if $\xi \in \mathrm{dom}\, D$ then:

$$Dab\xi - abD\xi = Dab\xi - aDb\xi + aDb\xi - abD\xi = [D, a]b\xi + a[D, b]\xi.$$

Thus, as operators on $\mathrm{dom}\, D$, we conclude that $[D, ab] = a[D, b] + [D, a]b$. So $ab \in \mathrm{Lip}_D(A)$. Moreover, we have

$$
\begin{aligned}
\|ab\|_D &\overset{(5.42)}{=} \|[D, ab]\|_{X \to X} = \|[D, a]b + a[D, b]\|_{X \to X} \\
&\leqslant \|[D, a]\|_{X \to X} \|\pi(b)\|_{B(X)} + \|\pi(a)\|_{B(X)} \|[D, b]\|_{X \to X} \\
&= \|a\|_D \|\pi(b)\|_{B(X)} + \|\pi(a)\|_{B(X)} \|b\|_D.
\end{aligned}
$$

2. If $\lambda \in \mathbb{C}$ and $a \in \mathrm{Lip}_D(A)$, we have $\|\lambda a\|_D \overset{(5.42)}{=} \|[D, \lambda a]\|_{X \to X} = \|\lambda[D, a]\|_{X \to X} = |\lambda| \|[D, a]\|_{X \to X} = |\lambda| \|a\|_D$. If $a, b \in \mathrm{Lip}_D(A)$, we have

$$
\begin{aligned}
\|a + b\|_D &\overset{(5.42)}{=} \|[D, a + b]\|_{X \to X} = \|[D, a] + [D, b]\|_{X \to X} \\
&\leqslant \|[D, a]\|_{X \to X} + \|[D, b]\|_{X \to X} \overset{(5.42)}{=} \|a\|_D + \|b\|_D.
\end{aligned}
$$

3. Let $a \in \mathrm{Lip}_D(A)$. If $\xi \in \mathrm{dom}\, D$ and $\zeta \in \mathrm{dom}\, D^*$ then:

$$
\begin{aligned}
\langle \pi(a)^* \zeta, D(\xi) \rangle_{X^*, X} &= \langle \zeta, \pi(a) D(\xi) \rangle_{X^*, X} \\
&= \langle \zeta, D\pi(a)\xi \rangle_{X^*, X} - \langle \zeta, [D, \pi(a)]\xi \rangle_{X^*, X} \\
&= \langle D^*(\zeta), \pi(a)\xi \rangle_{X^*, X} - \langle \zeta, [D, \pi(a)]\xi \rangle_{X^*, X}.
\end{aligned}
$$

Now, the linear map $\mathrm{dom}\, D \to \mathbb{C}, \xi \mapsto \langle D^*(\zeta), \pi(a)\xi \rangle_{X^*, X}$ is continuous. Since $[D, \pi(a)]$ is bounded, the linear map $\mathrm{dom}\, D \to \mathbb{C}, \xi \mapsto \langle \zeta, [D, \pi(a)]\xi \rangle_{X^*, X}$ is also continuous. Hence $\mathrm{dom}\, D \to \mathbb{C}, \xi \mapsto \langle \pi(a)^* \zeta, D(\xi) \rangle_{X^*, X}$ is continuous. Hence $\pi(a)^* \zeta$ belongs to $\mathrm{dom}\, D^*$.

For any $a \in \mathrm{Lip}_D(A)$, any $\xi \in \mathrm{dom}\, D$ and any $\zeta \in \mathrm{dom}\, D^*$, we have

$$
\begin{aligned}
\langle [D, \pi(a)]\xi, \zeta \rangle_{X,X^*} &= \langle (D\pi(a) - \pi(a)D)\xi, \zeta \rangle_{X,X^*} \\
&= \langle D\pi(a)\xi, \zeta \rangle_{X,X^*} - \langle \pi(a)D\xi, \zeta \rangle_{X,X^*} \\
&= \langle \pi(a)\xi, D^*\zeta \rangle_{X,X^*} - \langle D\xi, \pi(a)^*\zeta \rangle_{X,X^*} \\
&= \langle \xi, \pi(a)^*D^*\zeta \rangle_{X,X^*} - \langle \xi, D^*\pi(a)^*\zeta \rangle_{X,X^*} \\
&= \langle \xi, -[D^*, \pi(a)^*]\zeta \rangle_{X,X^*}.
\end{aligned}
$$

Hence the operators $[D, \pi(a)]$ and $-[D^*, \pi(a)^*]$ are formally adjoint to each other in the sense of [136, p. 167]. We infer that $-[D^*, \pi(a)^*] \subset [D, \pi(a)]^*$. Since this latter operator is bounded and since the domain $\mathrm{dom}\, D^*$ of $-[D^*, \pi(a)^*]$ is dense by [136, Theorem 5.29 p. 168], we obtain the result.

4. Suppose that $\pi : A \to B(X)$ is continuous when A is equipped with the topology $\mathcal{T}$ and when $B(X)$ is equipped with the weak operator topology. We shall show that $\|\cdot\|_D$ is lower semicontinuous on $\mathrm{Lip}_D(A)$ when $\mathrm{Lip}_D(A)$ is equipped with the induced topology by the topology $\mathcal{T}$ of A. Let $a \in A$ and (a_j) be a net of elements of $\mathrm{Lip}_D(A)$ such that (a_j) converges for $\mathcal{T}$ to a and $\|a_j\|_D \leqslant 1$. We have to show that a belongs to $\mathrm{Lip}_D(A)$ and that $\|a\|_D \leqslant 1$. Note that

$$
\left\| [D^*, \pi(a_j)^*] \right\|_{X^* \to X^*} \overset{(5.43)}{=} \left\| [D, \pi(a_j)] \right\|_{X \to X} \leqslant 1 \tag{5.44}
$$

and that $\pi(a_j) \to \pi(a)$ converges for the weak operator topology. Let $\xi \in \mathrm{dom}\, D$ and let $\zeta \in \mathrm{dom}\, D^*$. Note that $\pi(a_j)^*(\zeta)$ belongs to $\mathrm{dom}\, D^*$. Then for any index j

$$
\begin{aligned}
\langle \pi(a_j)\xi, D^*(\zeta) \rangle_{X,X^*} &= \langle \xi, \pi(a_j)^*D^*(\zeta) \rangle_{X,X^*} \\
&= \langle \xi, D^*\pi(a_j)^*(\zeta) \rangle_{X,X^*} - \langle \xi, [D^*, \pi(a_j)^*]\zeta \rangle_{X,X^*} \\
&= \langle D(\xi), \pi(a_j)^*\zeta \rangle_{X,X^*} - \langle \xi, [D^*, \pi(a_j)^*]\zeta \rangle_{X,X^*}.
\end{aligned}
$$

Passing to the limit and using the point 3, we deduce that

$$
\begin{aligned}
\left| \langle \pi(a)\xi, D^*(\zeta) \rangle_{X,X^*} \right| &= \lim_j \left| \langle \pi(a_j)\xi, D^*(\zeta) \rangle_{X,X^*} \right| \\
&\leqslant \liminf_j \left[\left| \langle D\xi, \pi(a_j)^*\zeta \rangle_{X,X^*} \right| \right. \\
&\qquad \left. + \left| \langle \xi, [D^*, \pi(a_j)^*]\zeta \rangle_{X,X^*} \right| \right] \\
&\overset{(5.44)}{\leqslant} \left| \langle D\xi, \pi(a)^*\zeta \rangle_{X,X^*} \right| + \|\xi\|_X \|\zeta\|_{X^*} \\
&\leqslant \|\zeta\|_{X^*} \left(\|D(\xi)\|_X \|\pi(a)\|_A + \|\xi\|_X \right).
\end{aligned}
$$

So the function $\text{dom}\,D^* \to \mathbb{C},\ \zeta \mapsto \langle \pi(a)\xi, D^*(\zeta)\rangle_{X,X^*}$ is continuous, and thus $\pi(a)\xi \in \text{dom}\,D^{**} = \text{dom}\,D$ by [136, Theorem 5.29]. We conclude that $\pi(a) \cdot \text{dom}\,D \subset \text{dom}\,D$. If $\xi \in \text{dom}\,D$ and $\zeta \in \text{dom}\,D^*$ with $\|\xi\|_X \leqslant 1$ and $\|\zeta\|_{X^*} \leqslant 1$ then

$$
\left| \langle \pi(a_j)\xi, D^*(\zeta)\rangle_{X,X^*} - \langle D(\xi), \pi(a_j)^*\zeta\rangle_{X,X^*}\right|
$$

$$
= \left| \langle D\pi(a_j)(\xi), \zeta\rangle_{X,X^*} - \langle \pi(a_j)D(\xi), \zeta\rangle_{X,X^*}\right|
$$

$$
= \left| \langle (D\pi(a_j) - \pi(a_j)D)(\xi), \zeta\rangle_{X,X^*}\right| = \left| \langle [D, \pi(a_j)](\xi), \zeta\rangle_{X,X^*}\right|
$$

$$
\leqslant \left\| [D, \pi(a_j)]\right\|_{X \to X} \|\xi\|_X \|\zeta\|_{X^*} \leqslant 1.
$$

Passing to the limit, we obtain

$$
\left| \langle [D, \pi(a)]\xi, \zeta\rangle_{X,X^*}\right| = \left| \langle (D\pi(a) - \pi(a)D)\xi, \zeta\rangle_{X,X^*}\right|
$$

$$
= \left| \langle \pi(a)\xi, D^*\zeta\rangle_{X,X^*} - \langle D\xi, \pi(a)^*\zeta\rangle_{X,X^*}\right| \leqslant 1.
$$

Hence $[D, \pi(a)]$ is bounded on $\text{dom}\,D$ and thus extends to X to a bounded operator and $\|[D, \pi(a)]\| \leqslant 1$.

$\square$

Remark 5.7 In the particular case of a spectral triple, we have $D = D^*$ and $\pi(a)^* = \pi(a^*)$. So the previous result says that $\text{Lip}_D(A)$ is a $*$-subalgebra of A.

The following is a Banach space generalization of [87, Proposition 2.1] proved with a similar method.

Proposition 5.12 *Let X be a Banach space and $D \colon \text{dom}\,D \subset X \to X$ be a closed operator.*

1. *Suppose that C is a core of D and that $a \in \text{B}(X)$ satisfies*

   (a) *$a \cdot C \subset \text{dom}\,D$*
   (b) *$[D, a]|C \colon C \to X$ is bounded on C.*

   Then $a \cdot \text{dom}\,D \subset \text{dom}\,D$ and the operator $[D, a] \colon \text{dom}\,D \to X$ is well-defined.
2. *If in addition D is densely defined and if there exists an adjoint[13] $T \colon \text{dom}\,T \subset X^* \to X^*$ of D, and a subspace Y of $\text{dom}\,T$ which is dense in X^* such that $a^* \cdot Y \subset \text{dom}\,T$, then $[D, a] \colon \text{dom}\,D \to X$ extends to a bounded operator on X.*

---

[13] In the sense of [136, p. 167].

Proof

1. Let $x \in \text{dom } D$. By (2.3) there exists a sequence (x_n) of elements of C such that $x_n \to x$ and $D(x_n) \to D(x)$ in X. Since $a \in B(X)$, we have $a(x_n) \to a(x)$ in X. Moreover, for any integer n, m, we have

$$\left\| Da(x_n) - Da(x_m) \right\|_X$$
$$= \left\| aD(x_n) - aD(x_m) + Da(x_n) - aD(x_n) - Da(x_m) + aD(x_m) \right\|_X$$
$$= \left\| aD(x_n) - aD(x_m) + [D, a](x_n) - [D, a](x_m) \right\|_X$$
$$\leqslant \|a\|_{X \to X} \left\| D(x_n) - D(x_m) \right\|_X + \left\| [D, a] \right\|_{C \to X} \|x_n - x_m\|_X.$$

We infer that $(Da(x_n))$ is a Cauchy sequence in X hence converges. Since D is closed, we conclude that $a(x)$ belongs to dom D.

2. Let $x \in \text{dom } D$ and $y \in Y$. Then

$$\langle [D, a](x), y \rangle_{X,X^*} = \langle (Da - aD)(x), y \rangle_{X,X^*} = \langle Da(x), y \rangle_{X,X^*} - \langle aD(x), y \rangle_{X,X^*}$$
$$= \langle a(x), T(y) \rangle_{X,X^*} - \langle D(x), a^*(y) \rangle_{X,X^*}$$
$$= \langle x, a^* T(y) \rangle - \langle x, T a^*(y) \rangle_{X,X^*}$$
$$= \langle x, -[T, a^*](y) \rangle_{X,X^*}.$$

We infer that $y \in \text{dom}[D, a]^*$. Hence $Y \subset \text{dom}[D, a]^*$. Since Y is dense in X^*, we deduce that $[D, a]^*$ is densely defined. By [136, Theorem 5.28 p. 168], this implies that $[D, a]$ is closable. From $[D, a]|_C \subset [D, a]$, we have $\overline{[D, a]|_C} \subset \overline{[D, a]}$ and the closure $\overline{[D, a]|_C}$ is the bounded extension of $[D, a]|_C$ on X (note that C is dense in X since dom D is dense). Hence the closed operator $\overline{[D, a]}$ is defined on X hence bounded by [136, Theorem 5.20 p. 166]. □

Next, we try to define the notion of a locally compact *Banach* spectral triple. When X is a Hilbert space, it is interesting to compare to the ones in [49], [88, p. 588] and [214].

Definition 5.11 Let X be a reflexive Banach space, D a densely defined closed operator on X and A a subalgebra of some Banach algebra. Let $\pi : A \to B(X)$ be a homomorphism. We call (A, X, D) a locally compact Banach spectral triple, provided that

1. D is bisectorial on X and has a bounded H^∞ functional calculus on some bisector.
2. We have $A = \text{Lip}_D(A)$.
3. For any $a \in A$, $\pi(a)|D|^{-1}|\overline{\text{Ran } D}$ is a compact operator $\overline{\text{Ran } D} \to X$.

The condition "for any $a \in A$, $\pi(a)(\text{iId} + D)^{-1}|\overline{\text{Ran}\,D}$ is a compact operator $\overline{\text{Ran}\,D} \to X$" could be a substitute for the last condition in some situations.

Recall that a spectral triple (A, H, D) is even if there exists a selfadjoint unitary operator $\gamma \colon H \to H$ such that $\gamma^2 = \text{Id}_H$, $\gamma D = -D\gamma$ and $\gamma \pi(a) = \pi(a)\gamma$ for all $a \in A$.

Definition 5.12 We say that a (locally compact or compact) Banach spectral triple (A, X, D) is even if there exists a surjective isometry $\gamma \colon X \to X$ such that $\gamma^2 = \text{Id}_X$, $\gamma D = -D\gamma$ and $\gamma\pi(a) = \pi(a)\gamma$ for all $a \in A$.

5.8   Spectral Triples Associated to Semigroups of Fourier Multipliers I

In this section, we consider a markovian semigroup of Fourier multipliers as in Proposition 2.3, together with the noncommutative gradient $\partial_{\psi,q,p}$ and its adjoint. Now, we generalize the construction of [131, p. 587-589] which corresponds to the case $q = 1$ and $p = 2$ below. So we obtain a scale of L^p-Banach spectral triples associated to these semigroups. In particular, here we shall complete the picture. Indeed, we prove the compactness axiom (point 6 below) of Connes' spectral triples. Finally, we shall compare our result Theorem 5.4 with the result of [131, p. 588] in Remark 5.11.

Suppose $1 < p < \infty$ and $-1 \leqslant q < 1$. Recall that the Hodge-Dirac operator is defined by

$$\mathcal{D}_{\psi,q,p} \overset{\text{def}}{=} \begin{bmatrix} 0 & (\partial_{\psi,q,p^*})^* \\ \partial_{\psi,q,p} & 0 \end{bmatrix} \tag{5.45}$$

on the subspace $\text{dom}\,\mathcal{D}_{\psi,q,p} = \text{dom}\,\partial_{\psi,q,p} \oplus \text{dom}(\partial_{\psi,q,p^*})^*$ of the Banach space $L^p(\text{VN}(G)) \oplus_p L^p(\Gamma_q(H) \rtimes_\alpha G)$. If $a \in \text{VN}(G)$, we define the bounded operator $\pi(a) \colon L^p(\text{VN}(G)) \oplus_p L^p(\Gamma_q(H) \rtimes_\alpha G) \to L^p(\text{VN}(G)) \oplus_p L^p(\Gamma_q(H) \rtimes_\alpha G)$ by

$$\pi(a) \overset{\text{def}}{=} \begin{bmatrix} L_a & 0 \\ 0 & \tilde{L}_a \end{bmatrix}, \quad a \in \text{VN}(G) \tag{5.46}$$

where $L_a \colon L^p(\text{VN}(G)) \to L^p(\text{VN}(G))$, $x \mapsto ax$ is the left multiplication operator and where $\tilde{L}_a \colon L^p(\Gamma_q(H) \rtimes_\alpha G) \to L^p(\Gamma_q(H) \rtimes_\alpha G)$ is the left action of the bimodule (2.85). It is easy to check that $\pi \colon \text{VN}(G) \to \text{B}(L^p(\text{VN}(G)) \oplus_p L^p(\Gamma_q(H) \rtimes_\alpha G))$ is continuous when $\text{VN}(G)$ is equipped with the weak* topology and when $\text{B}(L^p(\text{VN}(G)) \oplus_p L^p(\Gamma_q(H) \rtimes_\alpha G))$ is equipped with the weak operator topology. We will also use the restriction of π on $\text{C}_r^*(G)$.

We will use the following proposition.

Proposition 5.13 *Let* $1 < p < \infty$. *Let* G *be a discrete group. Consider the associated function* $b_\psi : G \to H$ *from (2.34). Assume that the Hilbert space* H *is finite-dimensional, that* b_ψ *is injective and that* $\mathrm{Gap}_\psi > 0$. *Finally assume that the von Neumann crossed product* $\Gamma_q(H) \rtimes_\alpha G$ *has QWEP.*

1. *The operator* $A_p^{-\frac{1}{2}} : \overline{\mathrm{Ran}\, A_p} \to \overline{\mathrm{Ran}\, A_p}$ *is compact.*

2. *Suppose* $-1 \leqslant q \leqslant 1$. *Then the operator* $B_p^{-\frac{1}{2}}$ *is compact where* B_p *is defined in (4.7).*

Proof

1. As in the proof of Theorem 5.1 the operator $A_2^{-\frac{1}{2}} : \overline{\mathrm{Ran}\, A_2} \to \overline{\mathrm{Ran}\, A_2}$ is compact.

   We show that $A_\infty^{-\frac{1}{2}} : \overline{\mathrm{Ran}\, A_\infty} \to \overline{\mathrm{Ran}\, A_\infty}$ is bounded. Note that $(T_{t,\infty})_{t \geqslant 0}$ is a bounded weak* continuous semigroup with weak* generator A_∞. Using a mild adaptation of Lemma 5.6 to weak* Markov semigroups, it suffices to establish the bound $\|T_t\|_{\mathrm{Ran}\, A_\infty \to \mathrm{Ran}\, A_\infty} \lesssim \frac{1}{t^d}$ for some $d > \frac{1}{2}$ (if dim $H = 1$ we need to embed H in a larger Hilbert space before). We conclude with [129, Lemma 5.8] and the contractive inclusion $\mathrm{VN}(G) \subset \mathrm{L}^1(\mathrm{VN}(G))$. Thus, $A_\infty^{-\frac{1}{2}} : \overline{\mathrm{Ran}\, A_\infty} \to \overline{\mathrm{Ran}\, A_\infty}$ is bounded.

   Now, assume that $p > 2$. We will use complex interpolation. Since the resolvents of the operators A_p are compatible for different values of p, the complementary projections associated to the ones of (2.17) onto the spaces $\overline{\mathrm{Ran}\, A_p}$ are compatible. Hence, the $\overline{\mathrm{Ran}\, A_p}$'s form an interpolation scale. Observe that $\overline{\mathrm{Ran}\, A_2}$ is a Hilbert space, hence a UMD space. Then we obtain the compactness of $A_p^{-\frac{1}{2}} : \overline{\mathrm{Ran}\, A_p} \to \overline{\mathrm{Ran}\, A_p}$ by means of complex interpolation between a compact and a bounded operator with Theorem 2.2.

   If $p < 2$, we conclude by duality and Schauder's Theorem [167, Theorem 3.4.15], since $\overline{\mathrm{Ran}\, A_p}$ is the dual space of $\overline{\mathrm{Ran}\, A_{p^*}}$ (use (2.8) and [167, page 94]) and $A_p^{-\frac{1}{2}}$ defined on the first space is the adjoint of $A_{p^*}^{-\frac{1}{2}}$ defined on the second space.

2. Since $A_2 = (\partial_{\psi,q,2})^* \partial_{\psi,q,2}$ and $B_2 = \partial_{\psi,q,2}(\partial_{\psi,q,2})^* | \overline{\mathrm{Ran}\, \partial_{\psi,q,2}}$, by Theorem 2.1 and Proposition 4.8 together with a functional calculus argument, we obtain that the operators $A_2^{-\frac{1}{2}} |_{\mathrm{Ker}(\partial_{\psi,q,2})^\perp}$ and $B_2^{-\frac{1}{2}}$ are unitarily equivalent. Since $A_2^{-\frac{1}{2}} : \overline{\mathrm{Ran}\, A_2} \to \overline{\mathrm{Ran}\, A_2}$ is compact and $\mathrm{Ker}(\partial_{\psi,q,2})^\perp = \overline{\mathrm{Ran}\, A_2}$, $A_2^{-\frac{1}{2}} |_{\mathrm{Ker}(\partial_{\psi,q,2})^\perp}$ and finally $B_2^{-\frac{1}{2}} : \overline{\mathrm{Ran}\, \partial_{\psi,q,2}} \to \overline{\mathrm{Ran}\, \partial_{\psi,q,2}}$ is compact.

   Recall that by (4.7), B_p is the generator of the (strongly continuous) semigroup $(\mathrm{Id}_{\mathrm{L}^p(\Gamma_q(H))} \rtimes T_t | \overline{\mathrm{Ran}\, \partial_{\psi,q,p}})_{t \geqslant 0}$. Now, we use [129, Lemma 5.8] and Proposition 2.8 to obtain the bound $\|\mathrm{Id} \rtimes T_t\|_{\infty \to \infty} \lesssim \frac{1}{t^d}$ for some $d > \frac{1}{2}$. According to Proposition 4.8, we have $\overline{\mathrm{Ran}\, B_p} = \overline{\mathrm{Ran}\, \partial_{\psi,q,p}}$. By Lemma 5.6, we obtain that $B_p^{-\frac{1}{2}} : \overline{\mathrm{Ran}\, B_p} \to \overline{\mathrm{Ran}\, B_p}$ is bounded.

Now, it suffices to interpolate with Theorem 2.2 the compactness of the operator $B_2^{-\frac{1}{2}} \colon \overline{\mathrm{Ran}\, \partial_{\psi,q,2}} \to \overline{\mathrm{Ran}\, \partial_{\psi,q,2}}$ and the boundedness of $B_{p_0}^{-\frac{1}{2}} \colon \overline{\mathrm{Ran}\, \partial_{\psi,q,p_0}} \to \overline{\mathrm{Ran}\, \partial_{\psi,q,p_0}}$ to obtain the compactness of $B_p^{-\frac{1}{2}} \colon \overline{\mathrm{Ran}\, \partial_{\psi,q,p}} \to \overline{\mathrm{Ran}\, \partial_{\psi,q,p}}$ for $2 \leqslant p < p_0 < \infty$ and the fact that on the L^2-level, we have a Hilbert space hence a UMD space. Note that the spaces $\overline{\mathrm{Ran}\, \partial_{\psi,q,p}} \subset L^p(\Gamma_q(H)\overline{\otimes}\mathrm{B}(\ell_I^2))$ interpolate by the complex interpolation method since we have bounded projections

$$P_p \colon L^p(\Gamma_q(H) \rtimes G) \to \overline{\mathrm{Ran}\, \partial_{\psi,q,p}} \subset L^p(\Gamma_q(H) \rtimes G)$$

which are compatible for different values of p according to (2.17). We use duality if $p < 2$.

$\square$

In the following theorem, recall the weak* closed operator $\partial_{\psi,q,\infty} \colon \mathrm{dom}\, \partial_{\psi,q,\infty} \subset \Gamma_q(H) \rtimes_\alpha G \to \Gamma_q(H) \rtimes_\alpha G$ from Proposition 3.5. The latter is valid if G has AP and $q \neq 1$. Finally, the QWEP assumption of the following result is satisfied if G is amenable or G is a free group and $q = -1$.

Theorem 5.4 *Suppose $1 < p < \infty$ and $-1 \leqslant q < 1$. The triple*

$$(C_r^*(G), L^p(\mathrm{VN}(G)) \oplus_p L^p(\Gamma_q(H) \rtimes_\alpha G), \mathcal{D}_{\psi,q,p})$$

is a Banach spectral triple in the case where G is weakly amenable, $b_\psi \colon G \to H$ is injective, $\mathrm{Gap}_\psi > 0$, H is finite-dimensional and $\Gamma_q(H) \rtimes_\alpha G$ has QWEP. Moreover, we have the following properties.

1. *We have $(\mathcal{D}_{\psi,q,p})^* = \mathcal{D}_{\psi,q,p^*}$. In particular, the operator $\mathcal{D}_{\psi,q,2}$ is selfadjoint.*
2. *We have*

$$\mathcal{P}_G \subset \mathrm{Lip}_{\mathcal{D}_{\psi,q,p}}(\mathrm{VN}(G)). \tag{5.47}$$

3. *For any $a \in \mathcal{P}_G$, we have*

$$\big\| [\mathcal{D}_{\psi,q,p}, \pi(a)] \big\|_{L^p(\mathrm{VN}(G)) \oplus_p L^p \to L^p(\mathrm{VN}(G)) \oplus_p L^p} \leqslant \big\| \partial_{\psi,q}(a) \big\|_{\Gamma_q(H) \rtimes_\alpha G}. \tag{5.48}$$

4. *Suppose that G has AP. We have*

$$\mathrm{dom}\, \partial_{\psi,q,\infty} \subset \mathrm{Lip}_{\mathcal{D}_{\psi,q,p}}(\mathrm{VN}(G)). \tag{5.49}$$

5. *Suppose that G has* AP. *For any* $a \in \operatorname{dom} \partial_{\psi,q,\infty}$, *we have*

$$\left\| [\mathcal{D}_{\psi,q,p}, \pi(a)] \right\|_{L^p \oplus_p L^p(\Gamma_q(H) \rtimes_\alpha G) \to L^p \oplus_p L^p(\Gamma_q(H) \rtimes_\alpha G)} \tag{5.50}$$

$$\leqslant \left\| \partial_{\psi,q,\infty}(a) \right\|_{\Gamma_q(H) \rtimes_\alpha G}.$$

6. *Assume that G is weakly amenable and that $\Gamma_q(H) \rtimes_\alpha G$ has* QWEP. *If $b_\psi : G \to H$ is injective,* $\operatorname{Gap}_\psi > 0$ *and if H is finite-dimensional, then the operator*

$$|\mathcal{D}_{\psi,q,p}|^{-1} : \overline{\operatorname{Ran}(\partial_{\psi,q,p^*})^*} \oplus \overline{\operatorname{Ran} \partial_{\psi,q,p}} \to \overline{\operatorname{Ran}(\partial_{\psi,q,p^*})^*} \oplus \overline{\operatorname{Ran} \partial_{\psi,q,p}}$$

is compact.

Proof

1. An element (z,t) of $L^{p^*}(\mathrm{VN}(G)) \oplus_{p^*} L^{p^*}(\Gamma_q(H) \rtimes_\alpha G)$ belongs to $\operatorname{dom}(\mathcal{D}_{\psi,q,p})^*$ if and only if there exists $(a,b) \in L^{p^*}(\mathrm{VN}(G)) \oplus_{p^*} L^{p^*}(\Gamma_q(H) \rtimes_\alpha G)$ such that for any $(x,y) \in \operatorname{dom} \partial_{\psi,q,p} \oplus \operatorname{dom}(\partial_{\psi,q,p^*})^*$ we have

$$\left\langle \begin{bmatrix} 0 & (\partial_{\psi,q,p^*})^* \\ \partial_{\psi,q,p} & 0 \end{bmatrix} \begin{bmatrix} x \\ y \end{bmatrix}, \begin{bmatrix} z \\ t \end{bmatrix} \right\rangle = \left\langle \begin{bmatrix} x \\ y \end{bmatrix}, \begin{bmatrix} a \\ b \end{bmatrix} \right\rangle,$$

that is

$$\left\langle (\partial_{\psi,q,p^*})^*(y), z \right\rangle + \left\langle \partial_{\psi,q,p}(x), t \right\rangle = \langle y, b \rangle + \langle x, a \rangle. \tag{5.51}$$

If $z \in \operatorname{dom} \partial_{\psi,q,p^*}$ and if $t \in \operatorname{dom}(\partial_{\psi,q,p})^*$ the latter holds with $b = \partial_{\psi,q,p^*}(z)$ and $a = (\partial_{\psi,q,p})^*(t)$. This proves that $\operatorname{dom} \partial_{\psi,q,p^*} \oplus \operatorname{dom}(\partial_{\psi,q,p})^* \subset \operatorname{dom}(\mathcal{D}_{\psi,q,p})^*$ and that

$$(\mathcal{D}_{\psi,q,p})^*(z,t) = \left((\partial_{\psi,q,p})^*(t), \partial_{\psi,q,p^*}(z) \right)$$

$$= \begin{bmatrix} 0 & (\partial_{\psi,q,p})^* \\ \partial_{\psi,q,p^*} & 0 \end{bmatrix} \begin{bmatrix} z \\ t \end{bmatrix} \overset{(5.45)}{=} \mathcal{D}_{\psi,q,p^*}(z,t).$$

Conversely, if $(z,t) \in \operatorname{dom}(\mathcal{D}_{\psi,q,p})^*$, choosing $y = 0$ in (5.51) we obtain $t \in \operatorname{dom}(\partial_{\psi,q,p})^*$ and taking $x = 0$ we obtain $z \in \operatorname{dom} \partial_{\psi,q,p^*}$.

2. Recall that the subspaces $\mathcal{P}_G$ and $\mathcal{P}_{\rtimes,G}$ are dense in $L^p(\mathrm{VN}(G))$ and $L^p(\Gamma_q(H) \rtimes_\alpha G)$ and are contained in the domains of $\partial_{\psi,q,p}$ and $(\partial_{\psi,q,p^*})^*$. So $\mathcal{P}_G \oplus \mathcal{P}_{\rtimes,G}$ is contained in $\operatorname{dom} \mathcal{D}_{\psi,q,p}$. For any $a \in \mathcal{P}_G$, we have $L_a(\mathcal{P}_G) \subset \mathcal{P}_G$ and $\tilde{L}_a(\mathcal{P}_{\rtimes,G}) \subset \mathcal{P}_{\rtimes,G}$. We infer that $\pi(a) \cdot (\mathcal{P}_G \oplus \mathcal{P}_{\rtimes,G}) \subset \operatorname{dom} \mathcal{D}_{\psi,q,p}$. Note also that $\pi(a)^* \cdot (\mathcal{P}_G \oplus \mathcal{P}_{\rtimes,G}) \subset \operatorname{dom} \mathcal{D}_{\psi,q,p^*} = \operatorname{dom}(\mathcal{D}_{\psi,q,p})^*$ (condition 2 of Proposition 5.12).

Let $a \in \mathcal{P}_G$. A simple computation shows that

$$\left[\mathcal{D}_{\psi,q,p}, \pi(a) \right]$$

$$\overset{(5.45)(5.46)}{=} \begin{bmatrix} 0 & (\partial_{\psi,q,p*})^* \\ \partial_{\psi,q,p} & 0 \end{bmatrix} \begin{bmatrix} L_a & 0 \\ 0 & \tilde{L}_a \end{bmatrix} - \begin{bmatrix} L_a & 0 \\ 0 & \tilde{L}_a \end{bmatrix} \begin{bmatrix} 0 & (\partial_{\psi,q,p*})^* \\ \partial_{\psi,q,p} & 0 \end{bmatrix}$$

$$= \begin{bmatrix} 0 & (\partial_{\psi,q,p*})^* \tilde{L}_a \\ \partial_{\psi,q,p} L_a & 0 \end{bmatrix} - \begin{bmatrix} 0 & L_a (\partial_{\psi,q,p*})^* \\ \tilde{L}_a \partial_{\psi,q,p} & 0 \end{bmatrix}$$

$$= \begin{bmatrix} 0 & (\partial_{\psi,q,p*})^* \tilde{L}_a - L_a (\partial_{\psi,q,p*})^* \\ \partial_{\psi,q,p} L_a - \tilde{L}_a \partial_{\psi,q,p} & 0 \end{bmatrix}.$$

We calculate the two non-zero components of the commutator. For the lower left corner, if $x \in \mathcal{P}_G$ and if we consider the canonical map $J \colon L^p(VN(G)) \to L^p(\Gamma_q(H) \rtimes_\alpha G)$, $x \mapsto 1 \rtimes x$, we have[14]

$$(\partial_{\psi,q,p} L_a - \tilde{L}_a \partial_{\psi,q,p})(x) = \partial_{\psi,q,p} L_a(x) - \tilde{L}_a \partial_{\psi,q,p}(x) \tag{5.52}$$

$$= \partial_{\psi,q,p}(ax) - a \partial_{\psi,q,p}(x) \overset{(2.86)}{=} \partial_{\psi,q}(a)x = L_{\partial_{\psi,q}(a)} J(x).$$

For the upper right corner, if $\mathbb{E}$ is the conditional expectation associated to J, note that for any $y \in \mathcal{P}_{\rtimes,G}$ and any $x \in \mathcal{P}_G$, (we recall that we have the duality brackets $\langle f, g \rangle$ antilinear in the first variable)

$$\langle ((\partial_{\psi,q,p*})^* \tilde{L}_a - L_a(\partial_{\psi,q,p*})^*)(y), x \rangle$$

$$= \langle (\partial_{\psi,q,p*})^* \tilde{L}_a(y), x \rangle - \langle L_a(\partial_{\psi,q,p*})^*(y), x \rangle$$

$$= \langle \tilde{L}_a(y), \partial_{\psi,q,p*}(x) \rangle - \langle (\partial_{\psi,q,p*})^*(y), L_{a*}(x) \rangle$$

$$= \langle y, \tilde{L}_{a*} \partial_{\psi,q,p*}(x) \rangle - \langle y, \partial_{\psi,q,p*} L_{a*}(x) \rangle$$

$$= \langle y, \tilde{L}_{a*} \partial_{\psi,q,p*}(x) - \partial_{\psi,q,p*} L_{a*}(x) \rangle = \langle y, a^* \partial_{\psi,q,p*}(x) - \partial_{\psi,q,p*}(a^*x) \rangle$$

$$\overset{(2.86)}{=} \langle y, -\partial_{\psi,q}(a^*)x \rangle = \langle y, -L_{\partial_{\psi,q}(a^*)}(1 \rtimes x) \rangle = \langle y, L_{(\partial_{\psi,q}(a))*}(1 \rtimes x) \rangle$$

$$= \langle L_{\partial_{\psi,q}(a)}(y), 1 \rtimes x \rangle = \langle \mathbb{E} L_{\partial_{\psi,q}(a)}(y), x \rangle_{L^p(VN(G)), L^{p*}(VN(G))}.$$

We conclude that

$$((\partial_{\psi,q,p*})^* \tilde{L}_a - L_a(\partial_{\psi,q,p*})^*)(y) = \mathbb{E} L_{\partial_{\psi,q}(a)}(y). \tag{5.53}$$

---

[14] Recall that the term $\partial_{\psi,q,p}(a)x$ is by definition equal to $\partial_{\psi,q,p}(a)(1 \rtimes x)$.

The two non-zero components of the commutator are bounded linear operators on $\mathcal{P}_G$ and on $\mathcal{P}_{\rtimes,G}$. We deduce that $[\mathcal{D}_{\psi,q,p}, \pi(a)]$ is bounded on $\mathcal{P}_G \oplus \mathcal{P}_{\rtimes,G}$. By Proposition 5.12, it extends to a bounded operator on $L^p(\mathrm{VN}(G)) \oplus_p L^p(\Gamma_q(H) \rtimes_\alpha G)$. Hence $\mathcal{P}_G$ is a subset of $\mathrm{Lip}_{\mathcal{D}_{\psi,q,p}}(\mathrm{VN}(G))$.

3. If $(x, y) \in \mathrm{dom}\,\mathcal{D}_{\psi,q,p}$ and $a \in \mathcal{P}_G$, we have

$$\left\| [\mathcal{D}_{\psi,q,p}, \pi(a)](x, y) \right\|_p \tag{5.54}$$

$$= \left\| \left(((\partial_{\psi,q,p^*})^* \tilde{L}_a - L_a(\partial_{\psi,q,p^*})^*) y, \, (\partial_{\psi,q,p} L_a - \tilde{L}_a \partial_{\psi,q,p}) x \right) \right\|_p$$

$$= \left(\left\| ((\partial_{\psi,q,p^*})^* \tilde{L}_a - L_a(\partial_{\psi,q,p^*})^*) y \right\|_p^p + \left\| (\partial_{\psi,q,p} L_a - \tilde{L}_a \partial_{\psi,q,p}) x \right\|_p^p \right)^{\frac{1}{p}}$$

$$\overset{(5.53)(5.52)}{=} \left(\left\| \mathbb{E} L_{\partial_{\psi,q}(a)}(y) \right\|_{L^p(\mathrm{VN}(G))}^p + \left\| \partial_{\psi,q}(a) J(x) \right\|_{L^p(\Gamma_q(H) \rtimes_\alpha G)}^p \right)^{\frac{1}{p}}$$

$$\leqslant \left\| \partial_{\psi,q}(a) \right\|_{\Gamma_q(H) \rtimes_\alpha G} \|(x, y)\|_p \,.$$

So we obtain (5.48).

4. Let $a \in \mathrm{dom}\,\partial_{\psi,q,\infty}$. Let (a_j) be a net in $\mathcal{P}_G$ such that $a_j \to a$ and $\partial_{\psi,q,\infty}(a_j) \to \partial_{\psi,q,\infty}(a)$ both for the weak* topology. The existence of such a net is guaranteed by Proposition 3.5. By Lemma 2.1, we can suppose that the nets (a_j) and $(\partial_{\psi,q,\infty}(a_j))$ are bounded. By the point 4 of Proposition 5.11, we deduce that $a \in \mathrm{Lip}_{\mathcal{D}_{\psi,q,p}}(\mathrm{VN}(G))$. By continuity of π, note that $\pi(a_j) \to \pi(a)$ for the weak operator topology. For any $\xi \in \mathrm{dom}\,\mathcal{D}_{\psi,q,p}$ and any $\zeta \in \mathrm{dom}(\mathcal{D}_{\psi,q,p})^*$, we have

$$\big\langle [\mathcal{D}_{\psi,q,p}, \pi(a_j)]\xi, \zeta \big\rangle_{L^p(\mathrm{VN}(G)) \oplus_p L^p(\Gamma_q(H) \rtimes_\alpha G), L^{p^*}(\mathrm{VN}(G)) \oplus_{p^*} L^{p^*}(\Gamma_q(H) \rtimes_\alpha G)}$$

$$= \big\langle (\mathcal{D}_{\psi,q,p}\pi(a_j) - \pi(a_j)\mathcal{D}_{\psi,q,p})\xi, \zeta \big\rangle$$

$$= \big\langle \mathcal{D}_{\psi,q,p}\pi(a_j)\xi, \zeta \big\rangle - \big\langle \pi(a_j)\mathcal{D}_{\psi,q,p}\xi, \zeta \big\rangle$$

$$= \big\langle \pi(a_j)\xi, (\mathcal{D}_{\psi,q,p})^*\zeta \big\rangle - \big\langle \pi(a_j)\mathcal{D}_{\psi,q,p}\xi, \zeta \big\rangle$$

$$\xrightarrow[j]{} \big\langle \pi(a)\xi, (\mathcal{D}_{\psi,q,p})^*\zeta \big\rangle - \big\langle \pi(a)\mathcal{D}_{\psi,q,p}\xi, \zeta \big\rangle$$

$$= \big\langle \mathcal{D}_{\psi,q,p}\pi(a)\xi, \zeta \big\rangle - \big\langle \pi(a)\mathcal{D}_{\psi,q,p}\xi, \zeta \big\rangle = \big\langle [\mathcal{D}_{\psi,q,p}, \pi(a)]\xi, \zeta \big\rangle.$$

Since the net $([\mathcal{D}_{\psi,q,p}, \pi(a_j)])$ is bounded by (5.48), we deduce that the net $([\mathcal{D}_{\psi,q,p}, \pi(a_j)])$ converges to $[\mathcal{D}_{\psi,q,p}, \pi(a)]$ for the weak operator topology by a "net version" of [136, Lemma 3.6 p. 151]. Furthermore, it is (really) easy to check that $L_{\partial_{\psi,q,p}(a_j)} J \to L_{\partial_{\psi,q,p}(a)} J$ and $-\mathbb{E} L_{\partial_{\psi,q}(a_j)} \to -\mathbb{E} L_{\partial_{\psi,q}(a)}$ both for the weak operator topology. By uniqueness of the limit, we deduce that the commutator is given by the same formula as that in the case of elements of $\mathcal{P}_G$.

5. We obtain (5.50) as in (5.54).

6. Recall that $\overline{\operatorname{Ran} A_p} = \overline{\operatorname{Ran}(\partial_{\psi,q,p*})^*}$ by Proposition 4.8 and that on the space $L^p(\operatorname{VN}(G)) \oplus_p \overline{\operatorname{Ran} \partial_{\psi,q,p}}$,

$$\mathcal{D}^2_{\psi,q,p} \overset{(4.19)}{=} \begin{bmatrix} A_p & 0 \\ 0 & (\operatorname{Id}_{L^p(\Gamma_q(H))} \rtimes A_p)|\overline{\operatorname{Ran} \partial_{\psi,q,p}} \end{bmatrix} \overset{(4.7)}{=} \begin{bmatrix} A_p & 0 \\ 0 & B_p \end{bmatrix}.$$

So on $\overline{\operatorname{Ran} \mathcal{D}_{\psi,q,p}}$,

$$|\mathcal{D}_{\psi,q,p}|^{-1} = \begin{bmatrix} A_p|\overline{\operatorname{Ran} A_p} & 0 \\ 0 & B_p \end{bmatrix}^{-\frac{1}{2}} = \begin{bmatrix} A_p^{-\frac{1}{2}}|\overline{\operatorname{Ran} A_p} & 0 \\ 0 & B_p^{-\frac{1}{2}} \end{bmatrix}.$$

With Proposition 5.13, we conclude that $|\mathcal{D}_{\psi,q,p}|^{-1}$ is compact.

$\square$

Remark 5.8 Suppose that G has AP. We do not know if $\operatorname{Lip}_{\mathcal{D}_{\psi,q,p}}(\operatorname{VN}(G)) = \operatorname{dom} \partial_{\psi,q,\infty}$. We will examine this question in a future work.

Remark 5.9 Consider the case that the assumptions of Theorem 5.4 are satisfied. Note that the (Banach) spectral triple $(C_r^*(G), L^p(\operatorname{VN}(G)) \oplus_p L^p(\Gamma_q(H) \rtimes_\alpha G), \mathcal{D}_{\psi,q,p})$ is even. Indeed, the Hodge-Dirac operator $\mathcal{D}_{\psi,q,p}$ anti-commutes with the involution

$$\gamma_p \overset{\text{def}}{=} \begin{bmatrix} -\operatorname{Id}_{L^p(\operatorname{VN}(G))} & 0 \\ 0 & \operatorname{Id}_{L^p(\Gamma_q(H) \rtimes_\alpha G)} \end{bmatrix} :$$

$$L^p(\operatorname{VN}(G)) \oplus_p L^p(\Gamma_q(H) \rtimes_\alpha G) \to L^p(\operatorname{VN}(G)) \oplus_p L^p(\Gamma_q(H) \rtimes_\alpha G).$$

(which is selfadjoint if $p = 2$), since

$$\mathcal{D}_{\psi,q,p}\gamma_p + \gamma_p\mathcal{D}_{\psi,q,p}$$

$$\overset{(5.45)}{=} \begin{bmatrix} 0 & (\partial_{\psi,q,p*})^* \\ \partial_{\psi,q,p} & 0 \end{bmatrix} \begin{bmatrix} -\operatorname{Id} & 0 \\ 0 & \operatorname{Id} \end{bmatrix} + \begin{bmatrix} -\operatorname{Id} & 0 \\ 0 & \operatorname{Id} \end{bmatrix} \begin{bmatrix} 0 & (\partial_{\psi,q,p*})^* \\ \partial_{\psi,q,p} & 0 \end{bmatrix}$$

$$= \begin{bmatrix} 0 & (\partial_{\psi,q,p*})^* \\ -\partial_{\psi,q,p} & 0 \end{bmatrix} + \begin{bmatrix} 0 & -(\partial_{\psi,q,p*})^* \\ \partial_{\psi,q,p} & 0 \end{bmatrix} = 0.$$

Moreover, for any $a \in \operatorname{VN}(G)$, we have

$$\gamma_p\pi(a) \overset{(5.46)}{=} \begin{bmatrix} -\operatorname{Id} & 0 \\ 0 & \operatorname{Id} \end{bmatrix} \begin{bmatrix} L_a & 0 \\ 0 & \tilde{L}_a \end{bmatrix} = \begin{bmatrix} -L_a & 0 \\ 0 & \tilde{L}_a \end{bmatrix} = \begin{bmatrix} L_a & 0 \\ 0 & \tilde{L}_a \end{bmatrix} \begin{bmatrix} -\operatorname{Id} & 0 \\ 0 & \operatorname{Id} \end{bmatrix} \overset{(5.46)}{=} \pi(a)\gamma_p.$$

Remark 5.10 The estimate (5.48) is in general not optimal. Indeed, already in the case $p = 2$ and $a = \lambda_s \in \mathcal{P}_G$ for some $s \in G$, we have according to (5.52) and (5.53),

$$\left\| [\mathcal{D}_{\psi,q,2}, \pi(a)] \right\|_{\mathrm{L}^2(\mathrm{VN}(G)) \oplus_2 \mathrm{L}^2(\Gamma_q(H) \rtimes_\alpha G) \to \mathrm{L}^2(\Gamma_q(H)) \oplus_2 \mathrm{L}^2(\Gamma_q(H) \rtimes_\alpha G)} \qquad (5.55)$$

$$\leqslant \max \left\{ \left\| \mathrm{L}_{\partial_{\psi,q}(a)} J \right\|_{\mathrm{L}^2(\mathrm{VN}(G)) \to \mathrm{L}^2(\Gamma_q(H) \rtimes_\alpha G)}, \right.$$

$$\left. \left\| \mathbb{E} \mathrm{L}_{\partial_{\psi,q}(a)} \right\|_{\mathrm{L}^2(\Gamma_q(H) \rtimes_\alpha G) \to \mathrm{L}^2(\mathrm{VN}(G))} \right\}.$$

Note that we have the Hilbert space adjoints $(\mathrm{L}_{\partial_{\psi,q}(a)} J)^* = J^* \mathrm{L}^*_{\partial_{\psi,q}(a)} = \mathbb{E} \mathrm{L}_{(\partial_{\psi,q}(a))^*} = -\mathbb{E} \mathrm{L}_{\partial_{\psi,q}(a^*)}$. Thus, in the maximum of (5.55), it suffices to consider the second term. For any element $x = \sum_t x_t \rtimes \lambda_t$ of $\mathrm{L}^2(\Gamma_q(H) \rtimes_\alpha G)$, we have

$$\mathbb{E} \mathrm{L}_{\partial_{\psi,q}(\lambda_s)}(x) = \mathbb{E}\left((s_q(b_\psi(s)) \rtimes \lambda_s)\left(\sum_t x_t \rtimes \lambda_t \right) \right)$$

$$\overset{(2.56)}{=} \mathbb{E}\left(\sum_t s_q(b_\psi(s))\alpha_s(x_t) \rtimes \lambda_{st} \right) = \sum_t \tau(s_q(b_\psi(s))\alpha_s(x_t))\lambda_{st}.$$

Thus we have

$$\left\| \mathbb{E} \mathrm{L}_{\partial_{\psi,q}(\lambda_s)}(x) \right\|_2^2 = \sum_t \left| \tau(s_q(b_\psi(s))\alpha_s(x_t)) \right|^2$$

$$\leqslant \left\| s_q(b_\psi(s)) \right\|_{\mathrm{L}^2(\Gamma_q(H))}^2 \sum_t \left\| \alpha_s(x_t) \right\|_{\mathrm{L}^2(\Gamma_q(H))}^2$$

$$= \left\| s_q(b_\psi(s)) \right\|_{\mathrm{L}^2(\Gamma_q(H))}^2 \sum_t \left\| x_t \right\|_{\mathrm{L}^2(\Gamma_q(H))}^2$$

$$\overset{(2.41)}{=} \left\| b_\psi(s) \right\|_H^2 \left\| x \right\|_{\mathrm{L}^2(\Gamma_q(H) \overline{\otimes} \mathrm{B}(\ell_I^2))}^2 .$$

We infer that

$$\left\| [\mathcal{D}_{\psi,q,2}, \pi(\lambda_s)] \right\|_{\mathrm{L}^2 \oplus_2 \mathrm{L}^2 \to \mathrm{L}^2 \oplus_2 \mathrm{L}^2} \leqslant \left\| \mathbb{E} \mathrm{L}_{\partial_{\psi,q}(\lambda_s)} \right\|_{2 \to 2} \leqslant \left\| b_\psi(s) \right\|_H . \qquad (5.56)$$

In the case where $-1 < q < 1$ and $b_\psi(s) \neq 0$, this quantity is strictly less than

$$\left\| \partial_{\psi,q}(\lambda_s) \right\|_{\Gamma_q(H) \rtimes_\alpha G} \overset{(2.84)}{=} \left\| s_q(b_\psi(s)) \rtimes \lambda_s \right\|_{\Gamma_q(H) \rtimes_\alpha G} = \left\| s_q(b_\psi(s)) \right\|_{\Gamma_q(H)}$$

$$\overset{[44,\ \mathrm{Th.}\ 1.10]}{=} \frac{2}{\sqrt{1-q}} \left\| b_\psi(s) \right\|_H .$$

Remark 5.11 In [131, p. 588], the authors obtain a similar estimate to (5.48) on the commutator for the case $q = 1$ and $p = 2$. The differences are that the space on which the commutator is regarded, is the smaller $L^2(VN(G)) \oplus_2 \overline{\Omega_\psi(G)} \subsetneq L^2(VN(G)) \oplus_2 L^2(\Gamma_1(H) \rtimes_\alpha G)$ [131, p. 587], where $\Omega_\psi(G) = \text{span}\{s_1(h) \rtimes \lambda_s : h \in H_\psi, s \in G\}$ and H_ψ is the Hilbert space spanned by the $b_\psi(s)$, $s \in G$ [131, Lemma C.1]. On the other hand, [131, p. 588] obtains an exact estimate (without proof)

$$\left\| [\mathscr{D}_{\psi,q=1,p=2}, \pi(a)] \right\|_{L^2 \oplus_2 \overline{\Omega_\psi(G)} \to L^2 \oplus_2 \overline{\Omega_\psi(G)}} \tag{5.57}$$

$$= \max \left\{ \|\Gamma(a,a)\|_{VN(G)}^{\frac{1}{2}}, \|\Gamma(a^*,a^*)\|_{VN(G)}^{\frac{1}{2}} \right\}.$$

Note that when $q = 1$ and $a \in \mathcal{P}_G$, then the right hand side of (5.57) is finite, whereas the right hand side of (5.48), that is, $\|\partial_{\psi,1,\infty}(a)\|_{\Gamma_1(H) \rtimes_\alpha G}$ is infinite. We do not know whether (5.57) holds on our full space $L^2(VN(G)) \oplus_2 L^2(\Gamma_q(H) \rtimes_\alpha G)$ or if some intermediate estimate out of (5.48) and (5.56) holds in that case. But we plan to examine this question.

5.9 Spectral Triples Associated to Semigroups of Fourier Multipliers II

We consider in this section a discrete group G and a cocycle $b_\psi : G \to H$. Now, we define another "Hodge-Dirac operator" by generalizing the construction of [131, p. 588] which corresponds to the case $q = 0$ and $p = 2$ below. Suppose $1 < p < \infty$ and $-1 \leqslant q < 1$. We let

$$\mathscr{D}_{\psi,q}(x \rtimes \lambda_s) \stackrel{\text{def}}{=} s_q(b_\psi(s))x \rtimes \lambda_s, \quad x \in \Gamma_q(H), s \in G. \tag{5.58}$$

We can see $\mathscr{D}_{\psi,q}$ as an unbounded operator acting on the subspace $\mathcal{P}_{\rtimes,G}$ of $L^p(\Gamma_q(H) \rtimes_\alpha G)$.

Finally, note that in [131, p. 588], the authors refer to a real structure but we warn the reader that the antilinear isometry $J : L^2(\Gamma_q(H) \rtimes_\alpha G) \to L^2(\Gamma_q(H) \rtimes_\alpha G)$, $x \mapsto x^*$ used in [131] does not[15] commute or anticommute with the Dirac operator $\mathscr{D}_{\psi,q}$.

---

[15] For any $x \in \Gamma_q(H)$ and any $s \in G$, we have

$$J\mathscr{D}_{\psi,q}(x \rtimes \lambda_s) \stackrel{(5.58)}{=} J\big(s_q(b_\psi(s))x \rtimes \lambda_s\big) = \big(s_q(b_\psi(s))x \rtimes \lambda_s\big)^*$$

$$\stackrel{(2.57)}{=} \alpha_{s^{-1}}(x^* s_q(b_\psi(s))) \rtimes \lambda_{s^{-1}}$$

Lemma 5.13 *Suppose* $1 < p < \infty$ *and* $-1 \leqslant q < 1$.

1. *For any* $a, b \in \mathcal{P}_{\rtimes,G}$ *we have*

$$\langle \mathscr{D}_{\psi,q}(a), b \rangle = \langle a, \mathscr{D}_{\psi,q}(b) \rangle \qquad (5.59)$$

where we use as usual the duality bracket $\langle x, y \rangle = \tau(x^* y)$.

2. *The operator* $\mathscr{D}_{\psi,q} \colon \mathcal{P}_{\rtimes,G} \subset \mathrm{L}^p(\Gamma_q(H) \rtimes_\alpha G) \to \mathrm{L}^p(\Gamma_q(H) \rtimes_\alpha G)$ *is closable.*

Proof

1. For any $s, t \in G$ and any $x, y \in \Gamma_q(H)$, we have

$$\langle \mathscr{D}_{\psi,q}(x \rtimes \lambda_s), y \rtimes \lambda_t \rangle \overset{(5.58)}{=} \langle s_q(b_\psi(s))x \rtimes \lambda_s, y \rtimes \lambda_t \rangle$$

$$= \tau_\rtimes \big((s_q(b_\psi(s))x \rtimes \lambda_s)^*(y \rtimes \lambda_t) \big)$$

$$\overset{(2.57)}{=} \tau_\rtimes \big((\alpha_{s^{-1}}(x^* s_q(b_\psi(s))) \rtimes \lambda_{s^{-1}})(y \rtimes \lambda_t) \big)$$

$$\overset{(2.56)}{=} \tau_\rtimes \big(\alpha_{s^{-1}}(x^* s_q(b_\psi(s)))\alpha_{s^{-1}}(y) \rtimes \lambda_{s^{-1}t} \big) = \tau \big(\alpha_{s^{-1}}(x^* s_q(b_\psi(s))y) \big) \delta_{s=t}$$

and

$$\langle x \rtimes \lambda_s, \mathscr{D}_{\psi,q}(y \rtimes \lambda_t) \rangle \overset{(5.58)}{=} \langle x \rtimes \lambda_s, s_q(b_\psi(t))y \rtimes \lambda_t \rangle$$

$$= \tau_\rtimes \big((x \rtimes \lambda_s)^*(s_q(b_\psi(t))y \rtimes \lambda_t) \big)$$

$$\overset{(2.57)}{=} \tau_\rtimes \big((\alpha_{s^{-1}}(x^*) \rtimes \lambda_{s^{-1}})(s_q(b_\psi(t))y \rtimes \lambda_t) \big)$$

$$\overset{(2.56)}{=} \tau_\rtimes \big(\alpha_{s^{-1}}(x^*)\alpha_{s^{-1}}(s_q(b_\psi(t))y) \rtimes \lambda_{s^{-1}t} \big)$$

$$= \tau \big(\alpha_{s^{-1}}(x^* s_q(b_\psi(s))y) \big) \delta_{s=t}.$$

Thus, (5.59) follows by linearity.

2. Since $\mathcal{P}_{\rtimes,G}$ is dense in $\mathrm{L}^{p^*}(\Gamma_q(H) \rtimes_\alpha G)$, this is a consequence of [136, Theorem 5.28 p. 168].

$\square$

---

and

$$\mathscr{D}_{\psi,q} J(x \rtimes \lambda_s) = \mathscr{D}_{\psi,q}\big(\alpha_{s^{-1}}(x^*) \rtimes \lambda_{s^{-1}} \big)$$

$$\overset{(5.58)}{=} s_q(b_\psi(s^{-1}))\alpha_{s^{-1}}(x^*) \rtimes \lambda_{s^{-1}} = -s_q(\pi_{s^{-1}}(b_\psi(s)))\alpha_{s^{-1}}(x^*) \rtimes \lambda_{s^{-1}}$$

$$= -\alpha_{s^{-1}}(s_q(b_\psi(s)))\alpha_{s^{-1}}(x^*) \rtimes \lambda_{s^{-1}} = -\alpha_{s^{-1}}(s_q(b_\psi(s))x^*) \rtimes \lambda_{s^{-1}}.$$

We denote by $\mathscr{D}_{\psi,q,p}\colon \operatorname{dom} \mathscr{D}_{\psi,q,p} \subset L^p(\Gamma_q(H) \rtimes_\alpha G) \to L^p(\Gamma_q(H) \rtimes_\alpha G)$ its closure. So $\mathcal{P}_{\rtimes,G}$ is a core of $\mathscr{D}_{\psi,q,p}$. We define the homomorphism $\pi \colon \mathrm{VN}(G) \to \mathrm{B}(L^p(\Gamma_q(H) \rtimes_\alpha G))$ by

$$\pi(a) \overset{\mathrm{def}}{=} \operatorname{Id}_{\Gamma_q(H)} \rtimes \mathrm{L}_a, \quad a \in \mathrm{VN}(G) \tag{5.60}$$

where L_a is the left multiplication operator by a. Note that $\pi(a)$ is equal to the map $\tilde{\mathrm{L}}_a$ of Sect. 5.8. It is not difficult to see that $\pi \colon \mathrm{VN}(G) \to \mathrm{B}(L^p(\Gamma_q(H) \rtimes_\alpha G))$ is continuous when $\mathrm{VN}(G)$ is equipped with the weak* topology of $\mathrm{VN}(G)$ and when $\mathrm{B}(L^p(\Gamma_q(H) \rtimes_\alpha G))$ is equipped with the weak operator topology. We will also use the restriction of π to $\mathrm{C}^*_r(G)$.

Theorem 5.5 *Suppose* $1 < p < \infty$ *and* $-1 \leqslant q < 1$.

1. *The operator* $\mathscr{D}_{\psi,q,2}$ *is selfadjoint.*
2. *Assume that* $L^{p^*}(\mathrm{VN}(G))$ *has* CCAP *and that* $\Gamma_q(H) \rtimes_\alpha G$ *has* QWEP. *For* $1 < p < \infty$, *the unbounded operators* $\mathscr{D}_{\psi,q,p}$ *and* $\mathscr{D}_{\psi,q,p^*}$ *are adjoint to each other (with respect to the duality bracket* $\langle x, y \rangle = \tau(x^* y)$*).*
3. *We have*

$$\mathcal{P}_G \subset \operatorname{Lip}_{\mathscr{D}_{\psi,q,p}}(\mathrm{VN}(G)). \tag{5.61}$$

4. *For any* $a \in \mathcal{P}_G$, *we have*

$$[\mathscr{D}_{\psi,q,p}, \pi(a)] = \mathrm{L}_{\partial_{\psi,q}(a)} \tag{5.62}$$

and

$$\left\| [\mathscr{D}_{\psi,q,p}, \pi(a)] \right\|_{L^p(\Gamma_q(H) \rtimes_\alpha G) \to L^p(\Gamma_q(H) \rtimes_\alpha G)} = \left\| \partial_{\psi,q}(a) \right\|_{\Gamma_q(H) \rtimes_\alpha G}. \tag{5.63}$$

5. *Suppose that* G *has* AP. *We have*

$$\operatorname{dom} \partial_{\psi,q,\infty} \subset \operatorname{Lip}_{\mathscr{D}_{\psi,q,p}}(\mathrm{VN}(G)). \tag{5.64}$$

6. *Suppose that* G *is weakly amenable. We have*

$$\operatorname{Lip}_{\mathscr{D}_{\psi,q,p}}(\mathrm{VN}(G)) = \operatorname{dom} \partial_{\psi,q,\infty}. \tag{5.65}$$

7. *Suppose that* G *has* AP. *For any* $a \in \operatorname{dom} \partial_{\psi,q,\infty}$, *we have*

$$[\mathscr{D}_{\psi,q,p}, \pi(a)] = \mathrm{L}_{\partial_{\psi,q,\infty}(a)}. \tag{5.66}$$

8. *Suppose that G has* AP. *For any* $a \in \mathrm{dom}\, \partial_{\psi,q,\infty}$, *we have*

$$\left\| [\mathscr{D}_{\psi,q,p}, \pi(a)] \right\|_{\mathrm{L}^p(\Gamma_q(H) \rtimes_\alpha G) \to \mathrm{L}^p(\Gamma_q(H) \rtimes_\alpha G)} = \left\| \partial_{\psi,q,\infty}(a) \right\|_{\Gamma_q(H) \rtimes_\alpha G}. \tag{5.67}$$

9. *Suppose* $p = 2$ *and* $q = -1$. *We have* $(\mathscr{D}_{\psi,-1,2})^2 = \mathrm{Id} \rtimes A_2$.

10. *If* $\Gamma_{-1}(H) \rtimes_\alpha G$ *has* QWEP, $b_\psi \colon G \to H$ *is injective,* $\mathrm{Gap}_\psi > 0$ *and if* H *is finite-dimensional then the operator* $|\mathscr{D}_{\psi,-1,2}|^{-1} \colon \overline{\mathrm{Ran}\, \mathscr{D}_{\psi,-1,2}} \to \overline{\mathrm{Ran}\, \mathscr{D}_{\psi,-1,2}}$ *is compact.*

Proof 1. It suffices to show that $\mathscr{D}_{\psi,q} \colon \mathcal{P}_{\rtimes,G} \subset \mathrm{L}^2(\Gamma_q(H) \rtimes_\alpha G) \to \mathrm{L}^2(\Gamma_q(H) \rtimes_\alpha G)$ is essentially selfadjoint. By (5.59), we infer that $\mathscr{D}_{\psi,q}$ is symmetric. By [195, Corollary p. 257], it suffices to prove that $\mathrm{Ran}(\mathscr{D}_{\psi,q} \pm i\mathrm{Id})$ is dense in $\mathrm{L}^2(\Gamma_q(H) \rtimes_\alpha G)$.

Let $z = \sum_s z_s \rtimes \lambda_s \in \mathrm{L}^2(\Gamma_q(H) \rtimes_\alpha G)$ be a vector which is orthogonal to $\mathrm{Ran}(\mathscr{D}_{\psi,q} + i\mathrm{Id})$ in $\mathrm{L}^2(\Gamma_q(H) \rtimes_\alpha G)$. For any $s \in G$ and any $x \in \mathrm{L}^2(\Gamma_q(H))$, we have

$$\left\langle s_q(b_\psi(s))x \rtimes \lambda_s, z \right\rangle_{\mathrm{L}^2(\Gamma_q(H) \rtimes_\alpha G)} - i\tau(x^* z_s)$$

$$\overset{(5.58)}{=} \left\langle \mathscr{D}_{\psi,q}(x \rtimes \lambda_s), z \right\rangle_{\mathrm{L}^2(\Gamma_q(H) \rtimes_\alpha G)} - i \langle x \rtimes \lambda_s, z \rangle_{\mathrm{L}^2(\Gamma_q(H) \rtimes_\alpha G)}$$

$$= \left\langle (\mathscr{D}_{\psi,q} - i\mathrm{Id})(x \rtimes \lambda_s), z \right\rangle_{\mathrm{L}^2(\Gamma_q(H) \rtimes_\alpha G)} = 0.$$

We infer that

$$\left\langle x, s_q(b_\psi(s))z_s \right\rangle_{\mathrm{L}^2(\Gamma_q(H))} = \tau(x^* s_q(b_\psi(s))z_s) = i\tau(x^* z_s) = \langle x, iz_s \rangle_{\mathrm{L}^2(\Gamma_q(H))}.$$

We deduce that $s_q(b_\psi(s))z_s = iz_s$. Recall that by [128, p. 96], the map $\Gamma_q(H) \to \mathcal{F}_q(H)$, $y \mapsto y(\Omega)$ extends to an isometry $\Delta \colon \mathrm{L}^2(\Gamma_q(H)) \to \mathcal{F}_q(H)$. For any $x \in \Gamma_q(H)$ and any $y \in \mathrm{L}^2(\Gamma_q(H))$, it is easy to check that $\Delta(xy) = x\Delta(y)$. We infer that

$$s_q(b_\psi(s))\Delta(z_s) = \Delta(s_q(b_\psi(s))z_s) = \Delta(iz_s) = i\Delta(z_s).$$

Thus the vector $\Delta(z_s)$ of $\mathcal{F}_q(H)$ is zero or an eigenvector of $s_q(b_\psi(s))$. Since $s_q(b_\psi(s))$ is selfadjoint, i is not an eigenvalue. So $\Delta(z_s) = 0$. Since Δ is injective, we infer that $z_s = 0$. It follows that $z = 0$. The case with $-i$ instead of i is similar.

2. By (5.59) and [136, Problem 5.24 p. 168], $\mathscr{D}_{\psi,q,p}$ and $\mathscr{D}_{\psi,q,p^*}$ are formal adjoints. Hence $\mathscr{D}_{\psi,q,p^*} \subset (\mathscr{D}_{\psi,q,p})^*$ by [136, p. 167]. Now, we will show the reverse inclusion. Let $z \in \mathrm{dom}(\mathscr{D}_{\psi,q,p})^*$. For any $y \in \mathrm{dom}\, \mathscr{D}_{\psi,q,p}$, we have

$$\left\langle \mathscr{D}_{\psi,q,p}(y), z \right\rangle = \left\langle y, (\mathscr{D}_{\psi,q,p})^*(z) \right\rangle. \tag{5.68}$$

Note that $(\mathrm{Id} \rtimes M_{\varphi_j})(z) \to z$ in $L^{p^*}(\Gamma_q(H) \rtimes_\alpha G)$, where $(M_{\varphi_j})_j$ is the approximating net of Fourier multipliers granted by the CCAP assumption. Now, for any $y \in \mathcal{P}_{\rtimes,G}$, we have

$$\langle y, \mathscr{D}_{\psi,q,p^*}(\mathrm{Id} \rtimes M_{\varphi_j})(z) \rangle = \langle \mathscr{D}_{\psi,q,p}(y), (\mathrm{Id} \rtimes M_{\varphi_j})(z) \rangle$$

$$= \langle (\mathrm{Id} \rtimes M_{\overline{\varphi_j}})\mathscr{D}_{\psi,q,p}(y), z \rangle = \langle \mathscr{D}_{\psi,q,p}(\mathrm{Id} \rtimes M_{\overline{\varphi_j}})(y), z \rangle$$

$$\overset{(5.68)}{=} \langle (\mathrm{Id} \rtimes M_{\overline{\varphi_j}})(y), (\mathscr{D}_{\psi,q,p})^*(z) \rangle = \langle y, (\mathrm{Id} \rtimes M_{\varphi_j})(\mathscr{D}_{\psi,q,p})^*(z) \rangle.$$

Hence

$$\mathscr{D}_{\psi,q,p^*}(\mathrm{Id} \rtimes M_{\varphi_j})(z) = (\mathrm{Id} \rtimes M_{\varphi_j})(\mathscr{D}_{\psi,q,p})^*(z) \underset{j}{\to} (\mathscr{D}_{\psi,q,p})^*(z)$$

By (2.1), we deduce that $z \in \mathrm{dom}\, \mathscr{D}_{\psi,q,p^*}$ and that

$$\mathscr{D}_{\psi,q,p^*}(z) = (\mathscr{D}_{\psi,q,p})^*(z).$$

3. and 4. For any $s, t \in G$ and any $x \in \Gamma_q(H)$, using the relation $\alpha_s = \Gamma_q^\infty(\pi_s)$ in the sixth equality, we have

$$\left[\mathscr{D}_{\psi,q,p}, \pi(\lambda_s)\right](x \rtimes \lambda_t) = \mathscr{D}_{\psi,q,p}\pi(\lambda_s)(x \rtimes \lambda_t) - \pi(\lambda_s)\mathscr{D}_{\psi,q,p}(x \rtimes \lambda_t)$$

$$\overset{(5.60)(5.58)}{=} \mathscr{D}_{\psi,q,p}(\alpha_s(x) \rtimes \lambda_{st}) - \pi(\lambda_s)(s_q(b_\psi(t))x \rtimes \lambda_t)$$

$$\overset{(5.60)}{=} \mathscr{D}_{\psi,q,p}(\alpha_s(x) \rtimes \lambda_{st}) - \alpha_s(s_q(b_\psi(t))x) \rtimes \lambda_{st}$$

$$\overset{(5.58)}{=} s_q(b_\psi(st))\alpha_s(x) \rtimes \lambda_{st} - \alpha_s(s_q(b_\psi(t))x) \rtimes \lambda_{st}$$

$$\overset{(2.33)}{=} s_q(b_\psi(s))\alpha_s(x) \rtimes \lambda_{st} + s_q(\pi_s(b_\psi(t)))\alpha_s(x) \rtimes \lambda_{st} - \alpha_s(s_q(b_\psi(t))x) \rtimes \lambda_{st}$$

$$= s_q(b_\psi(s))\alpha_s(x) \rtimes \lambda_{st} \overset{(2.56)}{=} (s_q(b_\psi(s)) \rtimes \lambda_s)(x \rtimes \lambda_t) \overset{(2.84)}{=} \partial_{\psi,q}(\lambda_s)(x \rtimes \lambda_t)$$

$$= \mathrm{L}_{\partial_{\psi,q}(\lambda_s)}(x \rtimes \lambda_t).$$

By linearity and density, we obtain $\left[\mathscr{D}_{\psi,q,p}, \pi(\lambda_s)\right] = \mathrm{L}_{\partial_{\psi,q}(\lambda_s)}$. It suffices to use Proposition 5.12. Finally by linearity, for any $a \in \mathcal{P}_G$ we obtain (5.62). In addition, we have

$$\left\|\left[\mathscr{D}_{\psi,q,p}, \pi(a)\right]\right\|_{L^p(\Gamma_q(H)\rtimes_\alpha G) \to L^p(\Gamma_q(H)\rtimes_\alpha G)} \tag{5.69}$$

$$\overset{(5.62)}{=} \left\|\mathrm{L}_{\partial_{\psi,q}(a)}\right\|_{L^p(\Gamma_q(H)\rtimes_\alpha G) \to L^p(\Gamma_q(H)\rtimes_\alpha G)} = \left\|\partial_{\psi,q}(a)\right\|_{\Gamma_q(H)\rtimes_\alpha G}.$$

5. We next claim that

$$\mathrm{dom}\, \partial_{\psi,q,\infty} \subset \mathrm{Lip}_{\mathscr{D}_{\psi,q,p}}(\mathrm{VN}(G)). \tag{5.70}$$

Let $a \in \mathrm{dom}\, \partial_{\psi,q,\infty}$ and consider with Proposition 3.5 a net (a_j) of elements of $\mathcal{P}_G$ such that $a_j \to a$ and $\partial_{\psi,q}(a_j) \to \partial_{\psi,q,\infty}(a)$ both for the weak* topology. By Lemma 2.1, we can suppose that the nets (a_j) and $(\partial_{\psi,q}(a_j))$ are bounded. So by (5.63), the net $([\mathscr{D}_{\psi,q,p}, \pi(a_j)])$ is also bounded. Now by the part 4 of Proposition 5.11, a belongs to $\mathrm{Lip}_{\mathscr{D}_{\psi,q,p}}(\mathrm{VN}(G))$. We have shown (5.70).

6. Next we pass to the reverse inclusion of (5.70) and claim $\mathrm{Lip}_{\mathscr{D}_{\psi,q,p}}(\mathrm{VN}(G)) \subset$ $\mathrm{dom}\, \partial_{\psi,q,\infty}$. To this end, let $a \in \mathrm{Lip}_{\mathscr{D}_{\psi,q,p}}(\mathrm{VN}(G))$ and denote $\hat{a}_r \overset{\mathrm{def}}{=} \langle a, \lambda_r \rangle = \tau(a\lambda_{r^{-1}})$ its Fourier coefficients for $r \in G$. We define a linear form $T_a \colon \mathcal{P}_{\rtimes,G} \to \mathbb{C}$ by

$$T_a(x \rtimes \lambda_s) \overset{\mathrm{def}}{=} \sum_{r \in G} \tau_{\rtimes}\big((\hat{a}_r s_q(b_\psi(r)) \rtimes \lambda_r) \cdot (x \rtimes \lambda_s)\big).$$

Since the trace vanishes for $r \neq s^{-1}$, the sum over r is finite. We will show that it extends to a bounded linear form on $\mathrm{L}^p(\Gamma_q(H) \rtimes_\alpha G)$. To this end consider the bounded net (a_j) in $\mathcal{P}_{\rtimes,G}$ defined by $a_j \overset{\mathrm{def}}{=} M_{\varphi_j}(a)$, where (φ_j) is the approximating net guaranteed by the fact that G has AP. By Proposition 2.9, we have $a_j \to a$ in the weak* topology. Then for any $\xi, \eta \in \mathcal{P}_{\rtimes,G}$, using the point 2 in the first equality, we obtain

$$\langle [\mathscr{D}_{\psi,q,p}, \pi(a)]\xi, \eta \rangle = \langle \pi(a)\xi, \mathscr{D}_{\psi,q,p^*}\eta \rangle - \langle \mathscr{D}_{\psi,q,p}\xi, \pi(a)^*\eta \rangle$$

$$= \lim_j \langle \pi(a_j)\xi, \mathscr{D}_{\psi,q,p^*}\eta \rangle - \langle \mathscr{D}_{\psi,q,p}\xi, \pi(a_j)^*\eta \rangle = \lim_j \langle [\mathscr{D}_{\psi,q,p}, \pi(a_j)]\xi, \eta \rangle$$

$$\overset{(5.62)}{=} \lim_j \langle \partial_{\psi,q}(a_j)\xi, \eta \rangle = \tau_{\rtimes}((\partial_{\psi,q}(a_j)\xi)^*\eta) = \lim_j \tau_{\rtimes}(\xi^*\partial_{\psi,q}(a_j)^*\eta)$$

$$= \lim_j \overline{\tau_{\rtimes}(\eta^*\partial_{\psi,q}(a_j)\xi)} = \lim_j \overline{\tau_{\rtimes}(\partial_{\psi,q}(a_j)\xi\eta^*)} = \overline{T_a(\xi\eta^*)},$$

where in the last equality, we used $\varphi_j(s) \to 1$ for each fixed $s \in G$ and the definition of T_a. Since $a \in \mathrm{Lip}_{\mathscr{D}_{\psi,q,p}}(\mathrm{VN}(G))$, for any $\xi, \eta \in \mathcal{P}_\rtimes$, we deduce the estimate $|T_a(\xi\eta^*)| \leqslant \big\| [\mathscr{D}_{\psi,q,p}, \pi(a)] \big\|_{p \to p} \|\xi\|_p \|\eta\|_{p^*}$. Choosing the element $\eta = 1$ of $\mathcal{P}_{\rtimes,G}$, we get $|T_a(\xi)| \lesssim \|\xi\|_p$, so that T_a induces an element of $(\mathrm{L}^p(\Gamma_q(H) \rtimes_\alpha G))^*$. We infer that the Fourier coefficients sequence $(\hat{a}_r s_q(b_\psi(r)))_{r \in G}$ belongs to an element $b \in \mathrm{L}^{p^*}(\Gamma_q(H) \rtimes_\alpha G)$. But then the previous calculation shows that for any $\xi, \eta \in \mathcal{P}_{\rtimes,G}$

$$\langle [\mathscr{D}_{\psi,q,p}, \pi(a)]\xi, \eta \rangle = \lim_j \langle \partial_{\psi,q}(a_j)\xi, \eta \rangle = \langle b\xi, \eta \rangle. \tag{5.71}$$

It follows from $a \in \mathrm{Lip}_{\mathscr{D}_{\psi,q,p}}(\mathrm{VN}(G))$ that

$$|\langle b\xi, \eta\rangle| \leqslant \left\|[\mathscr{D}_{\psi,q,p}, \pi(a)]\right\|_{p\to p} \|\xi\|_p \|\eta\|_{p^*},$$

so that $\|b\xi\|_p \leqslant \left\|[\mathscr{D}_{\psi,q,p}, \pi(a)]\right\|_{p\to p} \|\xi\|_p$. Thus, the element $b \in L^{p^*}(\Gamma_q(H) \rtimes_\alpha G)$ is a pointwise multiplier $L^p \to L^p$. It follows[16] that $b \in \Gamma_q(H) \rtimes_\alpha G$.

It remains to show that $a \in \mathrm{dom}\, \partial_{\psi,q,\infty}$. To this end, thanks to the AP property of G, we consider again the bounded net (a_j) defined by $a_j \overset{\mathrm{def}}{=} M_{\varphi_j}(a)$. By Proposition 2.9, we have $a_j \to a$ in the weak* topology. Moreover, we have

$$\left\|\partial_{\psi,q}(a_j)\right\|_{\Gamma_q(H)\rtimes_\alpha G} = \left\|\partial_{\psi,q}\left(\sum_r \hat{a}_r \varphi_j(r) \lambda_r\right)\right\| = \left\|\sum_r \hat{a}_r \varphi_j(r) s_q(b_\psi(r)) \rtimes \lambda_r\right\|$$

$$= \left\|\sum_r \hat{a}_r (\mathrm{Id} \rtimes M_{\varphi_j})(s_q(b_\psi(r)) \rtimes \lambda_r)\right\| = \left\|(\mathrm{Id} \rtimes M_{\varphi_j})(b)\right\|_{\Gamma_q(H)\rtimes_\alpha G}$$

$$\leqslant \left\|\mathrm{Id} \rtimes M_{\varphi_j}\right\|_{\Gamma_q(H)\rtimes_\alpha G \to \Gamma_q(H)\rtimes_\alpha G} \|b\|_{\Gamma_q(H)\rtimes_\alpha G}.$$

Using Proposition 2.9 in the last estimate (since G is weakly amenable), we deduce that $(\partial_{\psi,q}(a_j))$ is a bounded net. Now, using Banach-Alaoglu theorem, we can suppose that $\partial_{\psi,q}(a_j) \to y$ for the weak* topology for some y. Since the graph of the unbounded operator $\partial_{\psi,q,\infty}$ is weak* closed, we conclude that a belongs to the subspace $\mathrm{dom}\, \partial_{\psi,q,\infty}$.

7. Let $a \in \mathrm{dom}\, \partial_{\psi,q,\infty}$. Consider with Proposition 3.5 a net (a_j) of elements of $\mathcal{P}_G$ such that $a_j \to a$ and $\partial_{\psi,q}(a_j) \to \partial_{\psi,q,\infty}(a)$ both for the weak* topology. By Lemma 2.1, we can suppose that the nets (a_j) and $(\partial_{\psi,q}(a_j))$ are bounded. By continuity of π, $\pi(a_j) \to \pi(a)$ for the weak operator topology of $\mathrm{B}(L^p(\Gamma_q(H) \rtimes_\alpha G))$. For any $\xi, \zeta \in \mathcal{P}_{\rtimes, G}$, we have

$$\left\langle [\mathscr{D}_{\psi,q,p}, \pi(a_j)]\xi, \zeta\right\rangle_{L^p(\Gamma_q(H)\rtimes_\alpha G), L^{p^*}(\Gamma_q(H)\rtimes_\alpha G)}$$

$$= \left\langle (\mathscr{D}_{\psi,q,p}\pi(a_j) - \pi(a_j)\mathscr{D}_{\psi,q,p})\xi, \zeta\right\rangle$$

$$= \left\langle \mathscr{D}_{\psi,q,p}\pi(a_j)\xi, \zeta\right\rangle - \left\langle \pi(a_j)\mathscr{D}_{\psi,q,p}\xi, \zeta\right\rangle$$

$$= \left\langle \pi(a_j)\xi, \mathscr{D}_{\psi,q,p^*}\zeta\right\rangle - \left\langle \pi(a_j)\mathscr{D}_{\psi,q,p}\xi, \zeta\right\rangle$$

$$\underset{j}{\to} \left\langle \pi(a)\xi, \mathscr{D}_{\psi,q,p^*}\zeta\right\rangle - \left\langle \pi(a)\mathscr{D}_{\psi,q,p}\xi, \zeta\right\rangle = \left\langle [\mathscr{D}_{\psi,q,p}, \pi(a)]\xi, \zeta\right\rangle.$$

---

[16] For $t > 0$, let $\mu_t(b)$ denote the generalized singular number of $|b|$, so that $\tau_\rtimes(\phi(|b|)) = \int_0^\infty \phi(\mu_t(b))dt$ for any continuous function $\mathbb{R}_+ \to \mathbb{R}$ of bounded variation [227, p. 30]. If b were unbounded, then $\mu_t(b) \to \infty$ as $t \to 0$. Taking $\xi = \phi(|b|) \neq 0$ with ϕ a smoothed indicator function of an interval $[x, y]$ with large x, it is not difficult to see that $\|b\xi\|_p \geqslant x \|\xi\|_p$, which is the desired contradiction.

Since the net $([\mathscr{D}_{\psi,q,p}, \pi(a_j)])$ is bounded by (5.63), we deduce that the net $([\mathscr{D}_{\psi,q,p}, \pi(a_j)])$ converges to $[\mathscr{D}_{\psi,q,p}, \pi(a)]$ for the weak operator topology by a "net version" of [136, Lemma 3.6 p. 151]. Furthermore, it is (really) easy to check that $L_{\partial_{\psi,q,\infty}}(a_j) \to L_{\partial_{\psi,q,\infty}}(a)$ for the weak operator topology of $B(L^p(\Gamma_q(H) \rtimes_\alpha G))$. By uniqueness of the limit, we deduce that the commutator is given by the same formula as that in the case of elements of $\mathcal{P}_G$.

8. We obtain (5.67) as in (5.69).

9. If $x \in L^2(\Gamma_{-1}(H))$ and $s \in G$, note that

$$(\mathscr{D}_{\psi,-1,2})^2(x \rtimes \lambda_s) \overset{(5.58)}{=} \mathscr{D}_{\psi,-1,2}\big(s_{-1}(b_\psi(s))x \rtimes \lambda_s\big) \overset{(5.58)}{=} s_{-1}(b_\psi(s))^2 x \rtimes \lambda_s$$

$$\overset{(2.42)}{=} x \rtimes \big\|b_\psi(s)\big\|_H^2 \lambda_s = x \rtimes \psi(s)\lambda_s = (\mathrm{Id}_{L^2(\Gamma_{-1}(H))} \rtimes A_2)(x \rtimes \lambda_s).$$

By Hytönen et al. [110, Proposition G.2.4], note that $\mathcal{P}_{\rtimes,G}$ is a core of $\mathrm{Id}_{L^2(\Gamma_{-1}(H))} \rtimes A_2$. Then since $\mathscr{D}_{\psi,-1,2}^2$ is again selfadjoint, thus closed, we have $\mathrm{Id}_{L^2(\Gamma_{-1}(H))} \rtimes A_2 \subset \mathscr{D}_{\psi,-1,2}^2$. Now it follows from the fact that both $T_1 \overset{\mathrm{def}}{=} \mathrm{Id}_{L^2(\Gamma_{-1}(H))} \rtimes A_2$ and $T_2 \overset{\mathrm{def}}{=} \mathscr{D}_{\psi,-1,2}^2$ are selfadjoint that both operators are in fact equal. Indeed, consider some $\lambda \in \rho(T_1) \cap \rho(T_2)$ (e.g. $\lambda = i$). Then $T_1 - \lambda \subset T_2 - \lambda$, the operator $T_1 - \lambda$ is surjective and $T_2 - \lambda$ is injective. Now, it suffices to use the result [207, p. 5] to conclude that $T_1 - \lambda = T_2 - \lambda$ and thus $T_1 = T_2$.

10. By Proposition 5.13, the operator $A_2^{-\frac{1}{2}} : \overline{\mathrm{Ran}\, A_2} \to \overline{\mathrm{Ran}\, A_2}$ is compact. So there exists a sequence (T_n) of finite rank bounded operators $T_n : \overline{\mathrm{Ran}\, A_2} \to \overline{\mathrm{Ran}\, A_2}$ which approximate $A_2^{-\frac{1}{2}}$ in norm. Similarly to the proof of [121, Theorem 4.4], we may assume, without loss of generality, that the range of each T_n is contained in $\mathcal{P}_G$. Composing on the right with the orthogonal projection from $L^2(\mathrm{VN}(G))$ onto $\overline{\mathrm{Ran}\, A_2}$, we can see $A_2^{-\frac{1}{2}}$ and each T_n as operators on $L^2(\mathrm{VN}(G))$ which is an operator Hilbert space by Pisier [189, p. 139]. Moreover, $A_2^{-\frac{1}{2}}$ is then a Fourier multiplier (with symbol $\psi(s)^{-\frac{1}{2}}\delta_{s \neq e}$). Since the operator space OH is homogeneous by Pisier [189, Proposition 7.2 (iii)], each T_n and $A_2^{-\frac{1}{2}}$ are completely bounded. Proposition 2.8 says that $\mathrm{Id} \rtimes A_2^{-1/2}$ is bounded. Using the projection P_G of [17, Corollary 4.6], we obtain for any n a completely bounded Fourier multiplier $P_G(T_n) : L^2(\mathrm{VN}(G)) \to L^2(\mathrm{VN}(G))$ which has finite rank by the proof of [121, Theorem 4.4]. Moreover, by the contractivity of P_G we have for any n

$$\big\|A_2^{-1/2} - P_G(T_n)\big\|_{\mathrm{cb}, L^2(\mathrm{VN}(G)) \to L^2(\mathrm{VN}(G))} = \big\|P_G(A_2^{-1/2} - T_n)\big\|_{\mathrm{cb}, L^2 \to L^2}$$

$$\leqslant \big\|A_2^{-1/2} - T_n\big\|_{\mathrm{cb}, L^2(\mathrm{VN}(G)) \to L^2(\mathrm{VN}(G))}$$

$$= \big\|A_2^{-1/2} - T_n\big\|_{L^2(\mathrm{VN}(G)) \to L^2(\mathrm{VN}(G))} \xrightarrow[n \to +\infty]{} 0.$$

Now using the transference of Proposition 2.8, we obtain

$$\left\| \mathrm{Id} \rtimes A_2^{-1/2} - \mathrm{Id} \rtimes P_G(T_n) \right\|_{\mathrm{L}^2(\Gamma_q(H) \rtimes_\alpha G) \to \mathrm{L}^2(\Gamma_q(H) \rtimes_\alpha G)}$$

$$= \left\| \mathrm{Id} \rtimes \left(A_2^{-1/2} - P_G(T_n) \right) \right\|_{\mathrm{L}^2(\Gamma_q(H) \rtimes_\alpha G) \to \mathrm{L}^2(\Gamma_q(H) \rtimes_\alpha G)}$$

$$\leqslant \left\| A_2^{-1/2} - P_G(T_n) \right\|_{\mathrm{cb}, \mathrm{L}^2(\mathrm{VN}(G)) \to \mathrm{L}^2(\mathrm{VN}(G))} \xrightarrow[n \to +\infty]{} 0.$$

Since H is finite-dimensional, note that the fermionic space $\mathcal{F}_{-1}(H)$ is finite-dimensional. It is clear that $\mathrm{Id} \rtimes P_G(T_n)$ is again a finite rank operator.

Hence $\mathrm{Id} \rtimes A_2^{-1/2}$ is a norm-limit of finite rank bounded operators, hence compact. Thus, $\mathrm{Id} \rtimes A_2^{-\frac{1}{2}} \colon \overline{\mathrm{Ran}(\mathrm{Id} \rtimes A_2)} \to \overline{\mathrm{Ran}(\mathrm{Id}_{\mathrm{L}^2(\Gamma_{-1}(H))} \rtimes A_2)}$ is also compact. Finally, we note that $\overline{\mathrm{Ran}(\mathrm{Id} \rtimes A_2)} = \overline{\mathrm{Ran}(\mathscr{D}_{\psi,-1,2})^2} = \overline{\mathrm{Ran}\,\mathscr{D}_{\psi,-1,2}}$, where the last equality follows from the fact that $\mathscr{D}_{\psi,-1,2}$ is selfadjoint, see the first point. $\qquad\square$

5.10   Spectral Triples Associated to Semigroups of Schur Multipliers I

In this section, we consider a markovian semigroup of Schur multipliers on $\mathrm{B}(\ell_I^2)$ with associated gradient $\partial_{\alpha,q,p}$. Suppose $1 < p < \infty$ and $-1 \leqslant q < 1$. Recall that the (full) Hodge-Dirac operator $\mathcal{D}_{\alpha,q,p}$ with domain $\mathrm{dom}\,\mathcal{D}_{\alpha,q,p} = \mathrm{dom}\,\partial_{\alpha,q,p} \oplus \mathrm{dom}(\partial_{\alpha,q,p^*})^*$ is defined in (4.64) by the formula

$$\mathcal{D}_{\alpha,q,p} \overset{\mathrm{def}}{=} \begin{bmatrix} 0 & (\partial_{\alpha,q,p^*})^* \\ \partial_{\alpha,q,p} & 0 \end{bmatrix}. \tag{5.72}$$

We will see in the main results of this section (Propositions 5.14 and 5.16) how this Hodge-Dirac operator gives rise to a Banach spectral triple. Note that the compactness criterion needs particular attention and supplementary assumptions. The Banach spectral triple of this section will be locally compact.

Let us turn to the description of the homomorphism π. For any $a \in \mathrm{B}(\ell_I^2)$, we denote by $\tilde{L}_a \overset{\mathrm{def}}{=} \mathrm{Id}_{\mathrm{L}^p(\Gamma_q(H))} \otimes L_a \colon \mathrm{L}^p(\Gamma_q(H) \overline{\otimes} \mathrm{B}(\ell_I^2)) \to \mathrm{L}^p(\Gamma_q(H) \overline{\otimes} \mathrm{B}(\ell_I^2))$, $f \otimes e_{ij} \mapsto f \otimes a e_{ij}$ the left action. If $a \in \mathrm{B}(\ell_I^2))$, we define the bounded operator $\pi(a) \colon S_I^p \oplus_p \mathrm{L}^p(\Gamma_q(H) \overline{\otimes} \mathrm{B}(\ell_I^2)) \to S_I^p \oplus_p \mathrm{L}^p(\Gamma_q(H) \overline{\otimes} \mathrm{B}(\ell_I^2))$ by

$$\pi(a) \overset{\mathrm{def}}{=} \begin{bmatrix} L_a & 0 \\ 0 & \tilde{L}_a \end{bmatrix}, \quad a \in \mathrm{B}(\ell_I^2) \tag{5.73}$$

where $L_a \colon S_I^p \to S_I^p$, $x \mapsto ax$ is the left multiplication operator. Moreover, it is easy to check that $\pi(a)^* = \pi(a^*)$ in the case where $p = 2$. Finally, we will also use the restriction of π on S_I^∞.

Lemma 5.14 *Let* $1 < p < \infty$ *and* $-1 \leqslant q \leqslant 1$. *The map* π *is weak* continuous.*

Proof Indeed, we show that the map $\tilde{\pi} \colon \mathrm{B}(\ell_I^2) \to \mathrm{B}(S_I^p \oplus_p L^p(\Gamma_q(H) \overline{\otimes} \mathrm{B}(\ell_I^2)))$,

$a \mapsto \begin{bmatrix} L_a & 0 \\ 0 & \tilde{L}_a \end{bmatrix}$ is weak* continuous. Let (a_j) be a *bounded* net of $\mathrm{B}(\ell_I^2)$ converging

in the weak* topology to a. It is obvious that the nets (L_{a_j}) and $(\tilde{L}_{a_j})$ are bounded. If

$x \in S_I^p$ and if $y \in S_I^{p^*}$, we have $\langle L_{a_j}(x), y \rangle_{S_I^p, S_I^{p^*}} = \mathrm{Tr}((a_j x)^* y) = \mathrm{Tr}(yx^* a_j^*) \underset{j}{\to}$

$\mathrm{Tr}(yx^* a^*) = \langle L_a(x), y \rangle_{S_I^p, S_I^{p^*}}$ since $yx^* \in S_I^1$. So (L_{a_j}) converges to L_a in the

weak operator topology. Since S_I^p is reflexive, the weak operator topology and the weak* topology of $\mathrm{B}(S_I^p)$ coincide on bounded sets. We conclude that (L_{a_j}) converges to (L_a) in the weak* topology. If $\sum_k z_k \otimes t_k$ and if $\sum_l c_l \otimes b_l$ are elements of $\Gamma_q(H) \otimes \mathrm{M}_{I,\mathrm{fin}}$ we have

$$\left\langle \tilde{L}_{a_j}\left(\sum_k z_k \otimes t_k \right), \sum_l c_l \otimes b_l \right\rangle_{L^p(\Gamma_q(H) \overline{\otimes} \mathrm{B}(\ell_I^2)), L^{p^*}(\Gamma_q(H) \overline{\otimes} \mathrm{B}(\ell_I^2))}$$

$$= \sum_{k,l} \tau(z_k^* c_l) \mathrm{Tr}((a_j t_k)^* b_l) \underset{j}{\to} \sum_{k,l} \tau(z_k^* c_l) \mathrm{Tr}((a t_k)^* b_l)$$

$$= \left\langle \tilde{L}_a\left(\sum_k z_k \otimes t_k \right), \sum_l c_l \otimes b_l \right\rangle_{L^p(\Gamma_q(H) \overline{\otimes} \mathrm{B}(\ell_I^2)), L^{p^*}(\Gamma_q(H) \overline{\otimes} \mathrm{B}(\ell_I^2))}.$$

By a "net version" of [136, Lemma 3.6 p. 151], the *bounded* net $(\tilde{L}_{a_j})$ converges to $\tilde{L}_a$ in the weak operator topology. Once again, since $L^p(\Gamma_q(H) \overline{\otimes} \mathrm{B}(\ell_I^2))$ is reflexive, the weak operator topology and the weak* topology of $\mathrm{B}(L^p(\Gamma_q(H) \overline{\otimes} \mathrm{B}(\ell_I^2)))$ coincide on bounded sets. We infer that $(\tilde{L}_{a_j})$ converges to $(\tilde{L}_a)$ in the weak* topology. By Blecher and Le Merdy [38, Theorem A.2.5 (2)], we conclude that $\tilde{\pi}$ is weak* continuous. □

Recall in the following proposition the weak* closed operator

$$\partial_{\alpha,q,\infty} \colon \mathrm{dom}(\partial_{\alpha,q,\infty}) \subset \Gamma_q(H) \overline{\otimes} \mathrm{B}(\ell_I^2) \to \Gamma_q(H) \overline{\otimes} \mathrm{B}(\ell_I^2)$$

from the point 5 of Proposition 3.11. Note that the latter proposition is applicable since we suppose that $q \neq 1$ in the following.

Proposition 5.14 *Let $1 < p < \infty$ and $-1 \leqslant q < 1$.*

1. We have $(\mathcal{D}_{\alpha,q,p})^* = \mathcal{D}_{\alpha,q,p^*}$. In particular, the operator $\mathcal{D}_{\alpha,q,2}$ is selfadjoint.
2. We have

$$\operatorname{dom} \partial_{\alpha,q,\infty} \subset \operatorname{Lip}_{\mathcal{D}_{\alpha,q,p}}(\mathrm{B}(\ell_I^2)). \tag{5.74}$$

3. For any $a \in \operatorname{dom} \partial_{\alpha,q,\infty}$, we have

$$\left\| [\mathcal{D}_{\alpha,q,p}, \pi(a)] \right\|_{S_I^p \oplus_p L^p(\Gamma_q(H) \overline{\otimes} \mathrm{B}(\ell_I^2)) \to S_I^p \oplus_p L^p(\Gamma_q(H) \overline{\otimes} \mathrm{B}(\ell_I^2))} \tag{5.75}$$

$$\leqslant \left\| \partial_{\alpha,q,\infty}(a) \right\|_{\Gamma_q(H) \overline{\otimes} \mathrm{B}(\ell_I^2)}.$$

Proof 1. The proof is identical to the one of the first point of Theorem 5.4.

2. and 3. By Proposition 3.11 and Proposition 4.14, $M_{I,\mathrm{fin}}$ and $L^p(\Gamma_q(H)) \otimes M_{I,\mathrm{fin}}$ are cores of $\partial_{\alpha,q,p}$ and $(\partial_{\alpha,q,p^*})^*$. So $M_{I,\mathrm{fin}} \oplus (L^p(\Gamma_q(H)) \otimes M_{I,\mathrm{fin}})$ is a core of $\mathcal{D}_{\alpha,q,p}$. For any $a \in M_{I,\mathrm{fin}}$, we have $L_a(M_{I,\mathrm{fin}}) \subset M_{I,\mathrm{fin}}$ and $\tilde{L}_a(L^p(\Gamma_q(H)) \otimes M_{I,\mathrm{fin}}) \subset L^p(\Gamma_q(H)) \otimes M_{I,\mathrm{fin}}$. We infer that $\pi(a) \cdot (M_{I,\mathrm{fin}} \oplus (L^p(\Gamma_q(H)) \otimes M_{I,\mathrm{fin}})) \subset \operatorname{dom} \mathcal{D}_{\alpha,q,p}$. So the condition (a) of the first point of Proposition 5.12 is satisfied. Note also that $\pi(a)^* \cdot (M_{I,\mathrm{fin}} \oplus (L^p(\Gamma_q(H)) \otimes M_{I,\mathrm{fin}})) \subset \operatorname{dom} \mathcal{D}_{\alpha,q,p^*} = \operatorname{dom}(\mathcal{D}_{\alpha,q,p})^*$.

Let $a \in M_{I,\mathrm{fin}}$. A standard calculation shows that

$$\left[\mathcal{D}_{\alpha,q,p}, \pi(a) \right]$$

$$\overset{(5.72)(5.73)}{=} \begin{bmatrix} 0 & (\partial_{\alpha,q,p^*})^* \\ \partial_{\alpha,q,p} & 0 \end{bmatrix} \begin{bmatrix} L_a & 0 \\ 0 & \tilde{L}_a \end{bmatrix} - \begin{bmatrix} L_a & 0 \\ 0 & \tilde{L}_a \end{bmatrix} \begin{bmatrix} 0 & (\partial_{\alpha,q,p^*})^* \\ \partial_{\alpha,q,p} & 0 \end{bmatrix}$$

$$= \begin{bmatrix} 0 & (\partial_{\alpha,q,p^*})^* \tilde{L}_a \\ \partial_{\alpha,q,p} L_a & 0 \end{bmatrix} - \begin{bmatrix} 0 & L_a(\partial_{\alpha,q,p^*})^* \\ \tilde{L}_a \partial_{\alpha,q,p} & 0 \end{bmatrix}$$

$$= \begin{bmatrix} 0 & (\partial_{\alpha,q,p^*})^* \tilde{L}_a - L_a(\partial_{\alpha,q,p^*})^* \\ \partial_{\alpha,q,p} L_a - \tilde{L}_a \partial_{\alpha,q,p} & 0 \end{bmatrix}.$$

We calculate the two non-zero components of the commutator. For the lower left corner, if $x \in M_{I,\mathrm{fin}}$ we have[17]

$$(\partial_{\alpha,q,p} L_a - \tilde{L}_a \partial_{\alpha,q,p})(x) = \partial_{\alpha,q,p} L_a(x) - \tilde{L}_a \partial_{\alpha,q,p}(x)$$

$$= \partial_{\alpha,q,p}(ax) - a \partial_{\alpha,q,p}(x) \tag{5.76}$$

$$\overset{(2.96)}{=} \partial_{\alpha,q}(a)x = L_{\partial_{\alpha,q}(a)} J(x)$$

---

[17] Recall that the term $\partial_{\alpha,q,p}(a)x$ is by definition equal to $\partial_{\alpha,q}(a)(1 \otimes x)$.

where $J \colon S_I^p \to L^p(\Gamma_q(H) \overline{\otimes} B(\ell_I^2))$, $x \mapsto 1 \otimes x$. For the upper right corner, note that for any $y \in L^p(\Gamma_q(H)) \otimes M_{I,\mathrm{fin}}$ and any $x \in M_{I,\mathrm{fin}}$, (we recall that we have the duality brackets $\langle f, g \rangle$ antilinear in the first variable)

$$
\big\langle \big((\partial_{\alpha,q,p^*})^* \tilde{L}_a - L_a(\partial_{\alpha,q,p^*})^*\big)(y), x \big\rangle
$$

$$
= \big\langle (\partial_{\alpha,q,p^*})^* \tilde{L}_a(y), x \big\rangle - \big\langle L_a(\partial_{\alpha,q,p^*})^*(y), x \big\rangle
$$

$$
= \big\langle \tilde{L}_a(y), \partial_{\alpha,q,p^*}(x) \big\rangle - \big\langle (\partial_{\alpha,q,p^*})^*(y), L_{a^*}(x) \big\rangle
$$

$$
= \big\langle y, \tilde{L}_{a^*} \partial_{\alpha,q,p^*}(x) \big\rangle - \big\langle y, \partial_{\alpha,q,p^*} L_{a^*}(x) \big\rangle
$$

$$
= \big\langle y, \tilde{L}_{a^*} \partial_{\alpha,q,p^*}(x) - \partial_{\alpha,q,p^*} L_{a^*}(x) \big\rangle = \big\langle y, a^* \partial_{\alpha,q,p^*}(x) - \partial_{\alpha,q,p^*}(a^* x) \big\rangle
$$

$$
\overset{(2.96)}{=} -\big\langle y, \partial_{\alpha,q}(a^*) x \big\rangle = \big\langle y, -L_{\partial_{\alpha,q}(a^*)}(1 \otimes x) \big\rangle = \big\langle y, L_{(\partial_{\alpha,q}(a))^*}(1 \otimes x) \big\rangle
$$

$$
= \big\langle L_{\partial_{\alpha,q}(a)}(y), 1 \otimes x \big\rangle = \big\langle \mathbb{E} L_{\partial_{\alpha,q}(a)}(y), x \big\rangle_{S_I^p, S_I^{p^*}}.
$$

Here, $\mathbb{E} \colon L^p(\Gamma_q(H) \overline{\otimes} B(\ell_I^2)) \to S_I^p$, $t \otimes z \mapsto \tau(t) z$ denotes the canonical conditional expectation. We conclude that

$$
\big((\partial_{\alpha,q,p^*})^* \tilde{L}_a - L_a(\partial_{\alpha,q,p^*})^*\big)(y) = \mathbb{E} L_{\partial_{\alpha,q}(a)}(y). \tag{5.77}
$$

The two non-zero components of the commutator are bounded linear operators on $M_{I,\mathrm{fin}}$ and on $L^p(\Gamma_q(H)) \otimes M_{I,\mathrm{fin}}$. We deduce that $[\mathcal{D}_{\alpha,q,p}, \pi(a)]$ is bounded on the core $M_{I,\mathrm{fin}} \oplus (L^p(\Gamma_q(H)) \overline{\otimes} M_{I,\mathrm{fin}})$ of $\mathcal{D}_{\alpha,q,p}$. By Proposition 5.12, this operator extends to a bounded operator on $S_I^p \oplus_p L^p(\Gamma_q(H) \overline{\otimes} B(\ell_I^2))$. Hence $M_{I,\mathrm{fin}}$ is a subset of $\mathrm{Lip}_{\mathcal{D}_{\alpha,q,p}}(B(\ell_I^2))$. If $(x, y) \in \mathrm{dom}\, \mathcal{D}_{\alpha,q,p}$ and $a \in M_{I,\mathrm{fin}}$, we have in addition

$$
\big\| [\mathcal{D}_{\alpha,q,p}, \pi(a)](x, y) \big\|_p \tag{5.78}
$$

$$
= \Big\| \big(\big((\partial_{\alpha,q,p^*})^* \tilde{L}_a - L_a(\partial_{\alpha,q,p^*})^*\big)y, \big(\partial_{\alpha,q,p} L_a - \tilde{L}_a \partial_{\alpha,q,p}\big)x\big) \Big\|_p
$$

$$
= \Big(\big\| \big((\partial_{\alpha,q,p^*})^* \tilde{L}_a - L_a(\partial_{\alpha,q,p^*})^*\big)y \big\|_{S_I^p}^p + \big\| \big(\partial_{\alpha,q,p} L_a - \tilde{L}_a \partial_{\alpha,q,p}\big)x \big\|_p^p \Big)^{\frac{1}{p}}
$$

$$
\overset{(5.77)(5.76)}{=} \Big(\big\| \mathbb{E} L_{\partial_{\alpha,q}(a)}(y) \big\|_{S_I^p}^p + \big\| \partial_{\alpha,q}(a) J(x) \big\|_{L^p(\Gamma_q(H) \overline{\otimes} B(\ell_I^2))}^p \Big)^{\frac{1}{p}}
$$

$$
\leqslant \big\| \partial_{\alpha,q}(a) \big\|_{\Gamma_q(H) \overline{\otimes} B(\ell_I^2)} \big\| (x, y) \big\|_p.
$$

We conclude that

$$
\big\| [\mathcal{D}_{\alpha,q,p}, \pi(a)] \big\|_{S_I^p \oplus_p L^p(\Gamma_q(H) \overline{\otimes} B(\ell_I^2)) \to S_I^p \oplus_p L^p(\Gamma_q(H) \overline{\otimes} B(\ell_I^2))} \leqslant \big\| \partial_{\alpha,q}(a) \big\|_{\Gamma_q(H) \overline{\otimes} B(\ell_I^2)}. \tag{5.79}
$$

Let $a \in \text{dom}\,\partial_{\alpha,q,\infty}$. Let (a_j) be a net in $M_{I,\text{fin}}$ such that $a_j \to a$ and $\partial_{\alpha,q,\infty}(a_j) \to \partial_{\alpha,q,\infty}(a)$ both for the weak* topology. By Krein-Smulian Theorem (see Lemma 2.1), we can suppose that the nets (a_j) and $(\partial_{\alpha,q,\infty}(a_j))$ are bounded. By the point 4 of Proposition 5.11, we deduce that $a \in \text{Lip}_{\mathcal{D}_{\alpha,q,p}}(\mathrm{B}(\ell_I^2))$. By continuity of π, note that $\pi(a_j) \to \pi(a)$ for the weak operator topology. For any $\xi \in \text{dom}\,\mathcal{D}_{\alpha,q,p}$ and any $\zeta \in \text{dom}(\mathcal{D}_{\alpha,q,p})^*$, we have

$$
\langle[\mathcal{D}_{\alpha,q,p}, \pi(a_j)]\xi, \zeta\rangle_{S_I^p \oplus_p L^p(\Gamma_q(H)\overline{\otimes}\mathrm{B}(\ell_I^2)), S_I^{p^*} \oplus_{p^*} L^{p^*}}
$$

$$
= \langle(\mathcal{D}_{\alpha,q,p}\pi(a_j) - \pi(a_j)\mathcal{D}_{\alpha,q,p})\xi, \zeta\rangle
$$

$$
= \langle\mathcal{D}_{\alpha,q,p}\pi(a_j)\xi, \zeta\rangle - \langle\pi(a_j)\mathcal{D}_{\alpha,q,p}\xi, \zeta\rangle
$$

$$
= \langle\pi(a_j)\xi, (\mathcal{D}_{\alpha,q,p})^*\zeta\rangle - \langle\pi(a_j)\mathcal{D}_{\alpha,q,p}\xi, \zeta\rangle
$$

$$
\xrightarrow{j} \langle\pi(a)\xi, (\mathcal{D}_{\alpha,q,p})^*\zeta\rangle - \langle\pi(a)\mathcal{D}_{\alpha,q,p}\xi, \zeta\rangle
$$

$$
= \langle[\mathcal{D}_{\alpha,q,p}, \pi(a)]\xi, \zeta\rangle.
$$

Since the net $([\mathcal{D}_{\alpha,q,p}, \pi(a_j)])$ is bounded by (5.79), we deduce that the net $([\mathcal{D}_{\alpha,q,p}, \pi(a_j)])$ converges to $[\mathcal{D}_{\alpha,q,p}, \pi(a)]$ for the weak operator topology by a "net version" of [136, Lemma 3.6 p. 151]. Furthermore, it is (really) easy to check that $L_{\partial_{\alpha,q,\infty}(a_j)}J \to L_{\partial_{\alpha,q,\infty}(a)}J$ and $-\mathbb{E}L_{\partial_{\alpha,q,\infty}(a_j)} \to -\mathbb{E}L_{\partial_{\alpha,q,\infty}(a)}$ both for the weak operator topology. By uniqueness of the limit, we deduce that the commutator is given by the same formula as in the case of elements of $M_{I,\text{fin}}$. From here, we obtain (5.75) as in (5.78). $\qquad\square$

Remark 5.12 We do not know if $\text{Lip}_{\mathcal{D}_{\alpha,q,p}}(\mathrm{B}(\ell_I^2)) = \text{dom}\,\partial_{\alpha,q,\infty}$. We will investigate this question in subsequent work.

Remark 5.13 The estimate (5.75) is in general not optimal. Indeed, already in the case $p = 2$ and $a = e_{ij} \in M_{I,\text{fin}} \subset \text{dom}\,\partial_{\alpha,q,\infty}$ for some $i, j \in I$, we have according to (5.76) and (5.77),

$$
\left\|[\mathcal{D}_{\alpha,q,2}, \pi(a)]\right\|_{S_I^2 \oplus_2 L^2(\Gamma_q(H)\overline{\otimes}\mathrm{B}(\ell_I^2)) \to S_I^2 \oplus_2 L^2(\Gamma_q(H)\overline{\otimes}\mathrm{B}(\ell_I^2))} \tag{5.80}
$$

$$
\leqslant \max\left\{\left\|L_{\partial_{\alpha,q}(a)}J\right\|_{S_I^2 \to L^2(\Gamma_q(H)\overline{\otimes}\mathrm{B}(\ell_I^2))}, \left\|\mathbb{E}L_{\partial_{\alpha,q}(a)}\right\|_{L^2(\Gamma_q(H)\overline{\otimes}\mathrm{B}(\ell_I^2)) \to S_I^2}\right\}.
$$

Note that we have the Hilbert space adjoints $(L_{\partial_{\alpha,q}(a)}J)^* = J^*L_{\partial_{\alpha,q}(a)}^* = \mathbb{E}L_{(\partial_{\alpha,q}(a))^*} = -\mathbb{E}L_{\partial_{\alpha,q}(a^*)}$. Thus, in the maximum of (5.80), it suffices to consider the second term. We have for $x = \sum_{k,l} x_{kl} \otimes e_{kl} \in L^2(\Gamma_q(H)\overline{\otimes}\mathrm{B}(\ell_I^2))$

$$
\mathbb{E}L_{\partial_{\alpha,q}(e_{ij})}(x) = \mathbb{E}\left((s_q(\alpha_i - \alpha_j) \otimes e_{ij})\left(\sum_{k,l} x_{kl} \otimes e_{kl}\right)\right)
$$

$$
= \mathbb{E}\left(\sum_l s_q(\alpha_i - \alpha_j)x_{jl} \otimes e_{il}\right) = \sum_l \tau(s_q(\alpha_i - \alpha_j)x_{jl})e_{il}.
$$

Consequently

$$\left\|\mathbb{E}L_{\partial_{\alpha,q}(e_{ij})}(x)\right\|_2^2 = \sum_l \left|\tau(s_q(\alpha_i - \alpha_j)x_{jl})\right|^2$$

$$\leqslant \left\|s_q(\alpha_i - \alpha_j)\right\|_{L^2(\Gamma_q(H))}^2 \sum_l \left\|x_{jl}\right\|_{L^2(\Gamma_q(H))}^2$$

$$\leqslant \left\|s_q(\alpha_i - \alpha_j)\right\|_{L^2(\Gamma_q(H))}^2 \sum_{k,l} \left\|x_{kl}\right\|_{L^2(\Gamma_q(H))}^2$$

$$= \left\|\alpha_i - \alpha_j\right\|_H^2 \left\|x\right\|_{L^2(\Gamma_q(H)\overline{\otimes}B(\ell_I^2))}^2 .$$

We infer that

$$\left\|[\mathcal{D}_{\alpha,q,2}, \pi(a)]\right\|_{S_I^2\oplus_2 L^2(\Gamma_q(H)\overline{\otimes}B(\ell_I^2))\to S_I^2\oplus_2 L^2(\Gamma_q(H)\overline{\otimes}B(\ell_I^2))}$$

$$= \left\|\mathbb{E}L_{\partial_{\alpha,q}(e_{ij})}\right\|_{2\to 2} \leqslant \left\|\alpha_i - \alpha_j\right\|_H.$$

In the case where $-1 < q < 1$ and $\alpha_i - \alpha_j \neq 0$, this quantity is strictly less than

$$\left\|\partial_{\alpha,q}(e_{ij})\right\|_{\Gamma_q(H)\overline{\otimes}B(\ell_I^2)} \overset{(2.95)}{=} \left\|s_q(\alpha_i - \alpha_j) \otimes e_{ij}\right\|_{\Gamma_q(H)\overline{\otimes}B(\ell_I^2)}$$

$$= \left\|s_q(\alpha_i - \alpha_j)\right\|_{\Gamma_q(H)} \left\|e_{ij}\right\|_{B(\ell_I^2)}$$

$$= \left\|s_q(\alpha_i - \alpha_j)\right\|_{\Gamma_q(H)} \overset{[44,\, \text{Th. 1.10}]}{=} \frac{2}{\sqrt{1-q}}\left\|\alpha_i - \alpha_j\right\|_H.$$

Under additional assumptions on the family $(\alpha_i)_{i\in I}$ of our semigroup, we are able to prove that the triple from Proposition 5.14 is a locally compact spectral triple (see Definition 5.11). In the following, we shall need the restriction $A_{R_p} = A_p|_{R_p}$ of A_p to a row

$$R_p \overset{\text{def}}{=} \overline{\text{span}\{e_{0j} : j \in I\}}, \tag{5.81}$$

where 0 is some fixed element of I. Note that A_p leaves clearly R_p invariant.

Proposition 5.15 Let $1 < p < \infty$. Assume that $\alpha : I \to H$ is injective, where H is a Hilbert space of dimension $n \in \mathbb{N}$ and satisfies $\text{Gap}_\alpha > 0$ where Gap_α is defined in (5.16).

1. The operator $A_{R_p}^{-\frac{1}{2}} : \overline{\text{Ran}\, A_{R_p}} \to \overline{\text{Ran}\, A_{R_p}}$ is compact.
2. Suppose $-1 \leqslant q \leqslant 1$. Let $B_{R_p} = \partial_{\alpha,q,p}\partial_{\alpha,q,p^*}^*|_{\overline{\text{Ran}\, \partial_{\alpha,q,p}|R_p}} : \text{dom}\, B_{R_p} \subset \overline{\text{Ran}\, \partial_{\alpha,q,p}|R_p} \to \overline{\text{Ran}\, \partial_{\alpha,q,p}|R_p}$. Then B_{R_p} is sectorial and injective, and the operator $B_{R_p}^{-\frac{1}{2}}$ is compact.

Proof

1. We begin by showing that $A_{R_2}^{-\frac{1}{2}} : \overline{\operatorname{Ran} A_{R_2}} \to \overline{\operatorname{Ran} A_{R_2}}$ is compact. Note that it is obvious that A_{R_2} is selfadjoint and that the e_{0i}'s where $i \in I$, form an orthonormal basis of R_2 consisting of eigenvectors of A_{R_2}. Thus the e_{0i}'s where $i \in I \backslash \{0\}$ form an orthonormal basis of $\overline{\operatorname{Ran} A_{R_2}}$ consisting of eigenvectors of $A_{R_2}^{-\frac{1}{2}} : \overline{\operatorname{Ran} A_{R_2}} \to \overline{\operatorname{Ran} A_{R_2}}$ associated to the eigenvalues $\|\alpha_i - \alpha_0\|_H^{-1}$. It suffices to observe that the condition $\operatorname{Gap}_\alpha > 0$, the injectivity of α, together with the finite-dimensionality of H imply that any bounded subset of H meets the $\|\alpha_i - \alpha_0\|_H$ only for a finite number of $i \in I$. Hence theses eigenvalues $\|\alpha_i - \alpha_0\|_H^{-1}$ vanish at infinity.[18] We have proved the compactness.

   Next we show that $A_{R_\infty}^{-\frac{1}{2}} : \overline{\operatorname{Ran} A_{R_\infty}} \to \overline{\operatorname{Ran} A_{R_\infty}}$ is bounded where $R_\infty \subset S_I^\infty$. Since $(T_t|_{R_\infty})_{t \geqslant 0}$ is a semigroup with generator A_{R_∞}, it suffices by Lemma 5.6 to establish the bound $\|T_t|_{R_\infty}\|_{\operatorname{Ran} A_{R_\infty} \to R_\infty} \lesssim \frac{1}{t^d}$ for some $d > \frac{1}{2}$. We conclude with Lemma 5.4 and by restriction. Thus, $A_{R_\infty}^{-\frac{1}{2}} : \overline{\operatorname{Ran} A_{R_\infty}} \to \overline{\operatorname{Ran} A_{R_\infty}}$ is bounded.

   Now, assume that $p > 2$. We will use complex interpolation. Since the resolvents of the operators A_{R_p} are compatible for different values of p (in fact, they are equal since $R_{p_0} = R_{p_1}$ for $1 \leqslant p_0, p_1 \leqslant \infty$), the complementary projections of (2.17) onto the spaces $\overline{\operatorname{Ran} A_{R_p}}$ are compatible. Hence, the $\overline{\operatorname{Ran} A_{R_p}}$'s form an interpolation scale. Observe that $\overline{\operatorname{Ran} A_{R_2}}$ is a Hilbert space, hence a UMD space. Then we obtain the compactness of $A_{R_p}^{-\frac{1}{2}} : \overline{\operatorname{Ran} A_{R_p}} \to \overline{\operatorname{Ran} A_{R_p}}$ by means of complex interpolation between a compact and a bounded operator with Theorem 2.2.

   If $p < 2$, we conclude by duality and Schauder's Theorem [167, Theorem 3.4.15], since $\overline{\operatorname{Ran} A_{R_p}}$ is the dual space of $\overline{\operatorname{Ran} A_{R_{p^*}}}$ and $A_{R_p}^{-\frac{1}{2}}$ defined on the first space is the adjoint of $A_{R_{p^*}}^{-\frac{1}{2}}$ defined on the second space.

2. We will use the shorthand notation $\partial_p \overset{\text{def}}{=} \partial_{\alpha,q,p}|R_p$ and

$$\partial_p^* \overset{\text{def}}{=} (\partial_{\alpha,q,p^*})^* |\overline{L^p(\Gamma_q(H)) \otimes R_p}^p.$$

Note that ∂_p and ∂_p^* are again closed and densely defined.

   We begin with the case $p = 2$. The operators ∂_2 and ∂_2^* are indeed adjoints to each other (with the chosen domain). According to Theorem 2.1, the unbounded operators $\partial_2^* \partial_2 |(\operatorname{Ker} \partial_2)^\perp$ and $\partial_2 \partial_2^* |(\operatorname{Ker} \partial_2^*)^\perp$ are unitarily equivalent.

---

[18] Recall that a family $(x_i)_{i \in I}$ vanishes at infinity means that for any $\varepsilon > 0$, there exists a finite subset J of I such that for any $i \in I - J$ we have $|x_i| \leqslant \varepsilon$.

Note that we have $A_{R_2} = A_2|R_2 = \partial_2^* \partial_2$. Moreover, according to (2.8), $(\mathrm{Ker}\,\partial_2)^\perp = \overline{\mathrm{Ran}\,\partial_2^*}$, which in turn equals $\overline{\mathrm{Ran}(\partial_{\alpha,q,2})^*} \cap R_2 \overset{\text{Prop. 4.19}}{=} \overline{\mathrm{Ran}\,A_2} \cap R_2 = \overline{\mathrm{Ran}\,A_{R_2}}$. By the first part of the proof, $A_{R_2}^{-\frac{1}{2}} : \overline{\mathrm{Ran}\,A_{R_2}} \to \overline{\mathrm{Ran}\,A_{R_2}}$ is compact, hence $\partial_2^* \partial_2|(\mathrm{Ker}\,\partial_2)^\perp$ is invertible, and by functional calculus, the previous unitary equivalence also holds between $A_{R_2}^{-\frac{1}{2}}$ on $\overline{\mathrm{Ran}\,A_{R_2}}$ and $(\partial_2 \partial_2^*)^{-\frac{1}{2}}$ on $(\mathrm{Ker}\,\partial_2^*)^\perp \overset{(2.3)}{=} \overline{\mathrm{Ran}\,\partial_2}$. We infer that $B_{R_2}^{-\frac{1}{2}}$ is compact.

We turn to the case of general p and start by showing that $B_{R_p}^{-\frac{1}{2}}$ is bounded. Note that $B_{R_p} = \partial_{\alpha,q,p} \partial_{\alpha,q,p^*}^* | \overline{\mathrm{Ran}\,\partial_p} \overset{\text{Proposition 4.6}}{=} B_p | \overline{\mathrm{Ran}\,\partial_p}$. Since resolvents of B_p leave $\overline{\mathrm{Ran}\,\partial_p}$ invariant and B_p is sectorial by Lemma 4.14, by [83, Proposition 3.2.15], B_{R_p} is also sectorial. Since B_p is injective according to Lemma 4.14, also B_{R_p} is injective by restriction. Then again by [83, Proposition 3.2.15], $B_{R_p}^{-\frac{1}{2}} = B_p^{-\frac{1}{2}} | \overline{\mathrm{Ran}\,\partial_p}$. By Lemma 5.7 and restriction, we infer that $B_{R_p}^{-\frac{1}{2}}$ is bounded. To improve boundedness to compactness, we pick some $2 < p < \infty$ and fix some auxiliary $p < p_0 < \infty$. Finally, it suffices interpolate compactness of $B_{R_2}^{-\frac{1}{2}}$ and boundedness of $B_{R_{p_0}}^{-\frac{1}{2}}$ to conclude with Theorem 2.2 that $B_{R_p}^{-\frac{1}{2}}$ is compact. Details are left to the reader. For the case $p < 2$, we use duality. $\qquad\square$

Proposition 5.16 *Let* $1 < p < \infty$ *and* $-1 \leqslant q < 1$. *Assume that* H *is finite-dimensional, that* $\alpha \colon I \to H$ *is injective and that* $\mathrm{Gap}_\alpha > 0$. *Then* $(\mathrm{M}_{I,\mathrm{fin}}, S_I^p \oplus L^p(\Gamma_q(H) \overline{\otimes} \mathrm{B}(\ell_I^2)), \mathcal{D}_{\alpha,q,p})$ *is a locally compact Banach spectral triple. In other words, we have the following properties.*

1. $\mathcal{D}_{\alpha,q,p}$ *is densely defined and has a bounded* H^∞ *functional calculus on a bisector.*
2. *For any* $a \in \mathrm{dom}\,\partial_{\alpha,q,\infty}$, *we have* $a \in \mathrm{Lip}_{\mathcal{D}_{\alpha,q,p}}(\mathrm{B}(\ell_I^2))$.
3. *For any* $a \in S_I^\infty$, $\pi(a)(\mathrm{iId} + \mathcal{D}_{\alpha,q,p})^{-1}$ *and* $\pi(a)|\mathcal{D}_{\alpha,q,p}|^{-1}$ *are compact operators between the spaces* $\overline{\mathrm{Ran}\,\mathcal{D}_{\alpha,q,p}} \to S_I^p \oplus L^p(\Gamma_q(H) \overline{\otimes} \mathrm{B}(\ell_I^2))$.

Proof The first and second points are already contained in Theorem 4.9 and Proposition 5.14. We turn to the third point. Note that by bisectorial H^∞-calculus, we have a bounded operator $f(\mathcal{D}_{\alpha,q,p}) \colon \overline{\mathrm{Ran}\,\mathcal{D}_{\alpha,q,p}} \to \overline{\mathrm{Ran}\,\mathcal{D}_{\alpha,q,p}}$, where $f(\lambda) = \sqrt{\lambda^2}(i+\lambda)^{-1}$ belongs to $\mathrm{H}^\infty(\Sigma_\omega^\pm)$. Thus, recalling that $|\mathcal{D}_{\alpha,q,p}|^{-1}$ is the functional calculus of the function $n(\lambda) = \frac{1}{\sqrt{\lambda^2}}$, we obtain

$$\pi(a)(\mathrm{iId} + \mathcal{D}_{\alpha,q,p})^{-1} = \pi(a)|\mathcal{D}_{\alpha,q,p}|^{-1} f(\mathcal{D}_{\alpha,q,p})$$

as operators $\overline{\mathrm{Ran}\,\mathcal{D}_{\alpha,q,p}} \to S_I^p \oplus L^p(\Gamma_q(H) \overline{\otimes} \mathrm{B}(\ell_I^2))$. By composition, if the operator $\pi(a)|\mathcal{D}_{\alpha,q,p}|^{-1}$ is compact on $\overline{\mathrm{Ran}\,\mathcal{D}_{\alpha,q,p}}$, then so is $\pi(a)(\mathrm{iId}+\mathcal{D}_{\alpha,q,p})^{-1}$. It thus suffices to consider the former in the sequel.

Recall that $\overline{\operatorname{Ran} A_p} = \overline{\operatorname{Ran}(\partial_{\alpha,q,p^*})}^*$ by Proposition 4.19 and that as operators on $\overline{\operatorname{Ran} \mathcal{D}_{\alpha,q,p}}$,

$$\mathcal{D}_{\alpha,q,p}^2 \overset{(4.61)}{=} \begin{bmatrix} A_p|\overline{\operatorname{Ran} A_p} & 0 \\ 0 & (\operatorname{Id}_{L^p(\Gamma_q(H))} \otimes A_p)|\overline{\operatorname{Ran} \partial_{\alpha,q,p}} \end{bmatrix}.$$

So as operators on $\overline{\operatorname{Ran} \mathcal{D}_{\alpha,q,p}}$,

$$|\mathcal{D}_{\alpha,q,p}|^{-1} = \begin{bmatrix} A_p|\overline{\operatorname{Ran} A_p} & 0 \\ 0 & (\operatorname{Id}_{L^p(\Gamma_q(H))} \otimes A_p)|\overline{\operatorname{Ran} \partial_{\alpha,q,p}} \end{bmatrix}^{-\frac{1}{2}} = \begin{bmatrix} A_p^{-\frac{1}{2}} & 0 \\ 0 & B_p^{-\frac{1}{2}} \end{bmatrix}.$$

Hence we have

$$\pi(a)|\mathcal{D}_{\alpha,q,p}|^{-1} \overset{(5.73)}{=} \begin{bmatrix} L_a & 0 \\ 0 & \tilde{L}_a \end{bmatrix} \begin{bmatrix} A_p^{-\frac{1}{2}} & 0 \\ 0 & B_p^{-\frac{1}{2}} \end{bmatrix} = \begin{bmatrix} L_a A_p^{-\frac{1}{2}} & 0 \\ 0 & \tilde{L}_a B_p^{-\frac{1}{2}} \end{bmatrix}.$$

Now, it suffices to show that $L_a A_p^{-\frac{1}{2}}: \overline{\operatorname{Ran} A_p} \to S_I^p$ and $\tilde{L}_a B_p^{-\frac{1}{2}}: \overline{\operatorname{Ran} \partial_{\alpha,q,p}} \to L^p(\Gamma_q(H) \overline{\otimes} B(\ell_I^2))$ are compact. We start with the first operator. Recall that $A_p(M_{I,\operatorname{fin}})$ is a dense subspace of $\overline{\operatorname{Ran} A_p}$. If $P_j: S_I^p \to S_I^p$ is the Schur multiplier projecting onto the jth line associated with the matrix $[\delta_{j=k}]_{kl}$,[19] if we choose first $a = e_{ij}$, and if $e_{kl} \in A_p(M_{I,\operatorname{fin}})$ we have

$$L_a A_p^{-\frac{1}{2}}(e_{kl}) = L_{e_{ij}} A_p^{-\frac{1}{2}}(e_{kl}) = \|\alpha_k - \alpha_l\|^{-1} e_{ij} e_{kl}$$

$$= \delta_{j=k} \|\alpha_k - \alpha_l\|^{-1} e_{il} = \delta_{j=k} L_{e_{ij}} A_p^{-\frac{1}{2}}(e_{kl}) = L_a A_p^{-\frac{1}{2}} P_j(e_{kl}).$$

Then by span density of such e_{kl} in $\overline{\operatorname{Ran} A_p}$, we infer that $L_a A_p^{-\frac{1}{2}} = L_a A_p^{-\frac{1}{2}} P_j$ as bounded operators $\overline{\operatorname{Ran} A_p} \to S_I^p$. From Lemma 5.3, we infer that P_j is a completely contractive projection. It now suffices to show that $L_a A_p^{-\frac{1}{2}} P_j: P_j(\overline{\operatorname{Ran} A_p}) \to S_I^p$ is compact. We use the space R_p from (5.81), where we choose $0 = j$ there. Since $A_p^{-\frac{1}{2}} P_j = A_p^{-\frac{1}{2}}|\overline{\operatorname{Ran} A_{R_p}}$, we infer by Proposition 5.15 that $A_p^{-\frac{1}{2}} P_j$ is indeed compact on $P_j(\overline{\operatorname{Ran} A_p}) = \overline{\operatorname{Ran} A_{R_p}}$. We infer by composition that $L_a A_p^{-\frac{1}{2}} P_j = L_a A_p^{-\frac{1}{2}}$ is compact as an operator on $\overline{\operatorname{Ran} A_p}$, in case $a = e_{ij}$. By linearity in a, $L_a A_p^{-\frac{1}{2}}$ is also compact for $a \in M_{I,\operatorname{fin}}$. Finally, if $a \in S_I^\infty$ is a generic element, then $a = \lim_{n \to \infty} a_n$ in S_I^∞ for some sequence

---

[19] The entries are 1 on the j-row and zero anywhere else.

$a_n \in M_{I,\text{fin}}$, whence $L_a = \lim_{n\to\infty} L_{a_n}$ in $B(S_I^p)$ and $L_a A_p^{-\frac{1}{2}} = \lim_{n\to\infty} L_{a_n} A_p^{-\frac{1}{2}}$ in $B(\overline{\text{Ran } A_p}, S_I^p)$. Thus, $L_a A_p^{-\frac{1}{2}}$ is the operator norm limit of compact operators, and hence itself compact.

We turn to the second operator $\tilde{L}_a B_p^{-\frac{1}{2}}$. Let $x \otimes e_{kl} \in \partial_{\alpha,q,p}(M_{I,\text{fin}})$ (with $x = s_q(\alpha_k - \alpha_l)$), which is a dense subspace of $\overline{\text{Ran } \partial_{\alpha,q,p}}$ according to Proposition 3.11. We first fix $a = e_{ij}$ and turn to general $a \in S_I^\infty$ at the end.

$$\tilde{L}_a B_p^{-\frac{1}{2}}(x \otimes e_{kl}) = \|\alpha_k - \alpha_l\|^{-1}(1 \otimes e_{ij})(x \otimes e_{kl}) = \delta_{j=k} \|\alpha_k - \alpha_l\|^{-1} x \otimes e_{il}$$

$$= \delta_{j=k}(1 \otimes e_{ij})B_p^{-\frac{1}{2}}(x \otimes e_{kl}) = \tilde{L}_a B_p^{-\frac{1}{2}}(\text{Id}_{L^p} \otimes P_j)(x \otimes e_{kl}).$$

By the same arguments as previously, we infer that $\tilde{L}_a B_p^{-\frac{1}{2}} = \tilde{L}_a B_p^{-\frac{1}{2}}(\text{Id}_{L^p} \otimes P_j)$ as bounded operators $\overline{\text{Ran } \partial_{\alpha,q,p}} \to L^p(\Gamma_q(H)\overline{\otimes}B(\ell_I^2))$. We infer by Proposition 5.15 that $B_p^{-\frac{1}{2}}(\text{Id}_{L^p} \otimes P_j)$ is compact, thus by composition, also $\tilde{L}_a B_p^{-\frac{1}{2}} = \tilde{L}_a B_p^{-\frac{1}{2}}(\text{Id}_{L^p} \otimes P_j)$ is compact on $\overline{\text{Ran } \partial_{\alpha,q,p}}$. Then by linearity (resp. operator norm limit), $\tilde{L}_a B_p^{-\frac{1}{2}}$ is also compact for any $a \in M_{I,\text{fin}}$ (resp. for any $a \in S_I^\infty$). We finally deduce that $\pi(a)|\mathcal{D}_{\alpha,q,p}|^{-1}$ is compact on $\overline{\text{Ran } \mathcal{D}_{\alpha,q,p}}$. $\qquad\square$

Remark 5.14 Note that the locally compact (Banach) spectral triple

$$(M_{I,\text{fin}}, S_I^p \oplus_p L^p(\Gamma_q(H)\overline{\otimes}B(\ell_I^2)), \mathcal{D}_{\alpha,q,p})$$

is even. Indeed, the Hodge-Dirac operator $\mathcal{D}_{\alpha,q,p}$ anti-commutes with the involution

$$\gamma_p \overset{\text{def}}{=} \begin{bmatrix} -\text{Id}_{S_I^p} & 0 \\ 0 & \text{Id}_{L^p} \end{bmatrix} : S_I^p \oplus_p L^p(\Gamma_q(H)\overline{\otimes}B(\ell_I^2)) \to S_I^p \oplus_p L^p(\Gamma_q(H)\overline{\otimes}B(\ell_I^2))$$

(which is selfadjoint if $p = 2$) since

$$\mathcal{D}_{\alpha,q,p}\gamma_p + \gamma_p \mathcal{D}_{\alpha,q,p}$$

$$\overset{(5.72)}{=} \begin{bmatrix} 0 & (\partial_{\alpha,q,p*})^* \\ \partial_{\alpha,q,p} & 0 \end{bmatrix} \begin{bmatrix} -\text{Id}_{S_I^p} & 0 \\ 0 & \text{Id}_{L^p} \end{bmatrix} + \begin{bmatrix} -\text{Id}_{S_I^p} & 0 \\ 0 & \text{Id}_{L^p} \end{bmatrix} \begin{bmatrix} 0 & (\partial_{\alpha,q,p*})^* \\ \partial_{\alpha,q,p} & 0 \end{bmatrix}$$

$$= \begin{bmatrix} 0 & (\partial_{\alpha,q,p*})^* \\ -\partial_{\alpha,q,p} & 0 \end{bmatrix} + \begin{bmatrix} 0 & -(\partial_{\alpha,q,p*})^* \\ \partial_{\alpha,q,p} & 0 \end{bmatrix} = 0.$$

Moreover, for any $a \in \mathrm{B}(\ell_I^2)$, we have

$$
\gamma_p \pi(a) \overset{(5.73)}{=} \begin{bmatrix} -\mathrm{Id}_{\mathrm{S}_I^p} & 0 \\ 0 & \mathrm{Id}_{\mathrm{L}^p} \end{bmatrix} \begin{bmatrix} \mathrm{L}_a & 0 \\ 0 & \tilde{\mathrm{L}}_a \end{bmatrix} = \begin{bmatrix} -\mathrm{L}_a & 0 \\ 0 & \tilde{\mathrm{L}}_a \end{bmatrix}
$$

$$
= \begin{bmatrix} \mathrm{L}_a & 0 \\ 0 & \tilde{\mathrm{L}}_a \end{bmatrix} \begin{bmatrix} -\mathrm{Id}_{\mathrm{S}_I^p} & 0 \\ 0 & \mathrm{Id}_{\mathrm{L}^p} \end{bmatrix} \overset{(5.73)}{=} \pi(a)\gamma_p.
$$

5.11   Spectral Triples Associated to Semigroups of Schur Multipliers II

In this section, we shall investigate another triple, defined on the second component of the reflexive Banach space of the preceding section. Suppose $1 < p < \infty$ and $-1 \leqslant q < 1$. We define another "Hodge-Dirac operator" by letting

$$
\mathscr{D}_{\alpha,q}(x \otimes e_{ij}) \overset{\mathrm{def}}{=} s_q(\alpha_i - \alpha_j)x \otimes e_{ij}, \quad x \in \mathrm{L}^p(\Gamma_q(H)), i, j \in I. \tag{5.82}
$$

We can see $\mathscr{D}_{\alpha,q}$ as an unbounded operator acting on the subspace $\mathrm{L}^p(\Gamma_q(H)) \otimes \mathrm{M}_{I,\mathrm{fin}}$ of $\mathrm{L}^p(\Gamma_q(H)\overline{\otimes}\mathrm{B}(\ell_I^2))$.

Lemma 5.15 *Suppose* $1 < p < \infty$ *and* $-1 \leqslant q < 1$.

1. *For any* $a \in \mathrm{L}^p(\Gamma_q(H)) \otimes \mathrm{M}_{I,\mathrm{fin}}$ *and* $b \in \mathrm{L}^{p^*}(\Gamma_q(H)) \otimes \mathrm{M}_{I,\mathrm{fin}}$, *we have*

$$
\langle \mathscr{D}_{\alpha,q}(a), b \rangle = \langle a, \mathscr{D}_{\alpha,q}(b) \rangle \tag{5.83}
$$

where we use as usual the duality bracket $\langle x, y \rangle = \tau(x^*y)$.
2. *The operator*

$$
\mathscr{D}_{\alpha,q} : \mathrm{L}^p(\Gamma_q(H)) \otimes \mathrm{M}_{I,\mathrm{fin}} \subset \mathrm{L}^p(\Gamma_q(H)\overline{\otimes}\mathrm{B}(\ell_I^2)) \to \mathrm{L}^p(\Gamma_q(H)\overline{\otimes}\mathrm{B}(\ell_I^2))
$$

is closable.

Proof

1. For any $x, y \in \mathrm{L}^p(\Gamma_q(H))$ and any $i, j, k, l \in I$, we have

$$
\langle \mathscr{D}_{\alpha,q}(x \otimes e_{ij}), y \otimes e_{kl} \rangle \overset{(5.82)}{=} \langle s_q(\alpha_i - \alpha_j)x \otimes e_{ij}, y \otimes e_{kl} \rangle
$$
$$
= \tau(x^* s_q(\alpha_i - \alpha_j)y) \mathrm{Tr}(e_{ij}^* e_{kl}) = \delta_{ik}\delta_{jl}\tau(x^* s_q(\alpha_k - \alpha_l)y)
$$

$$= \mathrm{Tr}(e_{ij}^* e_{kl}) \tau \left(x^* s_q(\alpha_k - \alpha_l) y \right)$$

$$= \left\langle x \otimes e_{ij}, s_q(\alpha_k - \alpha_l) y \otimes e_{kl} \right\rangle \overset{(5.82)}{=} \left\langle x \otimes e_{ij}, \mathscr{D}_{\alpha,q}(y \otimes e_{kl}) \right\rangle.$$

Thus (5.83) follows by linearity.

2. Since $L^{p^*}(\Gamma_q(H)) \otimes M_{I,\mathrm{fin}}$ is dense in $L^{p^*}(\Gamma_q(H) \overline{\otimes} B(\ell_I^2))$, this is a consequence of [136, Theorem 5.28 p. 168]. □

We denote by $\mathscr{D}_{\alpha,q,p} \colon \mathrm{dom}\, \mathscr{D}_{\alpha,q,p} \subset L^p(\Gamma_q(H) \overline{\otimes} B(\ell_I^2)) \to L^p(\Gamma_q(H) \overline{\otimes} B(\ell_I^2))$ its closure. By definition, the subspace $L^p(\Gamma_q(H)) \otimes M_{I,\mathrm{fin}}$ is a core of $\mathscr{D}_{\alpha,q,p}$. We define the homomorphism $\pi \colon B(\ell_I^2) \to B(L^p(\Gamma_q(H) \overline{\otimes} B(\ell_I^2)))$ by

$$\pi(a) \overset{\mathrm{def}}{=} \mathrm{Id}_{L^p(\Gamma_q(H))} \otimes L_a, \quad a \in B(\ell_I^2) \tag{5.84}$$

where L_a is the left multiplication by a on S_I^p. Note that $\pi(a)$ is equal to the map $\tilde{L}_a$ of Sect. 5.10. It is not difficult to see that π is continuous when $B(\ell_I^2)$ is equipped with the weak* topology and when $B(L^p(\Gamma_q(H) \overline{\otimes} B(\ell_I^2)))$ is equipped with the weak operator topology, see also Lemma 5.14. Finally, we will also use the restriction of π on S_I^∞.

Theorem 5.6 *Suppose* $1 < p < \infty$ *and* $-1 \leqslant q < 1$.

1. *We have* $(\mathscr{D}_{\alpha,q,p})^* = \mathscr{D}_{\alpha,q,p^*}$ *with respect to the duality bracket* $\langle x, y \rangle = \tau(x^* y)$. *In particular, the operator* $\mathscr{D}_{\alpha,q,2}$ *is selfadjoint.*
2. *We have*

$$\mathrm{Lip}_{\mathscr{D}_{\alpha,q,p}}(B(\ell_I^2)) = \mathrm{dom}\, \partial_{\alpha,q,\infty}. \tag{5.85}$$

3. *For any* $a \in \mathrm{dom}\, \partial_{\alpha,q,\infty}$, *we have*

$$[\mathscr{D}_{\alpha,q,p}, \pi(a)] = L_{\partial_{\alpha,q,\infty}(a)}. \tag{5.86}$$

4. *For any* $a \in \mathrm{dom}\, \partial_{\alpha,q,\infty}$, *we have*

$$\left\| [\mathscr{D}_{\alpha,q,p}, \pi(a)] \right\|_{L^p(\Gamma_q(H) \overline{\otimes} B(\ell_I^2)) \to L^p(\Gamma_q(H) \overline{\otimes} B(\ell_I^2))} = \left\| \partial_{\alpha,q,\infty}(a) \right\|_{\Gamma_q(H) \overline{\otimes} B(\ell_I^2)}. \tag{5.87}$$

5. *Suppose* $p = 2$ *and* $q = -1$. *We have* $(\mathscr{D}_{\alpha,-1,2})^2 = \mathrm{Id}_{L^2(\Gamma_{-1}(H))} \otimes A_2$.

6. *If I is finite and if H is finite-dimensional then the operator*

$$|\mathcal{D}_{\alpha,-1,2}|^{-1} : \overline{\mathrm{Ran}\,\mathcal{D}_{\alpha,-1,2}} \to \overline{\mathrm{Ran}\,\mathcal{D}_{\alpha,-1,2}}$$

is compact.[20]

Proof 1. By (5.83) and [136, Problem 5.24 p. 168], $\mathcal{D}_{\alpha,q,p}$ and $\mathcal{D}_{\alpha,q,p^*}$ are formal adjoints. Hence $\mathcal{D}_{\alpha,q,p^*} \subset (\mathcal{D}_{\alpha,q,p})^*$ by [136, p. 167]. Now, we will show the reverse inclusion. Let $z \in \mathrm{dom}(\mathcal{D}_{\alpha,q,p})^*$. For any $y \in \mathrm{dom}\,\mathcal{D}_{\alpha,q,p}$, we have

$$\langle \mathcal{D}_{\alpha,q,p}(y), z \rangle = \langle y, (\mathcal{D}_{\alpha,q,p})^*(z) \rangle. \tag{5.88}$$

Note that $(\mathrm{Id}_{L^p(\Gamma_q(H))} \otimes \mathcal{T}_J)(z) \to z$. Now, for any $y \in L^p(\Gamma_q(H)) \otimes M_{I,\mathrm{fin}}$, we have

$$\langle y, \mathcal{D}_{\alpha,q,p^*}(\mathrm{Id}_{L^p(\Gamma_q(H))} \otimes \mathcal{T}_J)(z) \rangle = \langle \mathcal{D}_{\alpha,q,p}(y), (\mathrm{Id}_{L^p(\Gamma_q(H))} \otimes \mathcal{T}_J)(z) \rangle$$

$$= \langle (\mathrm{Id}_{L^p(\Gamma_q(H))} \otimes \mathcal{T}_J)\mathcal{D}_{\alpha,q,p}(y), z \rangle = \langle \mathcal{D}_{\alpha,q,p}(\mathrm{Id}_{L^p(\Gamma_q(H))} \otimes \mathcal{T}_J)(y), z \rangle$$

$$\stackrel{(5.88)}{=} \langle (\mathrm{Id}_{L^p(\Gamma_q(H))} \otimes \mathcal{T}_J)(y), (\mathcal{D}_{\alpha,q,p})^*(z) \rangle$$

$$= \langle y, (\mathrm{Id}_{L^p(\Gamma_q(H))} \otimes \mathcal{T}_J)(\mathcal{D}_{\alpha,q,p})^*(z) \rangle.$$

Hence

$$\mathcal{D}_{\alpha,q,p^*}(\mathrm{Id}_{L^p(\Gamma_q(H))} \otimes \mathcal{T}_J)(z) = (\mathrm{Id}_{L^p(\Gamma_q(H))} \otimes \mathcal{T}_J)(\mathcal{D}_{\alpha,q,p})^*(z) \to (\mathcal{D}_{\alpha,q,p})^*(z)$$

By (2.1), we deduce that $z \in \mathcal{D}_{\alpha,q,p^*}$ and that

$$\mathcal{D}_{\alpha,q,p^*}(z) = (\mathcal{D}_{\alpha,q,p})^*(z).$$

2. and 3. For any $i, j, k, l \in I$ and any $x \in L^p(\Gamma_q(H))$, we have

$$[\mathcal{D}_{\alpha,q,p}, \pi(e_{ij})](x \otimes e_{kl}) = \mathcal{D}_{\alpha,q,p}\pi(e_{ij})(x \otimes e_{kl}) - \pi(e_{ij})\mathcal{D}_{\alpha,q,p}(x \otimes e_{kl})$$

$$\stackrel{(5.84)(5.82)}{=} \mathcal{D}_{\alpha,q,p}(x \otimes e_{ij}e_{kl}) - \pi(e_{ij})(s_q(\alpha_k - \alpha_l)x \otimes e_{kl})$$

$$= \delta_{j=k}\mathcal{D}_{\alpha,q,p}(x \otimes e_{il}) - s_q(\alpha_k - \alpha_l)x \otimes e_{ij}e_{kl}$$

$$\stackrel{(5.82)}{=} \delta_{j=k}\big(s_q(\alpha_i - \alpha_l)x \otimes e_{il} - s_q(\alpha_k - \alpha_l)x \otimes e_{il}\big) = \delta_{j=k}s_q(\alpha_i - \alpha_k)x \otimes e_{il}$$

$$= s_q(\alpha_i - \alpha_j)x \otimes e_{ij}e_{kl} = \big(s_q(\alpha_i - \alpha_j) \otimes e_{ij}\big)(x \otimes e_{kl})$$

$$\stackrel{(2.95)}{=} (\partial_{\alpha,q}(e_{ij}))(x \otimes e_{kl}) = L_{\partial_{\alpha,q}(e_{ij})}(x \otimes e_{kl}).$$

---

[20] Note that since $\Gamma_{-1}(H)\overline{\otimes}B(\ell_I^2)$ is then finite-dimensional, the spaces $L^p(\Gamma_{-1}(H)\overline{\otimes}B(\ell_I^2))$ are all isomorphic for different values of p, $\overline{\mathrm{Ran}\,\mathcal{D}_{\alpha,-1,2}} = \overline{\mathrm{Ran}\,\mathcal{D}_{\alpha,-1,p}}$ for all $1 < p < \infty$ and thus, $|\mathcal{D}_{\alpha,-1,2}|^{-1}$ extends to a compact operator on $\overline{\mathrm{Ran}\,\mathcal{D}_{\alpha,-1,p}}$, too.

By linearity and density, we obtain $\left[\mathscr{D}_{\alpha,q,p}, \pi(e_{ij})\right] = L_{\partial_{\alpha,q}(e_{ij})}$. It suffices to use Proposition 5.12. Finally by linearity, for any $a \in M_{I,\mathrm{fin}}$ we obtain (5.86) and in addition

$$\left\|\left[\mathscr{D}_{\alpha,q,p}, \pi(a)\right]\right\|_{L^p(\Gamma_q(H)\overline{\otimes}B(\ell_I^2))\to L^p(\Gamma_q(H)\overline{\otimes}B(\ell_I^2))} \tag{5.89}$$

$$\overset{(5.86)}{=} \left\|L_{\partial_{\alpha,q,\infty}(a)}\right\|_{L^p(\Gamma_q(H)\overline{\otimes}B(\ell_I^2))\to L^p(\Gamma_q(H)\overline{\otimes}B(\ell_I^2))} = \left\|\partial_{\alpha,q,\infty}(a)\right\|_{\Gamma_q(H)\overline{\otimes}B(\ell_I^2)}.$$

We next claim that

$$\mathrm{dom}\,\partial_{\alpha,q,\infty} \subset \mathrm{Lip}_{\mathscr{D}_{\alpha,q,p}}(B(\ell_I^2)). \tag{5.90}$$

Let $a \in \mathrm{dom}\,\partial_{\alpha,q,\infty}$ and consider a net (a_j) of elements of $M_{I,\mathrm{fin}}$ such that $a_j \to a$ and $\partial_{\alpha,q}(a_j) \to \partial_{\alpha,q,\infty}(a)$ both for the weak* topology. By Lemma 2.1, we can suppose that the nets (a_j) and $(\partial_{\alpha,q}(a_j))$ are bounded. So by (5.89), the net $\left(\left[\mathscr{D}_{\alpha,q,p}, \pi(a_j)\right]\right)$ is also bounded. Now by the part 4 of Proposition 5.11, a belongs to $\mathrm{Lip}_{\mathscr{D}_{\alpha,q,p}}(B(\ell_I^2))$. We have shown (5.90).

Next we claim that for $a \in \mathrm{dom}\,\partial_{\alpha,q,\infty}$, we have

$$\left[\mathscr{D}_{\alpha,q,p}, \pi(a)\right] = L_{\partial_{\alpha,q,\infty}(a)}. \tag{5.91}$$

By continuity of π, $\pi(a_j) \to \pi(a)$ for the weak operator topology of the space $B(L^p(\Gamma_q(H)\overline{\otimes}B(\ell_I^2)))$. For any $\xi \in \mathrm{dom}\,\mathscr{D}_{\alpha,q,p}$ and any $\zeta \in \mathrm{dom}\,\mathscr{D}_{\alpha,q,p^*}$, we have

$$\left\langle\left[\mathscr{D}_{\alpha,q,p}, \pi(a_j)\right]\xi, \zeta\right\rangle_{L^p(\Gamma_q(H)\overline{\otimes}B(\ell_I^2)),L^{p^*}(\Gamma_q(H)\overline{\otimes}B(\ell_I^2))}$$

$$= \left\langle(\mathscr{D}_{\alpha,q,p}\pi(a_j) - \pi(a_j)\mathscr{D}_{\alpha,q,p})\xi, \zeta\right\rangle$$

$$= \left\langle\mathscr{D}_{\alpha,q,p}\pi(a_j)\xi, \zeta\right\rangle - \left\langle\pi(a_j)\mathscr{D}_{\alpha,q,p}\xi, \zeta\right\rangle$$

$$= \left\langle\pi(a_j)\xi, \mathscr{D}_{\alpha,q,p^*}\zeta\right\rangle - \left\langle\pi(a_j)\mathscr{D}_{\alpha,q,p}\xi, \zeta\right\rangle$$

$$\underset{j}{\to} \left\langle\pi(a)\xi, \mathscr{D}_{\alpha,q,p^*}\zeta\right\rangle - \left\langle\pi(a)\mathscr{D}_{\alpha,q,p}\xi, \zeta\right\rangle = \left\langle[\mathscr{D}_{\alpha,q,p}, \pi(a)]\xi, \zeta\right\rangle.$$

Since the net $([\mathscr{D}_{\alpha,q,p}, \pi(a_j)])$ is bounded by (5.89), we deduce that the net $([\mathscr{D}_{\alpha,q,p}, \pi(a_j)])$ converges to $[\mathscr{D}_{\alpha,q,p}, \pi(a)]$ for the weak operator topology by a "net version" of [136, Lemma 3.6 p. 151]. Furthermore, it is (really) easy to check that $L_{\partial_{\alpha,q,\infty}(a_j)} \to L_{\partial_{\alpha,q,\infty}(a)}$ for the weak operator topology of $B(L^p(\Gamma_q(H)\overline{\otimes}B(\ell_I^2)))$. By uniqueness of the limit, we deduce that the commutator is given by the same formula as that in the case of elements of $M_{I,\mathrm{fin}}$. From here, we obtain (5.87) as in (5.89).

Next we pass to the reverse inclusion of (5.90) and claim

$$\mathrm{Lip}_{\mathscr{D}_{\alpha,q,p}}(B(\ell_I^2)) \subset \mathrm{dom}\,\partial_{\alpha,q,\infty}. \tag{5.92}$$

To this end, let $a \in \mathrm{Lip}_{\mathscr{D}_{\alpha,q,p}}(\mathrm{B}(\ell_J^2))$. Consider the bounded net (a_J) defined by $a_J \overset{\mathrm{def}}{=} \mathcal{T}_J(a)$. We have $a_J(a) \to a$ in the weak* topology of $\mathrm{B}(\ell_J^2)$. The essential point will be to prove that

$$(\partial_{\alpha,q}(a_J)) \text{ is a bounded net.} \tag{5.93}$$

Indeed, once (5.93) is shown, using Banach-Alaoglu theorem, we can suppose that $\partial_{\alpha,q}(a_J) \to b$ for the weak* topology for some b. Since the graph of the unbounded operator $\partial_{\alpha,q,\infty}$ is weak* closed, we would conclude that a belongs to the subspace $\mathrm{dom}\, \partial_{\alpha,q,\infty}$ and (5.92) follows.

To show (5.93), we note that $\partial_{\alpha,q}(a_J)$ belongs to $\Gamma_q(H)\overline{\otimes}\mathrm{B}(\ell_J^2)$. So

$$\begin{aligned}
\left\| \partial_{\alpha,q}(a_J) \right\|_{\Gamma_q(H)\overline{\otimes}\mathrm{B}(\ell_J^2)} &= \left\| \partial_{\alpha,q}(a_J) \right\|_{\Gamma_q(H)\overline{\otimes}\mathrm{B}(\ell_J^2)} \\
&= \left\| \mathrm{L}_{\partial_{\alpha,q}(a_J)} \right\|_{\mathrm{L}^p(\Gamma_q(H)\overline{\otimes}\mathrm{B}(\ell_J^2)) \to \mathrm{L}^p(\Gamma_q(H)\overline{\otimes}\mathrm{B}(\ell_J^2))} \\
&= \sup_{\|\xi\|_{\mathrm{L}^p(\Gamma_q(H)\overline{\otimes}\mathrm{B}(\ell_J^2))} \leqslant 1,\ \|\eta\|_{\mathrm{L}^{p^*}(\Gamma_q(H)\overline{\otimes}\mathrm{B}(\ell_J^2))} \leqslant 1} \left| \langle \partial_{\alpha,q}(a_J)\xi, \eta\rangle \right|.
\end{aligned}$$

Now, for any $i,j,k,l \in J$, any $x \in \mathrm{L}^p(\Gamma_q(H))$ and any $y \in \mathrm{L}^{p^*}(\Gamma_q(H))$, we have

$$\begin{aligned}
\langle [\mathscr{D}_{\alpha,q,p}, &\pi(a)](x \otimes e_{ij}), y \otimes e_{kl}\rangle \\
&= \langle \pi(a)(x \otimes e_{ij}), \mathscr{D}_{\alpha,q,p^*}(y \otimes e_{kl})\rangle - \langle \pi(a)\mathscr{D}_{\alpha,q,p}(x \otimes e_{ij}), y \otimes e_{kl}\rangle \\
&= \langle x \otimes a e_{ij}, s_q(\alpha_k - \alpha_l)y \otimes e_{kl}\rangle - \langle \pi(a)(s_q(\alpha_i - \alpha_j)x \otimes e_{ij}), y \otimes e_{kl}\rangle \\
&= \langle x \otimes a e_{ij}, s_q(\alpha_k - \alpha_l)y \otimes e_{kl}\rangle - \langle s_q(\alpha_i - \alpha_j)x \otimes a e_{ij}, y \otimes e_{kl}\rangle \\
&= \tau\big(x^* s_q(\alpha_k - \alpha_l)y\big)\mathrm{Tr}(e_{ji}a^* e_{kl}) - \tau\big(x^* s_q(\alpha_i - \alpha_j)y\big)\mathrm{Tr}(e_{ji}a^* e_{kl}) \\
&= \tau\big(x^* s_q(\alpha_k - \alpha_l + \alpha_j - \alpha_i)y\big)\mathrm{Tr}(e_{ji}a^* e_{kl}) = \delta_{j=l}\overline{a_{ki}}\,\tau\big(x^* s_q(\alpha_k - \alpha_i)y\big)
\end{aligned}$$

and

$$\begin{aligned}
\langle [\mathscr{D}_{\alpha,q,p}, &\pi(a_j)](x \otimes e_{ij}), y \otimes e_{kl}\rangle \\
&= \left\langle \left(\sum_{r,s \in J} a_{rs} s_q(\alpha_r - \alpha_s) \otimes e_{rs}\right)(x \otimes e_{ij}), y \otimes e_{kl}\right\rangle \\
&= \sum_{r,s \in J} \overline{a_{rs}} \langle (s_q(\alpha_r - \alpha_s) \otimes e_{rs})(x \otimes e_{ij}), y \otimes e_{kl}\rangle \\
&= \sum_{r,s \in J} \overline{a_{rs}}\, \tau\big(x^* s_q(\alpha_r - \alpha_s)y\big)\mathrm{Tr}(e_{ji}e_{sr}e_{kl}) \\
&= \sum_{r,s \in J} \delta_{i=s}\delta_{r=k}\delta_{j=l}\overline{a_{rs}}\,\tau\big(x^* s_q(\alpha_r - \alpha_s)y\big) = \delta_{j=l}\overline{a_{ki}}\,\tau\big(x^* s_q(\alpha_k - \alpha_i)y\big).
\end{aligned}$$

Fixing $\xi \in L^p(\Gamma_q(H)) \otimes M_J$ and $\eta \in L^{p^*}(\Gamma_q(H)) \otimes M_J$, we deduce that

$$\langle \partial_{\alpha,q}(a_J)\xi, \eta \rangle \overset{(5.86)}{=} \langle [\mathscr{D}_{\alpha,q,p}, \pi(a_J)]\xi, \eta \rangle = \langle [\mathscr{D}_{\alpha,q,p}, \pi(a)]\xi, \eta \rangle.$$

We conclude by estimating the absolute value of the last term against the product $\left\| [\mathscr{D}_{\alpha,q,p}, \pi(a)] \right\| \|\xi\|_p \|\eta\|_{p^*}$.

4. For any $a \in \mathrm{dom}\, \partial_{\alpha,q,\infty}$, we have

$$\left\| [\mathscr{D}_{\alpha,q,p}, \pi(a)] \right\|_{L^p(\Gamma_q(H)\overline{\otimes}B(\ell_I^2)) \to L^p(\Gamma_q(H)\overline{\otimes}B(\ell_I^2))}$$

$$\overset{(5.86)}{=} \left\| L_{\partial_{\alpha,q,\infty}(a)} \right\|_{L^p(\Gamma_q(H)\overline{\otimes}B(\ell_I^2)) \to L^p(\Gamma_q(H)\overline{\otimes}B(\ell_I^2))} = \left\| \partial_{\alpha,q,\infty}(a) \right\|_{\Gamma_q(H)\overline{\otimes}B(\ell_I^2)}.$$

5. If $x \in L^2(\Gamma_{-1}(H))$ and $i, j \in I$, note that

$$(\mathscr{D}_{\alpha,-1,2})^2(x \otimes e_{ij}) \overset{(5.82)}{=} \mathscr{D}_{\alpha,-1,2}\big(s_{-1}(\alpha_i - \alpha_j)x \otimes e_{ij}\big) \overset{(5.82)}{=} s_{-1}(\alpha_i - \alpha_j)^2 x \otimes e_{ij}$$

$$\overset{(2.42)}{=} x \otimes \|\alpha_i - \alpha_j\|_H^2 e_{ij} = (\mathrm{Id}_{L^2(\Gamma_{-1}(H))} \otimes A_2)(x \otimes e_{ij}).$$

By Hytönen et al. [110, Proposition G.2.4], note that $L^2(\Gamma_{-1}(H)) \otimes M_{I,\mathrm{fin}}$ is a core of $\mathrm{Id}_{L^2(\Gamma_{-1}(H))} \otimes A_2$. Then since $\mathscr{D}_{\alpha,-1,2}^2$ is again selfadjoint, thus closed, we have $\mathrm{Id}_{L^2(\Gamma_{-1}(H))} \otimes A_2 \subset \mathscr{D}_{\alpha,-1,2}^2$. Now it follows from the fact that both $T_1 \overset{\mathrm{def}}{=} \mathrm{Id}_{L^2(\Gamma_{-1}(H))} \otimes A_2$ and $T_2 \overset{\mathrm{def}}{=} \mathscr{D}_{\alpha,-1,2}^2$ are selfadjoint that both operators are in fact equal. Indeed, consider some $\lambda \in \rho(T_1) \cap \rho(T_2)$ (e.g. $\lambda = \mathrm{i}$). Then $T_1 - \lambda \subset T_2 - \lambda$, the operator $T_1 - \lambda$ is surjective and $T_2 - \lambda$ is injective. Now, it suffices to use the result [207, p. 5] to conclude that $T_1 - \lambda = T_2 - \lambda$ and thus $T_1 = T_2$.

6. Since $\mathscr{D}_{\alpha,-1,2}$ is selfadjoint according to the first part, also $\mathscr{D}_{\alpha,-1,2}^2$ is selfadjoint. As H is finite-dimensional, $\Gamma_{-1}(H)$ is finite-dimensional. As also I is finite, $L^2(\Gamma_{-1}(H)\overline{\otimes}B(\ell_I^2))$ is finite-dimensional. Thus, $\overline{\mathrm{Ran}\, \mathscr{D}_{\alpha,-1,2}} = \mathrm{Ran}\, \mathscr{D}_{\alpha,-1,2}$ and $\mathscr{D}_{\alpha,-1,2}^2|\overline{\mathrm{Ran}\, \mathscr{D}_{\alpha,-1,2}} \colon \overline{\mathrm{Ran}\, \mathscr{D}_{\alpha,-1,2}} \to \overline{\mathrm{Ran}\, \mathscr{D}_{\alpha,-1,2}}$ is a bijective bounded operator. We infer that its inverse $\mathscr{D}_{\alpha,-1,2}^{-2}$ and the square root of it $|\mathscr{D}_{\alpha,-1,2}|^{-1}$ are also bounded operators on the same space. As $L^2(\Gamma_{-1}(H)\overline{\otimes}B(\ell_I^2))$ is finite-dimensional, $|\mathscr{D}_{\alpha,-1,2}|^{-1} \colon \overline{\mathrm{Ran}\, \mathscr{D}_{\alpha,-1,2}} \to \overline{\mathrm{Ran}\, \mathscr{D}_{\alpha,-1,2}}$ is compact.   □

Remark 5.15 In the case $p = 2$, we have an alternative proof of the point 1 of Theorem 5.6. It suffices to show that $\mathscr{D}_{\alpha,q} \colon L^2(\Gamma_q(H)) \otimes M_{I,\mathrm{fin}} \subset L^2(\Gamma_q(H)\overline{\otimes}B(\ell_I^2)) \to L^2(\Gamma_q(H)\overline{\otimes}B(\ell_I^2))$ is essentially selfadjoint.

Proof By linearity, (5.83) says that $\mathscr{D}_{\alpha,q}$ is symmetric. By [195, Corollary p. 257], it suffices to prove that $\mathrm{Ran}(\mathscr{D}_{\alpha,q} \pm \mathrm{iId})$ is dense in $L^2(\Gamma_q(H)\overline{\otimes}B(\ell_I^2))$.

Let $z = [z_{ij}] \in L^2(\Gamma_q(H)\overline{\otimes}B(\ell_I^2))$ be a vector which is orthogonal to $\mathrm{Ran}(\mathscr{D}_{\alpha,q} + \mathrm{iId})$ in $L^2(\Gamma_q(H)\overline{\otimes}B(\ell_I^2))$. For any $i, j \in I$ and any $x \in L^2(\Gamma_q(H))$, we have

$$\langle s_q(\alpha_i - \alpha_j)x \otimes e_{ij}, z\rangle_{L^2(\Gamma_q(H)\overline{\otimes}B(\ell_j^2))} - i\tau(x^*z_{ij})$$

$$\overset{(5.82)}{=} \langle \mathscr{D}_{\alpha,q}(x \otimes e_{ij}), z\rangle_{L^2(\Gamma_q(H)\overline{\otimes}B(\ell_j^2))} - i\langle x \otimes e_{ij}, z\rangle_{L^2(\Gamma_q(H)\overline{\otimes}B(\ell_j^2))}$$

$$= \langle (\mathscr{D}_{\alpha,q} - i\mathrm{Id})(x \otimes e_{ij}), z\rangle_{L^2(\Gamma_q(H)\overline{\otimes}B(\ell_j^2))} = 0.$$

We infer that

$$\langle x, s_q(\alpha_i - \alpha_j)z_{ij}\rangle_{L^2(\Gamma_q(H))} = \tau(x^* s_q(\alpha_i - \alpha_j)z_{ij}) = i\tau(x^*z_{ij}) = \langle x, iz_{ij}\rangle_{L^2(\Gamma_q(H))}.$$

We deduce that $s_q(\alpha_i - \alpha_j)z_{ij} = iz_{ij}$. Recall that by [128, p. 96], the map $\Gamma_q(H) \to \mathcal{F}_q(H)$, $y \mapsto y(\Omega)$ extends to an isometry $\Delta \colon L^2(\Gamma_q(H)) \to \mathcal{F}_q(H)$. For any $x \in \Gamma_q(H)$ and any $y \in L^2(\Gamma_q(H))$, it is easy to check that $\Delta(xy) = x\Delta(y)$. We infer that

$$s_q(\alpha_i - \alpha_j)\Delta(z_{ij}) = \Delta(s_q(\alpha_i - \alpha_j)z_{ij}) = \Delta(iz_{ij}) = i\Delta(z_{ij}).$$

Thus the vector $\Delta(z_{ij})$ of $\mathcal{F}_q(H)$ is zero or an eigenvector of $s_q(\alpha_i - \alpha_j)$. Since $s_q(\alpha_i - \alpha_j)$ is selfadjoint, i is not an eigenvalue. So $\Delta(z_{ij}) = 0$. Since Δ is injective, we infer that $z_{ij} = 0$. It follows that $z = 0$. The case with $-i$ instead of i is similar. $\square$

In the case where $-1 < q < 1$, we have the following equivalence with the norm of the commutator.

Lemma 5.16 *Suppose* $-1 < q < 1$. *For any* $x \in M_{I,\mathrm{fin}}$, *we have*

$$\|\partial_{\alpha,q}(x)\|_{\Gamma_q(H)\overline{\otimes}B(\ell_I^2)} \approx \max\left\{\|\Gamma(x,x)\|_{B(\ell_I^2)}^{\frac{1}{2}}, \|\Gamma(x^*,x^*)\|_{B(\ell_I^2)}^{\frac{1}{2}}\right\}. \tag{5.94}$$

Proof Here, we use an orthonormal basis $(e_k)_{k \in K}$ of H. Using the C*-identity, first note that the noncommutative Khintchine inequalities of [43, Theorem 4.1] can be rewritten under the following form for any element $f = \sum_{k \in K} s_q(e_k) \otimes x_k$ where $(x_k)_{k \in K}$ is a finitely supported family of elements of $B(\ell_I^2)$

$$\|f\|_{\Gamma_q(H)\overline{\otimes}B(\ell_I^2)} \approx \max\left\{\left\|\sum_{k \in K} x_k^* x_k\right\|_{B(\ell_I^2)}^{\frac{1}{2}}, \left\|\sum_{k \in K} x_k x_k^*\right\|_{B(\ell_I^2)}^{\frac{1}{2}}\right\} \tag{5.95}$$

$$\overset{(2.41)}{=} \max\left\{\left\|\sum_{k,j \in K} \langle s_q(e_k), s_q(e_j)\rangle_{L^2(\Gamma_q)} x_k^* x_j\right\|_{B(\ell_I^2)}^{\frac{1}{2}}, \right. \tag{5.96}$$

$$\left\| \sum_{k,j \in K} \langle s_q(e_k), s_q(e_j) \rangle_{L^2(\Gamma_q)} x_k x_j^* \right\|_{B(\ell_I^2)}^{\frac{1}{2}} \right\}$$

$$\overset{(2.68)}{=} \max \left\{ \|f\|_{L^\infty(B(\ell_I^2), L^2(\Gamma_q)_c)}, \|f\|_{L^\infty(B(\ell_I^2), L^2(\Gamma_q)_r)} \right\}$$

$$\overset{(2.74)(2.72)}{=} \max \left\{ \|\mathbb{E}(f^* f)\|_{B(\ell_I^2)}^{\frac{1}{2}}, \|\mathbb{E}(f f^*)\|_{B(\ell_I^2)}^{\frac{1}{2}} \right\}$$

where we use some easy extensions of the results of the second chapter to the case $p = \infty$. By weak* density, it is true for any element of $\mathrm{Gauss}_{q,\infty}(B(\ell_I^2))$. Replacing f by $\partial_{\alpha,q}(x)$ and using (2.97), we obtain

$$\left\| \partial_{\alpha,q}(x) \right\|_{\Gamma_q(H) \overline{\otimes} B(\ell_I^2)}$$

$$\overset{(5.95)}{\approx} \max \left\{ \left\| \mathbb{E}\left((\partial_{\alpha,q}(x))^* \partial_{\alpha,q}(x) \right) \right\|_{B(\ell_I^2)}^{\frac{1}{2}}, \left\| \mathbb{E}\left(\partial_{\alpha,q}(x)(\partial_{\alpha,q}(x))^* \right) \right\|_{B(\ell_I^2)}^{\frac{1}{2}} \right\}$$

$$\overset{(2.97)}{=} \max \left\{ \left\| \Gamma(x, x) \right\|_{B(\ell_I^2)}^{\frac{1}{2}}, \left\| \Gamma(x^*, x^*) \right\|_{B(\ell_I^2)}^{\frac{1}{2}} \right\}.$$

□

Remark 5.16 Suppose $2 \leqslant p < \infty$. If $x \in \mathrm{dom}\, A_p^{\frac{1}{2}}$, we have seen in the proof of Theorem 3.5 that

$$\left\| \partial_{\alpha,q,p}(x) \right\|_{L^p(\Gamma_q(H) \overline{\otimes} B(\ell_I^2))} \approx \max \left\{ \left\| \Gamma(x, x)^{\frac{1}{2}} \right\|_{S_I^p}, \left\| \Gamma(x^*, x^*)^{\frac{1}{2}} \right\|_{S_I^p} \right\}. \tag{5.97}$$

5.12 Bisectoriality and Functional Calculus of the Dirac Operator II

In the four preceding sections, we have investigated four Hodge-Dirac operators—one of which defines a compact Banach spectral triple and one defines a locally compact Banach spectral triple in the sense of Definitions 5.10 and 5.11 under suitable assumptions. The properties of bisectoriality and functional calculus of the Hodge-Dirac operators of type II were missing. In this section, we consider these operators, $\mathscr{D}_{\alpha,1,p}$ and $\mathscr{D}_{\psi,1,p}$ with parameter $q = 1$ and show that they are indeed bisectorial and do admit a bounded H^∞ functional calculus. We equally obtain partial results for the case $-1 \leqslant q < 1$ in Remark 5.17.

Bisectoriality of the Dirac Operator II for Fourier Multipliers on the Banach Space $L^p(L^\infty(\Omega) \rtimes_\alpha G)$ Consider the Dirac operator $\mathscr{D}_{\psi,1}$ defined by $\mathscr{D}_{\psi,1}(f \rtimes \lambda_s) = W(b_\psi(s)) f \rtimes \lambda_s$ for $f \in L^\infty(\Omega)$ and $s \in G$. Moreover, consider the subspace

$\mathrm{dom}_{\infty}(\mathscr{D}_{\psi,1}) \overset{\text{def}}{=} \mathrm{span}\{f \rtimes \lambda_s : s \in G, \ f \in \mathrm{L}^{\infty}(\Omega) : \mathrm{W}(b_{\psi}(s))f \in \mathrm{L}^{\infty}(\Omega)\}$ of $\mathcal{P}_{\rtimes,G}$ stable by "Gaussian multiplication" such that $\mathscr{D}_{\psi,1}(\mathrm{dom}_{\infty}(\mathscr{D}_{\psi,1})) \subset \mathcal{P}_{\rtimes,G}$.

Proposition 5.17

1. *There is a weak* continuous group $(U_t)_{t\in\mathbb{R}}$ of trace preserving *-automorphisms $U_t : \mathrm{L}^{\infty}(\Omega) \rtimes_{\alpha} G \to \mathrm{L}^{\infty}(\Omega) \rtimes_{\alpha} G$, such that its weak* generator $\mathrm{i}\mathscr{D}_{\psi,1,\infty}$ is an extension of the restriction $\mathrm{i}\mathscr{D}_{\psi,1}$ on $\mathrm{dom}_{\infty}(\mathscr{D}_{\psi,1})$.*
2. *Let $1 \leqslant p < \infty$. Suppose that $\mathrm{L}^p(\mathrm{VN}(G))$ has CCAP and that $\mathrm{L}^{\infty}(\Omega) \rtimes_{\alpha} G$ has QWEP. The operator $\mathrm{i}\mathscr{D}_{\psi,1} : \mathcal{P}_{\rtimes,G} \subset \mathrm{L}^p(\mathrm{L}^{\infty}(\Omega) \rtimes_{\alpha} G) \to \mathrm{L}^p(\mathrm{L}^{\infty}(\Omega) \rtimes_{\alpha} G)$ is closable and its closure $\mathrm{i}\mathscr{D}_{\psi,1,p}$ generates the strongly continuous group $(U_{t,p})_{t\in\mathbb{R}}$ of isometries on $\mathrm{L}^p(\mathrm{L}^{\infty}(\Omega) \rtimes_{\alpha} G)$.*

Proof We consider the weak* continuous group $(U_t)_{t\in\mathbb{R}}$ of trace preserving *-automorphisms from [13, Section 3] given by (an additional factor $\sqrt{2}$ has notational reasons there) $U_t(x \rtimes \lambda_s) \overset{\text{def}}{=} \mathrm{e}^{\mathrm{i}t\mathrm{W}(b_{\psi}(s))}x \rtimes \lambda_s$. Thus according to [174, Propositions 1.1.1 and 1.1.2 and Theorem 1.2.3], we have a weak* closed and weak* densely defined generator $\mathrm{i}\mathscr{D}_{\psi,1,\infty}$ of $(U_t)_{t\in\mathbb{R}}$ and a generator $\mathrm{i}\mathscr{D}_{\psi,1,p}$ for $(U_{t,p})_{t\in\mathbb{R}}$.

Now, let us show that $\mathrm{i}\mathscr{D}_{\psi,1,\infty}$ contains $\mathrm{i}\mathscr{D}_{\psi,1}$ restricted to $\mathrm{dom}_{\infty}(\mathscr{D}_{\psi,1})$. For that, let $s \in G$, $f \in \mathrm{L}^{\infty}(\Omega)$ such that $\mathrm{W}(b_{\psi}(s))f \in \mathrm{L}^{\infty}(\Omega)$, and y be an element of $\mathrm{L}^1(\mathrm{L}^{\infty}(\Omega) \rtimes_{\alpha} G)$. Using the differentiation under the integral sign by domination using $\mathrm{W}(b_{\psi}(s))f \in \mathrm{L}^{\infty}(\Omega)$, we obtain

$$\frac{1}{t}\langle y, (U_t - \mathrm{Id})f \rtimes \lambda_s\rangle = \frac{1}{t}\tau\left(y^*(\mathrm{e}^{\mathrm{i}t\mathrm{W}(b_{\psi}(s))} - 1)f \rtimes \lambda_s\right)$$

$$= \frac{1}{t}\int_{\Omega}(\mathrm{e}^{\mathrm{i}t\mathrm{W}(b_{\psi}(s))(\omega)} - 1)f(\omega)\overline{y_s(\omega)}\,\mathrm{d}\mu(\omega)$$

$$\xrightarrow[t\to 0]{}\int_{\Omega}\mathrm{i}\mathrm{W}(b_{\psi}(s))(\omega)f(\omega)\overline{y_s(\omega)}\,\mathrm{d}\mu(\omega) = \langle y, \mathrm{i}\mathrm{W}(b_{\psi}(s))f \rtimes \lambda_s\rangle.$$

We deduce that $f \rtimes \lambda_s$ belongs to $\mathrm{dom}\,\mathrm{i}\mathscr{D}_{\psi,1,\infty}$ and that $\mathrm{i}\mathscr{D}_{\psi,1,\infty}(f \rtimes \lambda_s) = \mathrm{i}\mathrm{W}(b_{\psi}(s))f \rtimes \lambda_s$.

Next we show that the generator $\mathrm{i}\mathscr{D}_{\psi,1,p}$ is the closure of $\mathrm{i}\mathscr{D}_{\psi,1} : \mathcal{P}_{\rtimes,G} \subset \mathrm{L}^p(\mathrm{L}^{\infty}(\Omega) \rtimes_{\alpha} G) \to \mathrm{L}^p(\mathrm{L}^{\infty}(\Omega) \rtimes_{\alpha} G)$. Let $f \rtimes \lambda_s$ be an element of $\mathcal{P}_{\rtimes,G}$. Using [85, Proposition 4.11 p. 32], we have

$$\frac{1}{t}(U_{t,p} - \mathrm{Id})(f \rtimes \lambda_s) = \frac{1}{t}\left(\mathrm{e}^{\mathrm{i}t\mathrm{W}(b_{\psi}(s))} - 1\right)f \rtimes \lambda_s$$

$$\xrightarrow[t\to 0]{}\mathrm{i}\mathrm{W}(b_{\psi}(s))f \rtimes \lambda_s.$$

We infer that $f \rtimes \lambda_s$ belongs to $\mathrm{dom}\,\mathscr{D}_{\psi,1,p}$ and that we have $\mathrm{i}\mathscr{D}_{\psi,1,p}(f \rtimes \lambda_s) = \mathrm{i}\mathrm{W}(b_{\psi}(s))f \rtimes \lambda_s$. So it suffices to prove that $\mathcal{P}_{\rtimes,G}$ is a core for $\mathrm{i}\mathscr{D}_{\psi,1,p}$.

According to Lemma 2.5 and Proposition 2.8, there exists a net (φ_j) of finitely supported functions $G \to \mathbb{C}$ converging pointwise to 1 such that

$\mathrm{Id}_{L^p(\Omega)} \rtimes M_{\varphi_j}$ converges to $\mathrm{Id}_{L^p(L^\infty(\Omega) \rtimes_\alpha G)}$ in the point norm topology and $\left\| \mathrm{Id}_{L^p(\Omega)} \rtimes M_{\varphi_j} \right\|_{\mathrm{cb}, L^p \to L^p} \leqslant 1$.

Note that $\mathrm{Id}_{L^p(\Omega)} \rtimes M_{\varphi_j}$ commutes with $U_{t,p}$. This is easy to see on elements of the form $f \rtimes \lambda_s$ and extends by a density argument to all of $L^p(L^\infty(\Omega) \rtimes_\alpha G)$. Hence, for $x \in \mathrm{dom}\, \mathscr{D}_{\psi,1,p}$, we have

$$\frac{1}{t}(U_{t,p} - \mathrm{Id})(\mathrm{Id}_{L^p(\Omega)} \rtimes M_{\varphi_j})(x) = \frac{1}{t}(\mathrm{Id}_{L^p(\Omega)} \rtimes M_{\varphi_j})(U_{t,p} - \mathrm{Id})(x)$$

$$\xrightarrow[t \to 0]{} \mathrm{i}(\mathrm{Id}_{L^p(\Omega)} \rtimes M_{\varphi_j}) \mathscr{D}_{\psi,1,p}(x).$$

We infer that $(\mathrm{Id}_{L^p(\Omega)} \rtimes M_{\varphi_j})(x)$ belongs again to $\mathrm{dom}\, \mathscr{D}_{\psi,1,p}$ and

$$\mathscr{D}_{\psi,1,p}(\mathrm{Id}_{L^p(\Omega)} \rtimes M_{\varphi_j})(x) = (\mathrm{Id}_{L^p(\Omega)} \rtimes M_{\varphi_j}) \mathscr{D}_{\psi,1,p}(x) \xrightarrow[j]{} \mathscr{D}_{\psi,1,p}(x).$$

Moreover, $(\mathrm{Id}_{L^p(\Omega)} \rtimes M_{\varphi_j})(x) \to x$ as $j \to \infty$. Therefore, to show the core property, it suffices to approximate an element $f \rtimes \lambda_s$ where $f \in L^p(\Omega)$ in the graph norm of $\mathscr{D}_{\psi,1,p}$ by elements in $\mathcal{P}_{\rtimes,G}$. To this end, it suffices to take $1_{|f| \leqslant n} f \rtimes \lambda_s$ and to use the same reasoning, noting that

$$\left(1_{|f| \leqslant n} \rtimes 1_{\mathrm{VN}(G)}\right) U_{t,p}(f \rtimes \lambda_s) = U_{t,p}\left(1_{|f| \leqslant n} f \rtimes \lambda_s\right).$$

$\square$

Corollary 5.2 *Let $1 < p < \infty$. Then the operator $\mathscr{D}_{\psi,1,p}$ is bisectorial and has a bounded $\mathrm{H}^\infty(\Sigma_\omega^\pm)$ functional calculus to any angle $\omega > 0$.*

Proof This follows at once from Proposition 5.17 together with the fact that a noncommutative L^p-space is UMD by [191, Corollary 7.7] together with the Hieber Prüss Theorem [110, Theorem 10.7.10] which says that the generator iA of a bounded strongly continuous group on a UMD-space has a bounded bisectorial H^∞ functional calculus of any angle $\omega > 0$. $\square$

Bisectoriality of the Dirac Operator II for Schur Multipliers on the Banach Space $L^p(L^\infty(\Omega) \overline{\otimes} \mathrm{B}(\ell_I^2))$ In the following, we consider the case $q = 1$ which corresponds to the classical Gaussian space. Consider the "Dirac operator" $\mathscr{D}_{\alpha,1}$ defined on $L^\infty(\Omega) \otimes \mathrm{M}_{I,\mathrm{fin}}$ with values in $L^0(\Omega) \otimes \mathrm{M}_{I,\mathrm{fin}}$ by

$$\mathscr{D}_{\alpha,1}(f \otimes e_{ij}) \overset{\mathrm{def}}{=} \mathrm{W}(\alpha_i - \alpha_j) f \otimes e_{ij}, \quad x \in L^\infty(\Omega), i, j \in I. \tag{5.98}$$

Moreover, we consider the subspace $\mathrm{dom}_\infty(\mathscr{D}_{\alpha,1}) \overset{\mathrm{def}}{=} \mathrm{span}\{ f \otimes e_{ij} : i, j \in I, f \in L^\infty(\Omega) : \mathrm{W}(\alpha_i - \alpha_j) f \in L^\infty(\Omega)\}$ of $L^\infty(\Omega) \otimes \mathrm{M}_{I,\mathrm{fin}}$ such that $\mathscr{D}_{\alpha,1}(\mathrm{dom}_\infty(\mathscr{D}_{\alpha,1})) \subset L^\infty(\Omega) \otimes \mathrm{M}_{I,\mathrm{fin}}$.

Proposition 5.18

1. *There is a weak* continuous group $(U_t)_{t \in \mathbb{R}}$ of trace preserving *-automorphisms*
   $U_t \colon \mathrm{L}^\infty(\Omega)\overline{\otimes}\mathrm{B}(\ell_I^2) \to \mathrm{L}^\infty(\Omega)\overline{\otimes}\mathrm{B}(\ell_I^2)$, *such that its weak* generator $\mathrm{i}\mathscr{D}_{\alpha,1,\infty}$ is*
   an extension of the restriction $\mathrm{i}\mathscr{D}_{\alpha,1}|_{\mathrm{dom}_\infty(\mathscr{D}_{\alpha,1})}$.
2. *Let $1 \leqslant p < \infty$. The operator $\mathrm{i}\mathscr{D}_{\alpha,1} \colon \mathrm{L}^\infty(\Omega) \otimes \mathrm{M}_{I,\mathrm{fin}} \subset \mathrm{L}^p(\Omega, S_I^p) \to$*
   $\mathrm{L}^p(\Omega, S_I^p)$ *is closable and its closure $\mathrm{i}\mathscr{D}_{\alpha,1,p}$ generates the strongly continuous*
   group $(U_{t,p})_{t \in \mathbb{R}}$ of isometries on $\mathrm{L}^p(\Omega, S_I^p)$.

Proof For $t \in \mathbb{R}$, we define the block diagonal operator $V_t \overset{\mathrm{def}}{=} \sum_{k \in I} \mathrm{e}^{\mathrm{i}t\mathrm{W}(\alpha_k)} \otimes$
$e_{kk} \in \mathrm{L}^\infty(\Omega)\overline{\otimes}\mathrm{B}(\ell_I^2)$. Note that $(V_t)_{t \in \mathbb{R}}$ is a group of unitaries. By the proof of [15, Theorem 5.2], $U_t \colon \mathrm{L}^\infty(\Omega)\overline{\otimes}\mathrm{B}(\ell_I^2) \to \mathrm{L}^\infty(\Omega)\overline{\otimes}\mathrm{B}(\ell_I^2), t \mapsto V_t x V_t^*$ defines a weak* continuous group $(U_t)_{t \in \mathbb{R}}$ of *-automorphisms. For any $t \in \mathbb{R}$, it is easy to check that U_t is trace preserving. So it induces for $1 \leqslant p < \infty$ a (uniquely) strongly continuous group $(U_{t,p})_{t \in \mathbb{R}}$ of complete isometries on $\mathrm{L}^p(\Omega, S_I^p)$. Thus according to [174, Propositions 1.1.1 and 1.1.2 and Theorem 1.2.3], we have a weak* closed and weak* densely defined generator $\mathrm{i}\mathscr{D}_{\alpha,1,\infty}$ of $(U_t)_{t \in \mathbb{R}}$ and a generator $\mathrm{i}\mathscr{D}_{\alpha,1,p}$ for $(U_{t,p})_{t \in \mathbb{R}}$.

Now, let us show that $\mathrm{i}\mathscr{D}_{\alpha,1,\infty}$ contains $\mathrm{i}\mathscr{D}_{\alpha,1}|_{\mathrm{dom}_\infty(\mathscr{D}_{\alpha,1})}$. To this end, let $i, j \in I$, $f \otimes e_{ij} \in \mathrm{dom}_\infty(\mathscr{D}_{\alpha,1})$ and y be an element of $\mathrm{L}^1(\Omega, S_I^1)$. Using differentiation under the integral sign by domination (note that $\mathrm{W}(\alpha_i - \alpha_j)f \in \mathrm{L}^\infty(\Omega)$), we obtain

$$
\frac{1}{t}\big\langle y, (U_t - \mathrm{Id})(f \otimes e_{ij})\big\rangle_{\mathrm{L}^1(\Omega, S_I^1), \mathrm{L}^\infty(\Omega)\overline{\otimes}\mathrm{B}(\ell_I^2)} = \frac{1}{t}\tau\left(y^*\big((\mathrm{e}^{\mathrm{i}t\mathrm{W}(\alpha_i - \alpha_j)} - 1)f \otimes e_{ij}\big)\right)
$$

$$
= \frac{1}{t}\int_\Omega \big(\mathrm{e}^{\mathrm{i}t\mathrm{W}(\alpha_i - \alpha_j)(\omega)} - 1\big)f(\omega)\overline{y_{ij}(\omega)}\,\mathrm{d}\mu(\omega)
$$

$$
\xrightarrow[t \to 0]{} \int_\Omega \mathrm{i}\mathrm{W}(\alpha_i - \alpha_j)(\omega)f(\omega)\overline{y_{ij}(\omega)}\,\mathrm{d}\mu(\omega) = \big\langle y, \mathrm{i}\mathrm{W}(\alpha_i - \alpha_j)f \otimes e_{ij}\big\rangle.
$$

$$\tag{5.99}$$

We deduce that $f \otimes e_{ij}$ belongs to $\mathrm{dom}\,\mathrm{i}\mathscr{D}_{\alpha,1,\infty}$ and that $\mathrm{i}\mathscr{D}_{\alpha,1,\infty}(f \otimes e_{ij}) = \mathrm{i}\mathrm{W}(\alpha_i - \alpha_j)f \otimes e_{ij}$.

Next we show that the generator $\mathrm{i}\mathscr{D}_{\alpha,1,p}$ is the closure of

$$
\mathrm{i}\mathscr{D}_{\alpha,1} \colon \mathrm{L}^\infty(\Omega) \otimes \mathrm{M}_{I,\mathrm{fin}} \subset \mathrm{L}^p(\Omega, S_I^p) \to \mathrm{L}^p(\Omega, S_I^p).
$$

For any $f \in \mathrm{L}^\infty(\Omega)$ and $i, j \in I$, using [85, Proposition 4.11 p. 32], we have

$$
\frac{1}{t}(U_{t,p} - \mathrm{Id})(f \otimes e_{ij}) = \frac{1}{t}\big(\mathrm{e}^{\mathrm{i}t\mathrm{W}(\alpha_i - \alpha_j)} - 1\big)f \otimes e_{ij}
$$

$$
\xrightarrow[t \to 0]{} \mathrm{i}\mathrm{W}(\alpha_i - \alpha_j)f \otimes e_{ij}.
$$

We infer that $f \otimes e_{ij}$ belongs to $\operatorname{dom} i\mathscr{D}_{\alpha,1,p}$ and $i\mathscr{D}_{\alpha,1,p}(f \otimes e_{ij}) = iW(\alpha_i - \alpha_j)f \otimes e_{ij}$. So it suffices to prove that $L^\infty(\Omega) \otimes M_{I,\mathrm{fin}}$ is a core for $i\mathscr{D}_{\alpha,1,p}$. Note that $\operatorname{Id}_{L^p(\Omega)} \otimes \mathcal{T}_J$ commutes with $U_{t,p}$. This is easy to check on elements of the form $f \otimes e_{ij}$ and extends by a density argument to all of $L^p(\Omega, S_I^p)$. Thus, for $x \in \operatorname{dom} \mathscr{D}_{\alpha,1,p}$, we have

$$\frac{1}{t}(U_{t,p} - \operatorname{Id})(\operatorname{Id}_{L^p(\Omega)} \otimes \mathcal{T}_J)(x) = (\operatorname{Id}_{L^p(\Omega)} \otimes \mathcal{T}_J)\frac{1}{t}(U_{t,p} - \operatorname{Id})(x)$$

$$\xrightarrow[t \to 0]{} i(\operatorname{Id}_{L^p(\Omega)} \otimes \mathcal{T}_J)\mathscr{D}_{\alpha,1,p}(x).$$

We infer that $(\operatorname{Id}_{L^p(\Omega)} \otimes \mathcal{T}_J)(x)$ belongs to $\operatorname{dom} \mathscr{D}_{\alpha,1,p}$ and $\mathscr{D}_{\alpha,1,p}(\operatorname{Id}_{L^p(\Omega)} \otimes \mathcal{T}_J)(x) = (\operatorname{Id}_{L^p(\Omega)} \otimes \mathcal{T}_J)\mathscr{D}_{\alpha,1,p}(x) \xrightarrow[J \to I]{} \mathscr{D}_{\alpha,1,p}(x)$. Moreover, $(\operatorname{Id}_{L^p(\Omega)} \otimes \mathcal{T}_J)(x) \to x$ as $J \to I$. Therefore, to show the core property, it suffices to approximate an element $f \otimes e_{ij}$ where $f \in L^p(\Omega)$ in the graph norm of $\mathscr{D}_{\alpha,1,p}$ by elements in $L^\infty(\Omega) \otimes M_{I,\mathrm{fin}}$. To this end, it suffices to take $1_{|f| \leqslant n} f \otimes e_{ij}$ and to argue similarly as beforehand, noting that

$$\left(1_{|f| \leqslant n} \otimes 1_{B(\ell_I^2)}\right)U_{t,p}(f \otimes e_{ij}) = U_{t,p}\left(1_{|f| \leqslant n} f \otimes e_{ij}\right).$$

$\square$

Similarly to Corollary 5.2, we obtain the following result.

Corollary 5.3 *Let $1 < p < \infty$. Then the operator $\mathscr{D}_{\alpha,1,p}$ is bisectorial and has a bounded $H^\infty(\Sigma_\omega^\pm)$ functional calculus to any angle $\omega > 0$.*

Remark 5.17 Now we study the properties of the operator $\mathscr{D}_{\alpha,q}$ from (5.82) in the case where I is finite and $-1 \leqslant q < 1$. For any $i, j \in I$, we will use the maps $J_{ij} : L^p(\Gamma_q(H)) \to L^p(\Gamma_q(H)\overline{\otimes}B(\ell_I^2))$, $x \mapsto x \otimes e_{ij}$ and

$$Q_{ij} : L^p(\Gamma_q(H)\overline{\otimes}B(\ell_I^2)) \to L^p(\Gamma_q(H)), \quad x = \sum_{k,l \in I} x_{kl} \otimes e_{kl} \mapsto x_{ij}.$$

For any $k, l \in I$, we introduce the linear operator $\mathscr{L}_{kl} : L^p(\Gamma_q(H)\overline{\otimes}B(\ell_I^2)) \to L^p(\Gamma_q(H)\overline{\otimes}B(\ell_I^2))$, $x \otimes e_{ij} \mapsto \delta_{i=k,j=l}s_q(\alpha_i - \alpha_j)x \otimes e_{ij}$. For any $i, j, j, l, k', l' \in I$ and any $x \in L^p(\Gamma_q(H))$, we have

$$\mathscr{L}_{kl}\mathscr{L}_{k'l'}(x \otimes e_{ij}) = \delta_{i=k',j=l'}\mathscr{L}_{kl}\left(s_q(\alpha_i - \alpha_j)x \otimes e_{ij}\right)$$

$$= \delta_{i=k',j=l'}\delta_{i=k,j=l}s_q(\alpha_i - \alpha_j)^2 x \otimes e_{ij}$$

$$= \delta_{i=k,j=l}\mathscr{L}_{k'l'}\left(s_q(\alpha_i - \alpha_j)x \otimes e_{ij}\right) = \mathscr{L}_{k'l'}\mathscr{L}_{kl}(x \otimes e_{ij}).$$

So the operators $\mathscr{L}_{kl}$ commute. Moreover, we have

$$\mathscr{D}_{\alpha,q} = \sum_{k,l \in I} \mathscr{L}_{kl}. \tag{5.100}$$

For any $k, l \in I$, note that

$$\mathscr{L}_{kl} = J_{kl} L_{s_q(\alpha_k - \alpha_l)} Q_{kl}. \tag{5.101}$$

Note that $s_q(\alpha_k - \alpha_l)$ is selfadjoint. So by McIntosh and Monniaux [165, Remark 2.25] it is bisectorial of type 0 and admits a bounded bisectorial H^∞ functional calculus for all $\theta \in (0, \frac{\pi}{2})$ (with $K_\theta = 1$). By adapting [128, Proposition 8.4], the operator $L_{s_q(\alpha_k - \alpha_l)}$ is bisectorial of type 0 on $L^p(\Gamma_q(H))$ and has bounded bisectorial H^∞ functional calculus. By (5.101), the operator $\mathscr{L}_{kl}$ is also bisectorial of type 0 on $L^p(\Gamma_q(H) \overline{\otimes} B(\ell_I^2))$ and admits a bounded bisectorial H^∞ functional calculus. It is easy to check[21] that for any $\lambda \in \mathbb{C} \backslash \mathbb{R}$, we have $R(\lambda, \sum_{k,l \in I} \mathscr{L}_{kl}) = \sum_{k,l \in I} J_{kl} R(\lambda, L_{s_q(\alpha_k - \alpha_l)}) Q_{kl}$. Then the Cauchy integral formula yields for any $\theta \in (0, \frac{\pi}{2})$ and any $f \in H_0^\infty(\Sigma_\theta^\pm)$ that $f(\sum_{k,l \in I} \mathscr{L}_{kl}) = \sum_{k,l \in I} J_{kl} f(L_{s_q(\alpha_k - \alpha_l)}) Q_{kl}$. Thus by (5.100), we deduce that $\mathscr{D}_{\alpha,q}$ is bisectorial and admits a bounded bisectorial H^∞ functional calculus. The study of the operator (5.58) is similar and the verification is left to the reader. Here the assumption is that G is a finite group.

---

[21] We have

$$\sum_{k,l} J_{kl} R(\lambda, L_{s_q(\alpha_k - \alpha_l)}) Q_{kl} \left(\lambda - \sum_{i,j} J_{ij} L_{s_q(\alpha_i - \alpha_j)} Q_{ij} \right)$$

$$= \sum_{k,l} J_{kl} \lambda R(\lambda, L_{s_q(\alpha_k - \alpha_l)}) Q_{kl} - \sum_{k,l,i,j} J_{kl} R(\lambda, L_{s_q(\alpha_k - \alpha_l)}) Q_{kl} J_{ij} L_{s_q(\alpha_i - \alpha_j)} Q_{ij}$$

$$= \sum_{k,l} J_{kl} \lambda R(\lambda, L_{s_q(\alpha_k - \alpha_l)}) Q_{kl} - \sum_{k,l,i,j} \delta_{k=i} \delta_{l=j} J_{kl} R(\lambda, L_{s_q(\alpha_k - \alpha_l)}) L_{s_q(\alpha_i - \alpha_j)} Q_{ij}$$

$$= \sum_{k,l} J_{kl} R(\lambda, L_{s_q(\alpha_k - \alpha_l)})(\lambda - L_{s_q(\alpha_k - \alpha_l)}) Q_{kl}$$

$$= \sum_{k,l} J_{kl} \mathrm{Id}_{L^p(\Gamma_q(H))} Q_{kl} = \mathrm{Id}_{L^p(\Gamma_q(H) \overline{\otimes} B(\ell_I^2))},$$

and similarly the other way around.

Appendix A
Appendix: Lévy Measures and 1-Cohomology

Abstract In this short appendix, we describe a relation between 1-cohomology and Lévy-Khintchine decompositions in the case of an abelian group. This observation allows us to obtain with the Lévy measure an explicit description of the 1-cocycle associated to a markovian semigroup of Fourier multipliers.

The following observation describes the link between Lévy-Khintchine decompositions and 1-cocycles. Note that examples of explicit Lévy measures are given in [33, p. 184].

Let G be a locally compact *abelian* group. Recall that a 1-cocycle is defined by a strongly continuous unitary or orthogonal representation $\pi \colon G \to \mathrm{B}(H)$ on a complex or real Hilbert space H and a continuous map $b \colon G \to H$ such that $b(s+t) = b(s) + \pi_s(b(t))$ for any $s, t \in G$. We also recall the following fundamental connection between continuous functions $\psi \colon G \to \mathbb{R}$ which are conditionally of negative type in the sense of [34, 1.8 Definition p. 89], [33, 7.1 Definition] and quadratic forms $q \colon G \to \mathbb{R}_+$ together with Lévy measures μ from [33, 18.20 Corollary p. 184]. If ψ is such a function satisfying $\psi(0) = 0$, there exist a unique continuous quadratic form $q \colon G \to \mathbb{R}_+$ and a unique positive symmetric measure μ on $\widehat{G} - \{0\}$ such that

$$\psi(s) = q(s) + \int_{\widehat{G}-\{0\}} 1 - \mathrm{Re}\langle s, \chi \rangle \, \mathrm{d}\mu(\chi), \quad s \in G. \tag{A.1}$$

Conversely, any continuous function of this type is conditionally of negative type and $\psi(0) = 0$. Note that a quadratic form is a map satisfying $q(s+t) + q(s-t) = 2q(s) + 2q(t)$ for any $s, t \in G$ [33, 7.18 Definition]. The quadratic form can be computed by $q(s) = \lim_{n \to \infty} \frac{\psi(ns)}{n^2}$, and μ is the Lévy measure associated with ψ.

In the following result, note that π_s is the multiplication operator by the function $\langle s, \cdot \rangle_{G, \widehat{G}}$.

Proposition A.1 *Let G be a locally compact abelian group. Let $\psi \colon G \to \mathbb{R}$ be a continuous conditionally of negative type function such that $\psi(0) = 0$ and $\lim_{n \to \infty} \frac{\psi(ns)}{n^2} = 0$ for any $s \in G$. If μ is the Lévy measure of ψ on $\widehat{G} - \{0\}$ then $H = L^2(\widehat{G} - \{0\}, \frac{1}{2}\mu)$, $b \colon G \to H$, $s \mapsto (\chi \mapsto 1 - \langle s, \chi \rangle)$ and $\pi \colon G \to \mathrm{B}(H)$, $s \mapsto (f \mapsto \langle s, \cdot \rangle_{G, \widehat{G}} f)$ define a 1-cocycle on G such that*

$$\psi(s) = \|b(s)\|_H^2, \quad s \in G.$$

Proof For any $s \in G$, note that

$$|1 - \langle s, \chi \rangle|^2 = (1 - \langle s, \chi \rangle)\overline{1 - \langle s, \chi \rangle} = 1 - \langle s, \chi \rangle - \overline{\langle s, \chi \rangle} + 1 = 2 - 2\operatorname{Re}\langle s, \chi \rangle. \tag{A.2}$$

For any $s \in G$, using (A.1) in the last equality, we deduce that

$$
\begin{aligned}
\|b(s)\|_{L^2(\widehat{G} - \{0\}, \frac{1}{2}\mu)}^2 &= \frac{1}{2} \int_{\widehat{G} - \{0\}} |(b(s))(\chi)|^2 \, \mathrm{d}\mu(\chi) \\
&= \frac{1}{2} \int_{\widehat{G} - \{0\}} |1 - \langle s, \chi \rangle|^2 \, \mathrm{d}\mu(\chi) \\
&\overset{(A.2)}{=} \int_{\widehat{G} - \{0\}} \big(1 - \operatorname{Re}\langle s, \chi \rangle\big) \, \mathrm{d}\mu(\chi) = \psi(s).
\end{aligned}
$$

Moreover, for any $s, t \in G$ and any $\chi \in \widehat{G} - \{0\}$, we have

$$(b(s + t))(\chi) = 1 - \langle s + t, \chi \rangle = 1 - \langle s, \chi \rangle\langle t, \chi \rangle = 1 - \langle s, \chi \rangle + \langle s, \chi \rangle(1 - \langle t, \chi \rangle)$$

$$= b(s)(\chi) + \pi_s(b(t))(\chi) = \big(b(s) + \pi_s(b(t))\big)(\chi).$$

Hence, we conclude that $b(s + t) = b(s) + \pi_s(b(t))$. $\square$

In particular, by restriction of scalars, we can consider the real Hilbert space $H_{|\mathbb{R}} = L^2(\widehat{G} - \{0\}, \frac{1}{2}\mu)_{|\mathbb{R}}$, which comes with real scalar product $\langle f, g \rangle_{H_{|\mathbb{R}}} = \langle \operatorname{Re} f, \operatorname{Re} g \rangle_H + \langle \operatorname{Im} f, \operatorname{Im} g \rangle_H$. Then we take $b \colon G \to H_{|\mathbb{R}}$, $s \mapsto (\chi \mapsto 1 - \langle s, \chi \rangle)$ and $\pi \colon G \to \mathrm{B}(H_{|\mathbb{R}})$, $s \mapsto (f \mapsto \langle s, \cdot \rangle_{G, \widehat{G}} f)$. Thus if $\lim_{n \to \infty} \frac{\psi(ns)}{n^2} = 0$ for all $s \in G$, then we obtain a 1-cocycle on G as we considered in Proposition 2.3 suitable for our markovian semigroups of Fourier multipliers.

In the case of an arbitrary continuous function $\psi \colon G \to \mathbb{R}_+$ is with $\psi(0) = 0$ of conditionally negative type, we can use [129, Lemma 3.1] and [131, Lemma B1] on the quadratic part q of ψ with a direct sum argument to obtain a fairly concrete Hilbert space and an associated 1-cocycle.

References

1. G. Alexopoulos, An application of homogenization theory to harmonic analysis: harnack inequalities and Riesz transforms on Lie groups of polynomial growth. Canad. J. Math. **44**(4), 691–727 (1992)
2. E.M. Alfsen, *Compact Convex Sets and Boundary Integrals*. Ergebnisse der Mathematik und ihrer Grenzgebiete, Band 57 (Springer, New York, 1971)
3. E.M. Alfsen, F.W. Shultz, Geometry of state spaces of operator algebras, in *Mathematics: Theory & Applications* (Birkhauser Boston, Inc., Boston, 2003)
4. A. Amenta, New Riemannian manifolds with L^p-unbounded Riesz transform for $p > 2$. Math. Z. **297**(1–2), 99–112 (2021)
5. J.-P. Anker, Sharp estimates for some functions of the Laplacian on noncompact symmetric spaces. Duke Math. J. **65**(2), 257–297 (1992)
6. N. Arcozzi, Riesz transforms on compact Lie groups, spheres and Gauss space. Ark. Mat. **36**(2), 201–231 (1998)
7. N. Arcozzi, X. Li, Riesz transforms on spheres. Math. Res. Lett. **4**, 401–412 (1997)
8. W. Arendt, C.J.K. Batty, M. Hieber, F. Neubrander, *Vector-Valued Laplace Transforms and Cauchy Problems*, 2nd edn. Monographs in Mathematics, vol. 96 (Birkhäuser/Springer Basel AG, Basel, 2011)
9. C. Arhancet, Noncommutative Figà-Talamanca-Herz algebras for Schur multipliers. Integr. Equ. Oper. Theory **70**(4), 485–510 (2011)
10. C. Arhancet, On Matsaev's conjecture for contractions on noncommutative L^p-spaces. J. Operator Theory **69**(2), 387–421 (2013)
11. C. Arhancet, On a conjecture of Pisier on the analyticity of semigroups. Semigroup Forum **91**(2), 450–462 (2015)
12. C. Arhancet, Dilations of semigroups on von Neumann algebras and noncommutative L^p-spaces. J. Funct. Anal. **276**(7), 2279–2314 (2019)
13. C. Arhancet, Dilations of markovian semigroups of Fourier multipliers on locally compact groups. Proc. Amer. Math. Soc. **148**(6), 2551–2563 (2020)
14. C. Arhancet, Erratum to "Dilations of markovian semigroups of Fourier multipliers on locally compact groups". To appear in Proc. Amer. Math. Soc.
15. C. Arhancet, Dilations of markovian semigroups of measurable Schur multipliers. Preprint online on https://arxiv.org/abs/1910.14434
16. C. Arhancet, Quantum information theory and Fourier multipliers on quantum groups. Preprint online on https://arxiv.org/abs/2008.12019

17. C. Arhancet, C. Kriegler, Projections, multipliers and decomposable maps on noncommutative L^p-spaces. Submitted, Preprint online on https://arxiv.org/abs/1707.05591
18. N.H. Asmar, B.P. Kelly, S. Montgomery-Smith, A note on UMD spaces and transference in vector-valued function spaces. Proc. Edinburgh Math. Soc. (2) **39**(3), 485–490 (1996)
19. J. Assaad, E.M. Ouhabaz, Riesz Transforms of Schrödinger Operators on Manifolds. J. Geom. Anal. **22**(4), 1108–1136 (2012)
20. P. Auscher, On necessary and sufficient conditions for L^p-estimates of Riesz transforms associated to elliptic operators on $\mathbb{R}^n$ and related estimates. Mem. Amer. Math. Soc. **186**(871), xviii+75 pp. (2007)
21. P. Auscher, S. Hofmann, M. Lacey, A. McIntosh, P. Tchamitchian, The solution of the Kato square root problem for second order elliptic operators on $\mathbb{R}^n$. Ann. Math. (2) **156**(2), 633–654 (2002)
22. P. Auscher, A. Axelsson, A. McIntosh, On a quadratic estimate related to the Kato conjecture and boundary value problems, in *Harmonic Analysis and Partial Differential Equations*. Contemporary Mathematics, vol. 505 (American Mathematical Society, Providence, 2010), pp. 105–129
23. A. Axelsson, S. Keith, A. McIntosh, Quadratic estimates and functional calculi of perturbed Dirac operators. Invent. Math. **163**(3), 455–497 (2006)
24. N. Badr, E. Russ, Interpolation of Sobolev spaces, Littlewood-Paley inequalities and Riesz transforms on graphs. Publ. Mat. **53**(2), 273–328 (2009)
25. D. Bakry, *Étude des transformations de Riesz dans les variétés riemanniennes à courbure de Ricci minorée*. (French) [A study of Riesz transforms in Riemannian manifolds with minorized Ricci curvature]. Séminaire de Probabilités, XXI. Lecture Notes in Mathematics, vol. 1247 (Springer, Berlin, 1987), pp. 137–172
26. D. Bakry, I. Gentil, M. Ledoux, *Analysis and Geometry of Markov Diffusion Operators*. Grundlehren der Mathematischen Wissenschaften, vol. 348 (Springer, Berlin, 2014)
27. L. Bandara, Functional calculus and harmonic analysis in geometry. São Paulo J. Math. Sci. **15**(1), 20–53 (2021)
28. R. Banuelos, The foundational inequalities of D. L. Burkholder and some of their ramifications. Illinois J. Math. **54**(3), 789–868 (2010)
29. R. Banuelos, M. Kwasnicki, On the ℓ^p-norm of the discrete Hilbert transform. Duke Math. J. **168**(3), 471–504 (2019)
30. R. Banuelos, G. Wang, Sharp inequalities for martingales with applications to the Beurling-Ahlfors and Riesz transforms. Duke Math. J. **80**(3), 57–600 (1995)
31. B. Bekka, P. de la Harpe, Pierre, A. Valette, *Kazhdan's Property (T)*. New Mathematical Monographs, vol. 11 (Cambridge University Press, Cambridge, 2008)
32. J.V. Bellissard, M. Marcolli, K. Reihani, Dynamical Systems on Spectral Metric Spaces. Preprint online on https://arxiv.org/abs/1008.4617
33. C. Berg, G. Forst, *Potential Theory on Locally Compact Abelian Groups*. Ergebnisse der Mathematik und ihrer Grenzgebiete, Band 87 (Springer, New York, 1975)
34. C. Berg, J. Christensen, P. Ressel, *Harmonic Analysis on Semigroups. Theory of Positive Definite and Related Functions* (Springer, New York, 1984)
35. J. Bergh, J. Löfström, *Interpolation Spaces. An Introduction* (Springer, Berlin, 1976)
36. E. Berkson, T.A. Gillespie, P.S. Muhly, Generalized analyticity in UMD spaces. Ark. Mat. **27**(1), 1–14 (1989)
37. R. Bhatia, *Positive Definite Matrices*. Princeton Series in Applied Mathematics (Princeton University Press, Princeton, 2007)
38. D. Blecher, C. Le Merdy, *Operator Algebras and Their Modules-An Operator Space Approach*. London Mathematical Society Monographs. New Series, vol. 30 (Oxford Science Publications, The Clarendon Press, Oxford University Press, Oxford, 2004)
39. V.I. Bogachev, *Measure Theory*, vol. II (Springer, Berlin, 2007)
40. N. Bouleau, F. Hirsch, Dirichlet forms and analysis on Wiener space, in *De Gruyter Studies in Mathematics*, vol. 14 (Walter de Gruyter & Co., Berlin, 1991)

41. M. Bożejko, *Positive and Negative Definite Kernels on Discrete Groups*. Lectures at Heidelberg University (1987)
42. M. Bożejko, Bessis-Moussa-Villani conjecture and generalized Gaussian random variables. Infin. Dimens. Anal. Quantum Probab. Relat. Top. **11**(3), 313–321 (2008)
43. M. Bożejko, R. Speicher, Completely positive maps on Coxeter groups, deformed commutation relations, and operator spaces. Math. Ann. **300**(1), 97–120 (1994)
44. M. Bożejko, B. Kümmerer, R. Speicher, q-Gaussian processes: non-commutative and classical aspects. Comm. Math. Phys. **185**(1), 129–154 (1997)
45. M. Bożejko, S.R. Gal, W. Mlotkowski, Positive definite functions on Coxeter groups with applications to operator spaces and noncommutative probability. Comm. Math. Phys. **361**(2), 583–604 (2018)
46. N.P. Brown, N. Ozawa, C^*-algebras and finite-dimensional approximations. *Graduate Studies in Mathematics*, vol. 88 (American Mathematical Society, Providence, 2008)
47. A.P. Calderon, A. Zygmund, On the existence of certain singular integrals. Acta Math. **88**, 85–139 (1952)
48. J. Cameron, R.R. Smith, Bimodules in crossed products of von Neumann algebras. Adv. Math. **274**, 539–561 (2015)
49. A.L. Carey, V. Gayral, A. Rennie, F.A. Sukochev, Index theory for locally compact noncommutative geometries. Mem. Amer. Math. Soc. **231**(1085) (2014)
50. G. Carron, T. Coulhon, A. Hassell, Riesz transform and L^p-cohomology for manifolds with Euclidean ends. Duke Math. J. **133**(1), 59–93 (2006)
51. M. Caspers, Harmonic analysis and BMO-spaces of free Araki-Woods factors. Studia Math. **246**(1), 71–107 (2019)
52. M. Caspers, Riesz transforms on compact quantum groups and strong solidity. Preprint online on https://arxiv.org/abs/2011.01609
53. A. Chamseddine, C. Consani, N. Higson, M. Khalkhali, H. Moscovici, G. Yu (eds.). Advances in Noncommutative Geometry. On the Occasion of Alain Connes' 70th Birthday (Springer, Berlin, 2019)
54. I. Chatterji, Introduction to the rapid decay property, in *Around Langlands Correspondences*. Contemporary Mathematics, vol. 691 (American Mathematical Society, Providence, 2017), pp. 53–72
55. L. Chen, T. Coulhon, B. Hua, Riesz transforms for bounded Laplacians on graphs. Math. Z. **294**, 397–417 (2020)
56. F. Cipriani, Noncommutative potential theory: a survey. J. Geom. Phys. **105**, 25–59 (2016)
57. F. Cipriani, J.-L. Sauvageot, Derivations as square roots of Dirichlet forms. J. Funct. Anal. **201**(1), 78–120 (2003)
58. F. Cipriani, D. Guido, T. Isola, J.-L. Sauvageot, Spectral triples for the Sierpinski gasket. J. Funct. Anal. **266**(8), 4809–4869 (2014)
59. A. Connes, Compact metric spaces, Fredholm modules, and hyperfiniteness. Ergodic Theory Dynam. Systems **9**(2), 207–220 (1989)
60. A. Connes, *Noncommutative Geometry* (Academic Press, Inc., San Diego, 1994)
61. A. Connes, A short survey of noncommutative geometry. J. Math. Phys. **41**(6), 3832–3866 (2000)
62. A. Connes, M. Marcolli, *A Walk in the Noncommutative Garden. An Invitation to Noncommutative Geometry* (World Scientific Publishing, Hackensack, 2008), pp. 1–128
63. T. Coulhon, *Heat Kernels on Non-compact Riemannian Manifolds: A Partial Survey*. Séminaire de Théorie Spectrale et Géométrie, No. 15, Année 1996–1997, Sémin. Théor. Spectr. Géom., vol. 15 (University Grenoble I, Saint-Martin-d'Hères, 1997), pp. 167–187
64. T. Coulhon, *Heat Kernel and Isoperimetry on Non-compact Riemannian Manifolds. Heat Kernels and Analysis on Manifolds, Graphs, and Metric Spaces* (Paris, 2002). Contemporary Mathematics, vol. 338 (American Mathematical Society, Providence, 2003), pp. 65–99
65. T. Coulhon, Heat kernel estimates, Sobolev-type inequalities and Riesz transform on noncompact Riemannian manifolds, in *Analysis and Geometry of Metric Measure Spaces*. CRM Proc. Lecture Notes, vol. 56 (American Mathematical Society, Providence, 2013), pp. 55–65

66. T. Coulhon, X.T. Duong, Riesz transforms for $1 \leqslant p \leqslant 2$. Trans. Amer. Math. Soc. **351**, 1151–1169 (1999)
67. T. Coulhon, X.T. Duong, Riesz transform and related inequalities on noncompact Riemannian manifolds. Comm. Pure Appl. Math. **56**(12), 1728–1751 (2003)
68. M. Cowling, I. Doust, A. McIntosh, A. Yagi, Banach space operators with a bounded H^∞ functional calculus. J. Austral. Math. Soc. Ser. A **60**(1), 51–89 (1996)
69. M. Cwikel, N. Kalton, Interpolation of compact operators by the methods of Calderon and Gustavsson-Peetre. Proc. Edinburgh Math. Soc. **38** (1995)
70. E.B. Davies, J.M. Lindsay, Noncommutative symmetric Markov semigroups. Math. Z. **210**(3), 379–411 (1992)
71. J. De Cannière, U. Haagerup, Multipliers of the Fourier algebras of some simple Lie groups and their discrete subgroups. Amer. J. Math. **107**(2), 455–500 (1985)
72. A. Defant, K. Floret, *Tensor Norms and Operator Ideals*. North-Holland Mathematics Studies, vol. 176 (North-Holland Publishing Co., Amsterdam, 1993)
73. B. Devyver, Heat kernel and Riesz transform of Schrödinger operators. Ann. Inst. Fourier (Grenoble) **69**(2), 457–513 (2019)
74. J. Dieudonné, *Treatise on Analysis*, vol. II. Enlarged and corrected printing. Translated by I. G. Macdonald. With a loose erratum. Pure and Applied Mathematics, 10-II (Academic Press [Harcourt Brace Jovanovich, Publishers], New York, 1976)
75. S. Dirksen, B. de Pagter, D. Potapov, F. Sukochev, Rosenthal inequalities in noncommutative symmetric spaces. J. Funct. Anal. **261**(10), 2890–2925 (2011)
76. K. Domelevo, S. Petermichl, Sharp L^p estimates for discrete second order Riesz transforms. Adv. Math. **262**, 932–952 (2014)
77. S.K. Donaldson, D.P. Sullivan, Quasiconformal 4-manifolds. Acta Math. **163**(3–4), 181–252 (1989)
78. R.M. Dudley, *Real Analysis and Probability* (CRC Press, Boca Raton, 2018)
79. N. Dungey, Riesz transforms on a solvable Lie group of polynomial growth. Math. Z. **251**(3), 649–671 (2005)
80. E.G. Effros, M. Popa, Feynman diagrams and Wick products associated with q-Fock space. Proc. Natl. Acad. Sci. USA **100**(15), 8629–8633 (2003)
81. E. Effros, Z.-J. Ruan, *Operator Spaces* (Oxford University Press, Oxford, 2000)
82. L.B. Efraim, F. Lust-Piquard, Poincaré type inequalities on the discrete cube and in the CAR algebra. Probab. Theory Related Fields **141**(3–4), 569–602 (2008)
83. M. Egert, On Kato's conjecture and mixed boundary conditions. PhD (2015)
84. T. Eisner, B. Farkas, M. Haase, R. Nagel, *Operator Theoretic Aspects of Ergodic Theory*. Graduate Texts in Mathematics, vol. 272 (Springer, Cham, 2015)
85. K.-J. Engel, R. Nagel, *One-Parameter Semigroups for Linear Evolution Equations*. Graduate Texts in Mathematics, vol. 194 (Springer, New York, 2000)
86. T. Ferguson, T. Mei, B. Simanek, H^∞-calculus for semigroup generators on BMO. Adv. Math. **347**, 408–441 (2019)
87. I. Forsyth, B. Mesland, A. Rennie, Dense domains, symmetric operators and spectral triples. New York J. Math. **20**, 1001–1020 (2014)
88. V. Gayral, J.M. Gracia-Bondía, B. Iochum, T. Schücker, J.C. Varilly, Moyal planes are spectral triples. Comm. Math. Phys. **246**(3), 569–623 (2004)
89. S. Ghorpade, B.V. Limaye, A course in multivariable calculus and analysis, in *Undergraduate Texts in Mathematics* (Springer, New York, 2010)
90. J.M. Gracia-Bondía, J.C. Varilly, H. Figueroa, Elements of noncommutative geometry, in *Birkhäuser Advanced Texts: Basler Lehrbücher* (Birkhäuser Boston, Inc., Boston, 2001)
91. L. Grafakos, Best bounds for the Hilbert transform on $L^p(\mathbb{R}^1)$. Math. Res. Lett. **4**(4), 469–471 (1997)
92. L. Grafakos, T. Savage, Best bounds for the Hilbert transform on $L^p(\mathbb{R}^1)$: a corrigendum. Math. Res. Lett. **22**(5), 1333–1335 (2015)
93. U. Haagerup, An Example of a nonnuclear C*-Algebra, which has the metric approximation property. Invent. Math. **50**(3), 279–293 (1978/79)

94. U. Haagerup, Group C*-algebras without the completely bounded approximation property. J. Lie Theory **26**(3), 861–887 (2016)
95. U. Haagerup, S. Knudby, A Lévy-Khinchin formula for free groups. Proc. Amer. Math. Soc. **143**(4), 1477–1489 (2015)
96. U. Haagerup, J. Kraus, Approximation properties for group C*-algebras and group von Neumann algebras. Trans. Amer. Math. Soc. **344**(2), 667–699 (1994)
97. U. Haagerup, M. Junge, Q. Xu, A reduction method for noncommutative L_p-spaces and applications. Trans. Amer. Math. Soc. **362**(4), 2125–2165 (2010)
98. M. Haase, The functional calculus for sectorial operators, in *Operator Theory: Advances and Applications*, vol. 169 (Birkhäuser Verlag, Basel, 2006)
99. M. Haase, Lectures on Functional Calculus. 21st International Internet Seminar (2018). https://www.math.uni-kiel.de/isem21/en/course/phase1
100. F. Hansen, G.K. Pedersen, Jensen's inequality for operators and Löwner's theorem. Math. Ann. **258**(3), 229–241 (1981/82)
101. S. Haran, Riesz potentials and explicit sums in arithmetic. Invent. Math. **1013**, 697–703 (1990)
102. S. Haran, Analytic potential theory over the p-adics. Ann. Inst. Fourier (Grenoble) **43**(4), 905–944 (1993)
103. A. Hassell, P. Lin, The Riesz transform for homogeneous Schrödinger operators on metric cones. Rev. Mat. Iberoam. **30**(2), 477–522 (2014)
104. M. Hellmich, Decoherence in Infinite Quantum Systems. PhD thesis, University of Bielefeld, Bielefeld (2009)
105. F. Hirsch, *Opérateurs carré du champ* (d'après J. P. Roth) (French). Séminaire Bourbaki, 29e année (1976/77), Exp. No. 501. Lecture Notes in Mathematics, vol. 677 (Springer, Berlin, 1978), pp. 167–182
106. S. Hofmann, M. Lacey, A. McIntosh, The solution of the Kato problem for divergence form elliptic operators with Gaussian heat kernel bounds. Ann. Math. (2) **156**(2), 623–631 (2002)
107. T. Hytönen, A. McIntosh, P. Portal, Kato's square root problem in Banach spaces. J. Funct. Anal. **254**(3), 675–726 (2008)
108. T. Hytönen, A. McIntosh, P. Portal, Holomorphic functional calculus of Hodge-Dirac operators in Lp. J. Evol. Equ. **11**(1), 71–105 (2011)
109. T. Hytönen, J. van Neerven, M. Veraar, L. Weis, *Analysis in Banach Spaces*, vol. I. Martingales and Littlewood-Paley Theory (Springer, Berlin, 2016)
110. T. Hytönen, J. van Neerven, M. Veraar, L. Weis, *Analysis in Banach Spaces*, vol. II. Probabilistic Methods and Operator Theory (Springer, Berlin, 2018)
111. T. Iwaniec, G. Martin, Quasiregular mappings in even dimensions. Acta Math. **170**(1), 29–81 (1993)
112. T. Iwaniec, G. Martin, Riesz transforms and related singular integrals. J. Reine Angew. Math. **473**, 25–57 (1996)
113. S. Janson, *Gaussian Hilbert Spaces*. Cambridge Tracts in Mathematics, vol. 129 (Cambridge University Press, Cambridge, 1997)
114. B. Jaye, F. Nazarov, M.C. Reguera, X. Tolsa, The Riesz transform of codimension smaller than one and the Wolff energy. Mem. Amer. Math. Soc. **266**(1293), v+97 pp. (2020)
115. R. Jiang, Riesz transform via heat kernel and harmonic functions on non-compact manifolds. Adv. Math. **377**, 50 pp. (2021). Paper No. 107464
116. M. Junge, Doob's inequality for non-commutative martingales. J. Reine Angew. Math. **549**, 149–190 (2002)
117. M. Junge, Fubini's theorem for ultraproducts of noncommmutative L_p-spaces. Canad. J. Math. **56**(5), 983–1021 (2004)
118. M. Junge, H.H. Lee, q-Chaos. Trans. Amer. Math. Soc. **363**, 5223–5249 (2011)
119. M. Junge, T. Mei, Noncommutative Riesz transforms–a probabilistic approach. Amer. J. Math. **132**(3), 611–680 (2010)
120. M. Junge, M. Perrin, Theory of $\mathcal{H}_p$-spaces for continuous filtrations in von Neumann algebras. Astérisque No. 362 (2014)

121. M. Junge, Z.-J. Ruan, Approximation properties for noncommutative L_p-spaces associated with discrete groups. Duke Math. J. **117**(2), 313–341 (2003)
122. M. Junge, Z.-J. Ruan, Decomposable maps on non-commutative L_p-spaces, in *Operator Algebras, Quantization, and Noncommutative Geometry*. Contemporary Mathematics, vol. 365 (American Mathematical Society, Providence, 2004), pp. 355–381
123. M. Junge, D. Sherman, Noncommutative L^p modules. J. Operator Theory **53**(1), 3–34 (2005)
124. M. Junge, Q. Xu, Noncommutative Burkholder/Rosenthal inequalities. II. Applications. Israel J. Math. **167**, 227–282 (2008)
125. M. Junge, Q. Zeng, Noncommutative Bennett and Rosenthal inequalities. Ann. Probab. **41**(6), 4287–4316 (2013)
126. M. Junge, Q. Zeng, Noncommutative martingale deviation and Poincaré type inequalities with applications. Probab. Theory Related Fields **161**(3–4), 449–507 (2015)
127. M. Junge, Q. Zeng, Subgaussian 1-cocycles on discrete groups. J. Lond. Math. Soc. (2) **92**(2), 242–264 (2015)
128. M. Junge, C. Le Merdy, Q. Xu, H^∞ functional calculus and square functions on noncommutative L^p-spaces. Astérisque No. 305 (2006)
129. M. Junge, T. Mei, J. Parcet, Smooth Fourier multipliers on group von Neumann algebras. Geom. Funct. Anal. **24**(6), 1913–1980 (2014)
130. M. Junge, C. Palazuelos, J. Parcet, M. Perrin, Hypercontractivity in group von Neumann algebras. Mem. Amer. Math. Soc. **249**(1183) (2017)
131. M. Junge, T. Mei, J. Parcet, Noncommutative Riesz transforms–dimension free bounds and Fourier multipliers. J. Eur. Math. Soc. (JEMS) **20**(3), 529–595 (2018)
132. M. Junge, S. Rezvani, Q. Zeng, Harmonic analysis approach to Gromov-Hausdorff convergence for noncommutative tori. Comm. Math. Phys. **358**(3), 919–994 (2018)
133. R.V. Kadison, A representation theory for commutative topological algebra. Mem. Amer. Math. Soc. (7), 39 (1951)
134. R.V. Kadison, J.R. Ringrose, *Fundamentals of the Theory of Operator Algebras*, vol. I. Elementary theory. Reprint of the 1983 original. Graduate Studies in Mathematics, 15 (American Mathematical Society, Providence, 1997)
135. N.J. Kalton, S. Montgomery-Smith, Interpolation of Banach spaces, in *Handbook of the Geometry of Banach Spaces*, vol. 2 (North-Holland, Amsterdam, 2003), pp. 1131–1175
136. T. Kato, *Perturbation Theory for Linear Operators*, 2nd edn. Grundlehren der Mathematischen Wissenschaften, Band 132 (Springer, Berlin, 1976)
137. F. King, *Hilbert Transforms*, vol. 1. Encyclopedia of Mathematics and Its Applications, 124 (Cambridge University Press, Cambridge, 2009)
138. F. King, *Hilbert Transforms*, vol. 2. Encyclopedia of Mathematics and its Applications, 125 (Cambridge University Press, Cambridge, 2009)
139. R.P. Kostecki, W*-algebras and noncommutative integration. Preprint online on https://arxiv.org/abs/1307.4818
140. P.C. Kunstmann, L. Weis, Maximal L_p-regularity for parabolic equations, Fourier multiplier theorems and H^∞-functional calculus, in *Functional Analytic Methods for Evolution Equations*. Lecture Notes in Mathematics, vol. 1855 (Springer, Berlin, 2004), pp. 65–311
141. K. Landsman, Notes on Noncommutative Geometry. https://www.math.ru.nl/~landsman/notes.html
142. L. Larsson-Cohn, On the constants in the Meyer inequality. Monatsh. Math. **137**(1), 51–56 (2002)
143. F. Latrémolière, Quantum locally compact metric spaces. J. Funct. Anal. **264**(1), 362–402 (2013)
144. F. Latrémolière, Convergence of fuzzy tori and quantum tori for the quantum Gromov-Hausdorff propinquity: an explicit approach. Münster J. Math. **8**(1), 57–98 (2015)
145. F. Latrémolière, The dual Gromov-Hausdorff propinquity. J. Math. Pures Appl. (9) **103**(2), 303–351 (2015)
146. F. Latrémolière, Equivalence of quantum metrics with a common domain. J. Math. Anal. Appl. **443**(2), 1179–1195 (2016)

147. F. Latrémolière, Quantum metric spaces and the Gromov-Hausdorff propinquity, in *Noncommutative Geometry and Optimal Transport*. Contemporary Mathematics, vol. 676 (American Mathematical Society, Providence, 2016), pp. 47–133
148. F. Latrémolière, The quantum Gromov-Hausdorff propinquity. Trans. Amer. Math. Soc. **368**(1), 365–411 (2016)
149. F. Latrémolière, Heisenberg Modules over Quantum 2-tori are metrized quantum vector bundles. Preprint online on https://arxiv.org/abs/1703.07073
150. F. Latrémolière, The Gromov-Hausdorff propinquity for metric Spectral Triples. Preprint online on https://arxiv.org/abs/1811.10843
151. C. Le Merdy, *H^∞-Functional Calculus and Applications to Maximal Regularity*. Semigroupes d'opérateurs et calcul fonctionnel (Besançon, 1998). Publ. Math. UFR Sci. Tech. Besançon, vol. 16 (Univ. Franche-Comté, Besançon, 1999), pp. 41–77
152. N. Lohoué, S. Mustapha, Sur les transformées de Riesz sur les espaces homogènes des groupes de Lie semi-simples. Bull. Soc. Math. de France **128**(4), 485–495 (2000)
153. F. Lust-Piquard, Inégalités de Khintchine dans C_p ($1 < p < \infty$). (French) [Khinchin inequalities in C_p ($1 < p < \infty$)]. C. R. Acad. Sci. Paris Sér. I Math. **303**(7), 289–292 (1986)
154. F. Lust-Piquard, Riesz transforms associated with the number operator on the Walsh system and the fermions. J. Funct. Anal. **155**(1), 263–285 (1998)
155. F. Lust-Piquard, Riesz transforms on deformed Fock spaces. Comm. Math. Phys. 205(3), 519–549 (1999)
156. F. Lust-Piquard, Dimension free estimates for discrete Riesz transforms on products of abelian groups. Adv. Math. **185**(2), 289–327 (2004)
157. F. Lust-Piquard, Riesz transforms on generalized Heisenberg groups and Riesz transforms associated to the CCR heat flow. Publ. Mat. **48**(2), 309–333 (2004)
158. F. Lust-Piquard, G. Pisier, Noncommutative Khintchine and Paley inequalities. Ark. Mat. **29**(2), 241–260 (1991)
159. Z.M. Ma, M. Röckner, Introduction to the theory of (nonsymmetric) Dirichlet forms. Universitext (Springer, Berlin, 1992)
160. J. Maas, J. van Neerven, Boundedness of Riesz transforms for elliptic operators on abstract Wiener spaces. J. Funct. Anal. **257**(8), 2410–2475 (2009)
161. P. Malliavin, *Stochastic Analysis* (Springer, Berlin, 1997)
162. C. Martinez Carracedo, M. Sanz Alix, The theory of fractional powers of operators, in *North-Holland Mathematics Studies*, vol. 187 (North-Holland Publishing Co., Amsterdam, 2001)
163. A. Mas, X. Tolsa, Variation for the Riesz transform and uniform rectifiability. J. Eur. Math. Soc. (JEMS) **16**(11), 2267–2321 (2014)
164. J. Mateu, X. Tolsa, Riesz transforms and harmonic Lip_1-capacity in Cantor sets. Proc. London Math. Soc. (3) **89**(3), 676–696 (2004)
165. A. McIntosh, S. Monniaux, Hodge-Dirac, Hodge-Laplacian and Hodge-Stokes operators in L^p spaces on Lipschitz domains. Rev. Mat. Iberoam. **34**(4), 1711–1753 (2018)
166. H.P. McKean, Geometry of differential space. Ann. Probab. 1, 197–206 (1973)
167. R.E. Megginson, An introduction to Banach space theory, in *Graduate Texts in Mathematics*, vol. 183 (Springer, New York, 1998)
168. F.G. Mehler, Über die Entwicklung einer Function von beliebig vielen Variablen nach Laplaschen Functionen höherer Ordnung. Crelles J. **66**, 161–176 (1866)
169. T. Mei, É. Ricard, Free Hilbert transforms. Duke Math. J. **166**(11), 2153–2182 (2017)
170. P.A. Meyer, *Démonstration probabiliste de certaines inégalités de Littlewood-Paley. II. L'opérateur carré du champ* (French). Séminaire de Probabilités, X (Première partie, Univ. Strasbourg, Strasbourg, année universitaire 1974/1975). Lecture Notes in Mathematics, vol. 511 (Springer, Berlin, 1976), pp. 142–161
171. P.A. Meyer, *Transformations de Riesz pour les lois gaussiennes* (French) [Riesz transforms for Gaussian laws]. Seminar on Probability, XVIII. Lecture Notes in Mathematics, vol. 1059, (Springer, Berlin, 1984), pp. 179–193

172. A. Naor, Discrete Riesz transforms and sharp metric X_p inequalities. Ann. Math. (2) **184**(3), 991–1016 (2016)
173. F. Nazarov, X. Tolsa, A. Volberg, On the uniform rectifiability of AD-regular measures with bounded Riesz transform operator: the case of codimension 1. Acta Math. **213**(2), 237–321 (2014)
174. J.v. Neerven, *The Adjoint of a Semigroup of Linear Operators*. Lecture Notes in Mathematics, vol. 1529 (Springer, Berlin, 1992), x+195 pp.
175. J.v. Neerven, *Stochastic Evolution Equations*. ISEM Lecture Notes (2007/08)
176. J.v. Neerven, R. Versendaal, L^p-analysis of the Hodge-Dirac operator associated with Witten Laplacians on complete riemannian manifolds. J Geom. Anal., 1–30 (2017)
177. A. Nou, Non injectivity of the q-deformed von Neumann algebra. Math. Ann. **330**(1), 17–38 (2004)
178. D. Nualart, *The Malliavin Calculus and Related Topics*, 2nd edn. (Springer, Berlin, 2006)
179. A. Osękowski, I. Yaroslavtsev, The Hilbert transform and orthogonal martingales in Banach spaces. Preprint online on https://arxiv.org/abs/1805.03948
180. E.M. Ouhabaz, *Analysis of Heat Equations on Domains*. London Mathematical Society Monographs Series, vol. 31 (Princeton University Press, Princeton, 2005)
181. N. Ozawa, About the QWEP conjecture. Internat. J. Math. **15**(5), 501–530 (2004)
182. N. Ozawa, M. Rieffel, Hyperbolic group C^*-algebras and free-product C^*-algebras as compact quantum metric spaces. Canad. J. Math. **57**(5), 1056–1079 (2005)
183. T.W. Palmer, *Banach Algebras and the General Theory of *-Algebras*, vol. 2. Encyclopedia of Mathematics and its Applications 79 (Cambridge University Press, Cambridge, 2001)
184. J. Parcet, K.M. Rogers, Twisted Hilbert transforms vs Kakeya sets of directions. J. Reine Angew. Math. **710**, 137–172 (2016)
185. V. Paulsen, *Completely Bounded Maps and Operator Algebras* (Cambridge University Press, Cambridge, 2002)
186. S.K. Pichorides, On the best values of the constants in the theorems of M. Riesz, Zygmund and Kolmogorov. Studia Math. **44**, 165–179 (1972)
187. G. Pisier, *Riesz Transforms: A Simpler Analytic Proof of P.-A. Meyer's Inequality*. Séminaire de Probabilités, XXII. Lecture Notes in Mathematics, vol. 1321 (Springer, Berlin, 1988), pp. 485–501
188. G. Pisier, Regular operators between non-commutative L_p-spaces. Bull. Sci. Math. **119**(2), 95–118 (1995)
189. G. Pisier, *Introduction to Operator Space Theory* (Cambridge University Press, Cambridge, 2003)
190. G. Pisier, Grothendieck's theorem, past and present. Bull. Amer. Math. Soc. (N.S.) **49**(2), 237–323 (2012)
191. G. Pisier, Q. Xu, Non-commutative L^p-spaces, in *Handbook of the Geometry of Banach Spaces*, vol. II, ed. by W.B. Johnson, J. Lindenstrauss (Elsevier, Amsterdam, 2003), pp. 1459–1517
192. N. Randrianantoanina, Hilbert transform associated with finite maximal subdiagonal algebras. J. Austral. Math. Soc. Ser. A **65**(3), 388–404 (1998)
193. N. Randrianantoanina, Non-commutative martingale transforms. J. Funct. Anal. **194**(1), 181–212 (2002)
194. J.G. Ratcliffe, *Foundations of Hyperbolic Manifolds*. Graduate Texts in Mathematics, vol. 149 (Springer, New York, 1994)
195. M. Reed, B. Simon, *Methods of Modern Mathematical Physics. I. Functional Analysis*, 2nd edn. (Academic Press, Inc. [Harcourt Brace Jovanovich, Publishers], New York, 1980)
196. M.A. Rieffel, Metrics on state spaces. Doc. Math. **4**, 559–600 (1999)
197. M.A. Rieffel, Compact quantum metric spaces, in *Operator Algebras, Quantization, and Noncommutative Geometry*. Contemporary Mathematics, vol. 365 (American Mathematical Society, Providence, 2004), pp. 315–330

198. M.A. Rieffel, Gromov-Hausdorff distance for quantum metric spaces. Appendix 1 by Hanfeng Li. Gromov-Hausdorff distance for quantum metric spaces. Matrix algebras converge to the sphere for quantum Gromov-Hausdorff distance. Mem. Amer. Math. Soc. **168**(796), 1–65 (2004)

199. M. Riesz, Sur les fonctions conjuguées. (French). Math. Z. **27**(1), 218–244 (1928)

200. L. Roncal, P.R. Stinga, Transference of fractional Laplacian regularity, in *Special Functions, Partial Differential Equations, and Harmonic Analysis.* Springer Proceedings in Mathematics & Statistics, vol. 108 (Springer, Cham, 2014), pp. 203–212

201. L. Roncal, P.R. Stinga, Fractional Laplacian on the torus. Commun. Contemp. Math. **18**(3), 1550033 (2016)

202. J.-P. Roth, Opérateurs dissipatifs et semi-groupes dans les espaces de fonctions continues. (French. English summary). Ann. Inst. Fourier (Grenoble) **26**(4), ix, 1–97 (1976)

203. E. Russ, Riesz transforms on graphs for $1 \leqslant p \leqslant 2$. Math. Scand. **87**(1), 133–160 (2000)

204. E. Russ, Racines carrées d'opérateurs elliptiques et espaces de Hardy. (French) [Square roots of elliptic operators and Hardy spaces]. Confluentes Math. **3**(1), 1–119 (2011)

205. J.-L. Sauvageot, Tangent bimodule and locality for dissipative operators on C*-algebras, in *Quantum Probability and Applications IV* (Rome, 1987). Lecture Notes in Mathematics, vol. 1396 (Springer, Berlin, 1989), pp. 322–338

206. J.-L. Sauvageot, Quantum Dirichlet forms, differential calculus and semigroups, in *Quantum Probability and Applications, V* (Heidelberg, 1988). Lecture Notes in Mathematics, 1442 (Springer, Berlin, 1990), pp. 334–346

207. K. Schmüdgen, Unbounded self-adjoint operators on Hilbert space, in *Graduate Texts in Mathematics*, vol. 265 (Springer, Dordrecht, 2012)

208. A. Sitarz, A friendly overview of noncommutative geometry. Acta Phys. Polon. B **44**(12), 2643–2667 (2013)

209. P. Sjögren, An estimate for a first-order Riesz operator on the affine group. Trans. Amer. Math. Soc. **351**(8), 3301–3314 (1999)

210. E.M. Stein, Topics in harmonic analysis related to the Littlewood-Paley theory, in *Annals of Mathematics Studies*, vol. 63 (Princeton University Press, Princeton; University of Tokyo Press, Tokyo, 1970)

211. E.M. Stein, Some results in harmonic analysis in $\mathbb{R}^n$, for $n \to \infty$. Bull. Amer. Math. Soc. (N.S.) **9**(1), 71–73 (1983)

212. S. Stratila, *Modular Theory in Operator Algebras.* Translated from the Romanian by the author. Editura Academiei Republicii Socialiste România, Bucharest (Abacus Press, Tunbridge Wells, 1981)

213. R.S. Strichartz, Analysis of the Laplacian on the complete Riemannian manifold. J. Funct. Anal. **52**(1), 48–79 (1983)

214. F. Sukochev, D. Zanin, The Connes character formula for locally compact spectral triples. Preprint online on https://arxiv.org/abs/1803.01551

215. V. Sunder, *An Invitation to von Neumann Algebras.* Universitext (Springer, New York, 1987)

216. M. Takesaki, Theory of operator algebras. II, in *Encyclopaedia of Mathematical Sciences*, vol. 125. Operator Algebras and Non-commutative Geometry, 6 (Springer, Berlin, 2003)

217. P. Tchamitchian, *The Solution of Kato's Conjecture* (after Auscher, Hofmann, Lacey, McIntosh and Tchamitchian). Journées "Équations aux Dérivées Partielles" (Plestin-les-Grèves, 2001), Exp. No. XIV (Univ. Nantes, Nantes, 2001), 14 pp.

218. A.F.M. ter Elst, D.W. Robinson, A. Sikora, Heat kernels and Riesz transforms on nilpotent Lie groups. Colloq. Math. **74**(2), 191–218 (1997)

219. B. Thaller, The Dirac equation, in *Texts and Monographs in Physics* (Springer, Berlin, 1992)

220. X. Tolsa, Principal values for Riesz transforms and rectifiability. J. Funct. Anal. **254**(7), 1811–1863 (2008)

221. A. van Daele, *Continuous Crossed Products and Type III von Neumann Algebras.* London Mathematical Society Lecture Note Series, vol. 31 (Cambridge University Press, Cambridge, 1978)

222. J.C. Varilly, Dirac operators and spectral geometry. Lecture notes on noncommutative geometry and quantum groups edited by P. M. Hajac. https://www.mimuw.edu.pl/~pwit/toknotes/
223. C. Villani, Optimal transport. Old and new, in *Grundlehren der Mathematischen Wissenschaften* [Fundamental Principles of Mathematical Sciences], vol. 338 (Springer, Berlin, 2009)
224. D. Voiculescu, The analogues of entropy and of Fisher's information measure in free probability theory. V. Noncommutative Hilbert transforms. Invent. Math. **132**(1), 189–227 (1998)
225. N. Weaver, Lipschitz algebras and derivations of von Neumann algebras. J. Funct. Anal. **139**(2), 261–300 (1996)
226. N. Weaver, *Lipschitz Algebras*, 2nd edn. (World Scientific Publishing Co. Pte. Ltd., Hackensack, 2018)
227. Q. Xu, Non-commutative L_p-spaces. Script not available

Index

© The Author(s), under exclusive license to Springer Nature Switzerland AG 2022
C. Arhancet, C. Kriegler, *Riesz Transforms, Hodge-Dirac Operators and Functional Calculus for Multipliers*, Lecture Notes in Mathematics 2304, https://doi.org/10.1007/978-3-030-99011-4

LECTURE NOTES IN MATHEMATICS   🐎 Springer

Editors in Chief: J.-M. Morel, B. Teissier;

Editorial Policy

1. Lecture Notes aim to report new developments in all areas of mathematics and their applications – quickly, informally and at a high level. Mathematical texts analysing new developments in modelling and numerical simulation are welcome.

   Manuscripts should be reasonably self-contained and rounded off. Thus they may, and often will, present not only results of the author but also related work by other people. They may be based on specialised lecture courses. Furthermore, the manuscripts should provide sufficient motivation, examples and applications. This clearly distinguishes Lecture Notes from journal articles or technical reports which normally are very concise. Articles intended for a journal but too long to be accepted by most journals, usually do not have this "lecture notes" character. For similar reasons it is unusual for doctoral theses to be accepted for the Lecture Notes series, though habilitation theses may be appropriate.

2. Besides monographs, multi-author manuscripts resulting from SUMMER SCHOOLS or similar INTENSIVE COURSES are welcome, provided their objective was held to present an active mathematical topic to an audience at the beginning or intermediate graduate level (a list of participants should be provided).

   The resulting manuscript should not be just a collection of course notes, but should require advance planning and coordination among the main lecturers. The subject matter should dictate the structure of the book. This structure should be motivated and explained in a scientific introduction, and the notation, references, index and formulation of results should be, if possible, unified by the editors. Each contribution should have an abstract and an introduction referring to the other contributions. In other words, more preparatory work must go into a multi-authored volume than simply assembling a disparate collection of papers, communicated at the event.

3. Manuscripts should be submitted either online at www.editorialmanager.com/lnm to Springer's mathematics editorial in Heidelberg, or electronically to one of the series editors. Authors should be aware that incomplete or insufficiently close-to-final manuscripts almost always result in longer refereeing times and nevertheless unclear referees' recommendations, making further refereeing of a final draft necessary. The strict minimum amount of material that will be considered should include a detailed outline describing the planned contents of each chapter, a bibliography and several sample chapters. Parallel submission of a manuscript to another publisher while under consideration for LNM is not acceptable and can lead to rejection.

4. In general, **monographs** will be sent out to at least 2 external referees for evaluation.

   A final decision to publish can be made only on the basis of the complete manuscript, however a refereeing process leading to a preliminary decision can be based on a pre-final or incomplete manuscript.

   Volume Editors of **multi-author works** are expected to arrange for the refereeing, to the usual scientific standards, of the individual contributions. If the resulting reports can be forwarded to the LNM Editorial Board, this is very helpful. If no reports are forwarded or if other questions remain unclear in respect of homogeneity etc, the series editors may wish to consult external referees for an overall evaluation of the volume.

5. Manuscripts should in general be submitted in English. Final manuscripts should contain at least 100 pages of mathematical text and should always include

   - a table of contents;
   - an informative introduction, with adequate motivation and perhaps some historical remarks: it should be accessible to a reader not intimately familiar with the topic treated;
   - a subject index: as a rule this is genuinely helpful for the reader.
   - For evaluation purposes, manuscripts should be submitted as pdf files.

6. Careful preparation of the manuscripts will help keep production time short besides ensuring satisfactory appearance of the finished book in print and online. After acceptance of the manuscript authors will be asked to prepare the final LaTeX source files (see LaTeX templates online: https://www.springer.com/gb/authors-editors/book-authors-editors/manuscriptpreparation/5636) plus the corresponding pdf- or zipped ps-file. The LaTeX source files are essential for producing the full-text online version of the book, see http://link.springer.com/bookseries/304 for the existing online volumes of LNM). The technical production of a Lecture Notes volume takes approximately 12 weeks. Additional instructions, if necessary, are available on request from lnm@springer.com.

7. Authors receive a total of 30 free copies of their volume and free access to their book on SpringerLink, but no royalties. They are entitled to a discount of 33.3 % on the price of Springer books purchased for their personal use, if ordering directly from Springer.

8. Commitment to publish is made by a *Publishing Agreement*; contributing authors of multiauthor books are requested to sign a *Consent to Publish form*. Springer-Verlag registers the copyright for each volume. Authors are free to reuse material contained in their LNM volumes in later publications: a brief written (or e-mail) request for formal permission is sufficient.

Addresses:
Professor Jean-Michel Morel, CMLA, École Normale Supérieure de Cachan, France
E-mail: moreljeanmichel@gmail.com

Professor Bernard Teissier, Equipe Géométrie et Dynamique,
Institut de Mathématiques de Jussieu – Paris Rive Gauche, Paris, France
E-mail: bernard.teissier@imj-prg.fr

Springer: Ute McCrory, Mathematics, Heidelberg, Germany,
E-mail: lnm@springer.com

Printed in the United States
by Baker & Taylor Publisher Services